90 0683387 2

D1756307

University of Plymouth Library

Subject to status this item may be renewed
via your Voyager account

http://voyager.plymouth.ac.uk

Exeter tel: (01392) 475049
Exmouth tel: (01395) 255331
Plymouth tel: (01752) 232323

Evolution Illuminated

EVOLUTION ILLUMINATED

Salmon and Their Relatives

Edited by

Andrew P. Hendry

Stephen C. Stearns

OXFORD
UNIVERSITY PRESS

2004

OXFORD

UNIVERSITY PRESS

Oxford   New York
Auckland   Bangkok   Buenos Aires   Cape Town   Chennai
Dar es Salaam   Delhi   Hong Kong   Istanbul   Karachi   Kolkata
Kuala Lumpur   Madrid   Melbourne   Mexico City   Mumbai   Nairobi
São Paulo   Shanghai   Taipei   Tokyo   Toronto

Copyright © 2004 by Oxford University Press, Inc.

Published by Oxford University Press, Inc.,
198 Madison Avenue, New York, New York 10016

www.oup.com

Oxford is a registered trademark of Oxford University Press

Library of Congress Cataloging-in-Publication Data
Evolution illuminated : salmon and their relatives / edited by Andrew P. Hendry
and Stephen C. Stearns.
p.   cm.
Includes bibliographical references and index.
ISBN 0-19-514385-X
1. Salmonidae—Evolution. 2. Fish populations. I. Hendry, A.P.
(Andrew P.) II. Stearns, S. C. (Stephen C.), 1946 III. Title.
QL638.S2E97 2003
597.5'5138—dc21   2002156410

9 8 7 6 5 4 3 2 1

Printed in the United States of America
on acid-free paper

PREFACE

How low on the scale of nature this law of battle descends, I know not; male alligators have been described as fighting, bellowing, and whirling round, like Indians in a wardance, for the possession of females; male salmons have been seen fighting all day long; male stag-beetles often bear wounds from the huge mandibles of other males. The war is, perhaps, severest between the males of polygamous animals, and these seem oftenest provided with special weapons. The males of carnivorous animals are already well armed; though to them and to others, special means of defence may be given through means of sexual selection, as the mane to the lion, the shoulder-pad to the boar, and the hooked jaw to the male salmon; for the shield may be as important for victory, as the sword or spear
—Charles Darwin, 1859

With this passage, salmonids made their entrée into evolutionary biology, appearing in no less than the first great work on the subject, Charles Darwin's *On the Origin of Species*. Darwin highlighted salmon again, this time with a beautiful illustration, in his *The Descent of Man and Selection in Relation to Sex*. From these auspicious beginnings, the contribution of salmonids to the development of evolutionary theory has waxed and waned. Recent years, in particular, have seen a resurgence in the use of salmonids to address evolutionary theory; this book attempts to synthesize some of that ongoing work.

The book begins with an introduction that reviews the various features of salmonids that commend (or do not commend) them to the testing of evolutionary theory. The introduction also highlights some of the areas to which salmonids are making substantial contributions. The first contributed chapter is by Bill Schaffer, whose pioneering work on life history theory helped launch a field that has remained vital to the present. Bill saw the value of using salmonids to test evolutionary theory and in doing so inspired much of the work that appears in this volume. Bill reviews and revisits his classic work in the light of new data from salmonids. He also provides personal anecdotes from a time when life history theory was just exploding onto the evolutionary scene.

The following chapters examine in detail the use of salmonids to address specific questions in evolutionary biology. Although each chapter focuses on a separate area of evolutionary theory, certain themes recur across chapters: e.g., the evolution of alternative mating tactics and adaptation to anadromy/non-anadromy. In their entirety, the chapters address only a subset of the many areas to which salmonids are actively contributing; the omission of other areas does not mean that we consider them to be of lesser importance.

Despite the affection that humans hold for salmonids, or perhaps because of it, many salmonid populations currently face the threat of extinction. For this reason, much of the current research on salmonids is motivated by their conservation and management. The final few chapters of the book highlight the contributions that evolutionary theory can make to such endeavors. By addressing these issues, we hope to encourage further integration of evolutionary theory and conservation biology, and more generally, basic and applied research.

We thank the numerous people who contributed to this book, especially the chapter authors and peer reviewers. We also acknowledge Heather Roffey and Vanessa Partridge, who helped with the otherwise thankless job of formatting the text and checking the references. Finally, we express our profound gratitude to Steve Johnson, whose drawings grace the frontispiece for each chapter. Steve took considerable time from his graduate work to produce these drawings and did so wonderfully with minimal monetary compensation. Salmon are remarkable photogenic animals and although we could not include color pictures in this book, we have posted them at http://www.oup-usa.org/sc/019514385X.

In hoping that you enjoy this book, we close with the immortal words of that most enthusiastic of fish fanciers, Gollum: "Fishes precious."

Andrew Hendry
McGill University

Steve Stearns
Yale University

CONTENTS

CONTRIBUTORS

Ole K. Berg
Norwegian University of Science and
  Technology
NTNU, N-7491
Trondheim, Norway
Email: Ole.Kristian.Berg@chembio.ntnu.no

Louis Bernatchez
Département de biologie
Pavillon Vachon
Université Laval
Ste-Foy, QC G1K 7P4 Canada
Email: louis.bernatchez@bio.ulaval.ca

Torgny Bohlin
Department of Zoology, Animal
  Ecology,
Göteborg University
Box 463, SE 405 30
Göteborg, Sweden
Email: torgny.bohlin@zool.gu.se

Vincent Castric
Laboratoire Genetique et Evolution des
  Populations Vegetales
Université des Sciences et Techniques de
  Lille
59 655 Villeneuve d'Ascq Cedex, France
Email: Vincent.Castric@univ-lille1.fr

Sigurd Einum
Norwegian Institute for Nature Research
Tungasletta 2, N-7485
Trondheim, Norway
Email: sigurd.einum@ninatrd.ninaniku.no

Ian A. Fleming
Coastal Oregon Marine Experiment Station
  and Department of Fisheries and Wildlife
Oregon State University
Hatfield Marine Science Center
Newport, OR 97365
Email: ian.fleming@oregonstate.edu

Michael J. Ford
National Marine Fisheries Service,
Northwest Fisheries Science Center
2725 Montlake Blvd E.
Seattle, WA 98125
Email: mike.ford@noaa.gov

Jeffrey J. Hard
National Marine Fisheries Service,
Northwest Fisheries Science Center
Conservation Biology Division
2725 Montlake Boulevard E.
Seattle, WA 98112
Email: Jeff.Hard@noaa.gov

Andrew P. Hendry
Redpath Museum and Department of
    Biology
McGill University
859 Sherbrooke St. W.
Montreal, PQ H3A 2K6 Canada
Email: andrew.hendry@mcgill.ca

Jeffrey A. Hutchings
Department of Biology
Dalhousie University
Halifax, NS B3H 4J1 Canada
Email: jhutch@mscs.dal.ca

Bror Jonsson
Norwegian Institute for Nature Research
Dronningensgt. 13
P.O. Box 736, N-0105
Oslo, Norway
Email: bror.jonsson@ninaosl.ninaniku.no

Michael T. Kinnison
Department of Biological Sciences
Murray Hall
University of Maine
Orono, ME 04469-5751
Email: michael.kinnison@umit.maine.edu

Thomas P. Quinn
School of Aquatic and Fishery Sciences
University of Washington
Seattle, WA 98195
Email: tquinn@u.washington.edu

John D. Reynolds
School of Biological Sciences
University of East Anglia
Norwich NR4 7TJ United Kingdom
Email: Reynolds@uea.ac.uk

William M. Schaffer
Department of Ecology and Evolutionary
    Biology
University of Arizona
Tucson, AZ 85721
Email: wms@u.arizona.edu

Stephen C. Stearns
Department of Ecology and Evolutionary
    Biology
Box 208106
Yale University
New Haven, CT 06520-8106
Email: stephen.stearns@yale.edu

Eric B. Taylor
Department of Zoology and Native Fish
    Research Group
University of British Columbia
6270 University Blvd.
Vancouver, BC V6T 1Z4 Canada
Email: etaylor@zoology.ubc.ca

Robin S. Waples
National Marine Fisheries Service,
Northwest Fisheries Science Center
2725 Montlake Boulevard East
Seattle, WA 98112
Email: robin.waples@noaa.gov

Kyle A. Young
Department of Biological Sciences
Simon Fraser University
8888 University Drive
Burnaby, BC V5A 1S6 Canada
Email: kayoung@sfu.ca

Evolution Illuminated

Introduction
The Salmonid Contribution to Key Issues in Evolution

Stephen C. Stearns

Andrew P. Hendry

Mature male sockeye salmon

In designing a biology book, two approaches are common. The first, question-oriented, organizes the book around a series of related questions and surveys the data, perhaps from many taxa, relevant to each question. Examples include books on sexual selection (Andersson 1994), adaptation (Rose and Lauder 1996), and life history evolution (Roff 2002). The second, taxon-oriented, organizes the book around a single taxon, and surveys many aspects of its biology. Examples include books on amphibians (Duellman and Trueb 1986), mammals (Eisenberg 1981), orchids (Dressler 1990), brown trout (Elliott 1994), and Pacific salmon (Groot and Margolis 1991; Groot et al. 1995). Here we take a hybrid approach, asking what one taxonomic group, the salmonids, contributes to our understanding of key questions in evolutionary biology.

Each approach has its advantages and disadvantages, and each contains an implicit focus strong enough to color an entire research program. The strategy used here is not common but is potentially very informative (e.g., sticklebacks: Bell and Foster 1994). We chose it for two reasons:

1. We want to inform evolutionary biologists about the wealth of information on salmonids that bears directly on evolutionary questions.
2. We want to increase awareness among salmon biologists of the relevance of their data for key issues in evolutionary biology.

Through improved communication we hope to advance both fields.

1. Important Features of Salmonids

Because some salmonid species have considerable value in recreational and commercial fisheries, they have been intensively studied for practical reasons. They have also been the subject of intensive aquaculture, a byproduct of which has been negative interactions between feral and wild fish in nature. This has yielded a wealth of data applicable to a variety of basic questions, data of a sort that is only rarely available in other groups. But because most research funding for salmonids has been for applied goals, the data generated has only sporadically been focused on basic evolutionary questions. This book reviews existing attempts along those lines, makes several new attempts, and suggests still further possibilities. Before embarking on this endeavor, it is useful to consider some of the advantages and disadvantages of salmonids for the study of evolution.

1.1. Advantages of Salmonids

1. Their relatively high fecundity (Table 1) generates considerable potential for mortality rates to vary among life history stages, populations, and species (Pacific salmon: Bradford 1995), just the kind of variation needed for comparative study of life history evolution. Note, however, that many organisms have higher fecundities than salmonids (e.g., plants and marine fishes) and that higher fecundities do not necessarily lead to greater variation in mortality (Mertz and Myers 1996).

Table I. Common names, scientific names, native geographic ranges, and life history information for some salmonids.

Common name[a]	Scientific name[a]	Native range	Semelparous or iteroparous	Anadromous or FW resident	Age (y)	Length (mm)	Egg #	Egg size (mm)
Chinook salmon	Oncorhynchus tshawytscha	N Pacific Ocean	S[b]	AN[c]	1–8	102–1490	1622–17,255	6.3–8.4
Coho salmon	Oncorhynchus kisutch	N Pacific Ocean	S	AN[c]	1–6	210–960	1724–7110	4.5–7.1
Chum salmon	Oncorhynchus keta	N Pacific Ocean, Arctic (N.A., E Asia)	S	AN[c]	2–7	441–1000	909–7779	5.5–9.5
Pink salmon	Oncorhynchus gorbuscha	N Pacific Ocean, Arctic (N.A., E Asia)	S	AN[c]	2	292–760	500–3000	5.5–6.2[d]
Sockeye salmon	Oncorhynchus nerka	N Pacific Ocean	S	AN & FW	3–8	170–840	139–5700	4.6–6.1
Masu salmon[e]	Oncorhynchus masou	Japan Sea and Okhotsk Sea	S[e]	AN & FW	2–7	88–743	78–5500	5.0–7.1
Rainbow trout[f]	Oncorhynchus mykiss	N Pacific Ocean, Central N.A.	I (5)	AN & FW	1–7	95–1220	326–7600	3.0–5.5
Cutthroat trout[g]	Oncorhynchus clarki	NE Pacific Ocean, Central N.A.	I (7)	AN & FW	2–11	175–550	2930	4.2–4.7
Gila trout[h]	Oncorhynchus gilae	SW N.A.	I (?)	FW	?	?–330	?	?
Atlantic salmon	Salmo salar	N Atlantic Ocean	I (11)	AN & FW	1–10	70–1150	33–18,847	4.5–7.0
Brown trout	Salmo trutta	Europe, W Asia	I (11)	AN & FW	2–20	125–706	68–10,588	3.2–5.9
Arctic charr	Salvelinus alpinus	N Pacific Ocean, N Atlantic Ocean, Arctic	I (18)	AN & FW	2–27	73–630	13–5400	4.0–5.0
Brook charr (brook trout)	Salvelinus fontinalis	E N.A.	I (8)	AN & FW	1–7	72–565	15–5920	2.9–5.0
Dolly Varden	Salvelinus malma	N Pacific Ocean	I (12)	AN & FW	5–11	91–619	38–4927	3.0–5.6
Bull trout (bull charr)	Salvelinus confluentus	N Pacific Ocean, Central N.A.	I (3+)	FW	3–20	250–1250	100–6000	?

(continued)

Table I. (continued)

Common name[a]	Scientific name[a]	Native range	Semelparous or iteroparous	Anadromous or FW resident	Age (y)	Length (mm)	Egg #	Egg size (mm)
Lake trout (lake charr)	*Salvelinus namaycush*	N N.A.	I (29)	FW	6–25	350–1260	500–18,000	5.0–6.0
White-spotted charr	*Salvelinus leucomaenis*	NE Pacific	I (?)	AN & FW	1–8	90–600	100–9000	4.5–6.5
Taimen (hucho)[i]	*Hucho hucho*	E Asia	I (?)	FW	4–55	368–2070	1600–35,040	3.6–6.0
Sakhalin taimen	*Hucho perryi*	NE Pacific	I (?)	AN & FW	4–15+	386–1260	2100–17,000	?
Arctic grayling	*Thymallus arcticus*	NW N.A., NW Asia	I (10)	FW	2–12	176–757	600–15,905	1.0–3.0
European grayling	*Thymallus thymallus*	Europe, NE Asia	I (20)	FW	2–28	220–600	421–36,000	2.0–4.6
Mountain whitefish	*Prosopium williamsoni*	W N.A.	I (16)	FW	3–18	197–572	1426–24,143	2.8–3.2
Round whitefish	*Prosopium cylindraceum*	N N.A., N Asia	I (12)	FW	4–14	310–561	2461–25,137	2.4–3.2
Pygmy whitefish	*Prosopium coulteri*	Great Lakes, NW N.A., Russia	I (?)	FW	2–9	130–272	440–600	2.0–2.6
Broad whitefish	*Coregonus nasus*	Arctic (N.A., Asia)	I (?)	AN & FW	4–20	180–800	?	?
Lake whitefish	*Coregonus clupeaformis*	N N.A.	I (22)	AN & FW	5–20	255–676	4800–110,500	2.3–3.0
Atlantic whitefish	*Coregonus canadensis*	Nova Scotia, Canada	?	AN & FW?	?	≤ 508	?	?
Inconnu (sheefish)	*Stenodus leucichthys*	NW N.A., N Asia	I (5)	AN & FW	5–21	540–1500	26,000–420,000	2.5
Shortjaw cisco	*Coregonus zenithicus*	Great Lakes, N N.A.	I	FW	2–10	100–368	?	2.1
Cisco	*Coregonus artedii*	N N.A.	I (12)	AN & FW	4–14	240–572	5000–29,000	1.8–2.1

6

Common name	Scientific name	Native range	Semelparous or iteroparous	Life history	Age	Length	Egg #	Egg size
Arctic cisco	*Coregonus autumnalis*	Arctic (N.A., Europe, Asia)	I (?)	AN	4–17	294–640	16,920–102,620	?
Bloater	*Coregonus hoyi*	Great Lakes	I (9)	FW	2–11	213–315	3116–18,768	2.0
Bering cisco	*Coregonus laurettae*	North Pacific Ocean	I	AN	3–8	254–480	?	?
European whitefish	*Coregonus lavaretus*	Europe, Asia	I (20)	FW	1–21	240–600	4000–66,100	0.5–1.7
Humpback whitefish	*Coregonus pidschian*	Alaska	I (?)	AN & FW	4–16	302–540	10,800–44,400	?
Vendace (European cisco)	*Coregonus albula*	Europe	I (20)	AN & FW	2–?	110–245	2000–20,000	?
Least cisco	*Coregonus sardinella*	NE Pacific, Arctic (N.A., Europe, Asia)	I (?)	AN & FW	5–26	229–419	2500–100,939	?

[a] We have excluded several North American Great Lakes coregonids that now appear to be extinct (Miller et al. 1989): longjaw cisco (*C. alpenae*), deepwater cisco (*C. johannae*), Lake Ontario kiyi (*C. kiyi*), blackfin cisco (*C. nigripinnis*), and shortnose cisco (*C. reighardi*). [b] Males that mature in their first year of life (mature parr) in a hatchery can sometimes be kept alive to mature again in subsequent years (Unwin et al. 1999). [c] May become freshwater resident when introduced outside of their native range, particularly in the North American Great Lakes. [d] These values are from a single study (Kaev and Kaeva 1987) and so the total range is undoubtedly higher. Most studies of pink salmon do not report the diameter of eggs that have not been water hardened. [e] Includes several subspecies: amago salmon (*O. masou ishikawae*, formerly *O. masou rhodurus*), masu salmon (*O. masou masou*), and Biwa salmon (*O. masou* subspecies). Males that mature in their first year of life (mature parr) in the wild can sometimes survive to mature in subsequent years (Tsiger et al. 1994). [f] Includes a complex of species/subspecies, such as steelhead trout, rainbow trout, redband trout, golden trout, and others (Behnke 1992). [g] Includes a complex of species/subspecies, such as coastal cutthroat trout (*O. clarki clarki*), westslope cutthroat trout (*O. clarki lewisi*), Yellowstone cutthroat trout (*O. clarki bouvieri*), and others (Behnke 1992). [h] Includes a complex of species/subspecies, such as Gila trout (*O. gilae gilae*), Apache trout (*O. gilae apache*), and others (Behnke 1992). [i] Includes the taimen (*Hucho hucho taimen*) and the huchen (*Hucho hucho hucho*), following Holčík et al. (1988).

Notes: This is not a complete listing and some taxonomic classifications remain fluid. Life history variation represents the range (minimum to maximum) observed within the species native range, and includes both anadromous and non-anadromous forms when they occur within a species. The values are approximate because maxima and minima will depend on the number of populations and individuals sampled. "Native range" gives the native range of the species with the following abbreviations: N = northern, E = eastern, W = western, NW = northwestern, NE = northeastern, SW = southwestern, and N.A. = North America. For the Arctic, additional regional distinctions are given in parentheses. "Semelparous or iteroparous" indicates whether each species is obligatorily semelparous (S) or has the potential for iteroparity (I), with the maximum recorded number of repeat spawnings in parentheses. "Age" is age at maturity. "Length" is length at maturity and is usually fork length but sometimes total length. "Egg #" is the number of eggs per breeding season (fecundity). "Egg size" is the diameter of fresh (i.e., not water-hardened) eggs. Data are from Scott and Crossman (1973), Groot and Margolis (1991), and various primary sources.

2. Populations and species differ dramatically in key life history traits such as semelparity versus iteroparity, migratory versus non-migratory, mono-morphic versus polymorphic, age and size at first reproduction, egg size, and egg number (Table 1). Among salmon species, for example, repeat breeding ranges from 0 to 65%, egg size ranges from 28 to 251 mg, and female GSI (ratio of gonad mass to body mass) ranges from 10.5 to 22.2% (Fleming 1998). Among populations of Atlantic salmon, the length of smolts (seaward migrants) ranges from 10.5 to 21.5 cm, the age of smolts ranges from 1.1 to 5.6 years, the length of mature fish that spent 2 years at sea ranges from 68.3 to 90.3 cm, and the length of time at sea ranges from 1.0 to 2.6 years (Hutchings and Jones 1998).

3. Salmonid species are complexes of populations that occupy dramatically different rearing habitats (lakes, streams, large rivers, ponds, estuaries, open ocean) and breeding habitats (lake beaches, ponds, streams, large rivers), and are at least partially reproductively isolated owing to natal homing (Quinn 1993; Quinn and Dittman 1990; Quinn et al. 1999). This type of population structure produces replicated sets of environment–phenotype gradients that are useful for the study of local adaptation (Taylor 1991b).

4. Much of their extant range was colonized from refugia after the last glaciation (Wood 1995), facilitating the study of adaptation and historical constraints: races from the same refugia colonized different environments and races from different refugia colonized similar environments. Salmonids have also been introduced all over the world (MacCrimmon and Gots 1979; Lever 1996), providing the opportunity to study the time scales of evolutionary change.

5. Salmonids exhibit a great diversity of trophic polymorphisms that persist in sympatry, and these polymorphisms have arisen multiple independent times within species and in parallel across species (Skúlason and Smith 1995; Jonsson and Jonsson 2001). This provides an opportunity to study how trophic specialization and competition can initiate and maintain divergence and speciation in sympatry.

6. Although populations tend to be isolated by natal homing, there is still enough movement among populations, called "straying," to allow studies of the evolution of dispersal. This analysis is facilitated by dramatic variation in straying rates. For example, an average of 41.6% of the chinook salmon released from Columbia River hatcheries stray to other locations (Pascual et al. 1995), whereas <1% of sockeye salmon stray between different lake systems (Quinn et al. 1987).

7. Because several species are commercially valuable and widely used in aquaculture, a great deal is known about the heritability and phenotypic plasticity of many life history traits (Gjerde and Schaeffer 1989; Crandell and Gall 1993; Su et al. 1997; Hebert et al. 1998; Quinn et al. 2000; Kinnison et al. 2001). Such information helps understand (a) how rapidly the traits that determine population dynamics can change, (b) the sources of variation in reproductive success that determine the strength

of selection, and (c) the capacity of populations to respond to selection through standing genetic variation.

1.2. Disadvantages of Salmonids

1. Their generation times are long, for females from 2 years (pink salmon) to more than 7 years (chinook salmon), and so they are not natural candidates for experimental evolution. However, this disadvantage can be compensated in part by intensive long-term study of the effects of fishing pressure and introductions to new environments (Bigler et al. 1996; Quinn and Adams 1996; Cox and Hinch 1997; Haugen and Vøllestad 2001).
2. They are often large at maturity and need large quantities of cold, clean water. This makes common-garden experiments difficult for adult traits, although not impossible (e.g., Wood and Foote 1996; Quinn et al. 2000, 2001b; Craig and Foote 2001; Kinnison et al. 2001; Haugen and Vøllestad 2001).
3. The sea phase of anadromous populations is difficult to observe. Again, this difficulty can be partly compensated by the intensity of fishing effort.
4. Their high fecundity, which leads to large numbers of juveniles, makes it hard to find the needle in the haystack if one needs to follow individuals or use mark-recapture techniques. Fortunately, the development of new tagging methods, including genetic markers (Bentzen et al. 2001; Potvin and Bernatchez 2001), has made it easier to track individuals.
5. Their large genomes make them less than ideal candidates for whole-genome sequencing followed by functional genomic studies of entire transcriptomes, but focused comparative genomics are certainly still possible.

No one model system is ideally suited to the study of all evolutionary questions but salmonids provide an excellent system for addressing certain questions that are difficult to examine using other taxa.

2. Key Evolutionary Issues to Which Salmonids Contribute

2.1. Life History Evolution

The striking contrast of closely related semelparous and iteroparous salmonids posed a classic evolutionary puzzle that helped motivate fundamental contributions from Cole (1954), Schaffer (1974a,b), Charnov and Schaffer (1973), Bell (1976), Stearns (1976), and others (see Schaffer 2003—*this volume*). Other classic and contemporary life-history themes represented by substantial work on salmonids include the evolution of age and size at maturity (Schaffer and Elson 1975; Healey 1987; Myers and Hutchings 1987a; Hutchings 1993a, 1996,

Hutchings 2003—*this volume*), semelparity versus iteroparity (Crespi and Teo 2002), alternative life histories and breeding tactics (Gross 1985; Hutchings and Myers 1994; Gross 1996), migratory tendency (Ricker 1940; Nordeng 1983; Jonsson 1985; Jonsson and Jonsson 1993; Bohlin et al. 2001; Hendry et al. 2003b—*this volume*), egg size versus number (Healey and Heard 1984; Hutchings 1991; Quinn et al. 1995; Einum and Fleming 1999, 2000a; Hendry et al. 2001b; Koops et al., 2003; Heath et al. 2003; Einum et al. 2003—*this volume*), and life history invariants (Charnov 1993; Mangel 1996). Although there is certainly room for more work on these classic themes, another area strikes us as warranting focused attention: the spreading of risk.

The evolution of risk-minimizing adaptations (bet-hedging) does not appear to have been given the attention in salmonids that it deserves. Iteroparous species have the potential to breed several times, spreading their risk temporally. They also tend to build several discrete redds in different areas, spreading their risk spatially (Barlaup et al. 1994). Pacific salmon, in contrast, are semelparous and build only a single redd, thus reducing their temporal and spatial spreading of risk. How then might Pacific salmon compensate for risk? (a) Individuals could mate with more than one partner, a pattern that is certainly common in natural populations (Foote 1990; Quinn et al. 1996; Garant et al. 2001). (b) Females could lay eggs in more than one nest within a redd but the nests of a Pacific salmon female are usually clustered near each other in a single redd, making the spatial spread minimal. (c) The offspring from a single clutch could display a mixture of ages at maturity, migration (freshwater residence versus anadromy, Hendry et al. 2003b—*this volume*), or homing (precise return to natal sites versus straying to non-natal sites, Hendry et al. 2003a—*this volume*). Some data can be brought to bear on these issues but have not yet been used to rigorously confront hypotheses derived from formal models of risk-spreading (but see Koops et al. 2003). The issue is both classic (Bernoulli 1738) and fundamental (Stearns 2000).

2.2. The Evolution of Reproductive Isolation: Ecological Speciation

The evolution of reproductive isolation ("speciation") is a key process in evolutionary biology because it straddles the boundary between micro-evolution and macro-evolution. Unfortunately, speciation is not well understood because it may often take longer than a human lifetime to complete, and is difficult to observe directly. A classical controversy contrasts allopatric scenarios (species originate in different locations) with sympatric scenarios (species originate in the same location). For many years, the weight of opinion favored allopatry as the primary mode of speciation (Mayr 1963). Recently, however, interest in sympatric speciation has surged because of new data, primarily on insect host races (Bush 1994) and cichlids in tiny crater lakes of Africa (Schliewen et al. 1994, 2001), and because of new models, primarily by Dieckmann and Doebeli (1999) and Kondrashov and Kondrashov (1999). Along with the renewed interest in sympatric speciation has come an increased appreciation of the role that divergent natural selection plays in the evolution of reproductive isolation. This view sees speciation as being driven by adaptation to different environments, competition for resources, and assortative mating ("ecological speciation"; Filchak et al. 2000; Rundle et al. 2000; Schluter 2000; Hawthorne and Via 2001; Bernatchez 2003—*this volume*).

Salmonids provide excellent systems for studying sympatric divergence and ecological speciation. First, sympatric but reproductively isolated forms with different patterns of resource use (benthic versus limnetic, freshwater resident versus anadromous) have arisen numerous times following the last glaciation (Atlantic salmon: Berg 1985; Arctic charr: Gíslason et al. 1999; Jonsson and Jonsson 2001; lake whitefish: Bernatchez et al. 1999; Lu and Bernatchez 1999; Bernatchez 2003—*this volume*; brown trout: Ferguson and Taggart 1991; sockeye salmon: Wood and Foote 1996; Taylor et al. 1996). In at least some cases, the divergence appears to have occurred entirely in sympatry, whereas in others the initial stages may have occurred in allopatry. Second, populations that are isolated only by natal homing (i.e., no physical barriers prevent their mixing) often colonize and adapt to dramatically different breeding environments (e.g., beaches versus streams for sockeye salmon: Hendry et al. 2000; Hendry 2001). Third, the role of natal homing in isolating salmonid populations in different selective environments is equivalent to the role of imprinting by insect larvae in isolating insect races on different host plants (Bush 1994). However, an interesting contrast ripe for theoretical exploration is that salmonids imprint on specific locations, whereas insect host races imprint on a specific type of plant that may be found in multiple locations.

In short, salmonids often occur as discrete populations or forms that are adapted to different environments but not isolated by geographic barriers. Moreover, many of these populations or forms have little or no intrinsic postzygotic barriers to reproduction (Taylor 2003—*this volume*), making them particularly well-suited to the study of "environment-dependent" (or "ecologically dependent") reproductive isolation (Rice and Hostert 1993; Rundle and Whitlock 2001). These are critical issues in research on speciation, and salmonids seem poised to make fundamental contributions to the development of this field.

2.3. Hybridization

Hybridization is the flip side of speciation and is thus thought to play a constraining evolutionary role by preventing the origin of new species (Mayr 1963). New species can only arise and remain distinct in sympatry when reproductive isolating mechanisms reduce hybridization. Conversely, hybridization may sometimes play a constructive role in evolutionary diversification (Arnold 1997). For example, hybridization can be a mechanism for acquiring potentially adaptive genetic variants in large quantities (Arnold 1997). Research on salmonids has proceeded from both of these perspectives (Taylor 2003—*this volume*).

Natural hybridization is common between several pairs of salmonid species, including rainbow trout and cutthroat trout (Campton and Utter 1985), brown trout and Atlantic salmon (Verspoor and Hammar 1991), and bull trout and Dolly Varden (Baxter et al. 1997). In some cases, this hybridization reduces genetic distinctiveness among species and places their conservation status in question (Allendorf et al. 2001). In other cases, species maintain their distinctiveness in sympatry, even in the presence of hybridization. A considerable amount of attention has therefore been placed on determining mechanisms of reproductive isolation between sympatric salmonids. Pre-mating isolating mechanisms include natal homing (Quinn and Dittman 1990; Quinn et al. 1999), spatial and temporal segregation during spawning (Leider et al. 1984; Fukushima and Smoker 1998), size-assortative mating (Foote and Larkin 1988), and species recognition. Post-mating isolating mechanisms include intrinsic genetic incompatibilities among distantly related species (Taylor 2003—*this volume*) and environment-dependent selection against hybrids in nature (Wood and Foote 1990; Hawkins and Foote 1998; Bernatchez 2003—*this volume*).

Other studies have suggested that hybridization may be advantageous because it can generate hybrid genomes that are sometimes superior to pure genomes. For example, some salmonid populations with nuclear loci fixed for one species can have high or even fixed frequencies of mitochondrial DNA from another species. Examples include Arctic charr mitochondria in brook charr (Bernatchez et al. 1995; Glémet et al. 1998) and lake trout (Wilson and Bernatchez 1998), and bull trout mitochondria in Dolly Varden (Taylor et al. 2001). Although this pattern could occur owing to past hybridization followed by genetic drift, some evidence suggests that mitochondrial introgression is adaptive. For example, Arctic charr mtDNA is better adapted for cold water and seems to introgress more frequently into cold water than warm water populations of brook charr (Glémet et al. 1998).

Arguing that selection can act both for and against hybridization might appear paradoxical, but in fact it is not. Under most circumstances, selection would be expected to act against hybridization, whether among locally adapted conspecific populations or among taxonomically recognized species. Under certain conditions, however, selection may tip the balance to favor hybridization, as has been observed in several bird species (Grant and Grant 1992; Veen et al. 2001). In salmonids, hybrid genotypes might be favored when (1) refugial populations are trapped in an isolated location during global climate change, (2) disturbance by humans causes rapid environmental change, (3) one species is rare and individuals have a hard time finding conspecific mates, and (4) costs of hybridization are low (e.g., for males that act as "sneaks").

2.4. Philopatry, Dispersal, and Metapopulations

The evolution of philopatry (homing) is of general interest because, among other things, individuals that choose specific habitats thereby choose the selection pressures to which they will be exposed (Bateson 1998). This link gives behavior an important causal role in shaping subsequent evolution. An increase in philopatry is conservative in some ways, contributing to evolutionary stasis and damping the evolution of adaptive phenotypic plasticity. In contrast, dispersal (straying) subjects a population to a variety of selection pressures. This promotes the evolution of adaptive phenotypic plasticity (although it is not sufficient to guarantee it) and increases the population's evolvability (ability of the population to respond to selection) by maintaining genetic variation.

Homing and straying vary dramatically within and among salmonid populations and species (Hendry et al. 2003a—*this volume*). This variation should be subject to many of the same selection pressures and genetic constraints as the evolution of dispersal in a metapopulation. A metapopulation is a set of local populations linked by movement (Hanski 1999), and metapopulation dynamics are driven in part by local extinction and recolonization. Even if all local populations are extinction-prone, a metapopulation can survive in a stochastic balance between extinction and colonization. Landscape features that affect these processes become important for regional persistence by contributing to a threshold ratio between colonization and extinction above which the metapopulation can persist.

There is considerable evidence for the existence of metapopulations in nature (Hanski 1999). Empirical work has shown that population size is significantly affected by migration; including source effects (isolated populations increase) and sink effects (isolated populations decrease). Population density is often affected by patch area and isolation, and population dynamics are often asynchronous among locations. Population turnover—extinction and recolonization—does occur, and empty habitats are often present. There is also strong evidence that extinction risk depends on patch size, that colonization rate depends on patch isolation, and that small, isolated patches are more likely to be empty than big, connected patches. However, there is as yet little evidence for the threshold condition—the balance between colonization and extinction—required for metapopulation persistence in studies of macroscopic, free-living plants and animals. Could data on salmonids be brought to bear on this issue?

Many salmonids are expected to exhibit metapopulation structure (Hansen and Mensberg 1996; Young 1999; Cooper and Mangel 1999), and so straying may evolve for the same reasons that dispersal evolves to intermediate levels in metapopulations (Hendry et al. 2003a—*this volume*). When local populations go extinct frequently, the metapopulation can only be maintained if dispersal rates are high. When local populations go extinct rarely, selection favors philopatry and dispersal rates decrease. Thus some of the variation in salmonid homing and straying may perhaps be understood as a adaptive response to local extinction rates in a metapopulation (Hendry et al. 2003a—*this volume*). Whether or not this variation can be used to reveal the threshold level between colonization and extinction remains to be seen.

2.5. Sexual Selection and Breeding Systems

Part of the received wisdom of behavioral ecology—ever subject to scrutiny—is that females find the resources they need to breed and males compete for females. If we ignore leks (Höglund and Alatalo 1995), this simple dictum does explain much of the variation in breeding systems, including that among salmonid species and populations (Fleming and Reynolds 2003—*this volume*). Females need well-aerated beds of gravel substrate in which to lay their eggs, substrate that they are capable of excavating and that will provide their offspring with an appropriate incubation environment (Chapman 1988). Females evolve to select such places for spawning and males join them (Foote 1990), sometimes arriving first (Morbey 2000), but then things get complicated.

One major complication is the conflict between female choice of males and male competition for females. From the female point of view, it would be best to choose a mate who would yield maximum fitness for her offspring. Because male salmon do not provide parental care, the benefits females seek must come primarily through the male's genes, not through any direct contributions to offspring survival. The genes could be for a particular life history, for parasite and pathogen resistance, or for male attractiveness to females (and therefore the attractiveness of a female's sons). In salmonids, the opportunity for mate choice by females is limited: active choice seems mostly related to the rapidity with which females spawn when attended by large or small males (de Gaudemar et al. 2000b; Berejikian et al. 2000), and female choice can be circumvented by male competition (Petersson et al. 1999). Recent evidence suggests that wild females mate with multiple males (Garant et al. 2001) and that mating may be dissassortative with respect to MHC genotype (Landry et al. 2001). These latter results suggest a role for genetic benefits in mate choice, but it isn't yet known whether the patterns result from male–male competition, male choice, female choice, or a combination of these factors.

From the male point of view, it would also be good to choose a mate with good genes or large size (because large females have more eggs; Foote 1988) but it might be even better simply to mate with as many females as possible (which could involve courting females that are close to oviposition; Hamon et al. 1999). In either case, males must be ready to fight with other males for access to females (Foote 1990; Quinn et al. 1996; Morbey 2002). Males that adopt an aggressive, fighting strategy will be selected for large body size and secondary sexual characters, such as hooked jaws, that make them more effective fighters (Foote 1990; Quinn and Foote 1994; Fleming and Gross 1994). The existence of fighter males, in turn, creates an opportunity for an alternative mating tactic, that of the small male who "sneaks" in to steal fertilizations (Gross 1985; Hutchings and Myers 1988; Gross 1996; Foote et al. 1997). The sneaker tactic may be characterized by sperm that are of higher quality, leading to potentially higher proportional fertilization success (Vladić and Järvi 2001). The balance between the two male tactics is probably under frequency-dependent selection favoring the rarer type and may be condition-dependent (Repka and Gross 1995; Gross and Repka 1998).

It has recently been suggested that in certain systems sneakers have higher fitness than fighters (Gross 1996) and that females may prefer to mate with sneakers (M. Gross unpublished). In at least some systems, however, females are more aggressive toward sneakers than toward fighters (L. Weir unpublished), suggesting that females actually prefer to have their eggs fertilized by fighters. It remains possible, however, that this aggression may arise because sneakers sometimes eat eggs (Blanchfield and Ridgway 1999), which might present females with a tradeoff between offspring number and offspring quality. Another possibility is that females may prefer sneaker genes but chase sneakers away because they do not want to pay the costs associated with coercion by fighter males (J. Watters unpublished). The relative fitness of sneakers versus fighters, as well as female preferences for each, likely vary among species, life histories, and ecological conditions. In particular, results may vary considerably between Pacific salmon, where semelparity is the rule and males are usually anadromous, and Atlantic salmon, where iteroparity is possible and fighters are anadromous but sneakers are non-anadromous. Two fundamental, unanswered questions thus face evolutionary biologists in this field: what is the relative fitness of fighters and sneakers under different conditions, and do females prefer fighters or sneakers under those conditions? Salmonids are on the verge of making important contributions to answering these questions, continuing the tradition of their seminal contributions to mating systems theory.

Under the salmonid mating system, where females try, with imperfect information and ability, to choose among potential mates, and males try, with imperfect efficiency, to gain access to fertilizations by fighting or sneaking, conflicts between the sexes over mate choice naturally arise ("sexual conflict," Parker 1979; Chapman et al. 2003). The result, probably better than random choice but worse than optimal choice for both sexes, determines the component of variation in reproductive success that gets attributed to sexual selection. There have not yet been any studies of sexual conflict in salmonids but this area seems ripe for exploration.

There are a host of unsolved problems associated with sexual selection to which salmonids can contribute (Fleming and Reynolds 2003—*this volume*). When alternative male tactics exist, do they have equal fitness on average? What factors in the environment influence the frequency of alternative mating tactics? What attributes in males are being actively chosen by females, and are those attributes associated with pathogen resistance, migratory strategy, sexy sons, or some other factor? To what degree does selection for performance in mating compromise or oppose selection for other traits, such as growth rates, body size, and age at maturity? Or, to turn that question around, if a population comes under selection for smaller body size because of fishing pressure, what consequences will that have for the performance of males in mating and for the ongoing evolution of female preferences? Is selection for non-anadromous populations driven primarily by mortality at sea, or is it aided and abetted by female choice? In the next few decades, work on salmonids should be able to address all of these questions.

2.6. Rates of Evolution

A major shift in evolutionary biology in the last quarter century is due to the insight that evolution can be very rapid when large populations containing ample genetic variation encounter strong selection (Hendry and Kinnison 1999; Stockwell and Weeks 1999; Bone and Farres 2001; Kinnison and Hendry

2001; Reznick and Ghalambor 2001; Stockwell et al. 2003). Salmonids have contributed substantially to this literature, primarily through the study of recent introductions to new locations (sockeye salmon in Lake Washington: Hendry et al. 2000; Hendry 2001; European grayling in Norway: Haugen and Vøllestad 2001; Koskinen et al. 2002b; chinook salmon in New Zealand: Kinnison et al. 2001; Quinn et al. 2001b; Unwin et al. 2003). Other potentially profitable analyses would be rates of evolution following post-glacial expansion (Kinnison and Hendry 2003—*this volume*), and in response to environmental change and size-selective fisheries (Bigler et al. 1996; Quinn and Adams 1996; Cox and Hinch 1997; Haugen and Vøllestad 2001; Hard 2003—*this volume*).

Many other questions about evolutionary rates remain open (Kinnison and Hendry 2001). Do some traits evolve more rapidly than others: at a broad level (life history versus morphology) or a specific level (e.g., egg size versus egg number)? If so, do the fastest evolving traits tend to be the same across populations and species, or is no pattern detectable? If certain traits consistently evolve the fastest, is it because they (1) are subject to stronger selection (Kingsolver et al. 2001), (2) have higher heritabilities (Mousseau and Roff 1987), (3) have higher additive genetic variation (Houle 1992), or (4) are not constrained by tradeoffs and opposing selection on correlated traits (Merilä et al. 2001)? Are evolutionary changes best explained by natural selection or can some be attributed to founder effects, genetic drift, and gene flow (Hendry 2001; Koskinen et al. 2002b)? Are the micro-evolutionary changes observed in contemporary populations the stuff of macro-evolutionary transitions (i.e., speciation)? Research on salmonids could address all the questions.

Rates of evolution are also relevant from the perspective of conservation and management (Stockwell et al. 2003; Ashley et al. 2003). Humans are influencing salmonid environments at unprecedented rates, primarily through habitat alterations (logging, urbanization, and dams), fishing, introductions of exotic species, and global climate change. Salmonids must evolve to meet these new challenges if they are to persist in a particular location. Indeed, some theoretical work has suggested the existence of a maximum sustainable rate of evolution that determines the critical rate of environmental change (Bürger and Lynch 1995). If environmental change exceeds this rate, populations are unlikely to persist. The vast number of salmon populations and the large number of recent population extirpations in impacted areas (CPMPNAS 1996) make this a good system for estimating maximum sustainable rates of evolution and critical rates of environmental change.

2.7. Conservation Biology and Conservation Genetics

Research on salmonids has had a substantial impact on conservation biology and conservation genetics. Perhaps the most obvious of these impacts is the application of the U.S. Endangered Species Act (ESA). The ESA mandates full protection of "any distinct population segment of any species of vertebrate fish or wildlife which interbreeds when mature." The interpretation of this clause is particularly critical for salmonid species because of their strong tendency to form thousands of populations that are at least partially isolated from each other through natal homing, geographic barriers, and selection against hybridization. To require protection of every such population would be intractable in practice and indefensible in a political arena. Waples (1991a) therefore interpreted "distinct population segments" as the more conservative "evolutionarily significant unit" (ESU). To be an ESU, a population or group of populations (1) "must be substantially reproductively isolated from other conspecific population units," and (2) "must represent an important component in the evolutionary legacy of the species." This definition has since guided the listing of populations of vertebrates as endangered under the ESA in the United States and elsewhere (Waples 1995; Ford 2003—*this volume*).

Salmonids have also made fundamental contributions to the estimation and interpretation of effective population size (N_e), which is the size of an "ideal" population that would have the same genetic properties as the actual population (Wright 1931). An ideal population is considered to have random mating, equal sex ratio, discrete generations, constant population size, and random variation in reproductive success (Wright 1931; Waples 2003—*this volume*). N_e is of critical importance in conservation biology because it strongly influences inbreeding and the rate at which genetic variation is lost from populations. Inbreeding and reduced genetic variation can depress population fitness owing to (1) the expression of recessive deleterious mutations, (2) the fixation of mildly deleterious alleles, (3) reduced heterozygosity, and (4) a reduced ability to respond to future environmental change (Lande 1994; Lynch et al. 1995). Pacific salmon have directly contributed to the development of theory on effective population size because traditional estimation approaches are inappropriate for their atypical life history, wherein different individuals from the same brood can mature at different ages but then reproduce only once (Waples 1990a, Waples 2003—*this volume*). More recently, biologists working on salmon have clarified the theoretical relationships between N_e and the census size of populations (Kalinowski and Waples 2002). Recent empirical work has examined how N_e is influenced by anadromous males (fighters) and non-anadromous males (sneakers) within populations (Jones and Hutchings 2001, 2002).

Another area of conservation and management to which salmonids have contributed concerns the effects of captive propagation on the genetics of populations (Young 2003—*this volume*). Several theoretical treatments have examined the effect of "supportive breeding" (rearing a fraction of the population in captivity and then releasing them into the wild) on N_e and the loss of genetic variation within populations (Ryman and Laikre 1991; Wang and Ryman 2001; Duchesne and Bernatchez 2002). As a result of this and other work, some supplemental breeding programs for salmonids tailor their protocols so as to reduce inbreeding and maximize N_e (Hedrick et al. 2000b). On a more empirical note, work on salmonids has been at the forefront of demonstrations that captive rearing can cause evolutionary changes that make populations maladapted for the natural environment (Reisenbichler and Rubin 1999; Heath et al. 2003). Releasing captive individuals into the wild can depress the fitness of natural populations (Reisenbichler and Rubin 1999; Fleming et al. 2000), despite the increase in census population size. This a particularly critical issue in Norway where escaped farmed Atlantic salmon can represent large proportions of the individuals spawning in rivers (Gausen and Moen 1991). Atlantic salmon are also farmed outside their native range, where they escape in large numbers (e.g., British Columbia: McKinnell et al. 1997; Chile: Soto et al. 2001), sometimes breed successfully (Volpe et al. 2000), and may introduce novel evolutionary pressures on native fishes.

2.8. Where a Salmonid Contribution Seems Less Likely

In several areas of intense evolutionary research, it does not seem likely that salmonids will make big contributions. These include intragenomic conflict (Haig 2000; Hurst et al. 1996), selection arenas (Stearns 1987; Mock and Parker 1997), the evolution of sex (Hurst and Peck 1996; West et al. 1999), the allocation of reproductive effort to male versus female offspring (Charnov 1982; salmonids, like humans, have chromosomal sex determination), and the estimation of mutation rates (Keightley and Eyre-Walker 2000). It would please us greatly to discover that we were wrong about these points.

2.9. Some Areas in Active Development or on the Verge of Completion

1. A reliable phylogenetic tree for the whole salmonid clade, including intraspecific populations and estimates of divergence times along the entire tree. Numerous trees have been proposed using a variety of methods but a consensus has yet to emerge (Kinnison and Hendry 2003—*this volume*). A recently developed phylogeny is shown in Figure 1 (Crespi and Fulton, unpublished).
2. Reliable data on the ancestral origins (i.e., glacial refugia) and major life history traits for replicate populations in the major habitat types. This work has been ongoing for many decades and is nearing the point where certain aspects of life history and genetic variation have been assayed in a fairly standard fashion in many hundreds of populations within *Oncorhynchus* and *Salmo* (e.g., Waples et al. 2001). This breadth of knowledge probably exceeds that for most other taxa, and thus represents an unparalleled opportunity to analyze the determinants of evolutionary diversification.
3. Long-term studies of variation in reproductive success (and the traits that influence reproductive success) on a large enough set of replicate populations to span the low-to-high variation continuum. This endeavor has been greatly assisted by the recent development of large numbers of hyper-variable microsatellite markers that are well suited for parentage analysis in natural populations (Bentzen et al. 2001; Garant et al. 2001, 2003a; Taggart et al. 2001).
4. Quantitative estimates of how rapidly populations can evolve to meet new evolutionary challenges as a function of key population parameters.

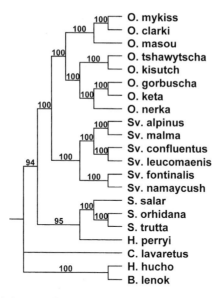

Figure I. A recent phylogeny of some of the salmonid species discussed in this book (for alternative phylogenies and discussions of them see Kinnison and Hendry 2003—*this volume*). Shown are results from Bayesian maximum-likelihood analysis of nuclear DNA data for salmonid fishes, from growth hormone genes 1 and 2, ITS1 and 2, 18S, vitellogenin, and major histocompatibility class II genes. Values at nodes represent *a posteriori* probabilities, derived from 5000 trees after likelihood stabilization (B. Crespi and M. Fulton, unpublished data).

This work is proceeding most rapidly in considering the response of populations to selective harvest (Hard 2003—*this volume*).

3. Concluding Statement

Salmonids are diverse, intriguing, beautiful—and very well studied. They offer a lot to evolutionary biologists in search of wonderful puzzles to solve, just as evolutionary biology offers a lot to salmonid biologists trying to understand why these magnificent fish live and behave as they do. It is our hope that this book will introduce to each other people and puzzles who might otherwise not have met.

Acknowledgments   Detailed comments were provided by Ole Berg, Louis Bernatchez, and Jeff Hutchings. Help with the salmonid life history table was provided by Eugene Balon, Bob Behnke, Ole Berg, Louis Bernatchez, Torgny Bohlin, Jeff Hard, Thrond Haugen, Jeff Hutchings, Bror Jonsson, Jason Ladell, Steve Latham, Koji Maekawa, Tony Ricciardi, Rick Taylor, Robin Waples, and Kyle Young. The table is intended as a consensus, but acknowledgement of the foregoing individuals does not mean that they all agree with all the table entries. The phylogeny was provided by B. Crespi and M. Fulton (unpublished).

Life Histories, Evolution, and Salmonids

William M. Schaffer

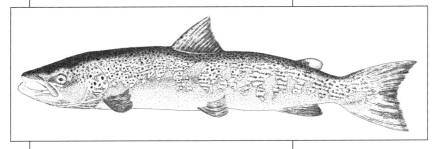

Mature male Atlantic salmon

1

This book is about salmonid fishes and the ways in which studying salmonids can inform general problems in evolution. My own involvement with salmon dates to the fall of 1969 when, at the urging of Professor J. T. Bonner, I attempted to apply my then nascent ideas on life history evolution to the genus *Salmo*. The occasion was the weekly ecology brown bag at which I had been scheduled to speak, and the ensuing discussion focused on the adaptive merits of semelparous versus iteroparous reproduction. Dr. Bonner, it transpired, spent his summers in Nova Scotia where he enjoyed angling for salmon. He was impressed by the tremendous obstacles elsewhere[1] encountered by these fish when they returned from the ocean to spawn. Even though *S. salar* is technically iteroparous, it was apparent that individuals often had but a single crack at reproduction (Belding 1934). I should go to the Maritimes, he proposed; solicit the assistance of P. F. Elson, a fisheries biologist at the research station in St. Andrews; and learn about salmon. Perhaps, there were data that would inform the theory I was developing. Perhaps I would gain insight as to why *Oncorhynchus* is obligatorily semelparous and *Salmo* is not.[2] And so, when the spring semester ended, I drove north, knowing little enough about the animals I had chosen to study, but convinced, as only a graduate student can be, that they would doubtless conform to my expectations. In due course, I met Dr. Elson who took me under a fatherly wing. With his assistance and encouragement, data were gathered, the theory refined, and results published (Schaffer and Elson 1975; Schaffer 1979a). Thereafter, I moved on to other taxa (Schaffer and Schaffer 1977, 1979) and other questions (Schaffer 1985). Thirty years later, an invitation arrived to write this chapter. Let me begin by thanking the editors for affording me an opportunity to revisit a formative stage in my scientific career.

1. Theory of Life History Evolution

1.1. Nature of the Problem

According to Merriam-Webster (http://www.refdesk.com/), demography is "the statistical study of human populations especially with reference to size and density, distribution, and vital statistics." Such studies need not be restricted to humans. One can just as well construct life tables for animals (Deevey 1947) and plants (Harper 1977), although in the latter case, especially, the situation may be complicated by ambiguity as to just what constitutes an individual. Such technicalities aside, one of the more interesting aspects of non-human demography is the fact that age-specific schedules of reproduction and mortality can vary dramatically among closely related species and even among populations of the same species. Some of this variation is nothing more than the proximate reflection of differences in environmental quality. But this is not the whole story. Animal and plant life histories are also the products of evolution, having been molded to a greater or lesser degree by natural selection. Regarding this second source of demographic variation, it is important to bear in mind that characters such as egg size and number, gestation period, generation time, and so on, are often correlated with traits such as body size (Bonner 1965; Calder 1984; Schmidt-Nielsen 1984), upon which selection also can act. As a result, inferring the phenotypic targets of past selection can be problematic. It follows that the biologist seeking adaptive explanations for particular life history phenomena is wise to focus on closely allied forms that differ principally with regard to the characters in question. Put another way, the sensible adaptationist, as opposed to ideologically inspired caricatures (Gould and Lewontin 1979) thereof, is guided by the expectation that *"related* species will differ in the direction of their respective optima."[3]

1.2. Early Investigations

The late 1960s and early 70s were years of heady optimism for evolutionary ecologists who believed that simple mathematical models could offer a useful accounting of the broad brushstroke patterns observable in nature (Kingsland 1985). Forever associated with this conviction will be the name of Robert Helmer MacArthur whose Mozartian career spanned a scant 15 years.[4] Best known for his interest in biogeography (MacArthur and Wilson 1967; MacArthur 1972) and competitive exclusion (May and MacArthur 1972; MacArthur 1958, 1969, 1970; MacArthur and Levins 1967), MacArthur also contributed (1962; MacArthur and Wilson 1967) to the *corpus theoretica* now known as the theory of life history evolution (Roff 1992, 2002; Stearns 1976, 1977, 1992). This line of inquiry traces to the work of Lamont Cole (1954) who was among the first to advocate a comparative approach to the study of animal demographics. Especially, Cole sought to understand the adaptive significance of divergent schedules of reproduction and mortality, in which regard, he presumed that fitness, the elusive "stuff" of evolution, is usefully approximated by the rate at which an asexual individual and its identical descendants multiply under the assumption of unchanging probabilities of reproduction and survival. As discussed by Leslie (1945, 1948), Keyfitz (1968), and others, this rate is determined by the so-called "stable age equation,"

$$1 = \sum_{i=0}^{n} \lambda^{-(i+1)} \ell_i B_i \tag{1}$$

where, ℓ_i is the probability of living to age i, and B_i is the *effective* fecundity of an i-year-old.[5,6]

Equation (1) is a polynomial equation of degree $n + 1$ from which it follows by the Fundamental Theorem of Algebra that there are $n + 1$ (possibly complex and not necessarily distinct) values of λ for which the equality holds. These values are called "roots," and we may arrange them in order of decreasing magnitude, that is, $|\lambda_1| \geq |\lambda_2| \geq \ldots \geq |\lambda_{n+1}|$.[7] If one neglects post-reproductive age classes, the Peron–Frobenius theorem (Gantmacher 1959) guarantees that λ_1 is real and positive. Moreover, we can usually replace the inequalities, $\lambda_1 \geq |\lambda_2| \geq \ldots \geq |\lambda_{n+1}|$, with *strict* inequalities, $\lambda_1 > |\lambda_2| > \ldots > |\lambda_{n+1}|$, in which case, λ_1 is the rate at which the population ultimately multiplies.[8] It is this final fact that justifies identification of λ_1 with fitness.

Equating fitness with λ_1 enabled Cole to inquire as to the circumstances favoring alternative schedules of birth and death. In particular, he argued that delaying the onset of reproduction would generally be disadvantageous, in which regard, he posed his famous paradox (text box) concerning the ubiquity of the perennial habit. Implicit in Cole's analysis was the notion of *tradeoffs*: over the course of evolution, he imagined, current fecundity can be exchanged for subsequent survival and reproduction, an idea that George Williams (1966a) later made explicit via the concept of "reproductive effort." Of course, many mutations result in fitness diminutions all round, that is, reductions in *both* reproduction and survivorship. *Modulo* the effects of small population size, such variations are eliminated by selection—which is what Levins (1968) meant when he observed that it is only those genotypes corresponding to points on the margins of a fitness set that matter.

What about potentially beneficial variations? To first approximation (Crow and Kimura 1970; Charlesworth and Williamson 1975), the likelihood of a favorable mutation being fixed depends on its selective advantage, which, in the present case, reflects the values of the partial derivatives,

$$\partial \lambda_1 / \partial B_i = \left(\ell_i / \lambda_1^i \right) / V_T \qquad (2a)$$

Cole's Paradox

Cole asserted that "for an annual species, the absolute gain in intrinsic population growth ... achieved by changing to the perennial reproductive habit would be exactly equivalent to adding one individual to the average litter." His argument was as follows:

Consider an annual plant, which produces B_a seeds and then dies, and a corresponding perennial which produces B_p seeds and itself survives with probability one to reproduce the following year and again with probability one, the year after that, *ad infinitum*. The annual and its descendants multiply yearly at rate $\lambda_a = B_a$; the perennial and its descendants, at rate $\lambda_p = B_p + 1$ Energetically, seeds are "cheap" compared to adaptations that permit plants to overwinter. Why then, asked Cole, aren't perennials replaced by annual mutants which "trade in" the energy required for long life for a small increment in seed production?

The answer (Charnov and Schaffer 1973) is that the B's are "effective fecundities," that is, the numbers of seeds that germinate and themselves survive to set seed. That is, $B_a = cb_a$, and $B_p = cb_p$, where b's are the numbers of seeds produced, and c is the aforementioned germination-survival probability. For the annual population to grow more rapidly than the perennial, it is therefore necessary that $b_a > b_p + 1/c$, or, in the case that adults survive with probability $p < 1$, that $b_a > b_p + p/c$. Often, $c \ll p$, in which case, $(p/c) \gg 1$. Then, superiority of the annual requires $b_a \gg b_p$, which is to say, an increase in fecundity substantially in excess of a single seed.

and

$$\partial \lambda_1 / \partial p_i = \left((\ell_i / \lambda_1^i) / V_T \right) (v_{i+1} / v_0) \tag{2b}$$

(Hamilton 1966; Emlen 1970; Caswell 1979).[9] Here,

$$v_i / v_0 = (\lambda_1 / \ell_i) \sum_{j=1}^{\infty} \lambda_1^{-(j+1)} \ell_j B_j \tag{3}$$

is the *reproductive value* (Fisher 1930) of an i-year-old, which is simply the expectation of current and future offspring discounted by appropriate powers of λ_1.[10] Correspondingly,

$$V_T = \sum_{j=i}^{\infty} (i+1) \lambda_1^{-(i+1)} \ell_i B_i \tag{4}$$

is the *total reproductive value* of a population that has achieved stable age distribution.

Equations (2a) and (2b) suggest two principal conclusions:

1. Increased reproduction at earlier ages confers greater benefits to fitness than comparable increases at later ages. This reflects the fact that, so long as $\lambda_1 \geq 1$, the term, (ℓ_i / λ_1^i) necessarily declines with age.
2. The benefits of increased survival are often greatest at intermediate ages. This is because $(\partial \lambda_1 / \partial p_i)$ depends on reproductive value, which generally peaks at, or shortly after, the age of first reproduction (Fisher 1930).

1.3. Fixation of Advantageous Mutants in the Presence of Density Dependence

Thus far, we have assumed constant rates of reproduction and survival. This is appropriate for populations growing exponentially. Of course, real-world populations do not multiply indefinitely. Rather, their growth rates are arrested by a variety of limiting factors, the effect of which is to reduce reproduction and increase mortality. It is important therefore to inquire as to what happens when life history parameters manifest *density dependence*, that is, when λ_1 depends on both population numbers and phenotype. To this end, we study the fate of an advantageous mutant arising in a population at equilibrium with respect to both age structure and density. In what follows, we denote the wild type phenotype φ_0, the mutant phenotype, φ' and the corresponding equilibrium densities (carrying capacities), K_0 and K'. Please refer to Figure 1.1.

When the mutant first appears, the wild type is in approximate equilibrium, and $\lambda_1(\varphi_0, K_0) \approx 1$. At the same time, the mutant's rate of increase, by virtue of its being advantageous, exceeds 1. That is, $\lambda_1(\varphi', K_0) > 1$. More precisely, the mutant manifests a selective advantage,

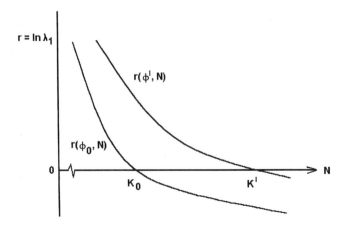

Figure 1.1. Fitness, $r(\varphi, N) = (1/N)(dN/dt)$, plotted against density for wild type and mutant phenotypes, φ_0 and φ'. In the course of evolving from φ_0 to φ', the population's equilibrium density increases from K_0 to K'.

$$s = \left[r(\varphi', K_0) - r(\varphi_0, K_0) \right] \approx \Delta\varphi(dr/d\varphi)|_{N=K_0} > 0 \qquad (5)$$

where $r = \ln \lambda_1$. Since $s > 0$, the number of mutants increases and, with it, the population's overall size. At the same time, increasing density forces reductions in the rates of increase of *both* phenotypes. Now $\lambda_1(\varphi_0, K') < 1$, and the number of wild type individuals begins to decline. Still, $N < K'$, and the number of mutants continues to increase, albeit more slowly. Eventually, the wild type dies out, and population size stabilizes at the new carrying capacity, $N = K'$, which is larger than the old. At this point, the mutant's frequency, $p = 1$, and its rate of multiplication, $\lambda_1(\varphi', K') = 1$.

Regarding this scenario, there are two principal caveats. In the first place, we assume the substitution process to be *slow*, by which we mean *quasi-static*, that is, slow enough that both density and age structure track their respective equilibria. We also require that the fitness of the mutant exceed that of the wild type on the entire range of population densities, $[K_0, K']$. According to this, substitution of a favorable mutant is formally equivalent to interspecific competition when the zero-growth isoclines do not cross. Absent this assumption, we are into the world of r- versus K-selection as these concepts are traditionally formulated (MacArthur and Wilson 1967; Pianka 1970).

When "phenotype," that is, φ, refers to the B's or p's above, we may use Equations (2a) and (2b) to evaluate $dr/d\varphi = (1/\lambda_1)(\partial\lambda_1/\partial\varphi)$. This ignores the possibility of tradeoffs, for example, between current fecundity and subsequent reproduction and survival. If there are tradeoffs, these must be taken into account, as discussed in Schaffer (1979b). To summarize, the machinery developed for the density-independent case can be applied to density-dependent situations given satisfaction of certain assumptions.

1.4. Life History Evolution as an Optimization Problem

The notion of reproductive effort was first linked to Cole's mathematics by Gadgil and Bossert (1970) who framed life history evolution in terms of costs and benefits. Specifically, they considered the set of age-specific reproductive expenditures, $E = [E_0, \ldots, E_n]$, and assumed explicit dependencies of the fecundities, B_i, and survival probabilities, $p_i = (\ell_{i+1}/\ell_i)$, on E. The computer was then used to determine the expenditures, $E^* = [E_0^*, \ldots, E_n^*]$, maximizing λ_1. As discussed by subsequent authors (Taylor et al. 1974; Leon 1976; Schaffer 1983), this is an *optimization problem* (Bellman 1957; Leitmann 1966; Intrilligator 1971) whereby a set of "controls," the E_i, are adjusted to maximize an "objective function," λ_1.

Because the dependence of λ_1 on the controls, $E = [E_0, \ldots, E_n]$, is via their effect on the B's and p's, viewing life history evolution as an optimization problem focuses our attention on the functional dependencies, $B_i(E_i)$, $p_i(E_i)$, and so on. Concerning these, we note the following:

1. Effective fecundity will generally be an *increasing* function of reproductive effort at the current age and possibly a decreasing function of expenditure at prior ages. That is, $\partial B_i / \partial E_i > 0$ and $\partial B_i / \partial E_j \leq 0$, for $j < i$.
2. In the special case that $B_{i+1} = g_i(E_i)B_i$, the rate, g_i, at which fecundity multiplies from one year to the next will be a *decreasing* function of reproductive effort at the current age, that is, $\partial g_i / \partial E_i < 0$.
3. Post-reproductive survival will also be a *decreasing* function of reproductive effort at the current age and possibly a decreasing function of expenditure at prior ages. That is, $\partial p_i / \partial E_i < 0$, and $\partial p_i / \partial E_j \leq 0$ for $j < i$.

This brings us to one of the more interesting subjects in life history theory, namely the adaptive significance of "big bang reproduction" (semelparity) whereby a single, often spectacular, bout of breeding is followed by the organism's obligate demise. Concerning this matter, Gadgil and Bossert (1970) suggested that such a strategy should be favored by concave (i.e., *accelerating,*) dependencies (positive second derivatives) of effective fecundity and post-breeding survival on reproductive effort. Conversely, they proposed that convex, (i.e., *decelerating*) dependencies (negative second derivatives), should favor more modest levels of reproduction and, hence, post-reproductive survival and iteroparity. The distinction is illustrated in Figure 1.2 for the special case in which the functions $B_i(E_i)$ and $p_i(E_i)$ are the same for all age classes, and there is a common reproductive effort, $E = E_0 = E_1 = \ldots = E_n$.

Under these assumptions, and in the limit that the number of age classes, $n \to \infty$, the multidimensional optimization problem collapses (Schaffer 1974a) to the simpler chore of maximizing

$$\lambda(E) = B(E) + p(E) \qquad (6a)$$

If effective fecundity, $B(E)$, multiplies at rate, $g(E)$, each year, this expression generalizes to

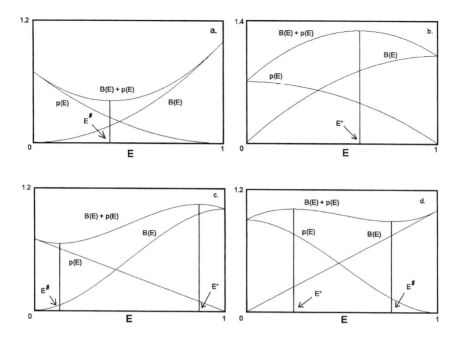

Figure 1.2. Optimal reproductive expenditure under different assumptions regarding the tradeoff between current effective fecundity, $B(E)$, and subsequent survival, $p(E)$, for populations growing according to Eq. (6a). Optimal expenditures are values of E at which $\lambda(E) = B(E) + p(E)$ is maximal. **a.** $B(E)$ and $p(E)$ are *accelerating* functions (*positive* second derivatives) of E. In this case, $\lambda(E)$ passes through a minimum, $E^{\#}$, and the optimal expenditure is 0 *or* 1. **b.** The dependencies, $B(E)$ and $p(E)$ are *decelerating* functions (*negative* second derivatives) of E with the consequence that the optimum expenditure, E^*, is often *between* 0 and 1. **c.** and **d.** More complex dependencies. In **c**, $E^* = 0$ or $0 < E^* < 1$; in **d**, $0 < E^* < 1$ or $E^* = 1$. Functions and parameter values used to compute these figures are as follows: **a.** $B(E) = \beta E^2$; $\beta = 1.0$; $p(E) = \pi(1 - E)^2$, $\pi = 0.7$. **b.** $B(E) = \beta(2E - E^2)$, $\beta = 1.0$; $p(E) = \pi(1 - E^2)$, $\pi = 0.7$. **c.** $B(E) = \beta(3E^2 - 2E^3)$, $\beta = 1.0$; $p(E) = \pi(1 - E)$, $\pi = 0.7$. **d.** $B(E) = \beta E$; $\beta = 1.0$; $p(E) = (1 - 2E^2 + 2E^3)$, $\pi = 0.9$.

$$\lambda(E) = B(E) + p(E)g(E) \tag{6b}$$

(Schaffer 1974a; Schaffer and Gadgil 1975). In either case, there is a tradeoff between current and future components of fitness: current fecundity and subsequent survival (Equation (6a)) or current fecundity and subsequent survival multiplied by the yearly rate at which fecundity multiplies (Equation (6b)).

Returning to Figure 1.2, we note that when $B(E)$ and $p(E)$ are concave (Figure 1.2a), the optimal expenditure, E^*, is either 0 or 1. Often, both of these so-called "boundary" solutions correspond to local maxima in $\lambda(E)$. Conversely, when $B(E)$ and $p(E)$ are convex (Figure 1.2b), E^* will often be intermediate between 0 and 1. That is, one has the "interior" solution, $0 < E^* < 1$.

The dependencies in Figures 1.2a and 1.2b are very simple. Figures 1.2c and 1.2d illustrate more complicated possibilities. Here, as in Figure 1.2a, there can be two expenditures corresponding to local maxima in fitness. Of these, one is always a boundary solution, that is, $E^* = 0$ (Figure 1.2c) or 1 (Figure 1.2d), while the other is often an interior solution, that is, $0 < E^* < 1$.

Of the four fecundity curves in Figure 1.2, the sigmoid function (Figure 1.2c) is arguably the one most often occurring in nature. Sigmoidal dependence combines the idea of "start-up costs" to reproduction with the notion of diminishing returns when expenditure is high. With regard to start-up costs, we note that in most organisms the commitment to breed entails an initial investment that, of itself, yields nothing in the way of viable offspring. In anadromous fishes, for example, this investment, involving physiological changes that facilitate returning to fresh water, an often costly upstream migration, and so on, can be substantial (Hendry et al. 2003b—*this volume*). Intraspecific competition—among males for females and among females for nesting sites—will also serve to increase the range of effort values for which fecundity evidences an accelerating dependence on expenditure. In extreme cases, such factors may have the effect of making the entire function concave as in Figure 1.2a.

Other factors can be expected to have the opposite effect, that is, to promote diminishing fecundity returns at high effort values (the decelerating portion of the curve). Examples include competition among sibs for resources (including parental attention) and the tendency of numerous, tasty offspring to attract natural enemies.

With regard to post-reproductive survival, the functional dependencies most frequently occurring in nature are arguably convex functions of the sort shown in Figure 1.2b. Biologically, such curves correspond to the idea that the consequences to post-breeding survival of increasing reproductive expenditure only become significant when investment exceeds some threshold. But here again, changing biological and environmental circumstances can mold the dependency. For example, organisms in poor condition to begin with may manifest abrupt declines in post-breeding survival at intermediate levels of expenditure—in which case the dependency will be convex–concave as shown in Figure 5 of Schaffer and Rosenzweig (1977). Conversely, the onset of reproduction may involve the initiation of morphological changes or behaviors that place the animal at immediate risk—think breeding coloration in birds—even though the energetic investment itself is modest. Then, survivorship may evidence an initial drop, in which case the curve will be concave–convex, as shown in Figure 4 of Schaffer and Rosenzweig (1977). If sufficiently pronounced, such effects can make the curve entirely concave (Figure 1.2a).

Regarding such considerations, we make two observations. The first is that they are difficult to quantify in the field (Schaffer and Schaffer 1977). Second, they can be confusing to think about, essentially because one is attempting to squeeze too much biology into a single variable. These are issues to which we will return in Sections 2 and 3. For the present, we reiterate that even the simplest model life histories can manifest alternative solutions to the same circumstances. With more than one age class (see below), the possibilities for multiple optima proliferate.

1.5. Characterizing an Optimal Life History

My own contribution to life history evolution with age structure was technical. As a student, I absorbed MacArthur's dictum that undue reliance on the computer is an invitation to counter-example. "What would you do," he once asked a visiting systems ecologist describing the results of an elaborate computer simulation, "if someone pulled the plug?" And so, I fretted about what was going on inside Gadgil's computer. Eventually, I convinced myself (Schaffer 1974a, 1979b) that selection acts to maximize reproductive value at all ages, a result which I later discovered had previously been conjectured by Williams (1966a).[11] More precisely, it turns out that the reproductive value, (v_i/v_0), of each age class is maximized with respect to the expenditure, E_i, at *that* age.[12] In symbols,

$$\underset{E}{\text{Max}}(\lambda_1) <=> \underset{E_i}{\text{Max}}(v_i/v_0), \text{ all } E_i \qquad (7)$$

which is read, "maximizing λ_1 by adjusting *all* of the expenditures is equivalent to maximizing each reproductive value, v_i/v_0, with respect to E_i." At about the same time, Taylor et al. (1974) and Pianka and Parker (1975) independently came to the same conclusion.

Intuitively, Equation (7) makes sense—reproductive value being the age-specific expectation of current and future reproduction discounted by appropriate powers of λ_1. One can further show (Schaffer 1974a) that these two components of fitness, that is, current and future reproduction, are exchangeable: maximizing (v_i/v_0) with respect to E_i also maximizes the sum, $B_i + p_i(v_{i+1}/v_0)$, where the second term, $p_i(v_{i+1}/v_0)$, is sometimes referred to as the "residual reproductive value," RRV, of an i-year-old. The complement to Equation (7) is thus

$$\underset{E}{\text{Max}}(\lambda_1) <=> \underset{E_i}{\text{Max}}[B_i + p_i(v_{i+1}/v_0)], \text{ all } E_i \qquad (8)$$

Equation (8) is the analog of Equations (6) appropriate to cases in which there are distinguishable age classes. Among other things, it allows for the definition (Schaffer and Gadgil 1975; Schaffer 1979a) of costs and benefits. In particular, with a small increase, ΔE_i, in reproductive expenditure at age i, we can associate the

$$\text{Benefit } (\Delta E_i) = \Delta B_i \approx \Delta E_i(\partial B_i/\partial E_i) \qquad (9a)$$

and the

$$\text{Cost } (\Delta E_i) = \Delta RRV_i \approx -\Delta E_i\{\partial[p_i(v_{i+1}/v_0)]/\partial E_i\} \qquad (9b)$$

1.6. Predictions of Simple Models

While Equations (7) and (8) speak to the nature of an optimal life history, they cannot be used to compute one directly. This is because the optimal reproductive expenditure for any *particular* age class depends on the reproductive expenditures at *all* ages.[13] As discussed below, this leads to the study of curves (surfaces) of *conditional optima* whereby one computes the optimal value of E_i for representative reproductive expenditures at the other ages (Schaffer and Rosenzweig 1977). Before turning to such matters, however, let us first review what can be learned from single age class models. In this instance, the following results can be deduced:

1. **Semelparity versus iteroparity (Schaffer 1974a; Schaffer and Gadgil 1975).** We have already remarked upon the fact that the evolution of "big-bang" reproduction is predicted under circumstances that make for concave, that is, accelerating dependencies of life table parameters on reproductive expenditure (Figures 1.2a, 1.2d). So far as I am aware, the existence of accelerating dependencies in a natural population has yet to be demonstrated convincingly. Still, one can speculate as to the biological circumstances that would have this effect. In the case of semelparous Agavaceae, Schaffer and Schaffer (1977, 1979) proposed (with some empirical support) that accelerating dependencies of effective fecundity on effort may result from competition among flowering plants for pollinators. Similarly, concavity of the fecundity function in Pacific salmon may result from competitive interactions among reproductives (see below). In such cases, reproductive success depends, not on an individual's reproductive effort per se, but on his/her effort relative to that of other individuals in the population. This suggests a connection between behavioral components of reproductive expenditure and dispersal, as discussed in Schaffer (1977).

2. **Allocation in proportion to relative returns (Schaffer 1974a; Schaffer and Gadgil 1975).** This is the theory's most general prediction: Increasing fecundity per unit investment favors greater reproductive effort (Figure 1.3a); increasing post-reproductive survival or year-to-year growth in fecundity per unit investment favors reduced reproductive expenditure, which is to say, greater allocation to maintenance and growth (Figure 1.3b).

3. **Effects of juvenile (pre-reproductive) versus adult (post-reproductive) mortality (Schaffer 1974a; Schaffer and Gadgil 1975).** This is a special case of the previous result. As such, it merits separate mention because it serves to emphasize the fact that the fecundities in Equation (2) are *effective* fecundities, that is, the numbers of offspring that survive to be counted as young-of-the year at census time. Accordingly, increasing (decreasing) survival rates among juveniles will select for greater (lesser) rates of reproductive expenditure. This is in contrast to changing survival rates among adults that select in the opposite direction.

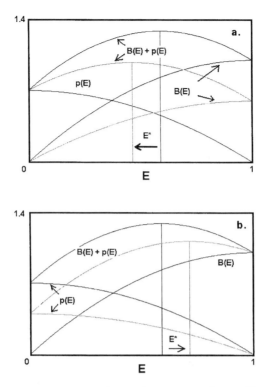

Figure 1.3. Consequences to optimal reproductive expenditure, E^*, of per unit effort changes in effective fecundity, $B(E)$, and post-reproductive survival, $p(E)$, in a model without age structure. **a.** Reducing $B(E)$ reduces E^*. **b.** Reducing $p(E)$ increases E^*.

4. **Under density dependence, optimal reproductive strategies maximize carrying capacity (Schaffer and Tamarin 1973).** This is the life historical realization of MacArthur's (1962) reformulation of Fisher's (1930) Fundamental Theorem of Natural Selection. Two examples are given in Figure 1.4. Here we plot optimal expenditure as a function, $E^*(N)$, of density and equilibrium density as a function, $N^*(E)$, of reproductive effort. In the first instance (Figure 1.4a), density is assumed to reduce juvenile survival and hence effective fecundity; in the second (Figure 1.4b), increasing population numbers are assumed to reduce post-reproductive survival. In both cases, the optimal expenditure changes with population size so as to maximize the carrying capacity, which is what MacArthur's more general analysis predicts. At the same time, the means by which this maximization is accomplished differ: in the first case, optimal reproductive effort declines with density; in the second, optimal expenditure increases. At first glance, the second result would seem to be at odds with the theory of r- and K-selection, which holds that r-selected species produce more young and are generally shorter lived than their K-selected counterparts. The discrepancy vanishes on consideration of the

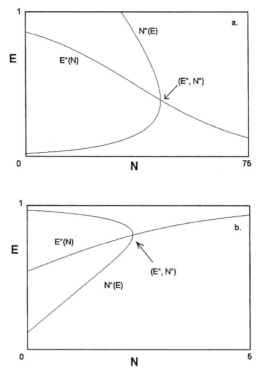

Figure 1.4. Optimal reproductive effort, $E^*(N)$, and equilibrium density, $N^*(E)$, under alternative assumptions regarding the effects of increasing population size. **a.** Increased density reduces pre-reproductive survival. Optimal effort declines with density. $\lambda(E, N) = B(E, N) + p(E) = \beta e^{-cN}(2E - E^2) + \pi(1 - E^2); \quad c = 0.5; \quad \beta = 5.0; \quad \pi = 0.8.$ **b.** Increased density diminishes post-reproductive survival. Optimal effort increases with density. $\lambda(E, N) = B(E) + p(E, N) = \beta(2E - E^2) + \pi e^{-cN}(1 - E^2); \quad c = 0.5; \quad \beta = 0.95; \quad \pi = 0.8.$

fact that under conditions of crowding, it is the youngest individuals that usually experience the most additional mortality.

5. **Year-to-year variation in relative returns is selectively equivalent to reducing the mean return in a constant environment (Schaffer 1974b).** Suppose the effective fecundity, $B(E)$, and survival, $p(E)$, functions are subject to year-to-year variations that result from environmental fluctuations. Choosing the population's *geometric* mean rate of increase,

$$\overline{\lambda_g} = \lim_{n \to \infty} \left(\prod_{i=0}^{n} \lambda_i \right)^{(1/n)}$$

as our criterion of optimality yields the following results:

(a) Variation in effective fecundity per unit investment selects for reduced reproductive expenditure;

(b) Variation in post-reproductive survivorship and growth favors increased reproduction.

These results are sometimes cited (Stearns 1976) as an evolutionary example of "bet-hedging." In fact, the arguments from which they follow have nothing to do with risk minimization (Schaffer 1974b). More generally, if variations in reproduction and survivorship per unit expenditure are predictable, selection will favor adaptations that allow organisms to "read" the environment and respond adaptively.

1.7. Optimal Allocation with Age Structure

With the addition of age structure, the problem of deducing general results becomes more difficult. As noted above, this is because the optimal reproductive expenditure for any *particular* age class depends on next year's reproductive value, and thus on the reproductive expenditures at *all* ages. Taking this complication into account yields the following results:

1. **Selective consequences of altered returns per unit investment depend on the age classes affected (Schaffer 1974a; Schaffer and Rosenzweig 1977).** As before, increasing effective fecundity for the *current age class* favors increased reproductive expenditure *at that age*, whereas increasing post-reproductive survival and/or growth per unit effort for the current age class selects for reduced expenditure (again, at that age). In addition, increasing fecundity *or* survivorship per unit investment at *earlier* age classes *increases* the optimal current expenditure. This is because these changes increase λ_1 and thereby reduce subsequent reproductive value (v_{i+1}/v_0). Conversely, increasing fecundity or survivorship at *later* age classes *reduces* the optimal current expenditure because such changes increase subsequent reproductive value. These results are summarized in Table 1.1.

Table 1.1. Consequences to current optimal reproductive effort, E_i^*, of increased effective fecundity and survival per unit investment.

Life history parameter affected					
Current age		Later ages		Earlier ages	
$B_i(E_i)$	$p_i(E_i)$	$B_i(E_i)$	$p_i(E_i)$	$B_i(E_i)$	$p_i(E_i)$
Consequence to E_i^*					
↑	↓	↓	↓	↑	↑

2. **Reduced reproductive effort in older individuals can be adaptive.** Gadgil and Bossert (1970) proposed that optimal effort invariably increases with age, a conclusion that suggests declining expenditures in older individuals are a maladaptive consequence of aging. Subsequently, Fagen (1972) produced a counter-example and observed that almost *any* age-specific pattern in optimal effort is possible depending on the way in which the functions $B_i(E_i)$ and $p_i(E_i)$ vary with age. In retrospect, one imagines that Gadgil and Bossert were misled by the fact that their model life histories assumed a maximum age following which death was obligate. Provided that fecundity increases monotonically with effort, the optimum expenditure for this final age class is necessarily 100%, that is, in terms of Equation (8), $(v_{i+1}/v_0) = 0$ for all values of E_i, which should thus be chosen to maximize B_i. It follows that optimal effort increases from the penultimate to the final age class, and, oftentimes, the pattern cascades back to the younger age classes.

From a computational point of view, assuming a fixed life span may be convenient. However, on biological grounds, it is often more realistic to suppose that even the oldest individuals have a non-zero probability of surviving another year. In such cases, there is no maximum age class per se, and optimal reproductive investment can decline in the later age classes. In particular, reduced per-unit effort fecundity in older individuals, as typically results from aging, can favor reductions in reproductive effort. This suggests a distinction between organisms in which body size and fecundity increase throughout life, and those in which body size and reproductive potential plateau. Some sample computations are presented in Table 1.2 wherein I report some results for a five-age-class model. Under the assumption of continuing age-dependent increases in potential fecundity, optimal effort increases with age regardless of what one assumes about the final age class.[14] Conversely, when fecundity per unit investment peaks at intermediate ages, the Gadgil–Bossert model only obtains if one assumes the obligate demise of individuals in the oldest age class. Absent this assumption, that is, if one regards the oldest age class as "adults" which survive from one year to the next with constant probability, optimal expenditure can manifest the up–down pattern noted above.[15] These results underscore the selective link between senescence, as the term is usually understood, that is, the onset of degenerative disorders of the sort targeted by geriatric medicine, and optimal life history evolution.[16]

1.8. Optimal Effort for Multiple Age Classes

Thus far, I have emphasized the evolution of reproductive expenditure at individual age classes, either in single age-class models, or in multi-age-class models, with expenditures at the other age classes implicitly held constant. To complete our discussion, we consider selection on the entire set of reproductive expenditures, $E = E_0, \ldots, E_n$. That is, we treat the life table as a whole. To

Table 1.2. Age-specific optimal effort, E_i^*.

	Age class				
	0	1	2	3	"Adults"

I. Potential fecundity increases with age

Maximum return on investment

B_i	0.50	1.00	2.00	3.00	4.00

Optimal effort

Case A. Adults die after one year

E_i^*	0.34	0.36	0.45	0.57	1.00

Case B. Adults survive with probability p_A

E_i^*	0.34	0.35	0.43	0.53	0.66

II. Potential fecundity declines in old age

Maximum return on investment

B_i	0.50	1.00	2.00	1.00	0.50

Optimal effort

Case A. Adults die after one year

E_i^*	0.37	0.44	0.73	0.76	1.00

Case B. Adults survive with probability p_A

E_i^*	0.37	0.44	0.72	0.72	0.63

$B_i(E_i) = \beta_i(2E_i - E_i^2); \pi_i(E_i) = \pi_i(1 - E_i^2); \pi_i = [0.50, 0.60, 0.70, 0.65, 0.50]$.

do this, we introduce the notion of *curves (or surfaces) of conditional optima* that specify the optimal reproductive effort at a particular age for some suitably chosen set of expenditures at the other ages (Schaffer 1974a; Schaffer and Rosenzweig 1977). We emphasize that this approach is neither esoteric nor difficult to implement. In practice, one identifies an age class of interest and computes the optimal reproductive effort at *that* age for a large number of expenditures at the *other* ages. For example, in the case of two-age-class models, one calculates the curve of optimal expenditures, $E_0^*(E_1)$, at age 0 for representative values of E_1 and the corresponding curve, $E_1^*(E_0)$, for representative values of E_0. The resulting curves are then plotted in the $E_0 - E_1$ plane and inspected. Points at which the curves intersect maximize λ_1 with respect to both expenditures and consequently correspond to optimal life histories. With three age classes, the curves become surfaces in $E_0 - E_1 - E_2$ space; and, with $n + 1 > 3$ age classes, hypersurfaces in $E_0 - \ldots - E_n$ space. In an entirely analogous fashion, one can compute curves (surfaces) of conditional *minima* in fitness. From the viewpoint of adaptive topographies (Wright 1968), intersections of curves (surfaces) of conditional optima correspond to *peaks* on an adaptive landscape; intersections of curves (surfaces) of conditional minima, to *valleys*; and intersections of conditional maxima with minima, to *saddles*.

Figure 1.5 shows the result of applying this approach to a two-stage life history in which there is a single juvenile age class followed by an infinite number of adult age classes wherein individuals survive with constant probability from one year to the next.[17] Four examples, corresponding to each of the single class models of Figure 1.2, are given. In each case, the thick white lines are curves of conditional optima, E_j^* (E_a) and E_a^* (E_j); and the thin white lines are curves of conditional minima, $E_j^\#$ (E_a) and $E_a^\#$ (E_j). In addition, reproductive effort pairs, (E_j, E_a), are color-coded according to fitness value as indicated by the different shades of gray. Color versions of Figures 1.5–1.7 and 1.9 are posted at http://www.oup-usa.org/sc/019514385X and at http://bill.srnr.arizona.edu, whereat ancillary material including an animation is also available. In Figure 1.5a, we consider *accelerating* dependencies of effective fecundity and post-reproductive survival on reproductive effort. In this case, there is a single global minimum in fitness with the consequence that optimal life histories, $(E_j, E_a) = (0, 0), (0, 1)$, and so on, are all boundary solutions corresponding to zero reproduction or the maximum possible reproduction at the different age classes. In Figure 1.5b, we consider *decelerating* dependencies of fecundity and survival on reproductive effort. Here, there is a single, *interior maximum* in fitness. Figures 1.5c and 1.5d treat cases in which effective fecundity and post-reproductive survival are more complex. Here, the fitness landscapes evidence multiple peaks and valleys.

As further discussed in the following section, studying curves of conditional optima allows one to think more clearly about the evolution of life histories as a whole. In particular, this approach facilitates the elaboration of falsifiable predictions regarding the selective consequences of changing environmental circumstances affecting reproduction, growth, and survival.

1.9. Optimal Age of First Reproduction

One consequence of the theory is that it generates predictions regarding the optimal age of first reproduction, a character more readily quantified than small variations in reproductive expenditure. In this regard, recall that when effective fecundity functions are sigmoidal (Figure 1.2c), there can be two values of reproductive effort corresponding to local maxima in λ_1. Of these, one is always zero, in which case, the organisms don't reproduce, while the second is often intermediate between 0 and 100%. In such cases, factors that *increase* optimal reproductive expenditure within an age class also act to *lower* the optimal age of first reproduction. Conversely, factors that *reduce* optimal expenditure within age classes favor *delays* in the onset of breeding. Computations illustrating these effects for the two-stage life history studied

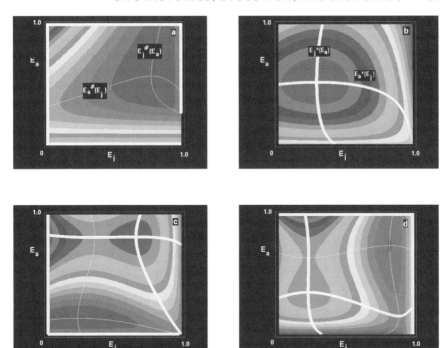

Figure 1.5. Adaptive topographies and optimal reproductive expenditures in a two-stage (juvenile–adult) life history. The coordinate axes, E_j and E_a, are juvenile and adult reproductive effort. Reproductive effort pairs, (E_j, E_a), are color coded by $\exp(\lambda_1)$. The thick white lines are curves of *conditional optima*, $E_j^*(E_a)$ and $E_a^*(E_j)$ whereon λ_1 is maximized. The thin white lines are curves of *conditional minima* whereon λ_1 is minimized. Color versions of these figures are posted at http:// www.oup-usa.org/sc/019514385X and at http://bill/srnr.arizona.edu, whereat ancillary material is also available. **a.** The effective fecundity and post-reproductive survival functions are *accelerating* functions of reproductive effort as in Figure 1.2a. In this case, there is a single global *minimum* in fitness and multiple optimal life histories. The latter correspond to zero or 100% reproduction at each age. **b.** The effective fecundity and post-reproductive survival functions are *decelerating* functions of reproductive effort as in Figure 1.2b. Here, there is a single global maximum in fitness. **c.** Effective fecundity is a sigmoid function of effort; post-reproductive survival declines linearly as in Figure 1.2c. In this case, there are two prominent adaptive peaks, the taller of which corresponds to no reproduction by juveniles and high effort by adults. **d.** Effective fecundity is a linear function of effort; post-reproductive survival, a reverse sigmoid as in Figure 1.2d. Effective fecundity and post-reproductive survival functions as in Figure 1.2.

in Figure 1.5 (but with the added possibility of year-to-year increases in effective fecundity among adults) are given in Figures 1.6 and 1.7. Figure 1.6 documents the fecundity effect. With increasing fecundity per unit expenditure in *both* stages of the life cycle, the optimal life history can

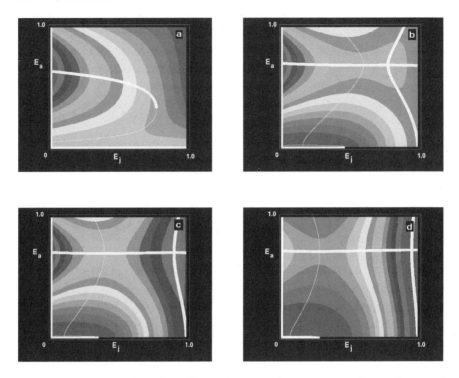

Figure 1.6. Effective fecundity effect. Changing adaptive topographies and optimal reproductive expenditures in response to increasing fecundity potential (effective fecundity per unit effort) in a two-stage life history. Curves of conditional optima and minima and reproductive effort pairs, (E_j, E_a), color coded as in Figure 1.5. Color versions of these figures are posted at http://www.oup-usa.org/sc/019514385X and at http://bill/srnr.arizona.edu, whereat ancillary material is also available. Life history functions, $B_i(E_i)$ and $p_i(E_i)$ as in Figure 1.2c (fecundity a sigmoidal function of effort; post-reproductive survival, a linear function with $\pi_j = 0.8$; $\pi_a = 0.9$). Post-reproductive growth function, $g_i(E_i) = 1 + \gamma_i(1 - E_i)$, with $\gamma_j = \gamma_a = 2.0$. **a.** $\beta_j = \beta_a = 0.5$. **b.** $\beta_j = \beta_a = 2.0$; **c.** $\beta_j = \beta_a = 3.0$; **d.** $\beta_j = \beta_a = 5.0$. With fecundity per unit effort low, juveniles should not reproduce. As potential fecundity increases, the optimal life history shifts to one in which reproductive effort is high at both stages in the life cycle.

shift from one in which only adults reproduce to one in which both age classes manifest high expenditures. Figure 1.7 documents the effect of increasing the rate—again, in *both* life cycle stages—at which fecundity multiplies from one age to the next. Here, the shift is in the opposite direction: increasing this rate selects for delayed reproduction. To complete the story, we note that increasing post-reproductive survival per unit expenditure also selects for delayed reproduction (not shown).

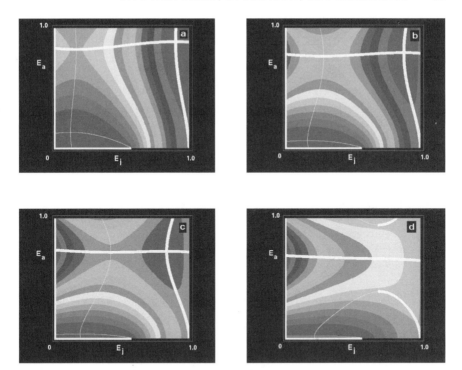

Figure 1.7. Post-reproductive growth rate effect. Changing adaptive topographies and optimal reproductive expenditures in response to increasing post-reproductive growth potential in a two-stage life history. Curves of conditional optima and minima and reproductive effort pairs, (E_j, E_a), color coded as in Figure 1.5. Color versions of these figures are posted at http://www.oup-usa.org/sc/019514385X and at http://bill/srnr.arizona.edu, whereat ancillary material is also available. Life history functions, $B_i(E_i)$ and $p_i(E_i)$ as in Figure 1.2c with $\beta_j = \beta_a = 1.5$ and $\pi_j = 0.8$; $\pi_a = 0.9$. Post-reproductive growth function, $g_i(E_i)$, as in Figure 1.7. **a.** $\gamma_j = \gamma_a = 0.0$. **b.** $\gamma_j = \gamma_a = 0.5$; **c.** $\gamma_j = \gamma_a = 1.0$; **d.** $\gamma_j = \gamma_a = 2.0$. For low growth potential, the optimal life history corresponds to large reproductive expenditures by juveniles and adults. With increasing growth potential, the optimal life history shifts to one in which the onset of reproduction is delayed, that is, juveniles no longer reproduce.

2. Limitations

As discussed, for example, by Roff (1992) and Stearns (1977, 1992), the theory as here elaborated ignores a great deal of biological detail. Rather than review this material, I here focus on the following, more fundamental deficiencies:

1. **Insufficiency of reproductive effort as a descriptor of allocation.** Use of a single set of control variables, $E = [E_1, \ldots, E_n]$, ignores the fact that there are multiple functions to which resources can be allocated. For example, in modeling plant life histories, it is useful to distinguish allocation to

reproduction, growth, and storage (Schaffer et al. 1982; Chiariello and Roughgarden 1983; Schaffer 1983). In short, the E_i in Equations (7) and (8) should really be E_{ij}, where the second subscript refers to the physiological function targeted.

2. **Insufficiency of reproductive effort as a descriptor of reproductive style.** This is an extension of #1, the point being that energy allocated to reproduction can itself be partitioned in various ways. We have already alluded to this fact in reviewing the functional dependence of fecundity on effort. Thus, in many species, one wants to distinguish energy expended on different aspects of reproduction, for instance, to distinguish mate competition from the actual production of offspring, the cost of birthing, from *post-partem* parental care, and so on.

3. **Total effort versus effort per offspring**. This is a further extension of # 1. Most organisms produce multiple young, with the consequence that selection can act on both the effort expended per offspring and the *total* effort.[18] In the absence of age structure, this leads to

$$\lambda(E, e) = B(E, e) + g(E)p(E) \tag{6c}$$

where E is total reproductive effort, and e is the caloric expenditure per offspring. With E held constant, maximizing $\lambda(E,e)$ collapses to the "clutch size problem" as it was studied in the 1950s and 60s by David Lack (1966) and his associates. In this case, and with a cap on juvenile survival, $c(e)$, the optimal expenditure per offspring, e^*, is the value of e at which $c(e)$ is tangent to the line,

$$c = ek \tag{10}$$

of greatest k as shown in Figure 1.8a. This so-called "marginal value solution" (Charnov 1976) was proposed independently by Smith and Fretwell (1974) and Schaffer and Gadgil (1975). The latter authors also considered the way in which competition among juveniles selects for increased parental expenditure. What happens if we allow both expenditures, that is, total and per offspring, to vary? One answer is shown in Figure 1.8 wherein we assume that offspring survival depends only on e. Then the optimal effort per offspring, e^*, is independent of the total expenditure. On the other hand, the optimal total expenditure, $E^*(e)$, depends on e. In particular, $E^*(e)$ rises and falls as shown in the figure, and, in fact, is maximal when $e = e^*$. There is no mystery to this: e^* maximizes effective fecundity, $B(E,e)$, and, so long as post-reproductive survival depends only on the total expenditure, the consequence of setting $e = e^*$ is to maximize $E^*(e)$.

As discussed by Parker and Begon (1986), McGinley (1989), and Hendry and Day (2003), more complex dependencies result when juvenile survival depends not only on effort per offspring but also on the total

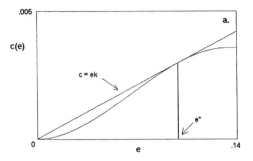

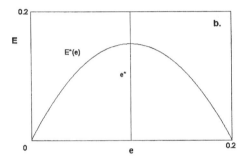

Figure 1.8. Evolution of effort per offspring, *e*, and the joint evolution of total effort, *E*, and *e*. **a.** Optimal value, e^*, of *e* is that for which juvenile survival, $c(e)$, is tangent to the straight line, $c = ek$, of greatest *k*. **b.** Covariation of effort per offspring and total reproductive expenditure. Because e^* maximizes effective fecundity, the optimal total expenditure, $E^*(e)$, is maximal at $e = e^*$. For these calculations, which are purely illustrative, it is assumed that $c(e) = 3\sigma e^2 - 2\tau e^3$; $e < \sigma/\tau$; $c(e) = \sigma^3/\tau^2$, $e \geq \sigma/\tau$; with $\sigma = 0.2$, and $\tau = 1.5$.

number of offspring produced. For additional discussion on such matters in the context of salmonids, see Einum et al. (2003—*this volume*).

4. **Insufficiency of age as a descriptor of potential reproduction, growth, and survival.** Especially in organisms with indeterminate growth, size is often of equal, if not greater, importance than age in determining components of fitness. If size is the principal determinant, one can replace Equation (1) with the corresponding equation that describes the growth of a population in which individuals are categorized by size or "stage." If both factors are important, an "age by stage" description is required.

5. **Dose–response relations.** In many species, patterns of reproductive expenditure depend on environmental circumstance. In such cases, what is selected among is not a set of age-specific reproductive efforts, but rather a set of "dose–response" relations, the "dose" being the environmental state, often as reflected by the organism's size or condition, and the "response," the age-specific expenditure. Such relationships—also called "norms of reaction" (Hutchings 2003—*this volume*)—vary

between and within species. For example, some desert birds (these notes are being written in Tucson, Arizona) are notoriously flexible with regard to the onset of breeding, in which regard they are influenced by variations in rainfall. Other species respond primarily to photoperiod and are locked into fixed breeding seasons. Likewise, a positive correlation between pre-reproductive growth rates and the age of first reproduction is widely observed in both plants (Harper 1977) and animals (Alm 1959).

6. **Insensitivity of fitness to variations in allocation.** As usually articulated, life history theory makes no reference to the expected degree of dispersion about a putative optimum. In fact, one expects that the selective pressures that mold life histories in nature are often minuscule, that is, adaptive topographies of the sort shown in Figures 1.5–1.7 will be quite flat. In this case, finite population effects *guarantee* that real-world populations will be suboptimal. The question then becomes not whether natural populations exhibit optimal behavior—they won't—but rather the degree to which deviations from optimality preclude validation of the theory. See also Einum et al. (2003—*this volume*).

7. **Life history evolution and nonlinear population dynamics.** The theory reviewed here antedates the realization (May 1976; Schaffer 1985) that ecological populations can manifest complex dynamics independent of environmental fluctuations. With regard to life history evolution, nonlinearity introduces the twist of deterministic fluctuations in population density and, hence, selective pressures which, over the short term, are predictable. There is also the contrapuntal issue as to the consequences to population dynamics of variations in life history.

3. The Theory Applied to Atlantic Salmon

3.1. Age of First Spawning

As discussed by Stearns and Hendry (2003—*this volume*), salmonid life histories make interesting grist for the theorist's mill. As a graduate student, I deduced predictions regarding the effects of environmental factors expected to influence the optimal (sea) age of first reproduction in anadromous North American Atlantic salmon. The predictions were tested by comparison with data from over 100 stocks obtained from, or with the assistance of, what was then the Fisheries Research Board of Canada. In particular, I considered the selective consequences of the following environmental factors: (a) the energetic demands of the upstream migration as indexed by river length; (b) the rate at which fish from different rivers gain weight in the ocean after their first summer at sea; and (c) the intensity of coastal and estuarine fishing as indexed by the numbers of commercial nets.

Because angling statistics were the principal source of data for most rivers, I relied largely on the mean weight of angled fish as a *proxy variable* for the age of first return. So long as the frequency of repeat spawners is low (generally true), this procedure is wholly defensible (see Table I of Schaffer and Elson 1975), the increase in weight resulting from each additional summer at sea being large compared to variations in weight among fish of the same (sea) age.

With regard to these factors, it was predicted that the optimal age of first spawning should (a) increase with river length—because increased energetic demands of migration depress effective fecundity per unit reproductive effort;[19] (b) increase with ocean growth rate—because this increases residual reproductive value per unit expenditure; and (c) decrease with the intensity of commercial fishing—because larger (= older) fish are more likely to be trapped than smaller (= younger) individuals.[20,21]

To confirm that these predictions do, in fact, derive from the theory, we refer back to the two-stage life history model of the preceding section. In the context of Atlantic salmon, think of these equations as modeling the adult (post-smolt) phase of the life cycle with everything that goes on in the river and during the first summer at sea being collapsed into the effective fecundity function of the first age class. In this spirit, we view the "juvenile" age class as "grilse," fish that spend a single summer in the ocean before returning to the river, and the "adult" age classes as "salmon," fish that spend two or more summers at sea before spawning.

Figure 1.6 addresses the river length effect on assumption that the more demanding the swim upstream, the greater the fraction of reproductive effort (*sensu lato*) that must be allocated to migration as opposed to reproduction per se. With migratory costs high (long rivers), fecundity per unit effort is reduced and fitness maximized by one sea-year fish not returning to spawn (Figures 1.6a and 1.6b). Conversely, with migratory costs low, fecundity per unit effort is high and selection favors reproduction by both age classes (Figures 1.6c and 1.6d).

Figure 1.7 addresses the ocean growth rate effect. Here the rate at which fecundity can increase from one year to the next, that is, the growth function, $g_i(E_i)$, is varied. For low rates of post-reproductive growth potential, reproduction by both age classes is favored (Figures 1.7a, 1.7b). For higher values of this function, an optimal life history is one in which only "salmon" reproduce (Figures 1.7c, 1.7d).

Finally, in Figure 1.9, we study the selective consequences of commercial fishing. Since commercial gear preferentially removes larger individuals, we model the effect of coastal and estuarine fishing by reducing *both* the effective fecundity and post-reproductive survival functions of older individuals. The not unexpected result is a shift from life histories in which one sea-year fish do not reproduce (Figures 1.9a, 1.9b) to those in which they do (Figures 1.9c, 1.9d).

To varying degrees, all three predictions were supported by the data. The results of subsequent studies (Thorpe and Mitchell 1981; Scarnecchia 1983; Fleming and Gross 1989; Myers and Hutchings 1987a; N. Jonsson et al. 1991a; L'Abée-Lund 1991; Hutchings and Jones 1998) of S. *salar* and other salmonids, however, were mixed. The follow-up investigations focused principally on the first prediction (but see Myers and Hutchings 1987a; Hutchings and Jones 1998), that is, on the effect of increasing costs of migration imposed as indexed by length of the river or mean annual discharge. This is something of a

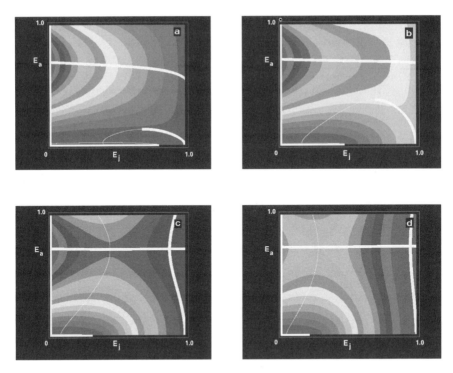

Figure 1.9. Commercial fishery effect. Changing adaptive topographies and optimal reproductive expenditures in response to increasing commercial fishing pressure as modeled by reductions in fecundity and post-reproductive survival per unit effort among adults in the two-stage life history model of Figures 1.5–1.7. Curves of conditional optima and minima and reproductive effort pairs, (E_j, E_a), color coded as in Figure 1.5. Color versions of these figures are posted at http://www.oup-usa.org/sc/019514385X and at http://bill/srnr.arizona.edu, whereat ancillary material is also available. Life history functions, $B_i(E_i)$ and $p_i(E_i)$ as in Figure 1.2c with $\beta_j = 1.0$ and $\pi_j = 0.8$. Post-reproductive growth function, $g_i(E_i)$, as in Figure 1.7 with $\gamma_j = \gamma_a = 2.0$. **a.** $\beta_a = 1.6$; $\pi_a = 0.9$. **b.** $\beta_a = 0.8$; $\pi_a = 0.45$. **c.** $\beta_a = 0.4$; $\pi_a = 0.225$. **d.** $\beta_a = 0.2$; $\pi_a = 0.1125$. Diminishing adult components of fitness lowers the optimal age of first reproduction.

pity, inasmuch as the original analysis suggested that river length effects are subject to masking or accentuation by the other variables.[22] In particular, river length, while strongly correlated with delayed reproduction in some regions (*Baie de Chaleur*, North Shore of the Gulf of St. Lawrence), evidenced no such relationship in others (New Brunswick and Newfoundland). In discussion, Elson and I suggested that this variation might reflect regional differences in commercial fishing pressure,[23] and growth rates in the ocean. With regard to the latter point, populations from regions manifesting a significant effect of river length also evidenced the highest post-grilse growth rates at sea, a finding subsequently confirmed by Hutchings and Jones (1998) in a study notable for its inclusion of data from both European and North American stocks. Such synergy is, in fact, to be expected, that is, if fish grow less rapidly at sea, the benefits of postponing reproduction are diminished.

3.2. Variation in Inter-spawning Intervals

One of the patterns reported by Schaffer and Elson (1975) was the observation (their Figure 7) that interspawning intervals, that is, the number of summers spent at sea between migrations into freshwater, was positively correlated with the mean age of first return. This finding was subsequently confirmed for Norwegian Atlantic salmon by N. Jonsson et al. (1991a) who ascribed it to the fact that larger (= older) fish require a longer period of ocean feeding to mend.[24] The twist that life history theory brings to the discussion is that factors (reduced fecundity per unit effort, increased post-reproductive growth rate) that select for delayed reproduction should also select for increasing intervals between reproductive episodes.

3.3. Life in the River

Contemporary science being the collective endeavor that it is, graduate students embarking upon what they believe to be a novel course of investigation often (and to their not inconsiderable chagrin!) discover as their work progresses that they are *not alone*. In my own case, a thesis (Schiefer 1971) relating heightened freshwater growth rates among parr to increased rates of pre-migratory sexual maturation and early return from the sea was submitted to the University of Waterloo even as I worked to complete my own dissertation. These observations were in accord with the widely held belief (Huntsman 1939) that growth rates in fresh water are an important determinant of the period salmon spend at sea before returning to the river. They are also consistent with the theory elaborated above: greater pre-reproductive growth rates increase fecundity per unit invest-ment for all age classes and thereby favor increased reproductive expenditure and an earlier age of first spawning. This point was first made in the paper with Elson. Later (Schaffer 1979a), I extended the argument by considering a model in which river and ocean stages of salmonid life histories were explicitly distin-guished. The latter analysis further predicts that the age at which young salmon go to sea should vary inversely with growth rates in the river, which prediction is consistent with the observation that the age of ocean-bound smolts increases with latitude (Power 1969; Schaffer and Elson 1975).

3.4. Males versus Females

One of the more interesting discussions in the salmonid life history literature (see Gross 1996; Fleming and Reynolds 2003—*this volume*; Hutchings 2003—*this volume*) revolves about the adaptive significance of precocious maturation by male parr. This phenomenon appears to reflect the fact that parr can slip past territorial adult males and fertilize some fraction of the female's eggs (Jones 1959; Fleming and Gross 1994). That comparable forces may also be at work among adult fish is suggested by the fact that, even though the mean age of returning adults varies from river to river, the average sea age of returning males is almost invariably less than that of returning females (Schaffer 1979a).

4. Pacific Salmon Versus Atlantic Salmon and Trout

Although post-spawning mortality approaches 100% in some populations of *S. salar*, there are other populations in which multiple returns to the river are the norm, and it is not unheard of for individuals to reproduce five times or more (Ducharme 1969). This is in contrast to Pacific salmon in which anadromous individuals are obligately semelparous.[25] During the upstream migration, apparently irreversible changes occur (Neave 1958; Foerster 1968; Groot and Margolis 1991): individuals stop feeding; the digestive tract is resorbed; and the jaws, especially in males, elongate to form a fearsome kype (Fleming and Reynolds 2003—*this volume*). In some species, males also develop a pronounced hump and there is the acquisition, to varying degrees depending on species, of characteristically gaudy spawning colors in both sexes. In thinking about the adaptive significance of these changes, I was first drawn to images of indomitable sockeye salmon fighting their way a thousand miles inland as they traverse rivers such as the Snake and Columbia in western North America. Under such circumstances, one can well imagine effective fecundity functions of the sort shown in Figure 1.2a, with anything less than a maximum effort resulting in the individual's death *before* it reaches the spawning grounds. Further reflection, however, suggests that such considerations do not explain the origin of semelparity in this group.

In the first place, anadromous rainbow trout or "steelhead" (*O. mykiss*), cohabit the same river systems and, in some cases, make the same grueling upstream migrations. Yet this species, like *S. salar*, with which it was previously grouped as *S. gairdneri*, is iteroparous. It follows that, if the adaptive significance of semelparity in *Oncorhynchus* relates to the rigors of migration, one must confront the fact that these factors, while operative on the ancestors of contemporary Pacific salmon, had no such effect on the line leading to steelhead. In addition, many Pacific salmon, for example, silver (coho) salmon in California, spawn in short, coastal streams (Shapovalov and Taft 1954). For these populations, getting back out to sea would seem to be no more of a problem than it is for multiple spawning *S. salar* in the rivers of southeastern Canada. Similarly, landlocked populations of *Oncorhynchus* generally undertake little in the way of spawning migrations. Yet, like their anadromous cousins, these populations are semelparous. In short, the distribution of reproductive and migratory habit suggests that the evolution of semelparity in *Oncorhynchus* is unrelated to migratory stress. This forces us to look to other possible explanations. Among them, are the following:

1. **Semelparity in contemporary *Oncorhynchus* is maladaptive, having been retained only because the ability to re-evolve repeated reproduction has**

been lost (Crespi and Teo 2002). This is the "hen's tooth" explanation. It holds that big bang reproduction in extant populations is reflective of selective regimes past. In this case, migratory stress may, indeed, have been the critical factor in the evolution of semelparity, but the ability to re-evolve the iteroparous habit was subsequently lost. This hypothesis raises the fascinating question as to whether or not heretofore unreported vestiges of *Oncorhynchus'* iteroparous heritage persist in extant populations of the Japanese species, *O. masou* and *O. rhodurus*, which are generally held (Neave 1958; Stearley and Smith 1993; McKay et al. 1996; Oakley and Phillips 1999; Osinov and Lebedev 2000; Kinnison and Hendry 2003—*this volume*) to be primitive for the group.

2. **Semelparity and iteroparity in salmonids represent multiple optima in allocation schedules (Schaffer 1979a).** According to this, both strategies are adaptive, that is, they correspond to alternative adaptive peaks in the here and now. The advantage of this explanation is that, with effective fecundity functions as in Figure 1.2d, alternative optima corresponding to iteroparous and semelparous reproduction are predicted. The problem is that with so many populations isolated to varying degrees by the tendency of individuals to return to the waters of their birth (Hendry et al. 2003a—*this volume*), one expects at least occasional jumps from one adaptive peak to the other. In short, the hypothesis of alternative evolutionary optima in contemporary populations is inconsistent with the phylogenetic distribution of parity modes.

3. **Semelparity in *Oncorhynchus* is adaptive by virtue of the acquisition of traits that increase effective fecundity per unit effort, thereby selecting for increased effort (Crespi and Teo 2002).** Recently this idea was advanced by Crespi and Teo (2002), who conjectured that increased egg size in *Oncorhynchus*, to the extent that it serves to increase effective fecundity, may have selected for greater reproductive expenditure. The question left unanswered is whether or not the effect would be sufficient to drive optimal levels of expenditure to the maximum. More precisely, increasing juvenile survival does not make for concavity in the effective fecundity function.

4. **Semelparity in *Oncorhynchus* is adaptive by virtue of the acquisition of traits that make post-reproductive survival unlikely regardless of reproductive expenditure (Crespi and Teo 2002).** Such traits would make for concave dependencies of residual reproductive value on current reproductive expenditure (Figure 1.2), thereby increasing the likelihood of selection for maximum reproductive expenditure. In this regard, Crespi and Teo (2002) suggest that Pacific salmon are characterized by more extensive ocean foraging than Atlantic salmon and trout, and that this would likely act to reduce post-reproductive survival. An alternative hypothesis, for which there is currently no evidence, is that physiological innovations in Pacific salmon may preclude post-spawning resumption of a saltwater existence. On the face of it, this hypothesis would seem to stumble on the fact that immature Pacific

salmon make the transition to marine environments as a matter of course. Still, experience teaches that the loss of plasticity with age is a common feature of biological systems.

5. **Semelparity in *Oncorhynchus* is adaptive by virtue of the acquisition of traits that make for accelerating dependencies of effective fecundity on reproductive expenditure, independent of the costs of migration into fresh water (Willson 1997; Crespi and Teo 2002).** This possibility is discussed by Willson (1997) and Crespi and Teo (2002), who emphasize the importance of competition for mates among males and nesting sites among females who further defend their eggs after fertilization. Among salmonids, these traits appear to be most highly developed in Pacific salmon, suggesting that semelparity in these species is essentially a consequence of sexual selection (Fleming and Reynolds 2003—*this volume*).

Of these five possibilities, which are not mutually exclusive, it is the last that is arguably the most intriguing. In the first place, intraspecific competition provides a mechanism that can keep driving populations toward higher and higher levels of reproductive investment, in effect manufacturing accelerating fecundity curves (Figure 1.2a) along the way. This is because reproductive success in such circumstances is a function of an individual's relative (as opposed to absolute) expenditure. Second, this hypothesis is consistent with the exaggerated secondary sexual characters (kype, hump, and gaudy spawning colors) which distinguish semelparous *Oncorhynchus* as a group. It further suggests a tie-in with the observation that semelparous salmonids generally breed under higher densities than their iteroparous allies. Specifically, high densities of spawners favor increased competition for mates and nesting sites, territorial defense of fertilized eggs, and increased *per offspring* allocation by females in the form of larger eggs—which is to say, the innovations of which we have been speaking. So perhaps the problem's solution is already in our grasp. And perhaps not. For, as A. Hendry has emphasized in correspondence, the situation is really quite complex, both with regard to the selective agents that may have triggered the process and the sequence in which traits were acquired. And of course, there is the not so minor consideration that, viewed across the animal kingdom, sexual selection doesn't always "run away." In short, the matter remains murky, for which reason, this essay concludes, not with an answer, but with an open question. As such, it is an appropriate way to end a prolegomenon, though clearly not so satisfying as being able to trumpet "Problem solved!" Maybe next time.

Acknowledgments I would like to dedicate this chapter to the memory of Paul F. Elson and Robert H. MacArthur, who helped me on my way.

The ideas reviewed here reflect the tutelage of John Bonner, Paul Elson, Henry Horn, Egbert Leigh, and Robert MacArthur. Mike Rosenzweig caught and helped repair a grievous error in my first attempt (Schaffer 1974a) to understand the life historical consequences of age structure. Characteristically, he suggested that we report the mistake and its correction together (Schaffer and Rosenzweig 1977). I am also grateful to contributors to the present volume: Bernard Crespi, Andrew Hendry, Michael Kinnison, Tom Quinn, and Eric Taylor, who kindly corresponded with me about salmonids and their evolution.

The work reviewed here was supported by Princeton University, the Universities of Utah and Arizona, and the National Science Foundation. More important was the support I received from my parents, Michael and Rose Schaffer, who gave me my start, taught me to distinguish right from wrong, and, in Dad's case, insisted that I learn to write. Other individuals have also made a difference: That I remain intellectually active at what I would have once considered an advanced age is due solely to Tatiana Valentinovna, light of my life, with whom I am privileged to work and cohabit. That I have had time these past ten years for intellectual activity beyond the mind-numbing, day-to-day grind at a university mired in mediocrity is due to the many teachers, trainers, and sitters who have worked with my son, Michael, or helped look after his interests. In particular, and with apologies to those whose names I have forgotten, I thank Nancy Sergeant Abate, Bink and Jack Campbell, Judy Cowgill, Ben Duncan, Cindy G., Alex, Ned and Jamie Gittings, Sharon Harrington, Linda Kuzman, Sherry Mulholland, Mary Ann Muratore, Steve Nagy, Ralph Schmidt, Suzie Speelman, and Eddie Vanture. Michael himself gives meaning to life in ways that only a parent who's "been there" can appreciate. G-d willing, he will yet learn to read and comprehend these words.

Notes

1. Nova Scotia's rivers themselves are short, gentle affairs.

2. At the time, only semelparous species (Pacific salmon) were assigned to *Oncorhynchus* (Neave 1958).

3. This is an amended version of Levins' (1966) "Principle of Optimality," the emendation being addition of the word, "related."

4. Struck down in his prime, MacArthur left the intellectual playing field, like the runner of Housman's poem, never having experienced defeat.

5. For purposes of exposition, I suppose a time step of a year, and further that the population is censused annually. The numbers of individuals in the various age classes are thus the numbers alive at census time. By this accounting, the probability that an individual survives from birth to first census is absorbed by the B's, which become "effective fecundities." Moreover, ℓ_0 necessarily equals 1.

6. In some texts, the rate at which the population multiplies, λ, is replaced by e^r, thereby emphasizing the relation of Equation (1) to its continuous analog, $1 = \int e^{-rx} l_x \, m_x \, d_x$ (Fisher 1930), in which case, the m's are (no longer effective) fecundities.

7. Recall that the magnitude of a complex number is the square root of the squares of its real and imaginary parts added together, that is, if $z = x \pm iy, |z| = (x^2 + y^2)^{1/2}$.

8. Equation (1) is the characteristic equation of the *linear* growth process, $N(t+1) = A \, N(t)$, where the N's are vectors, the elements of which are the numbers of individuals in the different age classes at times t and $t+1$, and A is the so-called "population projection" or "Leslie" matrix. Provided that λ_1 exceeds the other λ's in magnitude, it follows that as $t \to \infty$, $N(t) \to c_1 \lambda_1^t N_1$ where c_1 is a constant reflective of the population's initial age structure. According to this, N_1, which is an eigenvector defined by the relation $A N_1 = \lambda_1 N_1$, specifies the "stable age distribution."

9. For the reader uncomfortable with derivatives generally, and with partial derivatives in particular, we offer the following. If the value of y depends on a second quantity, x, the derivative, dy/dx, evaluated at $x = x_0$, is the rate at which y changes in response to a small variation in x when $x = x_0$. If y depends on a number of quantities, say, w, x, and z, then the *partial* derivative, $\partial y/\partial x$, is the same rate of change, but with the values of the other variables fixed.

10. Equation (3) is the definition one finds in the textbooks and, as such, has perplexed generations of students. A more useful rendering of this expression can be had as follows: Let p_i be the probability of surviving from age i to $i + 1$, that is, $p_i = \ell_{i+1}/\ell_i$. Now multiply the chicken scratches under the sum, that is, the terms following the "$\sum$," by the quantity in brackets outside it. Then $v_i/v_0 = B_i/\lambda_1 + p_i B_{i+1}/\lambda_1^2 + p_i p_{i+1} B_{i+2}/\lambda_1^3 + \ldots$, whence follows the verbal definition given above.

11. This result requires that varying reproductive effort at age i is without consequence to reproduction and survival at previous ages (Schaffer 1979b). On the face of it, this would seem always to be the case: time, after all, does not run backwards. In fact, there are instances when, so far as natural selection is concerned, the hand on the dial really does turn counter-clockwise. Specifically, when there is extended parental care, resources expended on this year's offspring can adversely affect the growth and survival of their older siblings. This reduces the parents' effective fecundity at previous ages. Similarly, if last year's offspring fail to disperse, as in the case of organisms that propagate "vegetatively" (plants, corals, etc.), they may compete with this year's progeny.

12. Equation (7) was misinterpreted (Schaffer 1981; Yodzis 1981) by Caswell (1980) who imagined the claim to be that the optimal expenditure at each age maximizes the reproductive value of *all* the age classes.

13. This is because E_i^*, for example, depends on (v_{i+1}/v_0) which, in turn, depends on λ_1 and hence on all the E's.

14. The set of optimal reproductive efforts, E^*, were computed using a "downhill simplex" algorithm (Press et al. 1992) to minimize $(1/\lambda_1)$. Note, however, that nothing having been proved, the proposed generality is purely conjectural.

15. With an infinite number of "adult" age classes, the stable age equation (1) becomes

$$1 = \sum_{i=0}^{a-1} \lambda^{-(i+1)} \ell_i B_i + (\lambda^{-a} \ell_a B_a)/(\lambda - p_a)$$

Here a is the age at which individuals achieve "adulthood," while B_a and p_a are respectively adult fecundity and the probability of survival from one year to the next.

16. While the mathematics of senescence (Hamilton 1966) and life history evolution are essentially equivalent, from a biological perspective the two topics are rather different. In the first case, one is concerned with adaptations that delay the onset of morbidity and mortality; in the latter, with the allocation of available resources to competing functions.

17. In terms of note 16, we consider the case in which $a = 1$.

18. In fishes, an important determinant of effort per offspring is egg size. To first order, the total caloric expenditure in females is thus $n \times e$ where n is the number of eggs.

19. River length is, of course, but the crudest of measures of the cost of migration. Subsequently, N. Jonsson et al. (1991a) suggested that mean annual discharge is a better indicator. For further discussion of migratory cost, see Hendry et al. (2003b—*this volume*) and references therein.

20. Obviously, if older fish are preferentially removed from the spawning run, the mean age of returning fish will drop, a prediction for which optimality theory is manifestly unnecessary. In fact, what is being predicted is that the numbers of fish returning to spawn at an early age, as indicated by census *before they hit the nets*, will increase. As discussed in (Schaffer and Elson 1975), this prediction is supported by increasing numbers of grilse taken in a research trap on the N. W. Miramichi from 1961 to 1969.

21. For discussion of the many other consequences of human exploitation, see Young (2003—*this volume*).

22. There was also some confusion. For example, regarding our use of angling statistics (fish weight) as a proxy variable for the age of first spawning, Fleming and Gross (1989) and, to a lesser extent, N. Jonsson et al. (1991a), imagined that what was being predicted was variation in fish size per se. Thus, the former authors reported as contradicting our results the observation that the size of returning female coho salmon varies *inversely* with river length. In fact, all of the fish in their study spent the same number of summers at sea with the consequence that the reported variations in body size are irrelevant to the life historical predictions we were trying to test.

23. As noted above, commercial nets and weirs preferentially remove larger (= older) fish from the spawning run and therefore select for an early age of first spawning. In fact, there is evidence that this has happened. For example, in the case of the Northwest Miramichi, the principal salmon river in New Brunswick, there is good reason (Hunstman 1939) to believe that the decades prior to those for which I was able to obtain data witnessed dramatic declines in the abundance of 2- and 3-sea-year virgin fish. By way of contrast, long-term angling records kept by the Moise Salmon Club (North Shore) indicate little change in the composition of the spawning run from 1930 to 1970 (Schiefer 1971).

24. My own take on this was to emphasize the fact that increased interspawning intervals were associated with increased losses in growth due to spawning which, in turn, correlated with increased migratory costs as evidenced by river length. Supporting this interpretation is the observation by N. Jonsson et al. (1991a) that post-spawning survival declines with the age of first return.

25. In what follows, I use the term "semelparity" to refer to the obligate demise of post-spawning *adults*, i.e., fish which have gone to sea and then returned to brackish or fresh water. This usage reflects the observation that, as in *S. salar*, male juvenile Pacific salmon can reproduce more than once as parr (Unwin *et al.* 1999) and further that some Pacific salmon returning to spawn have previously reproduced as parr (Tsiger et al. 1994). There *may* be occasional exceptions. For instance, Ricker (1972) cites published reports of purportedly "mending," that is, post-spawning chinook (O. *tshawytscha*) adults taken in salt water.

The Evolution of Philopatry and Dispersal
Homing Versus Straying in Salmonids

Andrew P. Hendry

Vincent Castric

Michael T. Kinnison

Thomas P. Quinn

Brown bear feeding on a sockeye salmon

2

Many animals reproduce at the same site as their parents, a phenomenon called philopatry. In some cases, philopatry can be the simple result of geographic isolation or limited movement ability. In other cases, philopatry can be strong even when migratory movements take individuals far from their natal site. We must therefore seek explanations for philopatry through selective factors that favor return to natal sites rather than non-natal sites. The tendency for philopatry in any population or species is balanced to varying degrees by some dispersal to non-natal sites, with some groups showing high philopatry and low dispersal and others low philopatry and high dispersal. Again, some of this variation may be the result of varying isolation or movement ability, whereas some can only be explained through natural selection. Our goal in the present chapter is to use data from salmonids to evaluate theories for how selection might influence the evolution of philopatry and dispersal.

Philopatry can be evaluated at a variety of spatial scales, ranging from precise nest locations to broad geographic areas. We focus on philopatry to spatially discrete locations that are predictably occupied by breeding aggregations. These aggregations often correspond to "populations": groups of individuals that breed together and are at least partially reproductively isolated from other such groups. By "reproductive isolation," we here mean that the different groups are not freely interbreeding within or between generations (i.e., not panmictic). Many factors can contribute to reproductive isolation (Dobzhansky 1937; Mayr 1963; Schluter 2000). "Extrinsic" isolation is caused by geographic barriers that severely limit movement between populations. This category does not apply to populations in separate physical locations as long as movement among them remains possible. Extrinsic isolation is of limited interest here because philopatry has been imposed on the organisms rather than evolving in response to selection. "Intrinsic" isolation is caused by features of organisms that limit gene flow between aggregations, even in the absence of strict geographic barriers. Intrinsic isolation can be the result of factors that reduce mating between individuals from different aggregations ("pre-zygotic") or reduce the success of hybrid progeny after mating ("post-zygotic"). Because philopatry reduces mixing even in the absence of strict geographic barriers, it is an intrinsic, pre-zygotic isolating mechanism.

"Dispersal of an individual vertebrate is the movement the animal makes from its point of origin to the place where it reproduces or would have reproduced if it had survived and found a mate" (Howard 1960). Although this is only one of many possible definitions of dispersal, it effectively encapsulates the phenomena considered in this chapter and is widely accepted (Johnson and Gaines 1990). Before proceeding, a number of features of dispersal warrant mention: (a) dispersal may or may not reflect individual choice ("voluntary" versus "enforced" dispersal, Greenwood 1980), (b) dispersal may take place before first breeding or between different breeding episodes ("natal" versus "breeding" dispersal, Greenwood 1980), (c) dispersal may or may not be influenced by environmental conditions ("environmental" versus "innate," Howard 1960), (d) dispersal may be density-dependent (Travis et al. 1999) or influenced by how close a population is to its carrying capacity ("saturation level," Kokko and Lundberg 2001), (e) effects of the environment and individual condition are often referred to as "condition-dependent" dispersal (Ims and Hjermann 2001), (f) dispersal can occur between spatially discrete populations or within continuous but spatially structured populations (e.g., Hastings 1983), (g) dispersal may be sex biased (Greenwood 1980), and (h) successful reproduction by a disperser results in gene flow (Endler 1977).

Dispersal is not the same as migration (Endler 1977), with the latter being the spatially and temporally predictable movement of individuals between breeding and foraging habitats. Individuals may thus migrate but not disperse (move from natal sites to foraging areas and then back to natal sites for breeding), disperse but not migrate (move from natal sites to non-natal sites without leaving breeding areas for separate foraging areas), migrate and disperse (move from natal sites to foraging areas and then to non-natal sites for breeding), or not migrate and not disperse (stay at natal sites for their entire lives). In the present chapter, we focus on the evolution of philopatry and dispersal, leaving migration for a separate chapter (Hendry et al. 2003b—*this volume*). However, it is important to remember that dispersal and migration may often interact. For example, the long-distance movements undertaken by migratory animals may make it more difficult to find the natal site. Conversely, philopatry may be increasingly favored in migratory individuals as an effective way of ensuring their return to suitable breeding areas.

In the following, we first outline how selection might favor philopatry or dispersal. This discussion is largely devoid of references to salmonids because our goal is to outline general theories, particularly those with mathematical formalization. Second, we describe the basic biology of salmonids and ways of gathering data on philopatry ("homing") and dispersal ("straying"). Third, we review basic patterns of homing, straying, and genetic diversity in salmonids. Finally, we consider each of the general theories in light of information from salmonids. Our overall goal is to provide a salmonid-based evaluation of existing general theories for the evolution of philopatry and dispersal.

1. Theory

Philopatry and dispersal are alternative behaviors for an individual but they represent opposing ends of a continuum for families, populations, or species. In fact, individuals can also be placed along this continuum by considering alternative behaviors as probabilities. Substantial variation in philopatry and dispersal among individuals, kin, and groups suggests that selection has the potential to act at each of these levels and, indeed, all have been invoked in mathematical models (Johnson and Gaines 1990; Table 2.1). Selection acting to increase the fitness of individuals (individual selection) is included in nearly all models, often appearing in the form of inbreeding depression or survival costs ("Inbreed" and "Costs" columns of Table 2.1). Selection acting to increase the fitness of closely related individuals (kin selection) also figures heavily ("Kin" column of Table 2.1). Selection acting to increase the success of unrelated groups (group selection) also appears in several models, particularly those examining effects of local extinctions ("Extinction" column of Table 2.1). In the following, we outline theories for how selection might favor philopatry ("Why philopatry?") and dispersal ("Why dispersal?"), listing empirically testable predictions of each. We typically focus on putative benefits of philopatry or dispersal but the benefits of one could just as easily be considered costs of the other. The theories are presented separately but are not mutually exclusive, and considering each alone assumes that "all else is equal." Ultimately, variation in philopatry and dispersal should reflect a balance between many interacting benefits and costs.

1.1. Why Philopatry?

Theory P1

Philopatry increases the likelihood of finding a suitable breeding habitat and mate. Suitable breeding habitats are distributed heterogeneously across space, and organisms whose foraging movements take them beyond natal areas may have difficulty finding suitable sites for breeding. If so, individuals predisposed to return to natal sites may have greater success than dispersing individuals because the former are more likely to find suitable habitat. Under this hypothesis, we predict that philopatry should increase with decreasing availability or accessibility of suitable alternative breeding sites (see also Travis and Dytham 1999) and with increasing temporal stability at natal sites (because successful reproduction by one's parents becomes an increasingly reliable predictor of successful reproduction by oneself at the same site). However, even if physically suitable alternative sites are easily discovered, reproductive success will still depend on the number of conspecifics. At least some individuals of the opposite sex are obviously necessary, and average individual reproductive success may increase with increasing density ("Allee effect," Courchamp et al. 1999), although it presumably declines again at higher densities. If dispersal is density-dependent, philopatry in such systems should be highest at intermediate densities.

Theory P2

Philopatry increases familiarity with local breeding conditions. The reproductive success of repeat breeders often increases with age (Martin 1995), implying benefits of prior experience. Some such benefits may result from first-hand knowledge of natal sites. For example, individuals that are familiar with an area may be better able to find high-quality nesting sites (Greenwood 1980; Pärt 1994). Social interactions, including mate choice, also influence reproductive success, and individuals breeding at natal sites may benefit from previously established relationships (Greenwood 1980, 1983). Competitive benefits of philopatry under this hypothesis parallel Greenwood's (1980, 1983) "resource enhancement" or "resource competition" mechanism (see also Perrin and Mazalov 1999). Under this hypothesis, we predict that philopatry will be higher

Table 2.1. Theoretical models examining the evolution of philopatry and dispersal, along with the major factors they consider.

Model	Spatial	Temporal	Extinction	Inbreeding	Kin	Costs	CD
Bengtsson 1978				Y		Y	
Bull et al. 1987	Y					Y	
Cohen and Levin 1991	Y	Y				Y	
Comins 1982			Y		Y	Y	
Comins et al. 1980			Y		Y	Y	
Crespi and Taylor 1990	Y		Y		Y	Y	Y
Frank 1986					Y	Y	
Gadgil 1971	Y	Y				Y	Y
Gandon and Michalakis 1999			Y		Y	Y	
Gandon and Rousset 1999					Y	Y	
Gandon 1999				Y	Y	Y	
Hamilton and May 1977					Y	Y	
Hastings 1983	Y						
Holt 1985	Y						Y
Holt and McPeek 1996	Y[a]	Y[a]				Y	
Irwin and Taylor 2000					Y	Y	
Jánosi and Scheuring 1997			Y				Y
Karlson and Taylor 1992, 1995	Y					Y	
Lemel et al. 1997	Y	Y				Y	Y
Levin et al. 1984	Y	Y	Y			Y	Y
Mathias et al. 2001	Y	Y					
McPeek and Holt 1992	Y	Y				Y	Y
Morris 1991	Y						Y
Motro 1982a, 1982b, 1983					Y	Y	
Motro 1991				Y	Y	Y	

Reference	Spatial	Temporal	Extinction	Inbreeding	Kin	Costs
Olivieri et al. (1995); Olivieri and Gouyon (1997); Ronce et al. (2000a)		Y			Y	
Perrin and Mazalov 1999		Y		Y		
Perrin and Mazalov 2000		Y	Y	Y		
Plantegenest and Kindlmann 1999	Y	Y	Y			
Pulliam 1988						Y
Roff 1975		Y				Y
Roff 1994		Y[e]			Y	
Ronce et al. 2000b		Y	Y			
Taylor 1988	Y[f]	Y	Y			
Travis and Dytham 1998						Y[b]
Travis and Dytham 1999						Y
Travis et al. 1999		Y[c]			Y	Y[b]
Van Valen 1971		Y			Y	
Waser et al. 1986		Y		Y		
Wilson 2001		Y[d]				Y

[a] Variation in patch quality is the result of chaotic population dynamics, not variation in carrying capacity. [b] Variation in patch quality is the result of random variation in the amount of movement among patches (the model is individual based and spatially explicit). [c] A cost of dispersal arises because some individuals disperse to patches that are unsuitable. [d] Individuals that disperse have a lower probability of becoming established in different habitats than in similar habitats. [e] Costs of dispersal include the energy cost of developing a dispersal phenotype (i.e., wings). [f] Dispersal rates of offspring may be related to maternal age.

Notes: Many theoretical models have examined factors influencing the evolution of philopatry and dispersal. This table summarizes some of the main effects considered in those models ("Y" indicates inclusion of an effect). "Spatial" refers to intrinsic variation among patches (sites) in quality (e.g., carrying capacity). "Temporal" refers to variation in patch quality across generations, which if asynchronous also causes spatial variation at any given time. "Extinction" refers to a non-zero probability that a given patch will go extinct. "Inbreeding" refers to effects of inbreeding depression owing to mating among kin. "Kin" refers to inclusive fitness effects owing to competition among kin. "Costs" refers to a mortality or fecundity cost of dispersing, specifically related to the movement period itself. "CD" refers to condition-dependent dispersal, usually considered as density-dependent dispersal.

for iteroparous species, which may breed multiple times at the same site, and for semelparous species that remain at natal sites for all (no migration) or a substantial fraction (delayed migration) of their lives. For such species, we might also expect higher philopatry in the sex that experiences higher competition for resources and receives greater benefits from prior experience (Greenwood 1980, 1983; Perrin and Mazalov 1999).

Theory P3

Philopatry returns locally adapted individuals to appropriate habitats. Many species are distributed across multiple breeding habitats that differ in biotic and abiotic features. Populations in different environments should experience divergent selection, which should lead to adaptive divergence (Endler 1986; Schluter 2000; Arnold et al. 2001; Reznick and Ghalambor 2001). As a result, individuals returning to natal sites should have higher average fitness than individuals dispersing to other sites: because dispersers are less likely to arrive at a site for which they are as well adapted. These conditions should strongly favor the evolution of philopatry (e.g., Balkau and Feldman 1973; Asmussen 1983). Accordingly, several studies have found that residents have higher fitness than immigrants (e.g., Verhulst and van Eck 1996; Orell et al. 1999), and some have implicated divergent adaptation as the reason (e.g., Hendry et al. 2000; Via et al. 2000). Under this hypothesis, we predict that philopatry will increase as the degree of divergent selection increases. However, it is important to recognize that (a) lower success of immigrants may arise for reasons other than maladaptation (e.g., site familiarity or assortative mating), and (b) immigrants may have higher success than residents when populations are inbred (Ingvarsson and Whitlock 2000; Ebert et al. 2002).

Theory P4

Philopatry is favored by spatial variation in habitat quality. Envision a group of populations linked by dispersal, where (a) the environments experienced by different populations differ in intrinsic quality (e.g., carrying capacity), (b) intrinsic site quality does not vary in time, (c) fitness within populations is density-dependent (higher when densities are lower), and (d) dispersal occurs at a fixed per-capita rate (a fixed percentage of each population disperses). Because high-quality sites have larger populations and therefore produce the most dispersers, emigration will be higher than immigration at high-quality sites but lower than immigration at low-quality sites. The resulting decrease in density at high-quality sites will increase average fitness at those sites, and the resulting increase in density at low-quality sites will decrease average fitness at those sites. These conditions favor philopatry (Gadgil 1971; Hastings 1983; Holt 1985) "because dispersal is basically moving individuals down gradients in fitness" (Holt and McPeek 1996). Under this hypothesis, we predict that philopatry will increase as spatial variation in site quality increases, relative to temporal variation at each site. If, however, philopatry is condition-dependent (higher when site quality is higher), spatial variation in habitat quality may favor some dispersal (Gadgil 1971; Pulliam 1988; Morris 1991; McPeek and Holt 1992; but see Lemel et al. 1997).

Theory P5

Philopatry improves access to parental resources. Philopatry may be favored if previous generations provide breeding resources that are used by subsequent generations (Waser and Jones 1983). Such resources may pass directly from specific parents to their offspring (e.g., inheritance of mates, nests, or nest locations; e.g., Brown and Brown 1984), and this should be particularly important when high-quality sites are limited (Pen and Weissing 2000). Alternatively, benefits may pass more generally from all parents to all offspring that breed in the same location (e.g., Stacey and Ligon 1987). Similarly, breeding adults may "condition" the environment, making it more suitable for subsequent reproduction. For example, penguins dig their nest burrows in the guano produced by previous generations, and this increases reproductive success (Paredes and Zavalaga 2001). The first of these mechanisms (benefits from a specific parent) directly favors philopatry because offspring must breed at the same location as their parents. The second mechanism (benefits from all previous breeders) does not directly favor philopatry because offspring simply need to breed at established sites, whether natal or non-natal. However, this mechanism may still indirectly favor philopatry as a way of increasing the likelihood that offspring return to an established site (invoking Theory P1). Under this hypothesis, we predict that philopatry should be higher when previous generations provide resources that increase the breeding success of subsequent generations.

Theory P6

Philopatry avoids costs of movement. In some species, philopatric individuals never leave the natal site, whereas dispersers may travel long distances. When this occurs, dispersers may incur costs that philopatric individuals do not. The preceding theories include some dispersal costs, such as failing to find a suitable breeding site, but not other costs, such as increased stress, predation, or energy expenditure. For example, some insects have winged morphs that disperse and wingless morphs that do not (Roff 1994), with the development of wings increasing energy costs and reducing reproductive output (Stirling et al. 1999). Bélichon et al. (1996) reviewed the literature and found that dispersal costs were often associated with the movement ("transience") and settlement phases of dispersal. This suggests that dispersal costs may be very important for the evolution of philopatry and, indeed, such costs figure prominently in most theoretical models (Table 2.1), usually represented as a decreased probability of survival. Under this hypothesis, we predict that philopatry should increase with increasing dispersal costs. This mechanism may be less important for migratory species, where philopatric and dispersing individuals may incur similar costs of movement.

1.2. Why Dispersal?

Several general issues regarding the evolution of dispersal must be considered before outlining theories for how it might be favored by selection. First, we consider how dispersal might be heritable, which is necessary for it to evolve in response to selection. Second, we consider how dispersal might be maintained within populations even though genes for dispersal tend to leave. Third, we consider how new populations might be established by dispersers, even though they produce offspring that might likewise be inclined to disperse (if dispersal is heritable).

The heritability of dispersal, or the *probability* of dispersal, might take several forms. First, dispersal could be a specific behavior (or related to a specific trait) coded by alleles at one or a few Mendelian loci. Second, dispersal could be a threshold trait related to some underlying continuous distribution of migratory "liability" (Roff and Fairbairn 2001). Third, dispersal could be a continuous (quantitative) trait such as "the duration of directed movement" (Roff and Fairbairn 2001). Fourth, dispersal could be condition-dependent (Ims and Hjermann 2001), wherein a reaction norm links dispersal to the condition of an individual (e.g., energy) or the environment (e.g., density). Fifth, dispersal might be a "mistake" made by individuals attempting to be philopatric. In this case, when selection for philopatry is weak, mutations that impair philopatric abilities will accumulate and dispersal should increase. When selection for philopatry is strong, such mutations will be purged and dispersal should decrease.

How might a heritable dispersal potential be maintained within a population when individuals with the greatest genetic proclivity for dispersal are always leaving? One possible way is through condition-dependent dispersal. If the conditions favoring dispersal are rare and dispersal is probabilistic under those conditions, the loss of heritable dispersal potential should be slow. However, this mechanism will not allow the permanent persistence of dispersal potential beyond that maintained by mutations compromising philopatric ability. Instead, dispersal might be maintained in perpetuity through metapopulation dynamics, where multiple populations are linked by dispersal (Hanski and Simberloff 1997). In this case, the loss of dispersal genes from a population is counterbalanced by the gain of dispersal genes from other populations (Bull et al. 1987; Morris 1991; Wilson 2001).

How can dispersal establish new populations when colonizing individuals had a genetic predisposition for dispersal and should therefore produce offspring that are similarly inclined? One possibility is that the *probability* of dispersal is heritable. In this case, some individuals colonizing new sites may actually have had a low genetic probability of dispersal, and will thus produce primarily philopatric offspring. Condition-dependent dispersal may also help. For example, competition may have been high in the ancestral environment, inducing dispersal, but low in the new environment, inducing philopatry. Yet another possibility is that individuals colonizing new sites may have been forced to leave their natal sites owing to inhospitable conditions or competitive interactions. In this case, dispersing individuals may have a strong genetic tendency for philopatry and will thus produce philopatric offspring.

Theory D1

Dispersal buffers against temporal variation in habitat quality. When habitat quality varies across years within sites, so too will the fitness of philopatric individuals. If this temporal variation is asynchronous among sites, the best site for breeding will vary among years. Under these conditions, an individual that disperses may often have higher fitness than an individual that does not. A number of theoretical models have examined this effect, with most assuming that temporal variation is caused by extrinsic conditions that influence carrying capacity (Table 2.1). However, temporal variation in site quality might also result from variation in population size relative to a fixed carrying capacity. This type of variation can result from chaotic population dynamics (Holt and McPeek 1996) or probabilistic dispersal in small populations: some sites will receive more dispersers than others just by chance (Travis and Dytham 1998). General conclusions of these models, and thus predictions under this hypothesis, are that dispersal rates should increase with increasing temporal variability and with increasing spatial asynchrony in that variation. We predict that these effects might also cause sex-biased dispersal if the relative amount of temporal variation in site quality differs between males (e.g., availability of mates) and females (e.g., availability of nest sites).

Theory D2

Dispersal allows colonization of new environments. Individual selection should often favor philopatry but group selection may favor dispersal as the only way of colonizing new sites. If philopatry is too strong, entire metapopulations may go extinct because local populations die out faster than vacated sites can be recolonized. Most theoretical models examining local extinctions (Table 2.1) predict that dispersal should increase with increasing extinction rates (e.g., Van Valen 1971; Comins et al. 1980; Levin et al. 1984; Olivieri et al. 1995; Gandon and Michalakis 1999, 2001). However, Ronce et al. (2000a) showed that incorporating realistic population growth into such models should lead to a peaked relationship between dispersal and extinction rates. This was the result of two opposing effects of increasing extinction: more empty sites favors increasing dispersal (as above) but lower competition within sites favors decreasing dispersal (populations may not reach carrying capacity). Karlson and Taylor (1992, 1995) also predict a peaked relationship between dispersal and extinction rates. Under this hypothesis, we predict that dispersal should increase as local extinctions increase, as long as new populations can reach their carrying capacity. Dispersal rates may also vary through time within local populations: starting high because colonists were dispersers but declining with time because philopatry is favored within populations (Roff 1994; Olivieri et al. 1995). We therefore predict that dispersal should be higher in younger than in older populations.

Theory D3

Dispersal reduces inbreeding depression. Inbreeding depression can reduce individual fitness and increase extinction risk (Charlesworth and Charlesworth 1987; Lande 1994; Lynch et al. 1995; Nieminen et al. 2001; Keller and Waller 2002). Selection might therefore favor behaviors that reduce mating between close kin, with dispersal being one such behavior. Furthermore, dispersal may be favored when populations are inbred because dispersers may have higher fitness than residents (Ingvarsson and Whitlock 2000; Ebert et al. 2002) and because gene flow can increase population fitness (Newman and Tallmon 2001). Theoretical models of dispersal under inbreeding depression (Table 2.1) often predict that one sex will be entirely philopatric, whereas the other will show some dispersal (Bengtsson 1978; Waser et al. 1986; Perrin and Mazalov 1999, 2000; Perrin and Goudet 2001). The sex that disperses is strongly influenced by initial conditions and sex-specific costs (Perrin and Mazalov 1999). Some models of inbreeding depression also include effects of competition among kin (Theory D4; Table 2.1), and these often predict that both sexes disperse at rates depending on the relative intensity of sex-specific dispersal costs, inbreeding depression, and competition (Motro 1991; Gandon 1999; Perrin and Mazalov 2000; Perrin and Goudet 2001). Under this hypothesis, we predict that species susceptible to inbreeding depression will show higher dispersal and that dispersal will be sex biased. An important caveat, however, is that this theory is based on inbreeding depression, rather than inbreeding per se. In some cases, a population that survives a prolonged period of inbreeding may have purged its deleterious mutations and no longer suffer strong inbreeding depression (Barrett and Charlesworth 1991; Keller and Waller 2002). If so, dispersal will no longer be favored under this theory, even if populations remain inbred.

Theory D4

Dispersal reduces competition among kin. This theory originates from the concept of inclusive fitness, in which fitness depends not only on an individual's reproductive success but also that of its kin, weighed by their relatedness (Hamilton 1964a). Inclusive fitness recognizes that related individuals are more likely to carry copies of the same alleles, and that selection should therefore favor altruistic behavior toward relatives. One such altruistic behavior might be dispersal that reduces competition, even in the face of dispersal costs (Hamilton and May 1977). Many theoretical models have explored this idea (Table 2.1), sometimes in conjunction with local extinctions (Theory D2) or inbreeding depression (Theory D3). Under this hypothesis, we predict that dispersal should increase as the potential for kin competition increases. Also, because iteroparity increases relatedness, dispersal should be higher as repeat breeding increases (Irwin and Taylor 2000), unless offspring dispersal varies with maternal age (Ronce et al. 2000b). Finally, the sex that experiences stronger competition among kin should disperse at higher rates (Perrin and Mazalov 2000).

The preceding theories each make qualitative predictions that can be subjected to empirical tests. However, some make similar predictions and may be difficult to distinguish. In addition, multiple selective factors probably act simultaneously, and the outcome will thus depend on the relative strengths of each factor. Experimental approaches would be the most powerful way to isolate individual selective factors but such experiments are logistically cumbersome, particularly for philopatry and dispersal. An alternative is to use comparative analyses. Our goal in the rest of this chapter is to use comparative analyses of salmonids for evaluating the relative merits of each theory. We start with some general information on homing and straying in salmonids, and then move to an explicit consideration of each theory.

2. The Salmonid System

2.1. Background

Here we provide a short review of the relevant salmonid life history, with detailed reviews appearing elsewhere (e.g., Balon 1980; Groot and Margolis 1991; Elliott 1994). Salmon, trout, and charr lay their eggs in the gravel of streams or lakes, where they incubate for several months and then hatch. Hatched embryos ("alevins") remain in the gravel for weeks to months, and then emerge as free-swimming "fry." In some cases, offspring remain for their entire life near their parent's original nest site (non-migratory), but in most, they move to other freshwater environments or the ocean (migratory). As they begin to mature, most salmonids return to the same general location where their parents bred ("homing"), whereas some disperse to non-natal sites ("straying"). Homing and straying will be strongly influenced by migration. For example, some wholly freshwater resident (non-anadromous) populations are isolated by physical barriers (landlocked) and so generate and receive few strays. Other non-anadromous populations are not isolated by barriers and can stray within their river system. However, straying between different river systems is greatly facilitated by migratory movements between fresh water and the ocean (anadromy).

Much of the early evidence for salmonid homing came from observations that phenotypic traits vary greatly among conspecific breeding aggregations (e.g., Robertson 1921). Presumably such variation could only arise if different aggregations represented different gene pools. Homing was later confirmed by experiments where juveniles tagged in a particular stream typically returned to that stream as adults (Scheer 1939; Ricker 1972; Stabell 1984). The principal mechanism of homing was then revealed by experiments where juveniles exposed to a particular odor overwhelmingly returned as adults to streams where that odor was released (Cooper et al. 1976; Scholz et al. 1976). These and other experiments, such as transplanting fish or cutting the olfactory nerve, confirmed that juveniles learn ("imprint") chemical characteristics of the water at their natal sites, and then use that information to guide their return migration

(Hasler and Scholz 1983). Accurate homing appears to be achieved by "sequential imprinting" (Harden Jones 1968), where "juvenile salmon learn a series of olfactory waypoints as they migrate through fresh water and later retrace this odor sequence as adults" (Dittman and Quinn 1996).

Most of the imprinted cues are probably abiotic, but Nordeng (1971, 1977) suggested that population-specific pheromones released by juveniles in fresh water might aid the return homing of adults. Although some juveniles and adults are preferentially attracted to odors from their own population (Groot et al. 1986; Brannon and Quinn 1990; Courtenay et al. 1997), the pheromone hypothesis appears to have little support as a general explanation for homing. As an obvious example, juveniles of many populations migrate to the ocean long before adults return. Some observations, however, suggest that pheromones may play a supplementary role: homing can be higher when population sizes are higher (Quinn and Fresh 1984; Hard and Heard 1999) and homing decreases when resident fish are removed (Tilzey 1977). In sum, olfactory imprinting, particularly on abiotic features, provides the best explanation for homing in fresh water. Homing from the open ocean to the mouth of a river system, however, must involve other mechanisms, such as magnetic or celestial orientation (Dittman and Quinn 1996).

The spatial scale of homing is currently unresolved. For example, do sockeye salmon home faithfully to a river system, to a lake within that system, to a tributary within that lake, to a riffle within that tributary, or to a location within that riffle? Evidence that sockeye salmon home faithfully to specific lakes is unequivocal (Quinn et al. 1987; Wood 1995) but *direct* evidence of homing to sites within lakes has been difficult to gather. Here, homing has been inferred *indirectly* from evidence that breeding aggregations show consistent differences in allelic frequencies at neutral genetic loci (Varnavskaya et al. 1994) and in population-specific adaptations (Quinn et al. 1995, 2001a). Recently, Quinn et al. (1999) used "natural tags" to provide direct evidence that sockeye salmon home faithfully to specific locations within lakes. At some scale, however, homing based on imprinting must break down because alternative sites will be too similar to distinguish. This threshold scale has yet to be determined and probably varies among populations and species. Also, heritable differences in breeding time could partially isolate adjacent aggregations even without imprinting (Hendry et al. 1995).

Homing and straying must be heritable for selection to drive their evolution, but we are unaware of any studies that have measured the heritability of homing or straying within a population (i.e., the degree to which homers beget homers and strays beget strays). However, several studies have compared homing and straying for different populations reared in a common environment. Hard and Heard (1999) found consistent differences in straying rates between two chinook salmon populations reared and released together at a non-natal site, suggesting a genetic difference in straying. In contrast, Labelle (1992) found no differences between three coho salmon populations reared and released at a non-natal site. We contend that homing *ability* (olfaction, imprinting, memory storage, memory recall) must have a genetic basis because juveniles return to non-natal release sites, even when no experienced individuals acted as guides and when their mothers never experienced that site.

Homing to *specific* sites, as opposed to overall homing ability, may also have a genetic basis. McIsaac and Quinn (1988) compared chinook salmon reared and released at natal sites (natal groups) to those reared and released at a non-natal site (experimental group). The natal groups homed accurately to their natal/ release sites but the experimental group showed an intermediate response: some returned to the non-natal release site and some (more than expected by chance) returned to the natal site, despite never having experienced that site. This suggests that salmonids have some genetic tendency to return to a specific natal site even if they did not imprint on that site. This inference is further supported by experiments that released pure natal, pure non-natal, and hybrid natal/non-natal fish into the same streams: homing was higher for hybrids than for pure non-natal fish (Bams 1976) and was higher for pure natal fish than for hybrids (Candy and Beacham 2000). All of these observations could be explained by a genetic tendency to return to a specific site or by a genetic tendency to return to a specific habitat type. In the later case, homing and straying would depend on the distribution of natal and non-natal sites with similar habitat. Regardless, homing to specific sites or habitats is clearly over-powered by imprinting because juveniles released at non-natal sites over-whelmingly return to those sites rather than to their natal site.

Even though homing ability is clearly heritable, it can also be influenced by individual condition and the environment. For example, straying may differ among fish of different ages (older > younger: Quinn and Fresh 1984; Quinn et al. 1991; Unwin and Quinn 1993; Labelle 1992; Pascual et al. 1995; older < younger: Hard and Heard 1999) and is higher for fish exposed to neurotoxic pesticides (Scholz et al. 2000). Also, changes in the physical characteristics of natal streams can influence homing: the eruption of Mt. St. Helens increased straying rates for steelhead (Leider 1989) and chinook salmon (Quinn et al. 1991). Of course, these patterns could still reflect underlying genetic variation, such as an adaptive reaction norm (Hutchings 2003—*this volume*) that links a threshold trait (straying) to the condition of an individual or the environment.

2.2. Estimating Homing and Straying

The two general methods for estimating homing and straying parallel the so-called "direct" and "indirect" methods for estimating gene flow (Slatkin 1987). Direct estimates can be obtained in several ways. First, juveniles can be tagged and the breeding adults then surveyed for the presence of tags. Some tags are administered to batches of fish: colored dyes for the skin, chemical dyes for scales or bones, and temperature-induced banding patterns on otoliths (Nielsen 1992; Schroder et al. 1995; Volk et al. 1999). Other tags are administered to individual fish: fin clips, dye marks, external tags, coded-wire tags, visual-implant tags, and passive integrated transponder tags (Nielsen 1992; Hughes et al. 2000). Some of these individual tags can be used to identify specific fish, and others are specific only to batches of fish. Second, breeding adults can be surveyed for the presence of "natural tags" induced by the environment: chemical signatures on scales (Kennedy et al. 2000), temperature-induced banding patterns on otoliths (Quinn et al. 1999), and parasites (Quinn et al. 1987). Third, adults can be characterized phenotypically or genetically and then assigned to putative source populations using discriminant functions, mixed-stock analyses (Pella and Masuda 2001), or "assignment tests" (Hansen et al. 2001).

Two general types of straying estimates are generated by tagging studies. First, all of the juveniles in a population can be tagged and then adults in that population can be surveyed for tags. Here, the number of untagged adults divided by the total number of adults estimates the proportion of fish at a site that are strays from elsewhere (e.g., Schroeder et al. 2001). Straying estimates using this method will be biased upward when some juveniles in the focal population are not tagged, but statistical adjustments can correct for this bias (e.g., Lindsey et al. 1959). Second, all or some of the juveniles in a population can be tagged and then that population and surrounding sites can be surveyed for tagged adults. Here, the number of tagged adults at all non-natal sites divided by the total number of tagged adults (natal and non-natal sites) estimates the proportion of fish that stray from a site. This method will underestimate straying when sampling effort is higher in the focal population than at other sites. However, this bias can be corrected by obtaining estimates of the proportion of adults sampled at each site (e.g., Mortensen et al. 2002). Even with this correction, however, straying rates may be underestimates because only a fraction of potential non-natal sites can be surveyed (Candy and Beacham 2000). However, most strays return to sites near the natal stream (see below) and complete sampling at nearby sites can therefore minimize any bias.

Several other direct methods have demonstrated strong site fidelity in salmonids but should not be used to estimate rates of homing and straying. First, adults can be captured at breeding sites and displaced to other sites. The extent to which these fish return to their capture site reveals site fidelity, and such return rates can be high (e.g., Hartman and Raleigh 1964). However, this approach should not be used to estimate straying because the displaced fish may already have been strays at their capture site, may not have had enough time to return to their capture location (see Lindsey et al. 1959), or may not have been exposed to the appropriate sequence of homing cues. Second, adults can be denied access to their natal site using a weir. Such fish will often die rather than breed at nearby sites (Hartman and Raleigh 1964) but the extent of such behavior will not reflect natural straying. Third, breeding adults of iteroparous species can be tagged and potential breeding sites can be surveyed in subsequent years (e.g., Lindsey et al. 1959). This last approach can be used to estimate *breeding* dispersal (switching sites between breeding seasons) but not *natal* dispersal (switching sites before the first breeding season).

Indirect methods rely on the expectation that the amount of among-population divergence at neutral genetic loci is negatively correlated with the amount of gene flow. The first step in this method is to obtain an accurate measure of genetic differentiation, which is facilitated by large sample sizes and many loci (Waples 1998b), as well as samples from multiple years or cohorts (E. Nielsen et al. 1999; Tessier and Bernatchez 1999; Garant et al. 2000; Heath et al. 2002b; Kinnison et al. 2002). The second step is to convert genetic differentiation to an estimate of gene flow. A traditional method is Wright's (1931) approximation: $F_{ST} = 1/[1 + 4N_e m]$, where F_{ST} is the proportion of the total genetic variation that is caused by differences among populations, $N_e m$ is the "effective number of migrants" per generation, m is the proportion of individuals that are "migrants," and N_e is the "effective population size." Effective population size is the size of an "ideal" population (random mating, equal sex ratio, non-overlapping generations, random variation in reproductive success) that would have the same genetic properties as the real population (Waples 2003—*this volume*). $N_e m$ thus represents gene flow in units of genetically effective individuals, which should correlate with straying rate.

Unfortunately, few natural systems meet the assumptions of Wright's approximation (Whitlock and McCauley 1999; Porter 2003). First, it assumes a genetic equilibrium, which requires thousands of generations when N_e is large, and will overestimate $N_e m$ when an equilibrium has not been reached (Nei and Chakravarti 1977; Figure 2.1A). Second, it assumes an infinite number of populations and will overestimate $N_e m$ when populations are few (Takahata 1983; Figure 2.1B). Third, it assumes that mutation rates are much lower than gene flow, which may not be the case for highly polymorphic markers such as microsatellites, and will overestimate $N_e m$ when this is not the case (Jin and Chakraborty 1995; Hedrick 1999; Figure 2.1C). Fourth, it assumes no selection on alternative alleles and will overestimate $N_e m$ when selection is purifying (removes variants) but underestimate $N_e m$ when selection is diversifying (favors different variants in different populations). Fifth, it assumes a particular population structure (island model), and may underestimate $N_e m$ for other types of population structure, such as "stepping-stone" and "isolation-by-distance" (Kimura and Weiss 1964; Slatkin 1993). Despite these concerns, it is important to recognize that F_{ST} remains a good descriptor of population structure (Whitlock and McCauley 1999; Balloux and Goudet 2002; Neigel 2002).

Some of the above concerns have been mitigated by recent developments, such as new genetic markers. The majority of early studies used allozymes, some of which are under selection (e.g., Jordan et al. 1997). In contrast, DNA microsatellites do not code for a product and should not be under direct selection, although they could be linked to a coding locus. With microsatellites, however, one must be wary of higher mutation rates (Hedrick 1999). Other improvements have focused on the relationship between genetic variation and $N_e m$. First, Takahata (1983) modified Wright's approximation to allow finite numbers of populations (Figure 2.1B), and this modification provides unbiased estimates in many cases (Balloux and Goudet 2002). Second, equations relating $N_e m$ to F_{ST} can be tailored to specific scenarios. For example, Hendry et al. (2000b) used a two-population model that allowed for different population sizes, asymmetric rates of gene flow, and non-equilibrium conditions. Third, Slatkin (1985) showed how $N_e m$ can be estimated from the distribution of rare alleles. Fourth, Beerli and Felsenstein (1999, 2001) showed how maximum likelihood and coalescent theory can be used to estimate gene flow with few assumptions.

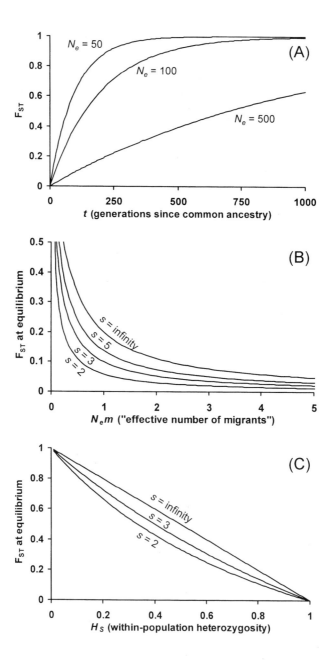

Fifth, simulation programs can accommodate diverse population structures in determining combinations of N_e, m, and divergence time that would generate observed patterns of divergence (Balloux 2001; Kinnison et al. 2002; Koskinen et al. 2002b).

Assuming an accurate $N_e m$ estimate has been obtained, several additional caveats are warranted. First, although many authors have argued that more than one migrant per generation will prevent genetic divergence (review: Mills and Allendorf 1996), divergence is still possible even if $N_e m$ is an order of magnitude larger (Allendorf and Phelps 1981). Second, $N_e m$ cannot by itself be used to argue that *adaptive* divergence is, or is not, constrained by gene flow; instead adaptive divergence is directly related to m (Hendry et al. 2001a). Third, estimates of straying produced by the direct and indirect methods may not be reliable surrogates for one another. For example, tagging work will often underestimate straying because few studies can survey all sites to which individuals might stray (Koenig et al. 1996; Candy and Beacham 2000). Also, tagging will underestimate gene flow when immigrants have higher reproductive success than residents (Ingvarsson and Whitlock 2000; Ebert et al. 2002) but overestimate gene flow when the converse is true (Hendry et al. 2000). Finally, when some individuals are migratory and some are not ("partial migration," Hendry et al. 2003b—*this volume*), tagging and genetic studies must include both migratory and non-migratory individuals if they are to provide accurate estimates of straying and gene flow.

Figure 2.1. Factors influencing the use of F_{ST} to estimate the effective number of migrants ($N_e m$). Panel (A) shows F_{ST} as a function of time (t, in generations) since a group of populations with a common ancestry started diverging (in the absence of ongoing gene flow: an isolation model). The rate of approach to unity depends heavily on N_e and when N_e is large it may take thousands of generations. The curves are calculated as $F_{ST} = 1 - (1 - 1/[2N_e])^t$, following Nei and Chakravarti (1977; see also Waples 1998b). If gene flow is taking place among the diverging populations, F_{ST} will asymptote below unity and will approach the asymptote more quickly than in the isolation model (Varvio et al. 1986). However, the rate of approach to the asymptote will still depend on N_e. Thus, estimating $N_e m$ from F_{ST} using equilibrium formulae will overestimate gene flow when diverging populations have not yet reached an equilibrium. Panel (B) shows equilibrium F_{ST} at different levels of gene flow ($N_e m$) and with different numbers of populations (s) exchanging migrants. Estimating $N_e m$ from F_{ST} using Wright's infinite island model will overestimate the true $N_e m$ if the actual number of populations is small. The curves are calculated as $F_{ST} = 1/(1 + 4N_e m[s/(s-1)]^2)$, following Takahata (1983). Once the number of populations exceeds five, the effect of further increases in the number of populations is minimal (at least in the island model). Panel (C) shows equilibrium F_{ST} at different levels of within-population heterozygosity (H_S) and different numbers of populations (s), in the absence of ongoing gene flow. Estimating $N_e m$ from F_{ST} without accounting for heterozygosity will overestimate the true $N_e m$. The curves were calculated as $F_{ST} = ([s-1][1-H_S])/(s-1-H_S)$, following Jin and Chakraborty (1995). When gene flow is ongoing among the populations, equilibrium F_{ST} will be low even at low heterozygosities (see panel B), and so the effect of heterozygosity on biasing $N_e m$ estimates may be small.

2.3. Patterns of Homing and Straying

A vast amount of data has been collected on homing and straying. Appendix 1 shows estimated straying rates from tagging studies of anadromous salmonids. The individual studies can be used to address specific questions (see below), but as a whole, the database is too small for detailed analyses of general trends. Broad generalizations, however, include (a) homing is more common than straying, (b) straying is variable within and among species, and (c) straying is more common among nearby populations than among distant populations.

Appendix 2 shows F_{ST} values from genetic studies, which should reflect the overall level of gene flow and straying. However, we did not convert F_{ST} estimates to $N_e m$ because of the aforementioned concerns. Several generalizations are possible. First, F_{ST} varies greatly among studies (0.000—0.645) and is strongly correlated with the maximum distance among populations (Figure 2.2). This matches the results of tagging studies in showing that straying and gene flow are negatively correlated with geographic distance. Second, wholly anadromous populations show less differentiation than wholly non-anadromous or mixed anadromous/non-anadromous populations (Figure 2.2; Appendix 2). Thus, species exhibiting life histories most conducive to straying do indeed stray at higher rates. Third, geographically isolated populations show greater differentiation than do populations among which straying remains possible (Appendix 2). Thus, in the absence of physically isolation, populations continue to exchange genes. Fourth, for a given geographic distance, differentiation is qualitatively greatest for coastal cutthroat trout, least for pink and chum salmon, and intermediate for the other species (Figure 2.3). This pattern likely reflects differences in their life history and migratory tendency.

Before using the empirical data to examine each theory, we here address two questions common to multiple theories. One question is whether or not straying is sex biased, and we suggest that it usually is not, at least not consistently. First, tagging studies of anadromous fish usually reveal similar straying rates in males and females (e.g., Unwin and Quinn 1993; Thedinga et al. 2000). An exception is Hard and Heard (1999), where 60.6% of the strays were males. Second, genetic assignment tests of anadromous brown trout show no evidence of sex biased straying (Hansen et al. 2001; Bekkevold 2002). Third, the dispersal of foraging juveniles is not sex biased (Dittman et al. 1998), except perhaps in masu salmon, where downstream movement is higher for newly emerged females than for males (Nagata and Irvine 1997). This last difference probably reflects variation in migration rather than dispersal because females are anadromous, whereas many males are not (Nagata and Irvine 1997). Fourth, although male-biased dispersal has recently been documented among locations within a landlocked population of brook charr (Hutchings and Gerber 2003), all of the fish remain part of the same population and the dispersal thus does not constitute straying. One context in which straying may be sex biased is partial migration, where females are anadromous more often than males (Hendry et al. 2003b—*this volume*) and should therefore stray at higher rates. However, this

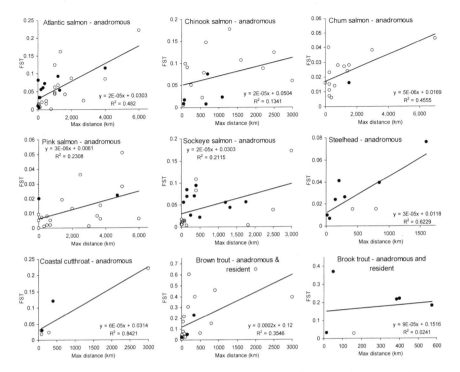

Figure 2.2. The amount of genetic variation among populations within species (F_{ST}) in relation to spatial scale, indexed as the shortest water distance between the two most distant populations. For Pacific and Atlantic salmon, only comparisons of exclusively anadromous populations (apart from mature parr) are shown. For brown trout and brook charr, the populations may be anadromous, resident, or a combination of the two, but they are not geographically isolated (i.e., they remain inaccessible to each other). Filled circles denote studies that included microsatellites or minisatellites, whereas open circles denote studies that used allozymes only. Note that the axes in the different panels have very different scales. The data are from Appendix 2.

difference probably reflects sex-specific costs and benefits of migration (Hendry et al. 2003b—*this volume*), rather than straying per se.

A second question is whether straying is density-independent or density-dependent (specifically, an increase in straying with increasing density), a distinction that is important because theoretical predictions can differ markedly between the two situations (Pulliam 1988; McPeek and Holt 1992; Travis et al. 1999). We suggest that salmonid straying typically does not increase with increasing density. First, tagging studies show that an increase in the number of adults returning to a site either causes decreased straying from that site (Quinn and Fresh 1984; Hard and Heard 1999) or has no effect on straying (Labelle 1992; Schroeder et al. 2001). Second, a dramatic reduction in the number of stream-resident brown trout led to reduced homing in the same year by repeat-spawning, lake-migratory brown trout (Tilzey 1977). In the absence of evidence

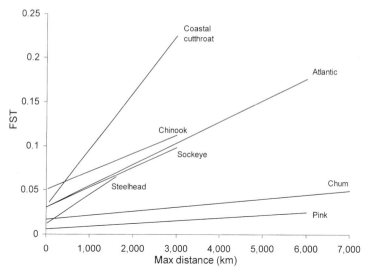

Figure 2.3. Comparison of genetic variation versus spatial scale (maximum shortest water distance between populations in a study) for the anadromous species in Figure 2.2. Shown are the best-fit linear relationships from Figure 2.2.

for increased straying with increasing density, our analyses will concentrate on the density-independent predictions. However, the trend in salmonids for decreased straying with increasing density suggests an interesting avenue for further theoretical work.

Although few of the empirical studies were specifically designed to evaluate alternative theories, many present data that we can use to test some of the theories' predictions. In the following, we attempt this task by examining variation in homing and straying in relation to life history (e.g., semelparity versus iteroparity, age at maturity, migratory tendency, duration of freshwater residence) and the environment (e.g., temporal and spatial variability).

3. Evaluating the Theories

3.1. Why Philopatry?

Theory P1

Philopatry increases the likelihood of finding a suitable breeding habitat and mate. This theory should be most important when strays have difficulty finding suitable alternative breeding sites or mates. It should be less important when strays typically enter established populations or find relatively vacant sites with suitable habitat and some conspecifics. Tagging and genetic studies clearly show that straying occurs among established populations (Appendices 1 and 2), a tendency that may be driven by several factors. First, most watersheds contain multiple populations, which will increase the likelihood that strays encounter established populations. Second, salmonids are attracted to water "conditioned" by conspecifics (e.g., Quinn and Busack 1985; Groot et al. 1986), which may increase the likelihood that strays will join established populations that they do encounter. Consistent with the idea that conspecifics influence adult returns, homing often increases as the number of returning adults increases (Quinn and Fresh 1984; Hard and Heard 1999). Straying among established populations is thus unequivocal but it is more difficult to determine the relative amount of straying to vacant sites because such sites are rarely surveyed.

Some straying probably does occur to sites without conspecifics because some hatcheries receive adults of species they do not propagate. Moreover, Hard and Heard (1999) found stray chinook salmon in several streams that lacked natural populations. In some cases, fish collected at non-natal sites may simply have been "proving" (Ricker 1972) or "exploring" (Griffith et al. 1999), and would have left the site had they been allowed to do so. Unambiguous evidence of straying to vacant sites comes from the colonization of areas made newly accessible by glacial recession (Milner and Bailey 1989; Milner et al. 2000), barrier removal (Bryant et al. 1999), or habitat improvement (Knutsen et al. 2001). Further evidence comes from the appearance of mature individuals outside their native range (McLeod and O'Neil 1983) and the rapid spread of introduced salmonids from introduction sites (Kwain and Lawrie 1981; Quinn et al. 2001b). These examples show that straying does occur to vacant sites but they also suggest that such straying is often successful. This implies that strays are not limited in their ability to find suitable alternative sites and mates. However, this conclusion must be tempered by the realization that straying to unsuitable sites, even if common, would rarely be documented because such sites are rarely surveyed.

In sum, homing probably does not evolve owing to direct selection for access to suitable breeding sites or mates: salmonids usually stray among established populations and strays to vacant sites are often successful. This theory may, however, play an important role in several contexts. First, the early stages of colonization may be tenuous if colonists are rare and poorly adapted. If so, returning adults may be few and homing may improve the chance that returning fish will find a suitable site with potential mates. Second, selection may generally favor returning to sites with conspecifics, and homing may simply be an efficient way to achieve this goal. Accordingly, this theory is indirectly invoked in some of the following theories.

Theory P2

Philopatry increases familiarity with local breeding conditions. The potential importance of this theory varies among salmonid species and life histories. It might seem most relevant for iteroparous species, where experiences gained during one breeding season might influence success during future breeding seasons. Although this would directly select against breeding dispersal (switching sites between breeding seasons), it would not directly select against natal dispersal (switching sites before first breeding). It might, however, indirectly select against natal dispersal if natal and breeding dispersal are correlated. Direct selection against natal dispersal requires that juveniles gain experiences at the natal site that later aid them while breeding at that site. This seems most likely for non-migratory salmonids, or for migratory salmonids with juveniles that remain at the natal site for an extended time. When juveniles leave natal sites soon after emergence (e.g., sockeye, chum, and pink salmon), it seems unlikely that they would be able to learn features of their environment that would later improve breeding success.

If this theory is important, we might therefore expect homing to be higher in iteroparous than semelparous salmonids, higher in non-migratory than migratory salmonids, and higher in migratory salmonids with juveniles that remain longer at natal sites. These predictions could be tested by comparing straying or genetic differences between these groups but this would be confounded by factors influencing the ease of homing. That is, homing should be easier for (a) iteroparous salmonids after their first return to a site (homing is higher for repeat breeders than for first-time breeders, Lindsey et al. 1959), (b) non-migratory salmonids because they do not leave the natal site, and (c) salmonids whose juveniles remain longer at natal sites because they have more time to imprint and may release pheromones that improve homing by adults (although Laikre et al. 2002 found that earlier emigration by brown trout did not decrease genetic differences among populations). Because of the directly confounding nature of this theory's predictions with the difficulty of homing, we do not further apply the comparative approach. Instead, we consider more generally whether juveniles might be able to gain experiences at the natal site that would improve their breeding success as adults.

Numerous studies have demonstrated that prior experience can improve competitive ability. For example, prior experience strongly influences the ability of size-matched juveniles to hold feeding territories (Rhodes and Quinn 1998; Cutts et al. 1999a). During breeding, prior residence increases the ability of females to defend nest sites and males to defend females (Foote 1990; Morbey 2002). To our knowledge, however, no study has tested whether experiences gained prior to breeding improve success during breeding. In sum, this theory is difficult to test but does not receive much support from salmonids. First, it does not provide a general explanation for homing because many species have life histories with which it is not compatible. Second, no evidence exists that experiences prior to breeding influence success during breeding.

Theory P3

Philopatry returns locally adapted individuals to appropriate habitats. In the following, we argue that this theory receives considerable support from salmonids. The stage is set by strong divergent selection among populations. For example, juvenile sockeye salmon usually occupy lakes for their first few years of life ("lake-type") and natal lakes differ dramatically in environmental conditions such as temperature and productivity (Burgner 1991; Edmundson and Mazumder 2001). In other sockeye salmon populations, juveniles remain in streams or migrate immediately to the ocean ("stream/ocean-type," Wood 1995). Sockeye salmon populations also differ dramatically in their breeding habitat: tiny streams, large rivers, spring-fed ponds, and lake beaches; with gravel size, water temperature, and water depth varying within and among these habitats (Quinn et al. 1995, 1999, 2001a). Many other examples of striking environmental variation among natal sites are found in all salmonid species (e.g., Groot and Margolis 1991; Taylor 1991b; Elliott et al. 1998).

Divergent environments favor divergent adaptation, which is thus a prominent feature of salmonids. For example, sockeye salmon show consistent among-population environment/phenotype correlations. First, adults in small streams are younger and smaller than those in large streams, probably owing to increased bear predation and "stranding" (Quinn et al. 2001a, 2001c). Second, adults on lake beaches have deeper bodies than those in streams, probably owing to sexual selection favoring this trait but natural selection acting against it in streams (Quinn et al. 2001a; Figure 2.4). Third, egg size and gravel size are positively correlated (Quinn et al. 1995). Fourth, fry from lake inlets show a negative response to current (move downstream), whereas fry from lake outlets show a positive response to current (move upstream), behaviors well suited for reaching their natal lake (e.g., Raleigh 1971). Some of these differences are known to have a genetic basis and the others are expected to. Numerous other examples of local adaptation have been described in salmonids (reviews: Ricker 1972; Taylor 1991b; Wood 1995).

The importance of divergent adaptation is best revealed by studying introduced populations (Kinnison and Hendry 2003—*this volume*). When such introductions are successful, new populations often become established in different habitats and have been found to exhibit adaptive divergence within at least 10–25 generations. Sockeye salmon introduced into Lake Washington, Washington, colonized beach and stream environments and now show apparently adaptive differences in body size, shape, and development (Hendry et al. 2000b; Hendry 2001). Chinook salmon introduced to New Zealand colonized multiple sites and diverged adaptively in many traits (Quinn et al. 2001b), particularly reproductive output in relation to migration distance (Kinnison et al. 2001). Adaptive divergence in introduced salmonids has also been documented for grayling in Norway (Haugen and Vøllestad 2001; Koskinen et al. 2002b).

If divergent adaptation has contributed to the evolution of homing, strays to a non-natal site should have lower reproductive success than homers to that site. One way to test this prediction is to compare the straying of adults (based on tagging) to the resulting gene flow (based on genetic divergence). If adult straying exceeds gene flow, strays must have reduced success relative to homers. Tallman and Healey (1994) tagged chum salmon juveniles and sampled adults from two adjacent streams. The tagging data suggested that adult straying averaged 37.9%, whereas the genetic data suggested that gene flow averaged only 5%. They concluded that "salmonids straying onto the spawning grounds of established popu-

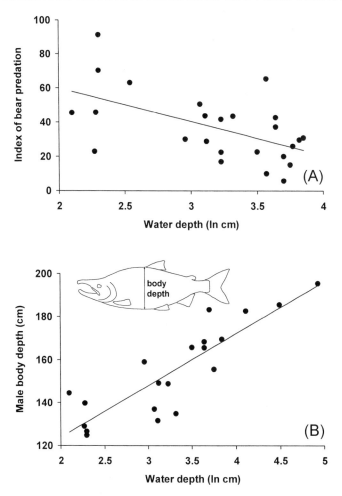

Figure 2.4. An example of probable local adaptation in sockeye salmon. Panel (A) shows that different populations in Bristol Bay (Alaska) breed in streams with very different average water depths, and that water depth is strongly correlated with the intensity of bear predation acting on the populations. Panel (B) shows that the body depth of males is highly correlated with water depth among populations, probably because large and deep-bodied males are more susceptible to stranding and predation in shallow water. Several studies have documented selection against large males owing to stranding and bear predation (e.g., Quinn et al. 2000c). Data are from Quinn et al. (2001a) and T. Quinn (unpublished).

lations of conspecifics had lower reproductive success than fish that return to their natal streams" (Tallman and Healey 1994). Unfortunately, the sample of adults they used to genetically characterize the groups was from a year when adults did not have tags. As a result, strays and homers were mixed in the genetic analysis, effectively affording them equal success and preventing an accurate estimate of gene flow.

Hendry et al. (2000b) conducted a similar study but were able to separate homers from strays. They examined two nearby populations that bred in different habitats (lake beach versus river) and had different population sizes (the river was two orders of magnitude larger). The physical proximity and disparate abundance of the two populations suggested that many adults breeding at the beach might be strays from the river, even if homing was strong. Breeding adults were sampled at the two sites and individual homers and strays were identified by their otolith banding patterns, which reflect site-specific incubation temperature. This analysis revealed that approximately 39% of the adults breeding at the beach were strays from the river. Microsatellites were then used to examine genetic divergence between beach residents (homed to the beach), river residents (homed to the river), and beach immigrants (strayed from the river to the beach). River residents and beach immigrants were genetically similar but river residents and beach residents were genetically distinct (Figure 2.5). This "pattern of genetic differentiation could only have arisen if beach immigrants have reduced reproductive success relative to beach residents" (Hendry et al. 2000b; see also Hendry 2001). A comparison of genetic introgression and physical straying for pink salmon in two adjacent creeks in Alaska yields the same conclusion: straying is higher than gene flow (Mortensen et al. 2002).

Voluminous circumstantial evidence further supports the contention that adapted homers have higher fitness than maladapted strays. First, introduced fish usually fail to establish new populations, particularly within their native range (Withler 1982; Wood 1995). Outside their native range, some introductions

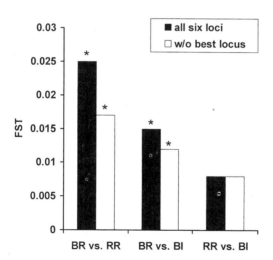

Figure 2.5. Genetic differentiation (F_{ST}) between beach residents (BR), river residents (RR), and beach immigrants (BI) for sockeye salmon introduced less than 14 generations previously into the Lake Washington system (Hendry et al. 2000b; Hendry 2001). Dark bars show F_{ST} over all six loci and white bars show F_{ST} after removing the locus that was best at discriminating between BR and RR (just in case that locus was linked to a locus under selection). Asterisks show comparisons that were statistically significant ($P < 0.05$).

have been successful (Lever 1996) but many more have failed (e.g., Harache 1992). Second, hatchery or farmed salmon released into the wild typically differ from wild fish in morphology and behavior, and have correspondingly lower reproductive success (Leider et al. 1990; Reisenbichler and Rubin 1999; Fleming et al. 2000). In fact, native gene pools often (but not always) retain their integrity despite large numbers of hatchery fish released into their midst (Hansen et al. 2000b; Utter 2000; Ruzzante et al. 2001).

Variation among species may also be explained by this theory. Relative to other anadromous species, chum and pink salmon populations typically use simple and similar freshwater habitats and have juveniles that do not remain long in fresh water (Quinn 1984). In this situation, divergent selection should be weak, divergent adaptation should be minor, strays should have similar success to homers, and straying rates should be high. As described below, tagging studies suggest that pink salmon do indeed stray at higher rates than the other species but comparable tagging data are not available for chum salmon. However, genetic variation among populations is considerably lower for both pink and chum salmon than for the other species (Figure 2.3). This difference is consistent with the present theory but it is also consistent with other explanations (Table 2.2).

We conclude that salmonid populations often show strong divergent adaptations that reduce the success of strays. At least two arguments might be raised in opposition to this as a general theory for the evolution of homing. The first is that divergent selection should be less effective at small spatial scales (review: Lenormand 2002). When populations are close together, it may be difficult for individuals to distinguish between natal and non-natal sites, which should increase straying and compromise adaptation. Indeed, most straying is to nearby sites (see above) and genetic differentiation is positively correlated with geographic distance (Figure 2.2). Divergent adaptation in salmonids will undoubtedly be minimal at very small spatial scales (Adkison 1995, Hansen et al. 2002) but these scales have not been determined and may be very small indeed. For example, populations within a few kilometers of each other often exchange few strays (Labelle 1992; Hard and Heard 1999; Quinn et al. 1999), are distinct at neutral markers (Carlsson et al. 1999; Garant et al. 2000; Hendry et al. 2000; Gharrett et al. 2001; Ruzzante et al. 2001), and show strong evidence of divergent adaptation (Hendry et al. 2000b; Hendry 2001; Quinn et al. 2001a).

A second criticism might be that this mechanism will only work when populations are already well adapted, and it may therefore be unimportant in new systems. We disagree because new populations should begin to adapt almost immediately, even in the presence of ongoing gene flow (Endler 1973; Hendry et al. 2001a). Accordingly, studies of introduced salmonids have shown that substantial adaptive divergence is already present after 10–25 generations (Haugen and Vøllestad 2001; Koskinen et al. 2002b), in some cases arising even with ongoing gene flow (Hendry et al. 2000b; Hendry 2001; Quinn et al. 2001b). Thus, strays should be at a slight disadvantage soon after colonization, and this disadvantage should increase as adaptation proceeds. In fact, the reduced reproductive success of strays in Lake Washington was already present after about a

Table 2.2. Plausible alternative explanations for some of the observed variation in the homing and straying of salmonids.

Observation	Key life history features	Alternative explanations
1. Pink salmon stray at higher rates than other anadromous Pacific salmon.	Pink salmon are semelparous, always mature at 2 years of age, often spawn in small streams close to the ocean, and have juveniles that migrate immediately to the ocean.	(a) Homing is more difficult for pink salmon because juveniles spend less time in fresh water and so have less opportunity for imprinting. (b) Homing is less important for pink salmon because local adaptation is weaker: natal streams are similar and juveniles spend little time in them (Theory P3). (c) Pink salmon cannot take advantage of parental resources because juveniles do not remain in fresh water and breeding sites have high flushing rates (Theory P5). (d) Semelparity and maturity at 2 years of age increase susceptibility of pink salmon to temporal variation in habitat quality—straying compensates (Theory D1).
2. Stream/ocean-type sockeye salmon show higher gene flow than lake-type sockeye salmon.	Both forms are semelparous, have multiple ages at maturity, and spawn in a diversity of environments.	(a) Homing is more difficult for stream/ocean type fish if streams are less distinctive than lakes. (b) Homing is less important for stream/ocean type fish because local adaptation is weaker: rivers may be more similar than lakes (Theory P3). (c) Lake-type juveniles receive greater benefits from parental resources (nutrients from carcasses) because flushing rates are slower (Theory P5). (d) Interannual variation may be greater for rivers than lakes, favoring increased straying in the former (Theory D1).

dozen generations (Hendry 2001). Adaptation and homing should reinforce each other in a positive feedback loop. As adaptive divergence begins, selection against strays will increase and homing should be favored. As homing then increases, adaptive divergence should also increase because gene flow is reduced. This process should in a positive feedback loop that continues until the population is maximally adapted.

Theory P4

Philopatry is favored by spatial variation in habitat quality. This theory predicts that homing is favored when fitness is negatively correlated with breeding density, straying rate is not positively correlated with breeding density, and habitat quality varies among sites but not among years. Several of these conditions are met by salmonids: (a) average reproductive success decreases with increasing density (Elliott 1994; Essington et al. 2000), (b) straying does not increase with increasing density (see above), and (c) habitat quality varies among sites as evidenced by variations in population sizes, breeding densities, and smolts-per-adult (Groot and Margolis 1991; Elliott 1994). One condition is not met: habitat quality varies substantially among years. Thus, the importance of this theory will depend on the relative magnitude of spatial and temporal variation. For example, if two metapopulations have similar levels of temporal variation but different levels of spatial variation, greater philopatry would be predicted in the case of greater spatial variation. We do not know of any data allowing a direct test of this prediction. Such data could consist of straying or genetic differences among populations in relation to spatial and temporal variation in habitat quality.

Theory P5

Philopatry improves access to parental resources. Salmonids probably do not inherit specific nest sites or mates, and so this theory would only be relevant through the general transfer of resources between generations. The nutrients released by dead and dying adults may be one such resource. Anadromous salmonids that die after breeding release large amounts of marine-derived nutrients into freshwater ecosystems (Kline et al. 1993; Schmidt et al. 1998). These nutrients increase primary productivity and invertebrate biomass (Bilby et al. 1996), and may improve the growth and survival of juvenile salmonids (Bilby et al. 1998; Schmidt et al. 1998; Wipfli et al. 2003). The chief beneficiaries of these nutrients are often heterospecifics but may sometimes include conspecifics. Among Alaskan lakes, for example, the number of sockeye salmon adults per unit of lake area (salmon-derived nutrient, SDN, loading) was positively correlated with marine-derived nitrogen in lake sediments, zooplankton, and juvenile sockeye salmon (Finney et al. 2000). Among years within lakes, SDN loading was positively correlated with the abundance of zooplankton, the primary food for sockeye salmon juveniles (Finney et al. 2000). This suggests "a positive feedback system, in which higher adult salmon abundance leads to increases in nutrient

(P and N) loadings. This enrichment, in turn, increases lake primary and second-ary productivity. Completing the cycle, the increase in lake carrying capacity for juvenile salmon may ultimately result in higher numbers of adult salmon" (Finney et al. 2000).

Whether or not these parental nutrients influence homing is unclear. One constraint is that parental nutrients will favor homing only when they improve the *breeding* success of offspring. It is not enough that parental nutrients improve the foraging success of their offspring because this would favor delayed migration rather than homing. Instead, nutrients must improve the foraging success of their grand offspring, and this will depend on the flushing rate of nutrients. Another constraint is that parental nutrients will not directly favor homing if straying is between established populations, where homers and strays would benefit equally from nutrients released by previous generations. However, parental nutrients may indirectly favor homing in at least two ways. The first ocurs if homing provides an efficient way for individuals to return to sites with abundant con-specifics (invoking Theory P1). The second is when potential breeding sites differ in population size and hence the availability of parental nutrients (invoking Theory P4). One prediction might be that homing should be highest for sockeye salmon because they are semelparous, abundant, and have juveniles that live in slowly flushing environments (lakes). Sockeye salmon do indeed have low stray-ing rates but so too do chinook salmon (Figure 2.3), which are much less abun-dant and have juveniles that live in quickly flushing habitats (streams). Moreover, strong homing in sockeye salmon is also consistent with other theories.

Another possible parental resource is improvements that breeding adults make to the incubation environment. Fine sediments drastically reduce the sur-vival of eggs and embryos (Chapman 1988), and nesting females purge the incubation environment of these suffocating materials (Kondolf et al. 1993). Much of this effect disappears before the next breeding season (Kondolf et al. 1993; Peterson and Quinn 1996) but frequently used and densely populated sites might nevertheless retain fewer fine sediments. (Of course, negative correlations between site use and fine sediments could also arise if salmonids prefer sites with fewer such sediments or have greater success in them.) Nest building also coar-sens the stream bed, which may reduce gravel scour (Montgomery et al. 1996). Homing may thus be favored if adults improve the long-term quality of incuba-tion sites. As above, this mechanism favors the return of adults to established sites rather than specific natal sites but might indirectly favoring homing through Theory P1 or·Theory P4.

Theory P6

Philopatry avoids costs of movement. This theory considers costs associated with movement itself, such as increased stress, energy use, or predation. Because migration also requires increased movement, this theory should be evaluated in three contexts: complete migration, complete non-migration, and partial migration. For complete migration, the theory may not be relevant because movement costs will be similar for homers and strays: most straying is to near the natal site (Appendix 1; Figures 2 and 3). For complete non-migration, the theory may be very relevant because strays should incur movement costs that homers do not. In this case, homing should be stronger when costs of movement are higher but no studies have yet related movement costs to homing. For partial migration, the theory may be relevant in indirect ways. For example, straying should be higher in anadromous than non-anadromous salmonids because anadromy increases the potential for movement between river systems. However, different populations experience different cost of migration and hence selection for or against anadromy (Hendry et al. 2003b—*this volume*). Direct selection for or against anadromy should therefore have strong indirect effects on homing and straying, and vice versa. Again, homing should be stronger when costs of movement are higher but, again, no empirical tests are available.

3.2. Why Dispersal?

Theory D1

Dispersal buffers against temporal variation in habitat quality. If straying evolves via this mechanism, site quality must vary among years and this variation must be asynchronous among sites. When these two conditions are met, at least two testable predictions follow: straying should increase as temporal variation increases and as variation in age at maturity decreases. This last prediction arises because variation in age at maturity also buffers against temporal variation in habitat quality. Selection should thus act more strongly to increase straying when age at maturity is less variable (Quinn 1984; Kaitala 1990). We will first consider the two conditions and then the two predictions.

The quality of natal sites clearly varies among years. One major source of variation is flooding, which mobilizes incubation gravel, destroys nests, and causes high mortality of eggs and alevins (Montgomery et al. 1996; Peterson and Quinn 1996; Lapointe et al. 2000). For example, the number of juveniles produced per breeder in the Cedar River, Washington, varied among years from 5.6 to 24.9, largely as a result of river discharge ($r^2 = 0.94$, Thorne and Ames 1987). Episodic floods can also have dramatic effects on stream-dwelling juveniles (e.g., Good et al. 2001). Predation is another factor that varies dramatically among years. For example, the number of adult sockeye salmon killed by bears is very low in some years but very high in others (7–100%, Quinn and Kinnison 1999; 16.4–92.3%, Ruggerone et al. 2000). Many other examples of temporal variation in habitat quality have been described (e.g., Giberson and Caissie 1998).

Some temporal variation probably has a spatially synchronous component: climate is usually regional and will vary across nearby sites in a similar manner. However, nearby sites often differ in physical features and will therefore respond differently to climate variation. For example, sockeye salmon breed at inlet streams, outlet streams, and lake beaches (Wood 1995), which differ in their susceptibility to temperature, gravel scour, and predation. Even within a specific habitat type, variation can be high in physical features such as basin area, stream gradient, water depth, ground-water influence, woody debris, and gravel size (Beechie and Sibley 1997; Quinn et al. 1995; 2001a). Each of these features will influence a site's response to climate variation. Moreover, some environmental catastrophes have localized effects, such as the damage to several streams caused by the eruption of Mount St. Helens. Although we do not know of any studies quantifying spatial asynchrony in temporal variation, we expect that it is probably high.

Do populations that experience greater temporal variation show higher rates of straying? Direct tests of this prediction are difficult because studies usually do not report both straying and temporal variation in habitat quality. However, some qualitative comparisons are possible, such as between lake-type and stream/ocean-type sockeye salmon. Temporal variation is presumably lower in lakes than in rivers, and straying should thus be lower for the former. Tagging studies have shown that sockeye salmon do home very precisely to natal lakes (Quinn et al. 1987) but no tagging studies have been performed for stream/ ocean-type populations. However, Gustafson and Winans (1999) found that genetic differentiation was higher among lake-type populations than among stream/ocean-type populations, consistent with higher straying among the latter. Although this difference supports the theory's prediction, several alternative explanations are possible (Table 2.2).

Another possible comparison is between spring breeders and fall breeders. L'Abée-Lund and Vøllestad (1985) argued that stream conditions are more variable in the spring than in the fall, and that spring species should therefore stray at higher rates. Quinn and Tallman (1987) countered that spring conditions are not necessarily more variable than fall conditions, and that spring species do not necessarily stray more than fall species. For example, spring-breeding steelhead strayed much less (2.5%) than fall-breeding coho salmon (21%) between the same two creeks (Shapovalov and Taft 1954). Our review of genetic variation suggests that spring-breeding steelhead show roughly similar levels of gene flow to fall-breeding species that also have stream-dwelling juveniles (coho, chinook, Atlantic salmon; Figure 2.3; Appendix 2). The prediction for spring breeders versus fall breeders thus lacks empirical support, and may not have been an appropriate prediction.

Do species with less variation in age at maturity show higher rates of straying? Pink salmon have a strict 2-year life cycle, making them the least variable Pacific salmon. Chinook salmon hold up the other end of the spectrum, with age at maturity ranging from 1 to 8 years (Healey 1991). Pink salmon should therefore stray the most, chinook salmon the least, and other Pacific salmon species at intermediate levels. Pink salmon have long been thought to stray more than other species (Heard 1991), but Quinn (1993) was "unable to locate a comprehensive study of straying by wild pink salmon populations" and felt "the conclusion that they stray more commonly than other salmon species seems premature." Recent evidence suggests that this conclusion is no longer premature. First, two tagging studies quantified straying rates in pink and chinook salmon in the same region (southeast Alaska) and in many of the same streams.

Sampling effort was similar for the two studies and yet straying was more than four times higher for pink salmon (5.1%, Thedinga et al. 2000) than chinook salmon (1.2%, Hard and Heard 1999). A caveat is that the release sites were surrounded by other pink salmon populations, which might attract strays, but not by other chinook salmon populations. Second, our review shows that pink salmon have the lowest levels of genetic variation in anadromous Pacific salmon (Figures 2.2 and 2.3; Appendix 2).

A counterpoint was argued by Gharrett et al. (2001), who genetically marked a subpopulation of pink salmon in Auke Lake, Alaska, and estimated gene flow into other Auke Lake subpopulations. Gene flow from the marked population was essentially 0% into two subpopulations that bred at different times and 8–9% into two subpopulations that bred at similar times. The marked subpopulation received 8% gene flow from the other subpopulations. Gharrett et al. (2001) concluded that "in some instances, the fidelity of homing in pink salmon is high." Although this is certainly true (e.g., Little Susitna River, see below), Gharrett et al.'s (2001) results should not be used to argue that pink salmon do not stray at higher rates than other species. First, gene flow was lowest among subpopulations with different breeding times, which is to be expected even if homing is weak (Tallman and Healey 1994; Varnavskaya et al. 1994; Hendry et al. 1995; Quinn et al. 2000; Woody et al. 2000). Second, a tagging study at Auke Creek revealed that physical straying of adults was actually fairly high (5.7%, Mortensen et al. 2002). For example, Gharrett et al. (2001) estimated that gene flow from Auke Creek into nearby Waydelich Creek was essentially zero, whereas Mortensen et al. (2002) found "that 3% of Waydelich Creek spawners had originated in Auke Creek." This difference between adult straying and gene flow implies selection against strays (Theory P3). Finally, gene flow among Auke Lake subpopulations with similar timing was actually fairly high (8–9%).

Pink salmon thus stray more than other salmonids, providing support for the theory that straying evolves to buffer temporal variation in habitat quality. A possible alternative is that homing is more difficult for pink salmon because they breed closer to the ocean and have juveniles that spend little time in fresh water (Table 2.2). Consistent with this idea, intertidal pink salmon stray more (Thedinga et al. 2000) and show lower levels of genetic differentiation (Seeb et al. 1999) than upstream populations. Another possibility that local adaptation is less important for pink salmon because they spend less time in fresh water (Table 2.2). These alternatives are plausible but not general because some pink salmon migrate long distances, and juveniles of other species sometimes leave natal sites immediately after emergence.

A related comparison is that between semelparous and iteroparous species, where straying should be lower in the latter because reproduction in multiple years buffers against temporal variation. The only study comparing adult straying rates between iteroparous (steelhead) and semelparous (coho) species in the same streams (Shapovalov and Taft 1954) found much higher straying in the later. Genetic differentiation is also greater in coastal cutthroat trout than in semelparous species but the same is not true of other iteroparous species

(Atlantic salmon and steelhead trout, Figures 2.2 and 2.3). Moreover, coastal cutthroat trout populations may contain some non-anadromous individuals that cause a reduction in total gene flow. In sum, salmonids provide some support for the theory that straying evolves to buffer temporal variation in habitat quality. The evidence is equivocal but is also wide-ranging and convincing.

Theory D2

Dispersal allows colonization of new environments. If extinction/recolonization dynamics are important in the evolution of straying, two conditions should be met: local populations should frequently go extinct and strays should rapidly recolonize vacant areas. Extinction/recolonization events have clearly influenced salmonid lineages because much of their native range has been covered periodically by glaciers. Each time the ice receded, newly exposed sites were colonized by fish whose ancestors had persisted in refugia (Wood 1995; McCusker et al. 2000; Bernatchez 2001). But how quickly does colonization take place? In Tustemena Lake, Alaska, sockeye salmon now breed in several inlet streams that were covered by a glacier as recently as 2000 years ago. These populations are distinct at neutral genetic loci (Burger et al. 1997) and at morphological traits that reflect local adaptation (Woody et al. 2000).

Milner and Bailey (1989) studied streams exposed by receding ice in Glacier Bay National Park, Alaska. They recorded 828 adult pink salmon in a stream accessible for 15 years and thousands in two streams accessible for 150 years. Other salmonid species in these new streams included sockeye salmon, chum salmon, coho salmon, and Dolly Varden (see also Milner et al. 2000). Similarly, Bryant et al. (1999) showed that the installation of a fish ladder allowing access to Margaret Lake, Alaska, resulted in natural colonization by pink salmon, cut-throat trout, steelhead trout, and Dolly Varden charr. Knutsen et al. (2001) studied Norwegian lakes where anadromous brown trout had gone extinct because of acidification, and found that the lakes were naturally recolonized soon after the addition of lime. The rapidity of colonization is also demonstrated by artificial introductions. In New Zealand, chinook salmon introduced to one stream quickly established self-sustaining populations in several other streams, and these populations are now genetically and phenotypically distinct (Kinnison et al. 2001, 2002; Quinn et al. 2001b). In the North American Great Lakes, pink salmon released into one stream in Lake Superior colonized at least four nearby streams within 4 years, and dozens in all the Great Lakes within 23 years (Kwain and Lawrie 1981).

These examples illustrate that salmonids rapidly colonize suitable habitats and establish distinct, self-sustaining populations. If, however, the present theory is to explain straying in extant populations, local extinctions and recolonizations must be quite frequent. Local extinctions are certainly common enough in modern times: "at least 106 major populations of salmon and steelhead on the West Coast have been extirpated" (Nehlsen et al. 1991; see also CPMPNAS 1996; Parrish et al. 1998). In most of these cases, human disturbances such as dams, habitat destruction, or fishing contributed to the extinctions, and so recolonization would not be expected until those impacts have been ameliorated. However, Hansen and Mensberg (1996) describe the natural recolonization of two streams in Jutland, Denmark, following the extinction of their original populations. Yet unless we invoke human-mediated disturbances as the primary cause of current straying rates, we must also seek evidence of frequent natural extinctions. Unfortunately, human impacts are so pervasive that it is difficult to know whether a given extinction was or was not influenced by humans. However, a variety of natural disturbances, such as volcanic eruptions (e.g., Mount St. Helens), landslides, sustained droughts, or debris flows, at least have the potential to cause frequent local extinctions.

An interesting prediction of this theory is that straying should be highest in the youngest populations and should decrease with population age (Olivieri et al. 1995). It might therefore be informative to compare straying among populations that differ in evolutionary age. For example, some populations persisted in refuges during the last glaciation, whereas others were established by post-glacial straying. Data are not yet available for a rigorous comparison of this nature but one study is suggestive. Pink salmon tend to stray at high rates (see above) but Churikov and Gharrett (2002) found one population (Little Susitna River, Cook Inlet, Alaska) that strayed at very low rates. "Not a single 'foreign' haplotype was found in the sample of 40 fish from Little Susitna and not a single haplotype from the Little Susitna lineage was detected among the samples from 6 other even-year Alaskan populations" (Churikov and Gharrett 2002). Based on a variety of evidence, the authors argued that the Little Susitna is a relict population that evolved in a glacial refuge, whereas the others were founded post-glacially. The difference in straying might thus be consistent with the above prediction. A broader speculation might be that the increase in anadromy with latitude (Hendry et al. 2003b—*this volume*) is caused by increased selection for straying because of the greater role of glaciers. In sum, straying does play an important role in recolonization, and local extinctions may therefore contribute to the evolution of straying rates in extant populations.

Theory D3

Dispersal reduces inbreeding depression. If inbreeding depression influences the evolution of straying, inbreeding should reduce fitness, wild populations should have the potential for inbreeding, other inbreeding avoidance mechanisms should be less effective, and males should disperse more than females. In the following, we consider each expectation in turn. Numerous studies of captive salmonids have examined the fitness effects of inbreeding (Su et al. 1996; Pante et al. 2001; Wang et al. 2001). For example, Myers et al. (2001) found no significant effects in coho salmon when each of five generations was founded by six males and six females from ten full-sib families. However, the progeny of highly inbred full-sib matings had 25% lower body mass than their outbred half-sibs. Heath et al. (2002a) found positive, but weak ($r^2 = 0.06$–0.08), correlations between genetic variation at microsatellite loci and reproductive traits in chinook salmon. Arkush et al. (2002) found that the offspring of full-sib inbred matings had lower survival than the offspring of outbred matings when challenged with one pathogen but not two others. In summary, inbreeding depression seems variable and often weak, except under intense inbreeding (Wang et al. 2001). Myers et al. (2001) suggested that the lack of strong inbreeding depression in salmonids was because they "have evolved through at least one tetraploid event, and the duplication of genes provides a buffer against inbreeding effects." As a caveat, studies of inbreeding depression in salmonids have thus far been for captive fish, whereas negative effects of inbreeding may be stronger in the wild (Wang et al. 2001). In the only study in the wild, Ryman (1970) released Atlantic salmon into Swedish rivers and recaptured more outbred than inbred individuals.

What is the potential for inbreeding in the wild? Unfortunately, inbreeding is usually quantified only for manipulated populations (Wang et al. 2001), wherein supportive breeding can reduce effective population sizes and increase inbreeding (Tessier et al. 1997; Hansen et al. 2000a). As a result, some breeding programs actively minimize inbreeding by equalizing individual contributions to the next generation (Hedrick et al. 2000b) or avoiding matings between genetically similar individuals (Letcher and King 2001). In unmanipulated populations, the potential for inbreeding will vary among species. In healthy populations of Pacific salmon, inbreeding is probably unlikely because population sizes are often large (thousands to hundreds of thousands), and long migrations would make it difficult for siblings to remain in close contact. In contrast, some trout and charr populations can be very small and have siblings that remain in close contact. For example, Laikre et al. (2002) estimated that effective population sizes were very low in some natural brown trout populations (e.g., $N_{e(females)} = 7.2, 15.7, 16.5,$ 17.6, 20.3, 24.0, 26.3, 27.8). Some salmonid populations appear to have experienced severe inbreeding, perhaps because of founder effects or genetic bottlenecks. For example, two populations of brown trout in northwestern Scotland were genetically monomorphic at 46 allozyme loci, six mtDNA restriction sites, and eight minisatellite loci, in sharp contrast to substantial genetic variation in other populations (Prodöhl et al. 1997). Despite the lack of genetic variation, growth rates were as high as in less inbred populations nearby (Prodöhl et al. 1997). Either inbreeding does not have strong fitness effects in these populations or deleterious mutations were purged during the bottleneck.

Do other inbreeding avoidance mechanisms operate in salmonids? One possibility is that individuals may be able to recognize close kin and avoid mating with them. For example, juvenile Arctic charr can distinguish between water scented by individuals with similar versus different major histocompatibility complex (MHC) genotypes (Olsén et al. 1998). Because MHC genotypes are highly polymorphic, mate choice favoring dissimilar genotypes would reduce mating between siblings, as appears to be the case in mice (Potts et al. 1991). Landry et al. (2001) studied mate choice in a population of 41 male and 35 female Atlantic salmon. They captured all adults at a weir, genotyped them at five microsatellite loci, and determined their genetic relatedness and MHC genotypes. They then sampled offspring from the stream and used multilocus genotypes to determine the matings that produced them. The results were suggestive of dissassortative mating by MHC genotype ($P = 0.049$) and related-ness ($P = 0.094$; Landry et al. 2001). In contrast, Ruzzante et al. (2001) exam-ined relatedness in anadromous brown trout and found that "in no case was the median individual inbreeding coefficient lower than the population inbreeding value, F, thus lending no support to the presence of inbreeding avoidance mechanisms." The evidence for inbreeding avoidance mechanisms in salmonids is thus equivocal.

Straying should be sex biased if it has evolved to reduce inbreeding, and the breeding system of salmonids would predict that males should be the sex with higher dispersal (see Section 1.2). Empirical evidence contradicts this prediction because dispersal is usually similar between the sexes, at least after controlling for migratory tendency (see above). When not controlling for migratory ten-dency, straying may be higher for females because they are anadromous more often than males (Bekkevold 2002; Hendry et al. 2003b—*this volume*). In sum-mary, inbreeding is probably rare for healthy populations of many species, espe-cially Pacific salmon, which have large populations and extensive migrations. Inbreeding is more likely in small, freshwater populations but here mate choice may reduce inbreeding (although empirical evidence is ambiguous). Even if inbreeding occurs, it may have only mildly deleterious effects (although studies in the wild have not been undertaken). Additional work is certainly warranted, especially studies of inbreeding depression in the wild, but the current evidence argues against a major role for inbreeding avoidance in the evolution of straying.

Theory D4

Dispersal reduces competition among kin. Kin selection might influence straying in two ways. First, kin might compete directly for mates, which should favor increased straying. This seems unlikely for many salmonids because large popu-lations and extensive migrations make it unlikely that siblings will remain in close contact. In small, non-migratory populations, kin might compete for mates but no study has directly tested for this possibility. If such competition does occur, straying should be higher for iteroparous than semelparous species and should be sex biased (see Section 1.2), probably higher in males because they may experience higher breeding competition. Neither of these predictions is supported by salmonids: iteroparous species do not stray more than semelpar-ous species (Figure 2.3, Appendices 1 and 2) and males usually do not stray more than females (see above). We conclude that competition among kin for mates is probably not an important factor driving the evolution of straying in most systems.

Second, kin might compete for resources as juveniles. This is certainly pos-sible because stream-dwelling juveniles often experience intense competition for territories and food, and may have reduced growth at high densities (Chapman 1966; Grant and Kramer 1990; Jenkins et al. 1999; Bohlin et al. 2002; Vøllestad et al. 2002). However, the type of dispersal considered in this chapter is move-ment from natal to non-natal sites, whereas juveniles may disperse while foraging and yet return to natal sites for breeding. This "foraging dispersal" might reduce competition among kin without increasing straying. To assess the relevance of foraging dispersal to straying, we here ask if juvenile salmonid kin might obtain benefits from dispersing, if foraging kin actively disperse to avoid competition, and if foraging dispersal might increase straying.

Do kin benefit from dispersal as juveniles? In laboratory studies, kin groups have some advantages over non-kin groups, including reduced aggression, increased growth, and reduced variance in growth (Brown and Brown 1993; Olsén and Järvi 1997; Brown et al. 1996). Some of these benefits, however, may be exaggerated because laboratory experiments often use recirculating water, which concentrates odors above natural levels (Griffiths and Armstrong 2000). The costs or benefits of kin association have rarely been examined in the wild but Griffiths and Armstrong (2001) found that groups of full-sib Atlantic salmon juveniles had lower densities and lower condition factors than mixed-family groups. They suggested that mixed family groups have greater genetic variation and therefore use the environment more efficiently ("heterogeneous advantage" theory).

Do kin actively disperse to avoid competition? In laboratory studies, juveniles spend more time in water conditioned by kin than by non-kin (Quinn and Busack 1985; Brown et al. 1993; Courtenay et al. 2001), suggesting that kin prefer to remain together. Studies in stream channels have shown that kin may remain in some degree of association with each other (e.g., Quinn et al. 1994) but evidence from the wild is mixed. A few studies of brown trout have found that juveniles from a specific stream location are more likely to be related than those from different locations (Hansen et al. 1997; Ruzzante et al. 2001). Other studies have found that within-river associations between genetic relatedness and geographic distance are either very weak (Mjølnerød et al. 1999) or entirely absent (Fontaine and Dodson 1999). Although these latter results might suggest active dispersal by kin, they might also reflect an inability of kin to remain together because of adverse conditions. Griffiths and Armstrong (2001) controlled for environmental conditions by releasing similar numbers of fish from full-sib and mixed-family groups into different sections of the same stream. Five months later, densities were lower in the full-sib sections than the mixed-family sections, which could reflect selection favoring dispersal of kin or simply lower performance when surrounded by kin. The latter seems more likely because condition factors were higher in the mixed groups despite their higher density.

Griffiths and Armstrong (2002) attempted to reconcile the results of laboratory studies, which find that kin remain together and benefit from doing so, with the results of field studies, which sometimes suggest that kin may avoid each other and benefit from doing so. They suggested that the heterogeneous advantage hypothesis leads kin to disperse in the wild but that kin selection leads dominant fish to be more tolerant of subordinate kin when they do come into contact. This idea is supported by work showing that juvenile Atlantic salmon have fluid spatial distributions with considerable home range overlap (Armstrong et al. 1999). For this hypothesis to work, however, kin cannot disperse very far because otherwise they would rarely come into contact with each other and kin selection would not be important. Considering all the above, it seems unlikely that salmonids actively disperse to avoid kin competition, except perhaps for short distances (see Hutchings and Gerber 2003). It therefore seems unlikely that such dispersal would have a large effect on straying rates. In sum, the theory that straying evolves to avoid competition among kin does not currently receive much support from salmonids.

4. Conclusions

Some general theories for the evolution of philopatry and dispersal receive strong support from salmonids. The most important factor favoring philopatry may be that it returns locally adapted individuals to appropriate environments (Theory P3). Salmonid populations in different locations show remarkably strong and rapid adaptive divergence, and strays should therefore have lower reproductive success than homers. A variety of evidence supports this assertion, most importantly the higher rates of adult dispersal than gene flow between populations in different habitats. The most important factor favoring dispersal may be that it buffers against temporal variation in habitat quality (Theory D1). Natal site quality varies dramatically among years, and populations in more variable environments appear to stray at higher rates, as do species with less variation in age at maturity.

Other theories receive moderate support from salmonids but their relative importance is not yet clear. For example, philopatry may sometimes be favored because it increases the likelihood of finding suitable breeding sites and mates (Theory P1). Although this is not a general explanation for philopatry (because dispersers often enter established populations), it may be a complementary mechanism to other theories because it is an efficient way for dispersers to find established populations. For dispersal, extinction/recolonization dynamics may be important, especially on longer time scales, because local salmonid populations often go extinct and strays rapidly establish new populations at suitable vacant sites (Theory D2). However, it is uncertain how often extinction/recolonization events take place in contemporary populations.

Other theories are clearly not general explanations for philopatry and dispersal in salmonids. Specifically, philopatry probably rarely evolves because (a) it increases familiarity with local environments and individuals (Theory P2—because most salmonids spend little time at natal sites before they start breeding), (b) it provides access to parental resources (Theory P5—because population sizes must be large, juveniles must spend extended periods in fresh water, and strays entering established populations would not be at a disadvantage over homers), (c) it compensates for spatial variation in habitat quality when temporal variation is absent (Theory P4—because temporal variation is prevalent), and (d) it avoids "costs" associated with leaving natal sites (Theory P6—because most movement costs are associated with migration rather than dispersal). Dispersal probably rarely evolves because (a) it reduces inbreeding (Theory D3—because strong inbreeding will not be common and will have limited effects when it occurs), and (b) it reduces competition among kin (Theory D4—because performance often seems higher in the presence of kin). However, each of these theories may still be relevant in specific contexts, suggesting the intriguing possibility that homing and straying evolve for different reasons in different situations.

Salmonids show great variation in philopatry (homing) and dispersal (straying), making them a good system for examining alternative theories for these phenomena. Our investigation is far from complete and we hope that future work will focus specifically on tests of alternative theories.

Acknowledgments Thorough reviews of the manuscript were provided by Michael Hansen, Fred Utter, and John Wenburg. Additional comments on specific points were provided by Dorte Bekkvold, Paul Bentzen, Louis Bernatchez, John Candy, Jeff Hard, Jeff Hutchings, Nicholas Perrin, Asbjorn Vøllestad, and Robin Waples. Help in compiling the appendices was provided by Fred Allendorf, Terry Beacham, John Candy, Jens Carlsson, Tony Gharrett, Kitty Griswold, Michael Hansen, Ken Kenaston, Marja-Liisa Koljonen, Steve Latham, Einar Nielsen, Núria Sanz, Eric Taylor, Martin Unwin, Fred Utter, and John Wenburg.

To Sea or Not to Sea?
Anadromy Versus Non-Anadromy in Salmonids

Andrew P. Hendry

Torgny Bohlin

Bror Jonsson

Ole K. Berg

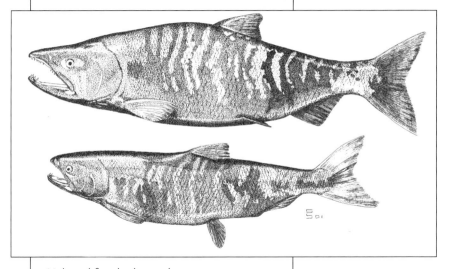

Male and female chum salmon

3

"Migration" has been defined in many ways that have a variety of connotations (Endler 1977; Dingle 1996). Here we follow Endler (1977) in considering migration to be "relatively long-distance movements made by large numbers of individuals in approximately the same direction at approximately the same time ... usually followed by a regular return migration." This definition is useful because it corresponds to salmonid migrations, and because it excludes dispersal, which is the subject of another chapter (Hendry et al. 2003a—*this volume*). Breeding often takes place at one terminus of the migration but not the other, and migrations may sometimes be undertaken multiple times over the course of an individual's life. Moreover migratory tendency varies at multiple levels: among species, populations, sexes, individuals, and years. In this chapter, we use salmonids (primarily salmon, trout, and charr) as a model system for examining evolutionary factors that may generate and maintain this variation.

Variation in migratory tendency is present at many levels in salmonids, making them a good system for examining how, when, and why migration will evolve, as well as how other traits will evolve in parallel. Salmonid migrations take place between a variety of contrasting environments (small streams vs. large rivers, streams vs. lakes, fresh water vs. the ocean) but we focus our analysis on migrations between fresh water and the ocean ("anadromy"; McDowall 1988). In salmonids, anadromy takes the form of individuals that hatched in fresh water migrating to the ocean, and later returning to breed in fresh water. In some instances, individuals may move back and forth between fresh water and the ocean several times in their life, and may enter and leave fresh water without breeding. In contrast, non-anadromous individuals remain their entire life in fresh water. In the present chapter, we typically (but not always) use the terms "anadromous" and "non-anadromous" rather than "migratory" and "resident" because the latter terms could also refer to migrations between different freshwater environments.

Whether salmonids had a freshwater or marine origin has long been debated (McDowall 2002), with some evidence suggesting an origin in fresh water (e.g., only some extant populations are anadromous whereas all breed in fresh water) and other evidence suggesting the opposite (e.g., non-anadromous populations seem to be the derived state in many species). Recently, McDowall (2002) has argued that the ancestral salmonid may already have been anadromous. In this chapter, we will not attempt to resolve this debate. Instead, we are interested in how the present diversity of salmonid migratory behavior has evolved. It is thus useful to know if extant anadromous populations can give rise to non-anadromous populations, and vice versa. Shifts from anadromy to non-anadromy are quite common in many species. For example, many of the non-anadromous populations of salmonids in Europe and North America arose after the last glaciation, owing to the colonization of newly exposed areas by anadromous fish (Berg 1985; Wood 1995). Moreover, anadromous salmonids have been transplanted all over the world, frequently giving rise to non-anadromous populations, often (but not always) because access to the ocean is difficult.

Well-documented examples of the opposite shift (from non-anadromy to anadromy) are less common, particularly within the native range of a species.

However, such shifts are physiologically possible because at least some populations that have been in fresh water for thousands of generations retain the ability to osmoregulate in salt water, although usually not as well as ancestral anadromous forms (e.g., Atlantic salmon, Staurnes et al. 1992; kokanee, Foote et al. 1994). Moreover, Foerster (1947) showed that kokanee (the non-anadromous, lake-resident form of *Oncorhynchus nerka*) could successfully undertake an anadromous migration when released at a non-natal site lacking access to a lake. Smolt to adult survival for these kokanee was more than an order of magnitude lower than for sockeye salmon released from the same site but this difference was confounded by different ages (kokanee were older when they returned) and different origins (sockeye were native but kokanee were non-native). Rounsefell (1958) summarizes additional unpublished kokanee release experiments that support Foerster's (1947) results. Finally, non-anadromous forms of some species have adopted anadromy following introduction to sites without native salmonids (e.g., rainbow trout in Argentina: Pascual et al. 2001). The apparent rarity of shifts from non-anadromy to anadromy within the native range of a species may simply reflect the rapid colonization by anadromous fishes of all sites with access to the ocean (e.g., Milner and Bailey 1989; Milner et al. 2000).

Our goal is to interpret variation in anadromy/non-anadromy in the context of "ultimate" considerations (i.e., as a function of relative costs and benefits) rather than proximate considerations (i.e., individual responses to internal and external stimuli). The latter approach is certainly useful (e.g., Thorpe et al. 1998) but is often context-specific and cannot be covered within the confines of the present chapter. In the following, we first outline theoretical expectations that (1) anadromy and non-anadromy have both benefits and costs, (2) anadromy (or non-anadromy) will increase as its benefits increase or its costs decrease, (3) populations will adapt to anadromy/non-anadromy so as to reduce the proximate costs, (4) populations will offset proximate costs that cannot be eliminated by compensatory adaptations in other life history traits, and (5) the tendency for anadromy/non-anadromy is likely to be condition-dependent, frequency-dependent, and density-dependent. These expectations are developed in the form of specific, testable predictions. We then outline variation in anadromy/non-anadromy in salmonids, and examine whether this variation supports the predictions. We also identify specific areas requiring further research.

1. Theory

The change in relative fitness conferred by a given behavior (e.g., migration) is determined by the influence of that behavior on survival to maturity, age at maturity, and reproductive output at maturity (extended to multiple reproductive episodes for iteroparous organisms). As we will show, migration has the potential to decrease survival to maturity, which would represent a cost, and increase reproductive output at maturity, which would represent a benefit. Migration might also influence age at maturity, with an increase possibly representing a cost and a decrease possibly representing a benefit. Variation in migratory tendency should therefore reflect its differential benefits and costs under different circumstances. In the following, we outline predictions regarding the benefits and costs of migration, and consider the evolution of migratory tendency as a function of those costs and benefits. Many other predictions could certainly be developed but the ones we present can be tested using salmonids. Although the hypotheses are framed around costs and benefits of migration, specifically anadromy, they could just as easily be reformulated around costs and benefits of non-anadromy. It is not our intention to imply a universal direction of evolutionary transitions between the two life histories.

When considering the fitness consequences of alternative behaviors, it is important to remember that fitness is best interpreted relative to other individuals within a given population or environment, not relative to other populations or environments. Consider an anadromous population in the lower reaches of a river and a non-anadromous population in the upper reaches of the same river. If these two populations are both stable in size, anadromous individuals within the downstream population might be interpreted as having equal fitness to non-anadromous individuals within the upstream population (because an individual in either population produces an average of one offspring that survives to maturity). The more relevant comparison, however, is between anadromous and non-anadromous individuals within each population. If a non-anadromous individual would have lower fitness than an anadromous individual in the downstream population (and vice versa for the upstream population), then the difference in life history between the populations is adaptive.

1.1. Benefits and Costs

Prediction 1

Migration should be beneficial to breeding adults. Benefits of migration may arise when the best breeding habitat is geographically separated from the best non-breeding habitat, or when certain locations are habitable for only part of the year. For example, high latitudes often have high productivity during the summer but low productivity during the winter. Accordingly, many migratory organisms spend their summers at high latitudes and their winters at low latitudes (Dingle 1996). With respect to anadromy, migration may be a way to avoid stressful conditions in fresh water or to take advantage of higher productivity in the ocean. Supporting this last supposition, anadromy is rare in tropical regions, where freshwater productivity exceeds ocean productivity, but is common in temperate regions, where ocean productivity exceeds freshwater productivity (Gross 1987; Gross et al. 1988). Greater productivity in the ocean clearly enables higher growth rates, larger size-at-age, and greater energy stores but what remains worthy of testing is whether or not these traits increase reproductive success. This should be the case if anadromy provides an evolutionary benefit.

Prediction 2

Migration should be costly during the migratory period. Such costs might include increased mortality or decreased reproductive output and might arise owing to increased energy expenditure, increased stress, increased predation, decreased foraging, or increased risk of not finding a suitable breeding site. With respect to anadromy, each of these costs might come into play. Anadromous individuals swim long distances, which should increase energy expenditure and time investment, and undertake stressful physiological transitions between fresh water and salt water. Anadromous individuals may experience higher predation if they are exposed to more predators or are less able to avoid them. Anadromy may also increase the difficulty of finding natal breeding habitats because individuals must return from the open ocean (McDowall 2001). We will examine these costs by comparing anadromous and non-anadromous salmonids, as well as salmonids that migrate different distances.

Prediction 3

Migration should become less common as its benefits decrease or its costs increase. A major benefit of migration should be higher growth opportunity in the ocean than in fresh water, which should translate into greater reproductive output for anadromous individuals. Thus, the tendency for anadromy should be greater as growth rates in fresh water decrease, or as growth rates in the ocean increase. Freshwater productivity will be more informative here because it varies more than ocean productivity across the native range of salmonids. Conversely, a major cost of migration should be increased mortality owing to energy loss, stress, or predation. Thus, the tendency for anadromy should be less for populations that would experience more difficult migrations. These predictions can be tested by jointly examining the degree of anadromy/non-anadromy among populations that differ in freshwater productivity (e.g., latitude) and migratory difficulty (e.g., distance or elevation). Another useful approach is to examine inter-annual or inter-individual variation in freshwater growth opportunity and migratory tendency. Specifically, non-anadromy should perhaps be more common in years and for individuals with higher freshwater growth.

A more direct, but logistically difficult, approach is to experimentally manipulate costs and benefits, such as the productivity of fresh water (e.g., through artificial fertilization or reductions in competition) or the difficulty of migration (e.g., by introducing a barrier or increasing mortality rates for migratory fish). In such manipulations, an immediate shift in migratory tendency would reflect an existing reaction norm linking migratory tendency to environmental conditions, where a reaction norm is defined as the phenotypic expression of a given genotype across a range of environments (Schlichting and Pigliucci 1998; Hutchings 2003—*this volume*). Such a reaction norm would likely reflect an adaptive plastic response that evolved to maximize fitness under variable costs and benefits. In contrast, demonstrating an evolutionary change in response to altered costs or benefits would require monitoring migratory tendency over multiple generations.

For example, one might increase and maintain fishing pressure while quantifying how migratory tendency changes through time.

Prediction 4

When costs and benefits of migration differ between the sexes, males and females should differ in migratory tendency (e.g., Adriaensen and Dhondt 1990; Jonsson and Jonsson 1993). In organisms with indeterminate growth, such as salmonids, larger females produce more eggs. The reproductive success of a female should therefore increase with body size, especially when all else is equal (which is not necessarily so—e.g., density dependence). Because anadromy increases body size, it should have strong positive effects on the reproductive success of females. The benefits of anadromy are less concrete for males. A male's success is a function of the number of eggs he fertilizes, which may not be strongly correlated with body size (see below). Moreover, the benefits of large size are often absolute in females (larger females produce more eggs, independent of other females) but relative in males (fertilization success depends on the size, condition, and behavior of other males). As a result, anadromy should be more common for females than for males in populations where only some individuals migrate.

1.2. Evolutionary Compensation

In the strictest sense, the above predictions assume that migratory and non-migratory individuals are similar in all respects, except for those imparted in a proximate sense by migration. It seems likely, however, that when reproductively isolated populations differ in migratory tendency, they should exhibit divergent adaptations that reduce the costs of each life history. If such compensatory adaptation takes place, we might expect the following predictions to be true. Each prediction relates to differences between separate populations, where selection has the potential to drive divergent adaptation. The predictions are not always relevant to migratory versus non-migratory individuals within populations.

Prediction 5

Non-migratory populations should show adaptations that decrease the proximate costs of not migrating. For example, growth opportunities are often lower for non-migratory individuals, and so selection may favor an evolutionary increase in their growth rate under those conditions. This prediction derives from evidence that large size (and hence fast growth) is usually beneficial for survival and reproduction (Arendt 1997; Blanckenhorn 2000). Fast growth in fishes may also carry some costs, such as reduced strength of skeletal elements and scales (Arendt and Wilson 2000; Arendt et al. 2001), reduced swimming ability (Billerbeck et al. 2001), and increased vulnerability to predation (Lankford et al. 2001). However, as long as the costs and benefits do not scale equally with a change in growth rate, a shift in environmental growth potential should favor a compensatory shift in growth rate. Such adaptation might proceed in two ways (Conover and Schultz 1995; Yamahira and Conover 2002). First, populations might shift their optimal conditions for growth: the peak of the reaction norm linking growth rate to environmental conditions should fall near the average conditions a population experiences in nature ("local adaptation"). Second, populations might change their "intrinsic" growth rate: i.e., the elevation of the reaction norm should be higher for populations adapted to environments with lower growth potential ("counter-gradient variation"). These two forms of compensation are not mutually exclusive: both the elevation and the peak of the reaction norm may vary among populations.

Environmental growth potential should be higher for anadromous salmonids, which have access to the productive ocean, than for non-anadromous salmonids. We might therefore predict that populations of the latter will increase their growth rate in fresh water. Ideally, empirical tests for such compensation would raise both forms under a range of environmental growth potentials (e.g., ration, day length, or water temperature), or would perform a reciprocal-transplant experiment. If non-anadromous fish show higher growth rates than anadromous fish under low-growth conditions but not under high-growth conditions, compensation has shifted the peak but not the elevation of the reaction norm (i.e., local adaptation). If non-anadromous fish show higher growth rates than anadromous fish under all conditions, then compensation has shifted the elevation of the reaction norm but not the location of the peak (i.e., counter-gradient variation). These predictions depend on the assumption that selection acting on size or growth is otherwise identical. One way to reduce possible variation in other factors is to compare genetically distinct anadromous and non-anadromous populations that are in sympatry for much of their life.

Prediction 6

Migratory populations should evolve traits and behaviors that decrease the proximate costs of migration. One approach is to store large amounts of fat, which can then be drawn upon to fuel migration. Another approach is to improve migratory efficiency by taking advantage of prevailing atmospheric winds (in birds and insects) or oceanic currents (in fishes and marine invertebrates; review: Dingle 1996). With respect to anadromy, adaptations that reduce migratory costs might relate to body size (if size influences migratory ability or efficiency), energy allocation (longer migrations should select for increased fat stores), or energy efficiency (longer migrations should select for greater swimming efficiency). For downstream migration, these traits can be compared between sympatric anadromous and non-anadromous populations, or among anadromous populations that vary in migratory difficulty (e.g., distance or elevation). For upstream migration, the comparison of anadromous and non-anadromous populations is not very useful because the traits in question are also influenced in a proximate sense by anadromy: anadromous fish are larger and have greater fat stores. Comparisons of populations that migrate different distances are more useful but should be made at the start of upstream migration, before migration has had its proximate effect.

Prediction 7

Anadromous populations should have adaptations that reduce the ultimate costs of migration. If increased stress or energy costs associated with migration have the proximate effect of reducing the expression of traits that influence reproductive success (e.g., egg size, egg number, reproductive life span), migratory populations should develop compensatory adaptations in these traits, and any such compensation should be greater for populations that experience greater migratory costs. In testing this prediction, common-garden experiments might be used to compare sympatric anadromous and non-anadromous populations, or to compare anadromous populations that migrate different distances. It is difficult to develop specific expectations for the first comparison because increased relative reproductive output might be predicted for anadromous fish (because of the proximate cost of migration) or for non-anadromous fish (because they have lower absolute reproductive output). For the second comparison, the specific prediction is more clear: anadromous populations with more difficult migrations should have a genetic tendency for greater relative reproductive output.

1.3. Other Predictions

Migratory tendency often varies among individuals within populations, and the relative frequency of migration may vary among years and among populations. Several hypotheses have been developed to explain this variation (Figure 3.1). First, migration may be density-dependent. If, for example, survival or growth is negatively correlated with density, increasing migration may be favored with an increase in density or a decrease in environmental quality (Kaitala et al. 1993). Interestingly, an increase in emigration will also benefit the remaining non-migrants through reduced competition. With respect to anadromy, we might predict that (1) increasing densities of residents at a site will decrease their average fitness and increase their anadromous tendency, (2) freshwater sites with greater carrying capacities will support higher densities of residents, and (3) the degree of anadromy should be positively correlated with the degree to which densities exceed freshwater carrying capacity. Density dependence may also have self-reinforcing effects on anadromy: an increase in the frequency of anadromy at a specific site may result in greater egg density, which should increase competition among juveniles and thereby favor anadromy to an even greater degree.

Second, migration may be frequency-dependent, wherein the relative fitness of a particular life history decreases as its relative frequency increases (Lundberg 1988; Figure 2.1). This might occur, for example, if migrants and non-migrants use alternative reproductive behaviors that are most successful when rare. With respect to anadromy, we might expect to see negative frequency dependence in

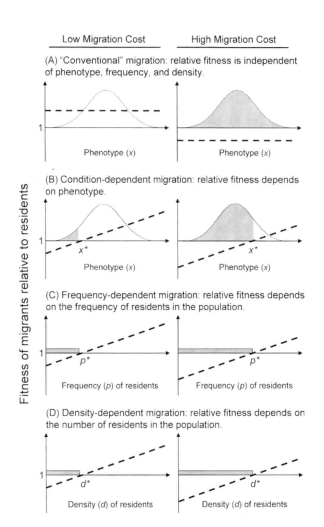

Figure 3.1. A graphical representation of how the evolution of migration may depend on migratory costs, individual phenotypes, and the density and frequency of non-migrants (residents). The cost of migration is shown in each panel by the dashed line, which represents the fitness of migrants relative to residents. Migration has a low cost (e.g., short distance) in the left column and a high cost (e.g., long distance) in the right column. Gray shading shows where residency should predominate and the absence of gray shading shows where migration should predominate. In Panel (A), the relative fitness of migrants is independent of phenotype, frequency, and density. The population is represented as a normal distribution of phenotypes taking the value x. The entire population should migrate when costs are low and not migrate when costs are high. In Panel (B), the relative fitness of migrants depends on their phenotype. Selection favors migration above some threshold phenotypic value (x^*) and residency below that value. Increasing migratory costs increase the threshold value, leading more of the population to remain resident. In Panel (C), the fitness of migrants depends on their frequency relative to residents (p). The equilibrium frequency of residents in the population (p^*) should occur where residents and migrants have equal average fitness. In Panel (D), the fitness of residents depends on their density, and migration should occur when densities exceed d^*. These are simplified representations because the different factors may act together (see text).

the mating success of anadromous males, which are large and typically fight for positions near breeding females, relative to non-anadromous males, which are small and typically act as "sneaks" during oviposition (Myers 1986; Hutchings and Myers 1994). If so, we might predict that (1) the average relative fitness of each life history will be highest in years when its relative frequency is lowest, and (2) an increase in the relative frequency of one life history will lead to a corresponding decrease in that life history in the next generation (assuming they are at least partially heritable). In testing this last prediction, it is important to also consider environmental factors (e.g., productivity) that have proximate effects on life history.

Third, migration may depend on an individual's phenotype, variously called its "condition," "status," or "state." For instance, the relative costs and benefits of migrating may depend on an individual's age, sex, size, energy stores, or social status (Lundberg 1988; Adriansen and Dhondt 1990). With respect to anadromy, predictions will be context-specific. For example, larger individuals may become anadromous because they are (1) closer to the asymptotic size in fresh water (Jonsson and Jonsson 1993) or (2) more likely to successfully complete migration (Bohlin et al. 1993). Alternatively, larger individuals, or those in better condition, may become non-anadromous if larger size increases the probability of successfully maturing in fresh water (Thorpe et al. 1998). A complicating factor is that condition dependence probably acts in concert with density dependence and frequency dependence. For example, an increase in density may cause increased migration and the individuals that migrate may be larger or smaller than those that do not migrate.

Two "tactics" (in this case, migrate or not) may be maintained within populations through a "conditional strategy" (Repka and Gross 1995; Gross 1996; Gross and Repka 1998). According to this theory, the individuals in a population share a genetically based strategy linking an individual's condition (or "state") to its choice of alternative tactics. The combination of individual condition and tactic then determine fitness. In the conditional strategy, the two tactics have equal fitness at a specific phenotypic state, or "evolutionary stable strategy (ESS) switchpoint," but they will typically not have equal average fitness (see also Hutchings and Myers 1994). Traditional game theory models would predict that the tactic with lower fitness would be lost from the population but this is not so in the conditional strategy. In the words of Gross and Repka (1998, p. 447): "The two tactics are regulated by the same strategy genotype, and although individuals compete with each other by using their tactics, the tactics do not compete in an evolutionary sense because they belong to the same conditional strategy. The conditional strategy evolves as a means of maximizing the fitness of the individual, essentially by allowing each individual to maximize its potential among alternative tactics." Frequency-dependent fitness may increase stability of the ESS switchpoint but frequency dependence is not necessary for maintenance of the conditional strategy (Gross and Repka 1998). Additionally, density-dependent fitness may influence the location of the ESS switchpoint (Gross 1996).

2. The Salmonid System

Salmonids show great variation in migratory tendency, and thus provide a good system for evaluating the evolution of migration (specifically anadromy), as well as its consequences for other aspects of life history. Differences in migratory tendency among species are substantial (Rounsefell 1958; Kinnison and Hendry 2003—*this volume*). For example, chinook, coho, chum, and pink salmon rarely (if ever) have non-anadromous populations within their native range. In contrast, golden, lake, and bull trout seldom (if ever) have anadromous populations. Other species, such as Atlantic salmon, sockeye salmon, masu salmon, brown trout, rainbow trout, and cutthroat trout have both anadromous and non-anadromous populations. This variation among species is intriguing but difficult to interpret in an adaptive context because it may reflect a strong phylogenetic signature. We therefore concentrate on variation within and among conspecific populations.

2.1. Landlocked Populations (i.e., Isolated Above Barriers)

Many river systems have obvious barriers to anadromous migration (such as waterfalls), and some of these systems have conspecific populations both above (landlocked) and below a barrier. In some cases, substantial genetic differences at allozyme, mtDNA, or microsatellite loci are present between the upstream and downstream populations (e.g., Skaala and Nævdal 1989; Vuorinen and Berg 1989; Appendix 3). These genetic differences presumably reflect mutation and random genetic drift during a period of allopatry caused by the barrier. Such populations also differ phenotypically in ways that reflect the proximate effect of (and adaptations to) anadromy/non-anadromy (Appendix 3).

Genetic differences between populations can only be maintained through reproductive isolation. Genetic integrity of landlocked populations is enforced by the physical barrier to upstream movement. Genetic integrity of the downstream populations, however, has the potential to be compromised by individuals that move over the barrier. Thus, reproductive isolation of downstream populations can only be maintained if few individuals actually move over the barrier, or if upstream individuals have reduced reproductive success when breeding in downstream populations. Some populations below barriers are a mix of anadromous and non-anadromous individuals but genetic studies have shown that the two forms are part of a common downstream gene pool, which is distinct from the upstream landlocked gene pool (Hindar et al. 1991a; Cross et al. 1992; Pettersson et al. 2001). These results suggest that individuals from landlocked populations do not contribute substantially to downstream populations.

How do landlocked populations originate? In some systems, areas above barriers may have been colonized by landlocked populations that arose in other systems. This can occur when geological processes cause dramatic shifts in drainage patterns (e.g., "stream capture"). In other systems, landlocked populations may have arisen directly from anadromous populations within that system. Here is one possible scenario. At the end of the last glaciation, ice receded from large areas of North America and Europe, opening up new streams to anadromous fishes. These streams were probably rapidly colonized, as has been found for Glacier Bay, Alaska (Milner and Bailey 1989). With time, areas formerly covered by large masses of ice would have risen through isostatic rebound, perhaps exposing barriers that had formerly been under the ocean (Berg 1985). These new barriers would progressively isolate upstream areas, making anadromous migrations more difficult. This increasing difficulty would steadily decrease the fitness of migratory individuals and ultimately lead to an evolutionary decrease in their frequency within the population (assuming migratory tendency has a genetic basis).

2.2. Sympatric Populations

Anadromous and non-anadromous populations can sometimes coexist without being separated by a strict physical barrier. A classic example occurs in O. *nerka*, where anadromous sockeye salmon and non-anadromous kokanee co-occur in numerous lake systems around the Pacific Rim (Ricker 1940; Nelson 1968; Wood 1995). In many of these systems, the two forms breed in the same streams at overlapping times, and yet maintain genetic differences at neutral loci and adaptive traits (Appendix 3). At this point, it is important to distinguish between kokanee, which are populations of O. *nerka* that have adapted to life entirely in fresh water, and residual sockeye salmon, which are non-anadromous progeny of anadromous fish (Ricker 1938). Kokanee presumably evolve through residuals but residuals are not specially adapted for freshwater life and show little if any genetic differences from sockeye salmon (Ricker 1940; Krogius 1982). Examples of sympatric anadromous and non-anadromous populations are also found in Atlantic salmon (Couturier et al. 1986; Verspoor and Cole 1989; Birt et al. 1991a, 1991b), rainbow trout (Neave 1944; Zimmerman and Reeves 2000, 2002; Docker and Heath 2003), and brook charr (Boula et al. 2002). Reproductive isolation in these latter examples may involve considerable (but not necessarily absolute) spatial and temporal separation during breeding.

Did currently sympatric anadromous and non-anadromous populations originate in sympatry, or have they come into secondary contact after originating in allopatry? A common way to address this question is to sample populations of both forms within multiple independent systems, and to test whether populations cluster genetically by system (suggesting independent sympatric origins within each system) or by form (suggesting allopatric origins followed by secondary contact). When sockeye salmon and kokanee are analyzed in this fashion, different populations from the same lake cluster together regardless of form, but different populations within lakes cluster by form and not by creek (Foote et al. 1989; Taylor et al. 1996). Most of these lakes were colonized following the recession of glaciers 10,000 years ago, and so it seems that (1) kokanee may be independently derived from anadromous sockeye salmon within each system but (2) may only have arisen once within each system. Unfortunately, a pattern of genetic clustering by system rather than by form could also arise if the two forms originated in allopatry but interbred following secondary contact.

Similarly, a pattern of clustering within lake systems by form rather than by creek could arise if kokanee originated independently in each creek and then interbred with kokanee from other creeks. The genetic data and geological evidence make it most likely that kokanee arose from sockeye salmon independently in many different lakes (Taylor et al. 1996) but it remains unknown whether kokanee had single versus multiple origins within lake systems.

How do anadromous and non-anadromous populations maintain their genetic integrity in sympatry: that is, without a geographic barrier that prevents their mixing? One possibility is that they breed at different times or places. Within the Deschutes River, Oregon, mean breeding date is 9–10 weeks earlier for anadromous steelhead than for non-anadromous rainbow trout, and steelhead tend to breed in deeper water with larger substrate (Zimmerman and Reeves 2000). These differences could cause substantial isolation because even very slight differences in breeding time (a few days) and location (15 m) can greatly reduce mixing within a season (Hendry et al. 1995; see also Quinn et al. 2000). But what about sockeye salmon and kokanee, which often breed in the same locations and at overlapping times (Foote and Larkin 1988; Wood and Foote 1996). In this case, interbreeding is reduced through size-assortative mating (kokanee are much smaller) and perhaps some additional "form-assortative" mating (Foote and Larkin 1988). However, kokanee males do sometimes act as "sneaks" on anadromous females and probably fertilize substantial numbers of eggs (see below). Thus, we must also invoke selection against hybrids or backcrosses to explain the persistent genetic differences. Sockeye salmon and kokanee lack intrinsic genetic incompatibilities (e.g., Wood and Foote 1990) and so post-zygotic reproductive isolation must have an ecological basis: that is, hybrids are inferior because they are poorly adapted for either parental environment (Bernatchez 2003—*this volume*). This seems eminently plausible because the two forms differ dramatically in traits that reflect adaptation to their different life histories (Appendix 3).

2.3. Variation Within Populations (Partial Migration)

Some populations with access to the ocean contain a mixture of anadromous and non-anadromous individuals that do not represent distinct gene pools but rather alternative life histories within a common population (Jonsson and Jonsson 1993). Confirmed or putative examples occur in sockeye salmon (Krogius 1982), brown trout (Hindar et al. 1991a; Cross et al. 1992; Pettersson et al. 2001), Arctic charr (Nordeng 1983; Reist 1989), white-spotted charr (Morita et al. 2000), brook charr (Wilder 1952; Jones et al. 1997), and Dolly Varden (Maekawa et al. 1993). Such populations are said to exhibit partial migration (Lundberg 1988; Adriaensen and Dhondt 1990; Jonsson and Jonsson 1993; Kaitala et al. 1993) and are often polymorphic in that the two forms differ phenotypically (Jonsson and Jonsson 1997; Appendix 3).

What determines whether or not a particular individual in such populations will migrate to the ocean in a given year? In many systems, individuals that will later migrate initially differ from those that will not in phenotypic traits such as size, growth rate, or energy status (e.g., Rowe et al. 1991; Prévost et al. 1992; Berglund 1992; Bohlin et al. 1994; Rikardsen and Elliott 2000). Opinions vary as to the evolutionary reasons for these differences. Jonsson and Jonsson (1993) suggested that fish will remain in a particular niche until they approach the asymptotic body size in that niche (i.e., until they become growth limited), at which time they either mature or switch to a new niche. The new niche may be new food resources, such as fish, or a new environment, such as the ocean. Based on this logic, downstream migrants may be larger than non-maturing residents because larger fish are nearer their asymptotic size. Variation in growth rate may have related effects: faster growing fish have higher metabolic requirements and so may be selected to undertake niche shifts at smaller sizes (Økland et al. 1993).

Thorpe (1994) has argued that an individual's evolutionary priority is to reproduce in any given year (even their first year of life), and that migration will take place only when energy resources are insufficient to mature. By this logic, "[anadromy] should be viewed as evidence of failure to meet the conditions necessary for maturation as parr in fresh water" (Thorpe 1994). If this is so, why do some fish remain in fresh water and yet not mature in a given year, and why are these fish initially smaller and in poorer condition than those becoming downstream migrants? Perhaps these non-maturing resident fish have not yet become growth limited (Jonsson and Jonsson 1993), or perhaps migration selects for fish that are larger or in better condition. Indeed, size- or condition-dependent survival during downstream migration may be an important reason for when (or if) an individual leaves fresh water for the ocean (Bohlin et al. 1993). For additional work see Morinville and Rasmussen (2003).

2.4. The Genetic Basis for Anadromy/Non-Anadromy

Anadromy/non-anadromy must have a genetic basis if it is to evolve in response to differential costs and benefits. Variation among species clearly has a genetic basis because some species are always anadromous whereas others are always non-anadromous, despite great variation in migratory difficulty and freshwater conditions. For example, chum salmon are invariably anadromous even though some migrate upstream for thousands of kilometers, whereas lake trout are invariably non-anadromous even though some occur very near the ocean. Variation between the sexes within a population presumably also has a genetic basis because they otherwise experience a common freshwater environment. Variation within and among populations within sexes, however, clearly has some environmental basis because it can be influenced by freshwater growth (Krogius 1982; Morita et al. 2000). The challenge is to detect genetic effects against this strong background of environmental influences.

Numerous studies have demonstrated genetic variation in some aspect of migratory behavior. First, adaptive variation in the directional response of juveniles to water current (rheotaxis) clearly has a genetic basis (e.g., Raleigh 1971; Kelso et al. 1981; N. Jonsson et al. 1994b). Second, populations that differ in migratory behavior show genetically based differences in swimming ability. For example, juvenile sockeye salmon are better swimmers than juvenile kokanee

(Taylor and Foote 1991) and juvenile coho from interior populations are better swimmers than those from coastal populations (Taylor and McPhail 1985b). Third, differences in the downstream migratory timing of Atlantic salmon populations has a genetic basis, at least in some cases (Riddell and Leggett 1981; C. Nielsen et al. 2001). Fourth, migrants and non-migrants may differ genetically within populations. For example, residual sockeye salmon in Lake Dal'nee have higher heterozygosities at an allozyme locus ($PGM-2^* = 62\%$) than do anadromous sockeye salmon (30%), perhaps because of positive associations between heterozygosity, growth, and maturity in fresh water (Altukhov and Salmenkova 1991; Thorpe 1993; Altukhov et al. 2000, pp. 241–242).

Despite the above, explicit tests for the inheritance of anadromy/non-anadromy have been rare and often inconclusive. Some evidence comes from Sr/Ca ratios in otoliths (Rieman et al. 1994; Doucett et al. 1999; Zimmerman and Reeves 2000, 2002). These ratios can be used to (1) identify anadromous and non-anadromous individuals (Sr is more common in salt water), and (2) determine whether or not an individual's mother was anadromous or non-anadromous (Sr/Ca ratios in an otolith's primordia reflect those in the maternal parent). Using this approach in the Deschutes River, Oregon, Zimmerman and Reeves (2000, 2002) found that anadromous steelhead had anadromous mothers and that non-anadromous rainbow trout had non-anadromous mothers. The same was generally true in the Babine River, British Columbia, although some exceptions were found (Zimmerman and Reeves 2000). This approach is informative but cannot by itself reveal the actual genetic basis for anadromy because it may reflect non-genetic maternal effects.

An alternative is to use common-garden, release, or transplant experiments. Neave (1944) raised and released the progeny of sympatric anadromous steelhead and non-anadromous rainbow trout into the Cowichan River, British Columbia. He subsequently caught more rainbow trout than steelhead progeny in fresh water, suggesting that the steelhead progeny had left for the ocean. Skrochowska (1969) raised and released the progeny of anadromous brown trout, non-anadromous brown trout, and their reciprocal hybrids. During subsequent recaptures, the progeny of non-anadromous trout were captured migrating to the ocean (likely anadromous) versus in the river (likely non-anadromous) in a ratio of 1.0:14.7. The analogous ratio for the progeny of anadromous trout ranged from 1.0:0.5 to 1.0:5.3, suggesting that the two forms usually bred true. Jonsson (1982) reciprocally transplanted juvenile brown trout between an upstream lake containing a landlocked population and a downstream lake containing an anadromous population. Subsequent recaptures revealed that downstream movement was more common in the anadromous-origin fish than the landlocked-origin fish, regardless of the lake of release. The differences in life history in these systems thus seem to have an at least partial genetic basis. In contrast, Nordeng (1983) raised the offspring of three sympatric forms of Arctic charr (anadromous, large non-anadromous, small non-anadromous) and found that each form readily gave rise to each other form in the hatchery and in the wild. He argued that this result suggested little if any genetic differences between the forms.

The above experiments demonstrate both genetic and environmental influences on migratory tendency, and suggest that these influences may vary among species and populations. One way in which these effects may be integrated is through thresholds for life history transitions. For example, individuals may migrate only when they exceed a threshold body size, energy status, or growth rate during a particular time window, with this threshold potentially varying among populations, ages, and growth rates (Myers et al. 1986; Økland et al. 1993). Alternatively, individuals that surpass a particular threshold may mature in fresh water without migration (see below). Such thresholds may be genetically determined and may vary adaptively among populations and species. For example, where the optimal size for migration is larger, populations should have a higher threshold size for migration. Thresholds may also show adaptive phenotypic plasticity (Hutchings 2003—*this volume*). For example, the optimal size for migration may be smaller (or larger) in years with lower freshwater growth opportunity, and so migration may be induced at a lower (or higher) threshold. A variety of methods have been developed for determining the genetic basis of thresholds (Roff 1997) but these have not yet been applied to anadromy/non-anadromy.

The genetic basis for anadromy is sometimes linked to the genetic basis for early maturity in fresh water (e.g., as mature parr), which is itself influenced by both environmental and genetic effects (Thorpe et al. 1983; Myers et al. 1986; Rowe et al. 1991; Berglund 1992; Prévost et al. 1992). For Atlantic salmon, Thorpe et al. (1998) has argued that the decision to initiate maturation is made in the fall (November) and is based on whether a fish exceeds a threshold condition. A fish then re-examines its condition at the end of the winter (April) and continues toward maturity if it exceeds a new threshold. If individual condition does not exceed the thresholds during both periods, maturation will not take place the following fall. A fish that will not mature in the fall then "decides" to initiate a trajectory toward smolting if it exceeds another threshold in the late summer (August). The decision to initiate the smolting trajectory can then be negated if the fish surpasses a new maturation threshold condition later that same fall (November). In this complex scenario, the genetic basis for anadromy depends on the genetic basis for several different thresholds.

3. Evaluation of the Theory

In the following, we interpret variation in anadromy/non-anadromy as an adaptive response to varying costs and benefits, but we do not intend to imply that all such variation is necessarily adaptive. First, several Pacific salmon species essentially never have non-anadromous forms, and this is probably the result of phylogenetic constraints. These species may have evolved in an environment that strongly favored anadromy, thereby eliminating alleles that would allow non-anadromy. The converse may be true for strictly freshwater populations or species, such as lake trout. Second, landlocked populations often contain some individuals that move downstream over barriers. This behavior is certainly not adaptive *within* the population because these individuals cannot return to their natal site. Selection should eliminate migratory behavior from such populations and yet it does not always do so. One possibility is that such behavior is maintained by selection for emigration when conditions deteriorate in a local area. In this case, movement over a barrier could be maintained by selection even though crossing the barrier itself is maladaptive.

3.1. Benefits and Costs

Prediction 1

The larger size-at-age and greater energy stores conferred by anadromy should increase success during breeding. For breeding females, the expected benefits of large size are several. First, larger females produce more eggs (Hendry et al. 2001b). For example, egg number averages 3149 for sockeye salmon and 143 for kokanee breeding sympatrically in Takla Lake, British Columbia (Wood and Foote 1996). Second, larger females can better acquire and defend preferred nesting sites (Foote 1990), which may increase the survival of their eggs (van den Berghe and Gross 1989). Third, larger females bury their eggs deeper in the gravel (Steen and Quinn 1999), which may protect them from disturbance by other females and by floods. Fourth, larger females produce larger eggs, which yield larger juveniles, which have higher growth and survival (Einum and Fleming 2000b; Einum et al. 2003—*this volume*). Fifth, larger females are preferred by males (Foote 1988), although this is probably a minor benefit because few if any females go without mates. In accord with these expected benefits, several studies of anadromous salmonids have found that female size is positively correlated with estimated reproductive success: coho salmon in the wild (van den Berghe and Gross 1989), coho salmon in stream channels (Fleming and Gross 1994), and Atlantic salmon in semi-natural tanks (Fleming et al. 1996).

Although large size thus seems to confer an overall benefit to breeding females, some of the above claims are tempered by ambiguities. First, increasing numbers of eggs may increase density-dependent mortality. Second, large eggs and deep burial may actually increase mortality if gravel sizes are fine or dissolved oxygen is low (Hendry et al. 2001b; but see Einum et al. 2002). Third, large females may be at a disadvantage under size-selective predation, such as that imposed by bears in small streams (Quinn and Kinnison 1999; Quinn et al. 2001c). Estimates of reproductive success in the wild hint at such effects. For example, Holtby and Healey (1986, p. 1946) "were unable to demonstrate that the reproductive success of large [coho salmon] females was consistently higher than that of small females." Similarly, Garant et al. (2001) found no significant correlation between female size and the number of offspring produced in Atlantic salmon. We are not aware of any studies directly comparing the reproductive success of anadromous and non-anadromous females within populations.

For breeding males, the expected benefits of large size stem from advantages during intrasexual competition: larger males are more likely to be dominant and to maintain positions closest to females (Fleming and Gross 1994; Quinn and

Foote 1994). Moreover, females spawn more quickly when courted by larger males (Berejikian et al. 2000; de Gaudemar et al. 2000b). These advantages should increase the reproductive success of large males but evidence from genetic parentage analyses is ambiguous. For anadromous fish, large males have more offspring than small males in some studies (Schroder 1981; Chebanov et al. 1984) but not in others (Foote et al. 1997; Garant et al. 2001). The ability of small males to fertilize a substantial number of eggs, despite their subordinate status, is due to their adoption of "sneaking" behavior, in which they hide near the female and dart in to release sperm at oviposition (Fleming and Reynolds 2003—*this volume*).

In polymorphic populations, dominant males tend to be large and anadromous, whereas sneaking males tend to be smaller and either anadromous or non-anadromous. In genetic parentage analyses comparing anadromous and non-anadromous males at 1:1 ratios, the former typically fertilize more eggs than the latter (Atlantic salmon: 92.5% vs. 7.5%, Hutchings and Myers 1988; 75.3% vs. 24.7%, Morán et al. 1996; 65% vs. 35%, Thomaz et al. 1997). However, a subordinate anadromous male may fertilize even fewer eggs than a non-anadromous male (Martinez et al. 2000). Only if an anadromous male is in a dominant position does his fertilization success seem guaranteed to be greater than that of a non-anadromous male. Adding to this uncertainty, the success of anadromous and non-anadromous males is expected to depend on their relative frequencies (Hutchings and Myers 1994). The mating benefits conferred by large size are thus ambiguous for males, and may be further diminished by size-selective predation, which targets males more strongly than females (Quinn and Kinnison 1999; Quinn et al. 2001c).

Anadromy may increase energy stores in males and females, both absolutely (larger body size) and relatively (greater mass-specific energy). Greater energy stores may then allow the production of more eggs, the construction of better nests, more effective nest defense, and greater competitive ability. In a polymorphic population of brown trout, anadromous fish had 10% greater mass-specific energy stores (lipid content was 40% higher) and anadromous females had a 19% greater gonadosomatic index (Jonsson and Jonsson 1997). Similarly, an anadromous population of brown trout in one stream invested 30–37% of their energy into egg production, whereas those in a nearby non-anadromous population invested only 16–17% (Elliott 1988).

In summary, anadromous salmonids are larger and have more energy than non-anadromous salmonids, which should increase their production of offspring. These benefits are more concrete for females than males but some uncertainty exists even for females. Most importantly, the magnitude of any benefits will depend on specific circumstances, such as the intensity of competition, gravel scour, predation, and the relative frequency of individuals adopting each life history. It is also important to remember that the above benefits apply only during breeding, and that anadromy/non-anadromy will impose other costs and benefits. For example, non-anadromous males are often younger than anadromous males and therefore have a much higher probability of survival to maturity (Hutchings and Myers 1994).

Prediction 2

Potential costs of anadromy may include increased mortality, stress, or energy expenditure during migration. Anadromous salmonids produce many more eggs than non-anadromous salmonids (Appendix 3) and so the former must have higher mortality rates, assuming their population sizes are not increasing at a much higher rate. The critical question thus becomes: when is the higher mortality manifested for anadromous fishes? If it occurs between the time that adults reach their breeding areas and the time that juveniles leave for the ocean, increased mortality is probably just a byproduct of density dependence in fresh water. If, however, increased mortality occurs between the time that juveniles leave for the ocean and the time that adults return to breed, anadromous migrations may impose a substantial cost.

Downstream migrants may suffer increased mortality owing to predation and stress. Predation may be particularly critical because (1) predacious fishes are more abundant lower in stream networks, (2) predators often gather at difficult points along migratory pathways, and (3) migrating salmonids may be less able to escape an attack. Accordingly, the proportion of migrating smolts eaten by predators can be very high: Arctic charr consume up to 66% of the sockeye salmon smolts in the Wood River system (Burgner 1991, p. 58), three fish species consume an average of 14% of the salmonid smolts in a single Columbia River reservoir (Rieman et al. 1991), and sea birds from two colonies consume 15% of steelhead smolts in the Columbia River estuary (Collis et al. 2001). Mortality owing to stress probably occurs because of osmoregulatory challenges associated with the transition from fresh water to salt water. For example, experiments with hatchery fish have shown that such mortality can be high if the fish are not well prepared for the transition (e.g., Staurnes et al. 1993).

Increased mortality probably also occurs during upstream migrations. Some migratory routes are interrupted by waterfalls or other physical obstructions that cause considerable delays. These sites may be attended by congregations of sea lions, seals, or bears that kill substantial numbers of migrants. They may also present insurmountable barriers when environmental conditions are poor (Hinch and Bratty 2000). Increased mortality may also be caused by energy depletion during upstream migration, and this may be manifested even after fish reach their breeding sites. For example, Gilhousen (1990) reported pre-breeding mortality rates of 0–90% (mean annual rates = 3.3–23.7%) for Fraser River sockeye salmon. These are just a few of many examples of how mortality can be high during downstream and upstream migration. We would ideally compare average daily mortality rates for migratory and non-migratory individuals throughout their lives but such comparisons are not currently available. One suggestive result is that iteroparous anadromous fish have lower post-breeding survival than non-anadromous conspecifies (Fleming and Reynolds 2003—*this volume*).

Another cost of upstream migration may be sublethal energy loss that reduces the energy available for egg production, secondary sexual development, or active metabolism during breeding. These costs may be partly offset by the higher energy stores of anadromous individuals but this offset will diminish as energy costs increase. Such costs have been assessed in several ways. First, somatic energy stores have been compared between the beginning and end of migration. Such methods suggest that sockeye, pink, and chum salmon expend 0.62–1.74 kcal per kg per km. This can represent a substantial cost: e.g., Early Stuart sockeye salmon expend 40% of their energy stores during their 1022 km upstream migration (Gilhousen 1980; Brett 1995). Second, ultrasonic telemetry can be used to measure heart rate, which correlates with swim speeds and energy expenditure. Such work has shown that energy use is higher when migratory conditions are less favorable (Rand and Hinch 1998; Hinch and Rand 1998). Third, salmon can be placed in swim chambers to measure energy depletion, oxygen consumption, and critical swimming velocities. Such work has shown that high water velocities lead to higher metabolic rates, greater energy consumption, and increased fatigue (review: Brett 1995). All of these results demonstrate that migration increases energy use per unit time. Migration may also be costly simply because it reduces the time available for other activities, such as foraging and breeding (Hedenström and Alerstam 1998).

Comparisons of energy use among natural populations may underestimate the cost of migration because populations that migrate greater distances may have adaptations that reduce migratory costs (see below). A powerful way to uncover the true cost of migration is through experimental manipulations. Kinnison et al. (2001) generated full sibling families for two populations of chinook salmon, reared them in a common environment to smolting, and then released them from two locations, one requiring a freshwater migration of 17 km (17 m elevation) and the other of 100 km (430 m elevation). When adults returned to the release sites, on which they had imprinted as smolts, those that migrated the longer distance had a 17% reduction in metabolizable energy stores and 13.7% smaller ovaries. The ovary mass was smaller because less energy had been allocated to eggs, which were smaller but not fewer (Figure 3.2). Kinnison et al. (in review) found comparable results for male chinook salmon in the same experiment: lower tissue energy reserves and smaller secondary sexual traits in fish migrating longer distances.

If migration negatively impacts fitness through increased energy costs, populations with more difficult migrations should be compromised for fitness-related traits. Kinnison et al. (2001) provides evidence for this cost because reproductive output (ovarian mass) decreased when migration was experimentally increased. Studies of unmanipulated anadromous populations similarly reveal lower reproductive output when migrations are longer (Beacham and Murray 1993; Healey 2001; Kinnison et al. 2001). A direct test for the negative fitness effects of migration is to ask whether the production of juveniles is lower in populations with more difficult migrations. Indeed, Bohlin et al. (2001) found a negative correlation between elevation and juvenile density in anadromous brown trout. This correlation was most likely the result of variation in migratory effort because juvenile density was not correlated with elevation in non-anadromous brown trout (Bohlin et al. 2001). In summary, numerous lines of evidence demonstrate that anadromous migrations are costly to the individuals that undertake them.

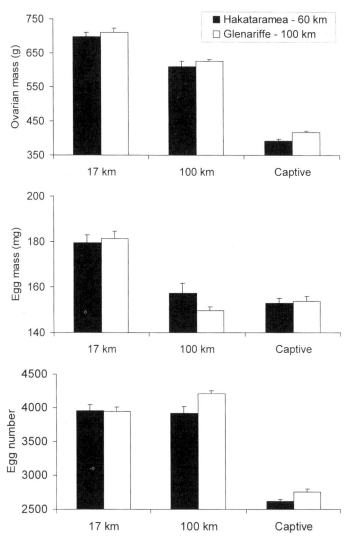

Figure 3.2. Proximate costs of migration on reproductive development, as well as evolutionary compensation for those costs in chinook salmon (data from Kinnison et al. 2001). Two New Zealand populations that had common ancestors 90 years previously but now migrate different distances (Hakataramea—60 km, Glenariffe—100 km) were compared under common-garden conditions (i.e., their entire lives in a common hatchery environment, "Captive"), and after returning from the ocean to a site with a short migration (Silverstream—17 km) and a site with a long migration (Glenariffe—100 km). The top panel shows that migration exacted a proximate cost by reducing total ovarian mass (compare ovary mass between the two experimental migration distances) and that the population adapting to longer migrations (Glenariffe) showed an evolutionary compensation by investing more energy in ovaries (compare ovary mass between the two populations under captive conditions). The bottom two panels show that the proximate reduction in ovary mass with increasing migration was the result of decreased egg size but that the evolutionary compensation was an increase in egg number.

Prediction 3

The tendency for anadromy should decrease as its benefits decrease, with the same true for non-anadromy. The relative benefits of anadromy, and therefore its prevalence, should decrease with increasing freshwater productivity (growth) or increasing migratory difficulty (distance or elevation). This hypothesis might be tested by (1) comparing populations that vary in freshwater growth or migratory difficulty, (2) comparing years that vary in freshwater growth or migratory difficulty, (3) comparing individuals that vary in freshwater growth, or (4) experimentally manipulating freshwater growth or migratory difficulty. In the following, we first present correlative evidence for effects of freshwater productivity, then correlative evidence for effects of migratory difficulty, and finally experimental evidence for the effects of both.

Several studies have examined the relationship between freshwater growth and anadromy/non-anadromy. McGurk (2000) showed that body size and egg number decrease dramatically with increasing latitude in kokanee but not sockeye salmon (Figure 3.3). This shows that the relative benefits of anadromy increase with latitude and lead to the prediction that kokanee should be less common, relative to sockeye salmon, at higher latitudes. No quantitative tests of this prediction have been made but the qualitative evidence strongly suggests just such a trend (Nelson 1968; Wood 1995; McGurk 2000). Trends toward increasing anadromy with increasing latitude have also been noted for other salmonid species (Rounsefell 1958). For example, L'Abée-Lund et al. (1990) studied the incidence of parr maturation in Atlantic salmon populations along the coast of Norway. The proportion of mature parr (relative to all mature males) varied from 0.6–60%, and was positively correlated with the mean length of parr but negatively correlated with mean smolt age. These patterns suggest that higher freshwater growth induces more males to forgo anadromy. Similarly, Myers et al. (1986) found a positive correlation between parr size and parr maturity among populations and among years within populations. Krogius (1982) showed that for cyclically abundant sockeye salmon in Lake Dal'nee, the frequency of residuals was highest in years when juvenile density was lowest and therefore freshwater growth highest. Supporting these among-population and among-year comparisons, many studies have shown that larger individuals or those with higher growth rates, greater energy stores, or better condition are more likely to mature in fresh water (e.g., Rowe et al. 1991; Berglund 1992; Økland et al. 1993; Bohlin et al. 1994).

An apparent contradiction to the above patterns is found in polymorphic Arctic charr, where increased growth in lakes *increases* the tendency for anadromy. In particular, the largest parr within populations tend to become anadromous and the smallest parr tend to remain as residents (Rikardsen and Elliott 2000). Moreover, populations in lakes with higher productivity have a greater propensity for anadromy (Kristoffersen et al. 1994; Rikardsen and Elliott 2000). What might explain this apparent exception to the general rule that high growth in fresh water leads to increased residency? One possibility is that selection strongly favors anadromy (the lakes are at very high latitudes) but also favors

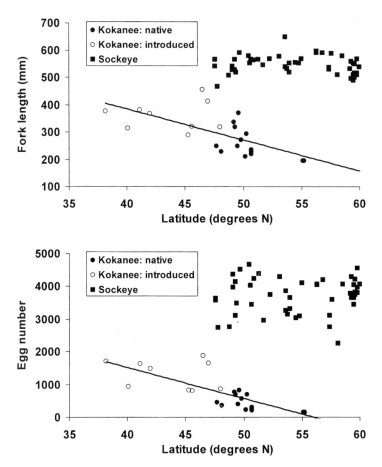

Figure 3.3. Variation in the size (fork length) and egg number (fecundity) of anadromous sockeye salmon and non-anadromous kokanee with latitude in North America (data from McGurk 2000). The top panel shows that fork length decreases with increasing latitude in kokanee ($r^2 = 0.55$, $P < 0.001$) but not sockeye salmon ($r^2 = 0.01$, $P = 0.53$). The bottom panel shows that egg number decreases with increasing latitude in kokanee ($r^2 = 0.63$, $P < 0.001$) but not sockeye salmon ($r^2 = 0.01$, $P = 0.87$). These egg numbers are not corrected for variation in body length among populations but when such a correction is performed, the negative correlation between egg number and latitude strengthens ($r^2 = 0.74$, $P < 0.001$). Note that the range of latitudes has been increased by the inclusion of kokanee populations introduced south of their native range.

large migrants. Indeed, the proportion of Arctic charr out-migrants that later return to fresh water is strongly correlated with their size (Finstad and Heggberget 1995; Gulseth and Nilssen 2000). Perhaps the only individuals that remain non-anadromous are those that don't grow large enough to success-fully migrate to the ocean.

Several studies have examined among-population correlations between ana-dromy and migratory difficulty. For northern Arctic charr, Kristoffersen (1994) showed that anadromy was less common in populations that had more difficult migrations. Similarly, kokanee are more common in interior lakes than coastal lakes (Wood 1995). Bohlin et al. (2001) recently provided a direct test of the hypothesis that variation in anadromy/non-anadromy is influenced by the costs of migration (Figure 3.4). They studied brown trout populations along the southern coast of Sweden that varied in elevation (a surrogate for migratory difficulty) and were either anadromous (migratory) or non-anadromous (resident). Some of the resident populations were isolated above barriers but near the ocean, where migration would have been easy had it not been for the barrier. At low elevations, juvenile density (a surrogate for population productivity) was higher for migratory than non-migratory populations, showing that migration provided positive fitness benefits when its costs were low (Figure 3.5). (Remember that fitness is relative to other individuals in the same population.) Juvenile density then decreased with increasing elevation in migratory but not resident trout, showing that increased migratory costs reduced the fitness benefits of anadromy. At an elevation of about 150 m, juvenile density was similar for migratory and resident populations, suggesting that the benefits and costs of anadromy were approximately equal at this elevation (Figure 3.5). As theory would predict, migration was rare above this elevation (Bohlin et al. 2001; Figure 3.5).

Experimental approaches would seem the most powerful way to test the prediction that variation in anadromy/non-anadromy is a function of relative costs and benefits. Dam construction represents one type of experiment. Morita et al. (2000) studied white-spotted charr populations above and below dams constructed 20–30 years previously. Juveniles below dams were more likely to smolt and migrate downstream than were juveniles above dams. This difference parallels the expectation that an increase in migratory costs causes an evolutionary reduction in anadromy: that is, migrants are lost from above-dam populations. However, Morita et al. (2000) then transplanted fry captured from above and below a dam in one stream to a fishless above-dam site in another stream. The transplanted fish showed very low rates of smolting and emigration regardless of whether they came from the below-dam or above-dam site. This suggests that reduced anadromy in the above-dam sites is a phenotypically plastic response to above-dam environments. The factor influencing this response appears to be growth rate: natural juvenile densities were much lower and juvenile growth rates much higher at above-dam sites than below-dam sites. Thus, increasing growth rates in fresh water cause a reduction in anadromy through a reaction norm linking migratory tendency to growth rate.

A second type of experiment is to increase fishing pressure on migrating individuals, which should select for reduced anadromy. This experiment is being conducted in many places around the world (Gross 1991), albeit inadvertently, and the effects have been evaluated in several Russian lakes (English reviews: Altukhov and Salmenkova 1991; Thorpe 1993; Altukhov et al. 2000, pp. 239–240). Fishing pressure dramatically reduced the abundance of adult sockeye salmon over a 40-year period (100,000 to 300 in Uyeginsk; 62,000 to 1600 in Dal'nee). This change was matched within lakes by a decrease in juvenile den-

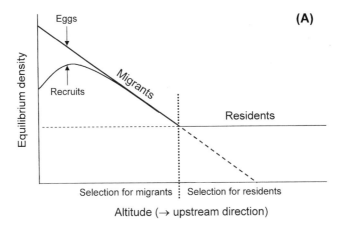

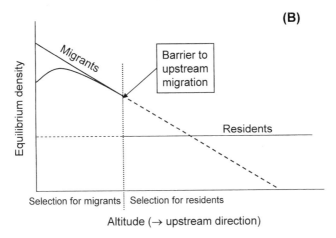

Figure 3.4. Theoretical predictions for the occurrence of anadromy (migrants) and non-anadromy (residents) in relation to migration difficulty (altitude). The equilibrium density of eggs (or recruits) remains constant with altitude for resident fishes because they do not incur any costs from migration (this assumes other factors influencing the equilibrium density do not vary with altitude). In contrast, the equilibrium density of recruits decreases with altitude for migrants because those migrating further distances incur higher costs. Panel (A) shows that migration is expected until the equilibrium density attained by residents exceeds that attained by migrants, after which residents are expected to predominate. Panel (B) shows that a barrier to migration may cause a higher equilibrium density of recruits below a barrier than above it. This figure is modified from Bohlin et al. (2001).

sities, an increase in juvenile growth, and a dramatic increase in the proportion of residuals among adult males (13% to 82% in Uyeginsk; 26% to 92% in Dal'nee). Thus, increases in the costs of anadromy (fishing pressure) and the benefits of non-anadromy (freshwater growth) led to the expected decrease in anadromy

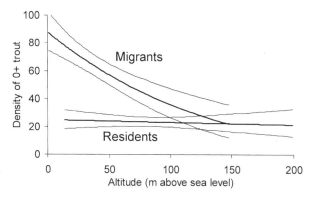

Figure 3.5. Relationship between the density of young-of-the-year (0+) brown trout (per m^2 × 100) in migratory populations and resident populations of brown trout (lines represent regressions with 95% confidence intervals, estimated from raw data). These data are consistent with the predictions outlined in Figure 3.4 that (1) the equilibrium density of recruits should decrease with increasing altitude for migratory populations but not for resident populations, (2) migratory populations should be absent above the elevation where the number of recruits produced by residents is equal to that produced by migrants, and (3) when barriers to migration isolate resident populations at low elevations they will have lower recruitment than migrant populations at similar elevations.

(see also Caswell et al. 1984). Unfortunately, it isn't known whether the changes are phenomenological, phenotypically plastic, or evolutionary. The change could be phenomenological simply because the proportion of residuals among breeding males would increase as fewer anadromous males returned from the ocean, even if the proportion of residuals among juveniles remained constant. The change could be phenotypically plastic because higher growth rates could mean that more individuals surpass a threshold for maturity in fresh water (see also Myers et al. 1986).

In summary, a number of studies have shown that anadromy decreases when its costs increase or its benefits decrease. Much of this variation can be explained by a reaction norm linking the tendency to migrate (or not) to some threshold condition influenced by the environment (e.g., growth rate). What remains to be determined is how much of the variation in anadromy/non-anadromy is the result of varying environmental conditions along a fixed reaction norm or to genetic variation in the reaction norm itself.

Prediction 4

Anadromy should be more common in females than males. A consequence of anadromy is increased adult size and energy stores, which may provide benefits during breeding. These benefits can be dramatic and absolute for females but subtle and relative for males (see above). Because the benefits of anadromy thus seem greater for females than males, the latter should be more likely to forgo anadromy at a given level of costs. In accord with this expectation, males are more likely than females to be non-anadromous in populations with partial migration (Jonsson and Jonsson 1993). For brown trout in 17 coastal streams of Norway, 48.9% of adult males but only 3.7% of adult females were non-anadromous (B. Jonsson et al. 2001a). The difference is even more striking in some Atlantic salmon populations, where all females are anadromous but most breeding males are non-anadromous "mature parr." In the 24 Atlantic salmon populations reviewed by Hutchings and Jones (1998), 0–95% of the adult males were mature parr. In ten populations, more that 50% of the adult males were mature parr (Hutchings and Jones 1998). Residual sockeye salmon are also over-whelmingly males (Ricker 1938, 1940; Krogius 1982). Partial migration is less common in other Pacific salmon species, but when it occurs, it is the males that may forgo anadromy (masu salmon, Tsiger et al. 1994; chinook salmon, Unwin et al. 1999).

3.2. Evolutionary Compensation

Prediction 5

Non-anadromous populations should compensate for the lower productivity of fresh water by evolving faster growth rates. To best test this hypothesis and determine the nature of any observed compensation (local adaptation, counter-gradient variation, or both), we would need multiple common-garden environments or reciprocal transplant experiments that compare fish from sympatric (but genetically distinct) anadromous and non-anadromous populations. No study has yet performed either of these tasks. However, one relevant study has used a unidirectional transplant experiment (Morita et al. 2000) and one has used a single common-garden environment (Wood and Foote 1996). These experiments can be used to test if non-anadromous forms show higher growth rates under the specific experimental conditions but they cannot determine the nature of such compensation.

Morita et al. (2000) studied white-spotted charr populations above and below recently constructed dams (see above). As part of this work, they collected fry from above and below a dam in one stream and transplanted them to an above-dam site in another stream. Recaptures over the next few years revealed that the below-dam origin fish grew faster than the above-dam origin fish. This pattern seems to counter the above prediction that non-anadromous fish should grow faster than anadromous fish. In this particular system, however, it is the above-dam fish that have the higher environmental growth potential in fresh water. This is because above-dam sites have lower charr densities and correspondingly higher growth rates. Thus, populations exposed to different growth conditions have evolved compensatory adaptations in a mere 20–30 years. Interestingly, most of the evolution must have taken place in the above-dam populations (because it was their environment that changed), and it thus represents a reduction in genetically based growth rate. The direction of evolution thus suggests selection against faster growth when environmental conditions already allow fast growth, supporting the evidence for costs of intrinsically high growth (see above).

Wood and Foote (1996) sampled adult sockeye salmon and kokanee breeding sympatrically in Narrows Creek, Takla Lake, and generated pure kokanee, pure sockeye, and hybrid crosses. When these crosses were raised under common conditions, pure sockeye were larger than pure kokanee at each of five sampling times (0, 76, 173, 377, and 640 days post-ponding), and hybrids were intermediate for the first four of these times. Similar results were obtained for a different brood year of Narrows Creek fish (Craig and Foote 2001) and for sympatric populations in Shuswap Lake (Wood and Foote 1990). Apparently, non-anadromous kokanee have not evolved higher growth rates than anadromous sockeye salmon, despite the reduced productivity of their environment. Why might this be so? The two sympatric forms share the same lake for the first year or two of life, and feed on the same food items (Wood et al. 1999). However, only juvenile sockeye salmon must subsequently smolt and migrate to the ocean. Perhaps sockeye salmon are under selection for faster growth because large size will increase survival during downstream migration (Wood and Foote 1996). Support for this hypothesis comes from Takla Lake where juvenile sockeye salmon are about 35% longer than juvenile kokanee by the end of their first summer (Wood et al. 1999).

In summary, very few studies have compared the growth rate of reproductively isolated, sympatric anadromous and non-anadromous forms, and those that have find the former have higher growth rates. These results seem to counter the above prediction, but perhaps only because they violate the implicit "all else being equal" assumption. Moreover, none of the studies conducted thus far has used an experimental design sufficient to distinguish between local adaptation and counter-gradient variation. This is not surprising because the adaptation of growth rate to environmental conditions has generally proven difficult to demonstrate in salmonids (B. Jonsson et al. 2001b).

Prediction 6

Anadromous populations should have adaptations that reduce the proximate costs of migration. In the following, we first consider adaptations to downstream migration and then upstream migration. Evidence of adaptations to downstream migration come from comparisons of sympatric sockeye salmon and kokanee: the former show greater saltwater adaptability (Foote et al. 1992), are larger at a common age (common-garden: Wood and Foote 1996; wild: Wood et al. 1999), and are better swimmers (Taylor and Foote 1991). Similar differences in saltwater adaptability and juvenile size have been documented between sympatric anadromous and non-anadromous Atlantic salmon populations (Birt et al. 1991b). High saltwater tolerance, large size, and strong swimming are obviously beneficial for fish that must migrate to the ocean. But why have these traits shown an evolutionary reduction in non-anadromous populations? One possibility is that selection in fresh water acts against some of the traits, most obviously physiological changes that prepare smolts for salt water (although some aspects of smolting may assist life in lakes, Foote et al. 1994). However, it seems less likely that selection in fresh water acts directly against large size and increased swimming ability. Here, the changes in non-anadromous populations may arise because selection favoring these traits is weaker in fresh water. When selection weakens, evolutionary changes can arise if the traits carry some indirect costs, such as those associated with high growth (see above). Alternatively, mutation and genetic drift could cause evolutionary change if population sizes are small and selection is weak (Adkison 1995).

Several adaptive differences have also been documented between anadromous juveniles that must migrate different distances. For example, Taylor and McPhail (1985a,b) used wild individuals, coupled with common-garden experiments, to compare juvenile body shape and swimming performance between coho salmon from coastal streams (short migrations) and interior streams (long migrations). They found that interior juveniles were more streamlined (Taylor and McPhail 1985a) and had greater swimming stamina (Taylor and McPhail 1985b), differences that would assist their long downstream migrations. In short, anadromous populations show a suite of adaptations that seem to reduce the proximate costs of downstream migration. For the rest of this section, we will examine complementary evidence for adaptations that reduce the proximate costs of upstream migration. Here, such adaptations are best revealed by comparing anadromous populations that migrate different distances.

Bernatchez and Dodson (1987) analyzed 15 populations of anadromous fishes (eight were salmonids), and found that populations migrating longer distances were more efficient in their energy use (kJ per kg per km). They also found that the populations most efficient in their energy use swam at the theoretically most efficient speeds. Hinch and Rand (2000) took a closer look at three of the Fraser River sockeye salmon populations, and found that they all swam at the theoretical optimum speed. Hinch and Rand (2000) did find one population (Chilko) that seemed to be "superoptimal migrators," showing the greatest ground speed at a given tail-beat frequency and using the least energy per unit of elevation. This observation supports the prediction that populations with more difficult migrations are more efficient because Chilko is one of the highest sockeye salmon lakes (1158 m) and is reached by a grade (1.9 m per km) more than twice as steep as any other Fraser River population (Gilhousen 1980). Recently, Crossin et al. (in review) found that sockeye salmon populations undertaking more difficult migrations in the Fraser River were considerably more efficient in their energy use. However, more populations must be studied before broad generalizations can be made about correlations between migratory difficulty and efficiency.

What traits might increase migratory ability and efficiency? Increased body size and increased fat stores have been suggested as possibilities. Potential advantages of large size include higher maximum and critical swimming speeds (to overcome obstacles), greater swimming efficiency (to conserve energy), and greater energy stores (to allow longer migrations). However, although fish with longer migrations are larger in some systems (Atlantic salmon: Schaffer and Elson 1975; N. Jonsson et al. 1991a; brown trout: L'Abée-Lund 1991), they are smaller in others (chum salmon: Beacham et al. 1988a; coho salmon: Fleming and Gross 1989; sockeye salmon: Moore 1996; Crossin et al. in review; Figure 3.6). At present, possible adaptive responses of body size to migration distance are unclear, and may depend on factors that have yet to be considered.

The potential advantage of increased fat stores is increased energy available for migration and breeding. Hendry and Berg (1999) found that a Bristol Bay population migrating only 98 km had considerably less mass-specific fat at the start of migration than did three Fraser River populations migrating 483–977 km. A similar trend is evident within the Fraser River: fat stores at river entry are higher for longer migrating populations (Gilhousen 1980; Crossin et al. in review). Fat stores at river entry are thus one of the strongest correlates of migration distance, and seemingly reflect adaptation to migration. Another potentially important trait is body shape, with longer migrating populations being more streamlined and having shorter snouts (Fleming and Gross 1989; Blair et al. 1993; Moore 1996; Crossin et al. in review; Figure 3.6).

Prediction 7

Anadromous populations should have adaptations that reduce the ultimate costs of migration (i.e., compensatory increases in reproductive output). The data available to test this prediction are sparse. Only one study has used a common-garden experiment to compare the reproductive output of sympatric anadromous and non-anadromous populations (Wood and Foote 1996). Unfortunately, maturation schedules and sampling dates in that study precluded a direct comparison of reproductive output. Only one study has used a common-garden experiment to compare the reproductive output of anadromous populations that migrate different distances (Kinnison et al. 2001). That study compared ovarian mass, egg number, and egg size (all standardized to a common body size) between two populations of New Zealand chinook salmon that migrate different distances (60 km vs. 100 km) to different elevations (200 m vs. 430 m). Full sibling families from each population were (1) raised under common conditions in a hatchery, and (2) released as smolts from a location with a long freshwater migration and a location with a short freshwater migration. The authors found evidence of adaptive compensation for the proximate cost of migration: the population from the site with a more difficult migration had larger ovaries under all three treatments (Kinnison et al. 2001; Figure 3.2). This evolutionary compensation was remarkable because the two populations shared common ancestors less than 100 years previously.

3.3. Other Predictions

Anadromy/non-anadromy may be density-dependent, frequency-dependent, or condition-dependent. The predictions here are more intricate, less well developed, and more difficult to test than those considered thus far, and the relevant data are correspondingly sparse. Here we briefly mention some relevant research and suggest potentially useful comparisons and experiments.

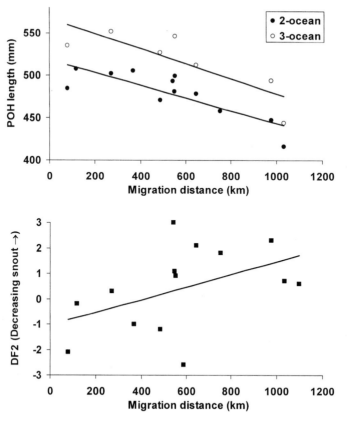

Figure 3.6. Variation in body size and shape among wild Fraser River sockeye salmon populations in relation to their migration distance (data from Moore 1996). The top panel shows variation in post-orbit to hypural (POH) length for adults that spent 2 years in the ocean (2-ocean, $r^2 = 0.68$, $P < 0.005$) or 3 years in the ocean (3-ocean, $r^2 = 0.67$, $P < 0.05$). The bottom panel shows variation in the second discriminant function (DF2) from truss measurements, on which measures of snout length load most heavily ($r^2 = 0.33$, $P < 0.05$).

First, a variety of studies have shown that survival and growth in fresh water are often density-dependent (e.g., Le Cren 1973; Burgner 1987; Jenkins et al. 1999). This suggests that the average fitness of non-anadromous individuals will decrease as they become more abundant, which should increasingly favor anadromy. Moreover, several studies have shown that increasing juvenile densities lead to increasing emigration from stream sections (e.g., Keeley 2001). With respect to anadromy, Morita et al. (2000) showed experimentally that a decrease in juvenile density for white-spotted charr led to an increase in freshwater growth rate and a decrease in anadromy. Similarly, a decrease in the density of juvenile sockeye salmon in lakes caused an increase in growth rate and an increase in the frequency of non-anadromous residuals (effects of cyclical abundance: Krogius 1982; effects of fishing: Thorpe 1993; Altukhov et al. 2000).

These studies demonstrate the existence of a reaction norm linking anadromy to freshwater density, but no studies have demonstrated an evolutionary response to variation in density.

Second, some evidence suggests that frequency-dependent mating success may play a role in the evolution of anadromy/non-anadromy (Myers 1986; Hutchings and Myers 1994). For example, genetic parentage analyses of salmon in experimental arenas have shown that non-anadromous males can fertilize many eggs, even in competition with anadromous males (see Fleming and Reynolds—*this volume*). Moreover, the *average* fertilization success of non-anadromous males may decrease as their abundance increases (Hutchings and Myers 1988; Thomaz et al. 1997). Unfortunately, these studies have yet to isolate frequency dependence from density dependence: that is, varying frequencies while holding densities constant. We also need more studies of fish in the wild because experimental arenas may not capture all of the factors influencing reproductive success. The few studies conducted in the wild have found that non-anadromous males can fertilize a substantial fraction of a female's eggs (Jordan and Youngson 1992; Taggart et al. 2001) but have yet to examine effects of frequency on fertilization success.

Third, many studies have shown that anadromy/non-anadromy is condition-dependent. In some systems, the body size of individuals is positively correlated with their likelihood of anadromy (e.g., Bohlin et al. 1994; Rikardsen and Elliott 2000), perhaps because smaller fish are less likely to survive migration (Finstad and Heggberget 1995; Gulseth and Nilssen 2000). In other systems, males that are larger (or faster growing, or higher in fat stores) are more likely to mature in fresh water as parr (Rowe et al. 1991; Berglund 1992; Prévost et al. 1992). Relative to anadromous individuals, these mature parr (1) have phenotypic characteristics suggesting they are of higher quality, (2) have higher survival to maturity (Hutchings and Myers 1994), and (3) are successful in fertilizing at least some eggs when competing with anadromous males (Fleming and Reynolds 2003—*this volume*). These observations suggest that non-anadromy actually generates higher average fitness than anadromy (Gross 1996), and that the two life histories may be maintained through a single conditional strategy (Gross and Repka 1998). If this proves true, it calls for a re-examination of earlier analyses that assumed each life history had equal average fitness through frequency dependence (Myers 1986; Hutchings and Myers 1994). However, no studies have yet directly tested for a conditional strategy in salmonids.

To fully evaluate the theories outlined in this section, we need empirical assessments of lifetime reproductive success (number of grandchildren produced) in nature. Such analyses would ideally examine the success of individuals of different condition (size, growth rate, age, energy stores) that adopt different life histories (anadromy vs. non-anadromy), across a range of densities and frequencies. Although difficult, such studies are feasible because hypervariable microsatellite markers for salmonids should be adequate for parentage (and even grandparentage) assignment in natural populations. Moreover, densities and frequencies of anadromy/non-anadromy vary widely among years within populations (e.g., sockeye salmon: Krogius 1982; Atlantic salmon: Myers et al. 1986; brown trout: Dellefors and Faremo 1988).

4. Conclusions

We examined evidence that variation in anadromy/non-anadromy is the result of variation in the benefits and costs of these alternative life histories. We find strong evidence that anadromy has both benefits and costs. Benefits come in the form of increased body size and energy stores, which may then increase reproductive success. These benefits tend to be concrete and absolute for females but variable and relative for males. Costs come in the form of increased mortality and increased energy expenditure during migration. These costs and benefits apply in an opposite manner to non-anadromy. Although these general conclusions seem robust, additional work is needed. For example, comparisons of stage-specific rates of mortality between the two life histories would provide a clearer picture of the actual fitness costs associated with migration per se.

Variation in anadromy/non-anadromy should evolve as a function of variation in costs and benefits. For example, the benefits of anadromy are greater for females than for males and, accordingly, males are more likely to forgo anadromy. Among populations, anadromy should decrease with increasing migratory difficulty and with increasing freshwater productivity. These predictions enjoy support from distribution patterns (Rounsefell 1958), direct correlative tests (Kristoffersen 1994; Bohlin et al. 2001), and experimental manipulations (Morita et al. 2000; Altukhov et al. 2000). What remains entirely unknown, is the extent to which variation in anadromy/non-anadromy is the result of phenotypic plasticity or genetic variation. We encourage further quantitative analyses of variation in anadromy with respect to costs and benefits, experimental manipulations of costs or benefits, and comparisons of migratory tendency in common-garden environments.

Evolution should favor adaptations that reduce the proximate costs of anadromy or non-anadromy. One prediction is that non-anadromous populations should have higher genetic growth rates to compensate for the lower productivity of their environment. Evidence for such compensation is currently lacking. In fact, anadromous fish grew faster than non-anadromous fish in a transplant experiment (Morita et al. 2000) and a common-garden experiment (Wood and Foote 1996). Post hoc explanations for these results are that environmental growth potential was actually lower for the anadromous population in Morita et al. (2000) and migration may be size-selective in Wood and Foote (1996). Other predictions are that longer anadromous migrations should lead to the evolution of (1) increased swimming ability (true for juveniles), (2) increased migratory efficiency (perhaps true for adults), (3) larger body sizes (true for juveniles, true for adults in some systems but not in others), (4) greater energy stores (true for adults), and (5) more efficient morphology (true for juveniles, perhaps true for adults).

Evolution should also favor adaptations that reduce the ultimate costs of anadromy or non-anadromy: that is, reproductive output should increase for populations with longer migrations. In the only relevant study thus far, two populations of chinook salmon that differed in migration difficulty showed clear evidence of such compensation: the longer migrating population had greater reproductive output when reared in a common environment (Kinnison et al. 2001). This latter result was remarkable because the two populations had a common origin about 100 years earlier, and because the evolutionary compensation for increased migratory difficulty was through a trait (egg number) that was different from the trait directly influenced by the proximate cost of migration (egg size).

Anadromy/non-anadromy should be influenced by density, frequency, and condition dependence. Research on these topics is as yet fragmentary but some preliminary generalizations are possible. First, density-dependent survival and growth is common, and can influence emigration from a local area. Moreover, several studies have shown that anadromy may indeed be density-dependent (Krogius 1982; Morita et al. 2000; Altukhov et al. 2000). Second, studies in experimental arenas have suggested that the mating success of anadromous and non-anadromous males may be frequency-dependent, but these have yet to remove potentially confounding effects of density dependence. Third, individual condition may influence migratory tendency in different ways. In some systems, the largest juveniles become anadromous, whereas in other systems, the largest juveniles remain non-anadromous. In any given system, the average fitness of the two life histories may not be equal and instead may be maintained within populations through a conditional strategy (Gross and Repka 1998). Fully testing these hypotheses will require studies of lifetime reproductive success in natural systems.

Acknowledgments   We thank Bob McDowall and John Thorpe for extensive comments on earlier drafts of the manuscript.

Evolution of Egg Size and Number

Sigurd Einum

Michael T. Kinnison

Andrew P. Hendry

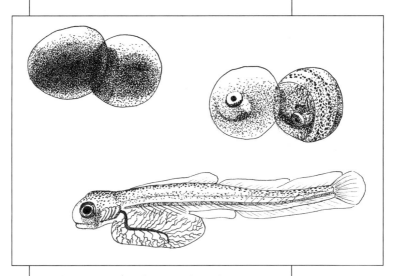

Early stage eggs, eyed eggs, and an alevin

4

Reproductive traits are strong determinants of fitness and should thus be associated with intense selection. One important component of this selection is the period following fertilization of the ova. In a wide range of organisms, particularly those with high numbers of offspring, early life stages experience massive mortality (e.g., Fortier and Leggett 1985; Brosseau and Baglivo 1988; Iverson 1991). Thus, if phenotypic variation is present, the potential for selection on juveniles may exceed that at any other stage of life. Although some phenotypic variation may be caused by genetic variation among the juveniles themselves, the importance of maternal effects (particularly through egg size) has become increasingly apparent (reviewed by Roff 1992; Bernardo 1996a; Mousseau and Fox 1998a,b). It is now clear that egg size is an important fitness component, with offspring fitness after hatching tending to increase with increasing egg size (e.g., Roff 1992, pp. 348–352; Sinervo et al. 1992; Hipfner and Gaston 1999; Einum and Fleming 2000a).

All else being equal, and barring density-dependent effects, the fitness of a female should increase with increasing egg number. However, egg size not only influences offspring fitness but also the number of offspring a female can produce. The traditional reasoning is that (1) a female has only a limited amount of energy resources available for egg production (or space to hold eggs in her body cavity), and (2) for a given energetic investment (or amount of space), an increase in egg size will inevitably cause a decrease in egg number. Due to this tradeoff, the size of eggs imposes a constraint on egg number (e.g., Rohwer 1988; Elgar 1990; Sinervo and Licht 1991), resulting in negative phenotypic and genetic correlations between these traits in many organisms (Roff 1992, p. 357), including salmonids (e.g., Jonsson and Jonsson 1999; Kinnison et al. 2001).

Egg size thus presents breeding females with an evolutionary dilemma, arising because maximization of both numbers of eggs and offspring fitness (which is size-dependent) is not attainable. Because egg size is inherently linked to egg number, the evolution of each cannot be considered in isolation. Here we outline general theoretical models for the evolution of egg size and number (Sections 1.1–1.5), and show how research on salmonids has provided critical empirical tests of these models (Sections 2.1–2.5). Due to our empirical focus on salmonids, we use terms such as "egg" and "hatching," but the theoretical framework also applies to other organisms. We close with a consideration of how egg size evolves on a macro-evolutionary scale (i.e., among species), and suggest that further progress can be made by simultaneously considering multiple features of an organism's biology that are likely to influence egg size evolution.

Several issues should be addressed before proceeding. First, our approach is primarily based on traditional optimization theories, because these have stimulated much empirical work, and have been pivotal in our understanding of egg size and number variation. Yet, alternative approaches may be relevant in some systems. For example, if selection is frequency-dependent or density-dependent, the optimal value may fluctuate over time, making an evolutionary stable strategy (ESS) approach appropriate (Sinervo et al. 2000). Furthermore, variation within or among populations may be non-adaptive if caused by mutational input and genetic drift. Fortunately, most optimization theories are built on explicit testable assumptions, and provide qualitative predictions regarding the nature of variation (e.g., correlations with environment conditions), enabling rigorous tests.

Second, the optimization approach assumes that optimal values are achieved, and that there are no constraints (e.g., mechanical or genetic) preventing adaptation (other than any constraints explicitly included in a model). Although these may be inaccurate assumptions in many cases, the relative success of the optimization approach, at least qualitatively, suggests that such constraints may rarely be strong enough to prevent trait values from approaching optimal values. We discuss such constraints in Sections 3.1 and 3.2.

Third, in empirical and theoretical studies of selection one must determine whether or not offspring fitness components should be assigned to parental fitness (Wolf and Wade 2001). If the component of offspring fitness is controlled largely by the mother (i.e., maternal effect), then it will be appropriate to incorporate this component into maternal fitness. This may not be true if both (1) the offspring fitness component is strongly influenced by the offspring's own genotype (i.e., direct effect), and (2) there is a genetic correlation between the maternal trait and the direct effect (Wolf and Wade 2001). For example, if juveniles originating from large eggs are also genetically superior, then incorporating the offspring fitness component into maternal fitness may yield somewhat incorrect conclusions regarding the evolutionary dynamics of egg size. In empirical studies this potential problem can be solved by manipulating the trait in question and thereby controlling for such correlations (Sections 2.1–2.2).

1. Theory

The idea that some tradeoff must exist between egg size and number can be traced back at least to Lack (1947a) and Svärdson (1949). Lack discussed how clutch size evolves in altricial birds (i.e., birds where offspring are fed by parents). His hypothesis was that clutch size is determined by the number of young that parents can raise to independence. Although he was mainly concerned with the level of resources provided to each young *after* hatching, the main argument remains the same for provisioning before hatching; an increase in number of offspring for a constant level of total investment cannot occur without a cost to individual offspring fitness. Similarly, Svärdson (1949) realized that there must be a balance between the number and size of eggs in fishes. Based on observations suggesting that larger eggs give rise to larger larvae, which have better survival prospects, he proposed that there must be a premium on large eggs, which must tend to decrease egg number. These verbal models made important contributions toward an increased interest in reproductive allocation (e.g., Kendeigh et al. 1956; Olsson 1960; Bagenal 1969; Blaxter 1969; Klomp 1970). Then, a mathematical model on reproductive allocation was published by Smith and Fretwell (1974). This model has been incredibly influential (more than 750 citations over 15 years, ISI Web of Science), in part by sparking an explosion of theoretical efforts to better understand the evolution of egg size and number. These efforts have remained a prominent area of life history theory (e.g., Brockelman 1975; Parker and MacNair 1978; Parker and Begon 1986; Lloyd 1987; McGinley et al. 1987; Morris 1987; Sargent et al. 1987; McGinley 1989; Geritz 1995; Hutchings 1997; Forbes 1999; Hendry et al. 2001b; Sakai and Harada 2001). We start our discussion of theoretical models by presenting the analysis of Smith and Fretwell (1974). We then consider how the basic theoretical framework has been extended by subsequent models.

1.1. The Smith–Fretwell Model

Smith and Fretwell (1974) set out to solve the problem of the allocation of a fixed quantity of resources into offspring. They assumed (1) a tradeoff between egg size and number, (2) the currency being maximized through selection is maternal fitness (i.e., number of grandchildren), and (3) the pattern of allocation does not influence survival or future reproduction of the mother. Thus, for a given breeding event, maternal fitness (W) depends on the number of offspring produced and the mean fitness of offspring. This can be expressed as:

$$W(M, m) = \frac{M}{m} \times f(m) \qquad (1)$$

where M represents the total amount of resources invested into egg production (e.g., total ovary mass in oviparous fishes), and m represents the amount of resources invested into each egg. Thus, M/m equals the number of eggs produced. $f(m)$ is a function describing how offspring fitness depends on egg size.

Smith and Fretwell (1974) assumed that $f(m)$ is an asymptotic function having some minimum viable egg size (Figure 4.1a). Two suggested explicit functions for $f(m)$ are:

$$f(m) = 1 - \exp[-(m - m_{min})] \qquad (2)$$

and

$$f(m) = 1 - \left(\frac{m_{min}}{m}\right)^a \qquad (3)$$

where m_{min} is the minimum viable egg size, and a is a constant determining the rate at which the function approaches the asymptote (Parker and Begon 1986; McGinley et al. 1987). Thus, offspring fitness is maximized (but with ever-diminishing proportional increases) as egg size approaches infinity. This is not the case for maternal fitness. The optimal egg size with respect to maternal fitness (m^*) can be found by setting $dW/dm = 0$ and solving for $m = m^*$:

$$m^* = \frac{f(m)}{f'(m)} \qquad (4)$$

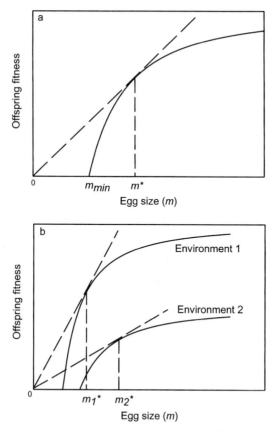

Figure 4.1. (a) The hypothetical asymptotic relationship between egg size and offspring fitness as suggested by Smith and Fretwell (1974). m^* denotes the optimal egg size that maximizes maternal fitness and m_{min} denotes the minimum viable egg size. The optimal egg size can be obtained by drawing a tangent to the function from the origin as done here, or by determining where the derivative of the function relating maternal fitness to egg size equals zero (see Figure 4.4). (b) Due to better environmental conditions, maximum relative fitness is higher and minimum viable egg size smaller in environment 1 than in environment 2. As a result, the optimal egg size in environment 1 (m_1^*) is smaller than in environment 2 (m_2^*).

This is the well-known marginal value solution (Charnov 1976), and the result can also be obtained graphically (Figure 4.1a). An important point is that because $f(m)$ is independent of M (i.e., no density-dependent effects on $f(m)$), so too is the optimal egg size (m^*), and the resulting ratio (M/m^*) determines the optimal egg number. As we shall see in Section 1.4, relaxing the assumption of no density dependence can change the outcome.

The simple structure of the Smith–Fretwell model makes it tractable for extensions beyond predicting a single optimal egg size and number in a single population. It has therefore had an enormous impact on life history theory but empirical tests have been rare. In Section 2.1, we outline such a test using a phenotypic manipulation of egg size in Atlantic salmon.

1.2. Optimal Egg Size and Environment Quality

According to the basic Smith–Fretwell model, all females reproducing in a particular environment should produce identical egg sizes. However, different environments may have different relationships between egg size and offspring fitness (i.e., variable $f(m)$). Any changes in this function have implications for Equation (4), and thus for optimal egg size. This means that the model has the potential to predict not only evolutionary stable egg sizes within a population, but also adaptive variation among populations. In general (although not always— see Section 1.3), the Smith–Fretwell framework predicts that as the quality of the environment experienced by offspring decreases, optimal egg size should increase (Figure 4.1b; McGinley et al. 1987). In Section 2.2, we outline empirical tests of this prediction using experiments on brook charr and brown trout.

1.3. Opposing Selection Stages

The Smith–Fretwell model has commonly been used to consider contributions to offspring fitness after hatching. In some organisms, however, egg size may also influence offspring fitness *before* hatching. For example, predation on pelagic eggs in marine environments has been suggested to select against small sizes (Rijnsdorp and Jaworski 1990). Egg size may also influence development, with larger eggs sometimes requiring more time from fertilization to hatching or yolk absorption (Sibly and Monk 1987; Einum and Fleming 2000a; but see Hutchings 1991; Hendry et al. 1998; Einum and Fleming 1999). Accordingly, if the egg stage is a "safe harbor" with low mortality, and an increase in egg size reduces the duration of the later high-risk juvenile stage, selection should favor large eggs (Williams 1966b; Shine 1978). Conversely, in organisms that experience high intrinsic mortality during the egg stage there may be selection for small, fast-developing eggs. These examples suggest selection for egg size before hatching, and that selection pressures may sometimes oppose each other during different developmental stages (e.g., large eggs may provide a negative contribution to offspring fitness before hatching, but a positive contribution after hatching). Such opposing selection pressures at different life history stages may be common in nature (Schluter et al. 1991).

A recent analysis examining how opposing selection pressures may shape egg size evolution is given by Hendry et al. (2001b). They considered a situation where offspring fitness and egg size were negatively correlated before hatching, but positively correlated after hatching. A negative pre-hatching survival/size relationship could occur for many reasons. In particular, egg size in aquatic organisms has been assumed to be negatively correlated with survival before hatching because levels of dissolved oxygen may constrain egg size (Krogh 1959). The reasoning here is that as egg size increases, the volume of the egg (and thus its oxygen demand) increases faster than the surface area (over which oxygen diffusion occurs). This assumption had been widely stated but never empirically tested at the time Hendry et al. (2001b) developed their model. A recent experiment designed to test the assumption, however, found the opposite effect: large eggs survived *better* than small eggs when challenged with low dissolved oxygen (Einum et al. 2002). Furthermore, egg surface area increased faster than oxygen consumption with increasing egg size, suggesting that the ability of eggs to acquire the necessary oxygen for metabolism actually increases with increasing egg size (Einum et al. 2002). Thus, modifications to existing expectations and theories appear necessary. Hendry and Day (2003) have shown that this new evidence can be accommodated within the general theoretical approach of Hendry et al. (2001b).

For now, we will continue with the traditional reasoning that large eggs suffer disproportionately in low dissolved oxygen, because it is instructive even if ultimately incorrect. For a given level of oxygen, a maximum viable egg size may exist, and with increasing oxygen this maximum size may increase. In the model of Hendry et al. (2001b), the overall individual offspring fitness function ($f(m)$) consisted of two components:

$$f(m) = f_{pre}(m) \times f_{post}(m) \qquad (5)$$

where the first and second components describe the relationship between egg size and fitness before (pre) hatching and after (post) hatching, respectively (Figure 4.2a). The post-hatching fitness function (f_{post}) followed Equation (3). The pre- hatching fitness function (f_{pre}) was modeled as:

$$f_{pre}(m) = k\left[1 - \left(\frac{m}{km_{max}}\right)^{b}\right] \qquad (6)$$

where m_{max} is the maximum viable egg size, b is a constant determining the shape of the function, and k is a constant related to the quality of the incubation environment. Thus, by varying the value of k, Hendry et al. (2001b) could solve for optimal egg sizes in different pre-hatching environments (Figure 4.2b).

One result emerging from this analysis was that an increase in pre-hatching environment quality (caused by differences among populations or female phenotypes) should cause an increase in the optimal egg size (Figure 4.2c). Thus, this model predicts that if the quality of the incubation environment influences pre-hatching fitness (all else being equal), populations experiencing low-quality incubation environments (i.e., low levels of dissolved oxygen) should produce small eggs. In Section 2.3, we outline a correlative study on sockeye salmon that provides support for this model. However, the above logic and results need to be reconsidered if the assumption that large eggs suffer more than small eggs in low dissolved oxygen proves incorrect (Einum et al. 2002; Hendry and Day 2003).

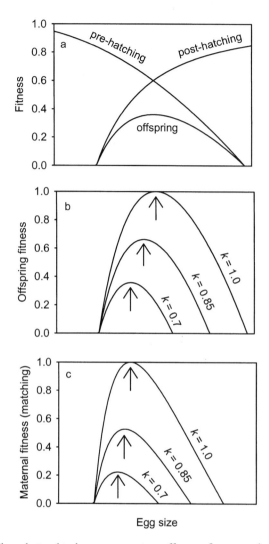

Figure 4.2. The relationship between egg size, offspring fitness, and maternal fitness in a situation of opposing selection pressures (e.g., negative relationship between egg size and offspring fitness before hatching, but a positive relationship after hatching). a) Pre-hatching and post-hatching fitness functions and the corresponding offspring fitness for a range of possible egg sizes. (b) Offspring fitness functions (rescaled to a maximum of 1.0) under three different incubation environment qualities (k, see Equation (6) in text; increasing k corresponds to increasing incubation environment quality). (c) Maternal fitness (rescaled to a maximum of 1.0) for females of three different sizes, when their eggs incubate in different pre-hatching environments (i.e., female size is positively correlated with incubation environment quality). Arrows indicate egg sizes that maximize offspring or maternal fitness. Modified from Hendry et al. (2001).

1.4. Phenotype-Specific Optimal Egg Size

Classic optimality models predict that a single optimal egg size maximizes maternal fitness in a given environment. Why then do we so commonly observe extensive egg size variation among individuals within populations? This observation is clearly at odds with the classic models, and calls for an explanation (Bernardo 1996b). Several competing hypotheses have been advanced. First, if environment quality varies through the breeding season, optimal egg size may vary accordingly (e.g., Bagenal 1971; Kerfoot 1974; Chambers 1997). This hypothesis cannot, however, explain the variation in a population at a given time. Second, egg size may be constrained in small females due to morphological features found in some taxa (e.g., pelvic girdle size; Congdon et al. 1983; Sinervo and Licht 1991). This constraint seems unlikely in organisms without such features, such as salmonids (see Section 3.1). Third, if females have imperfect information about optimal egg sizes, selection may favor increased variation in egg size both within and among females (Koops et al. 2003).

A quite different set of arguments, all based on extensions of the Smith–Fretwell model, rest on the assumption that shapes of the egg size/offspring fitness function depend on maternal phenotype. For example, larger females are expected to produce larger and/or more eggs due to higher total amounts of reproductive effort, and this in turn may influence the egg size/offspring fitness function. Parker and Begon (1986) developed a series of models examining how social interactions among siblings and non-siblings may influence rules of allocation among females of different sizes. McGinley (1989) took a different approach, wherein the number of offspring may influence the egg size/offspring fitness function through effects on levels of predation. These two models and others demonstrate the theoretical potential for adaptive variation in egg size within populations. More specifically, they predict that optimal egg size should increase with increasing gonad investment under negative density dependence caused by sibling competition or predation (i.e., larger clutches experience higher levels of competition or predation). In these cases, the reduction in egg number with an increase in egg size is more than compensated by the combined effects of reduced density dependence among offspring and their increased size-dependent fitness (see also Hendry and Day 2003).

A common feature of the Parker and Begon (1986) and McGinley (1989) models is that the selective pressures for variable egg sizes (i.e., mothers adjust their egg size to optimize fitness) occur after hatching. Sargent et al. (1987) suggested an alternative hypothesis. They assumed that survival of juveniles could be described by the Smith–Fretwell model (i.e., survival increases with egg size), but that increasing egg size increased egg or larval mortality. This negative effect of increased egg size could occur either by (1) an increase in instantaneous mortality, or (2) an increase in developmental time while keeping instantaneous mortality constant. Under both of these scenarios, females that were able to reduce egg or larval mortality through parental care should also produce large eggs.

The idea that optimal egg size can be shaped by egg mortality was further investigated by Hendry et al. (2001b). In Section 1.3 we discussed how their model identified pre-hatching environment quality as a potential evolutionary force shaping egg size. They suggested that this could have implications for variation in optimal egg size within populations in the case of a positive correlation between maternal phenotype and environment quality. Hendry et al. (2001b) referred to such situations as "phenotype/habitat matching," and pointed out that such matching can occur in certain organisms (e.g., when female phenotype influences incubation environments indirectly through interference competition, or directly through parental care). To examine if this could select for intrapopulation variation in egg size, they developed a model where matching occurs (larger females provide higher quality incubation environments), larger females have greater gonad investment, and offspring fitness is described by Equations (3), (5) and (6) (i.e., offspring fitness is related to egg size negatively before hatching and positively after hatching). Their model predicted that larger females should produce larger eggs (Figure 4.2c). In addition, it predicted that increasing female size should result in a greater proportional increase in egg number than in egg size because egg number shows a linear relationship with maternal fitness, whereas egg size does not. Finally, because it is female size and not gonad investment that determines environment quality, egg size should not increase with relative gonad investment (i.e., ovary mass adjusted for body size).

In Section 2.4, we outline a study that provides correlative evidence for these hypotheses in a variety of salmon populations. However, Hendry et al. (2001b) relied on the assumption that larger eggs are selected against in low dissolved oxygen conditions, whereas Einum et al. (2002) showed that the opposite may be true. Hendry and Day (2003) have recently shown that inverting the traditional assumption according to the new results of Einum et al. (2002) yields the following predictions. First, larger females should produce larger eggs if (1) they place their eggs in lower quality environments (i.e., less dissolved oxygen), which seems unlikely but has been suggested by Holtby and Healey (1986), or (2) they place their eggs in higher quality environments but this improvement in habitat quality is less than the decline habitat quality owing to the larger clutch sizes of larger females (i.e., negative density dependance). Testing these new predictions will require additional empirical study.

1.5. Non-Equilibrium and Non-Optimal States

We have thus far discussed the evolution of egg size and number from the perspective of optimality. The Smith–Fretwell model, and other models that derive from its basic premise, describe evolution under a simplified set of ideal conditions. For example, these models generally imply evolution over unlimited time frames, with consistent environment quality and selection through time, and with ample genetic variation. However, equilibrium and optimal conditions may not be achieved under some circumstances. For example, most new populations are likely established with trait values that are not at their optimal (or equilibrium) state, and many extant populations have been exposed to altered environment conditions that may influence $f(m)$. It should be possible to gain some insights into non-equilibrium and non-optimal states by examining perturbed or introduced populations where ancestral genotypes interact with new selective regimes. In Section 2.5, we discuss experimental and correlative work that provides insights into non-equilibrium/non-optimal aspects of egg size and number in chinook salmon.

2. Empirical Tests of the Theory

The theoretical treatment of egg size/number evolution has flourished in the wake of Smith and Fretwell (1974) but empirical tests have lagged behind, and some authors have questioned the validity of optimality models for examining egg size evolution (e.g., Sinervo et al. 1992; Bernardo 1996b). As in other fields of science, models are valuable for deriving testable predictions, but without rigorous tests, they remain purely theoretical.

Salmonids are particularly suitable for empirical tests of existing egg size/number theory. First, the predictive power of the Smith–Fretwell model and its extensions depend on multiple offspring among which resources are divided. The potentially high egg number and its great range in salmonids (hundreds to many thousands, see Table 1 in Hutchings and Morris 1985; Stearns and Hendry 2003—*this volume*) makes them a powerful system for testing these models. Second, salmonids lack post-hatching parental care, which would otherwise complicate patterns of optimal reproductive allocation (Trivers 1974). Third, salmonid development from fertilized egg to adult can be divided into discrete stages, and these stages may be exposed to different and even opposing selection pressures. Fourth, salmonids form well-defined populations with limited degrees of gene flow among them (Quinn 1993; Hendry et al. 2003a—*this volume*). These populations inhabit a range of biotic and abiotic environments, which sets the stage for adaptations to local conditions (review: Taylor 1991b). Fifth, salmonid eggs are large and variable relative to other fishes (Wootton 1984; Hutchings and Morris 1985; Beacham and Murray 1993; Fleming 1998), suggesting considerable potential for egg size variation to influence offspring and maternal fitness. Finally, salmonids are recreationally and commercially important and so (1) a vast amount of data have been collected on their life history and its genetic basis, and (2) numerous translocations have occurred providing seminatural experiments in the evolution of populations faced with altered selective regimes.

As a result of their suitability for studies of egg size and number, salmonids have been extensively used both in field studies and experiments. In the following, we show how each of the theoretical issues outlined in Sections 1.1–1.5 has been examined using research in salmonids.

2.1. The Smith–Fretwell Model: A Test Using Phenotypic Manipulations in Atlantic Salmon

Empirical tests of the Smith–Fretwell model have been rare for wild populations. One exception is a recent study of Atlantic salmon (Einum and Fleming 2000a). A previous study had demonstrated that juvenile body size at emergence from nests was an important determinant of early survival (Einum and Fleming 2000b), probably because larger juveniles have improved competitive ability, increased resistance to starvation, and/or reduced risk of predation (Bagenal 1969; Hutchings 1991; Kristjánsson and Vøllestad 1996; Einum and Fleming 1999; Heath et al. 1999). However, no manipulations had been performed to establish whether this was a causal effect of egg size, or due to correlations between size and other traits.

To isolate the effects of egg size per se, Einum and Fleming (2000a) manipulated egg size by rearing parents to adulthood in captivity, a procedure that can result in some females producing more variable egg sizes than in the wild (CV = 18.5% versus 4.0% for wild fish, Einum and Fleming 2000a). Such variation in egg size appears to be independent of genetic effects, or other characteristics of the oocytes themselves, and may arise owing to variation in the position of the egg relative to blood vessels in the female ovary (Einum and Fleming 1999). A sample of small eggs and large eggs was obtained from each of eight females, producing a total of 16 groups that differed substantially in egg size (small eggs were on average 31.4% lighter than their large siblings), and allowing a test for causal relationships between egg size and offspring fitness (e.g., Wade and Kalisz 1990; Sinervo and Svensson 1998). The different groups were allowed to develop naturally in artificial nests, and emerging juveniles were group marked and released into a stream. Body size and survival was assessed by extensive sampling 28 and 107 days after median emergence. The results were then used to test the assumptions and predictions of the Smith–Fretwell model, including: (1) egg size determines juvenile body size, (2) the relationship between egg size and juvenile fitness (measured as survival) will be positive but asymptotic, and (3) egg size will evolve to maximize maternal fitness. Thus, the predicted optimal egg size that would maximize maternal fitness, based on the experimental data, should match the mean egg size observed in the natural population.

Juvenile body size at emergence from the gravel, and at the two subsequent sampling periods, was positively related to egg size (Figure 4.3), supporting the first implicit assumption of the Smith–Fretwell model. Estimated selection intensities indicated strong directional selection toward larger egg size, with respect to offspring fitness. Furthermore, significant stabilizing selection for maternal fitness acted directly to reduce the variance in egg size. This means that females producing eggs that were too small or too large would have reduced fitness. The relationship between egg size and the proportion of fry recaptured (a measure of survival) was best described by asymptotic regressions in accordance with the second assumption of the Smith–Fretwell model (Figure 4.4). Finally, the optimal egg size estimated from these data corresponded to the mean egg

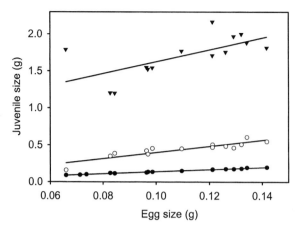

Figure 4.3. The relationship between Atlantic salmon egg and juvenile size (wet mass) at emergence (filled circles, $y = 1.4x, r^2 = 0.99$), at 28 days after median emergence (open circles, $y = 4.2x, r^2 = 0.81$), and at 107 days after median emergence (triangles, $y = 0.8 + 8.0x, r^2 = 0.43$). Each data point represents the mean value from one nest.

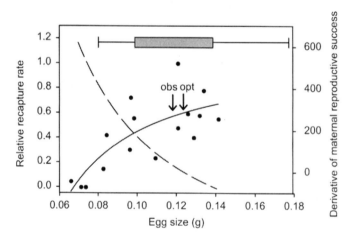

Figure 4.4. The relationship between egg size and the relative recapture rate of juvenile Atlantic salmon 28 days after median emergence. A similar relationship was found after 107 days. Each data point represents mean values for one nest. The asymptotic function relating offspring recapture rates to egg size (solid line) is estimated according to Equation (3), where $m_{min} = 0.068$ g and $a = 1.507$ $(r^2 = 0.59)$. The dashed line represents the derivative of the function relating maternal reproductive success to egg size. The derivative equals zero for the egg size that maximizes maternal reproductive success (opt). Observed mean (obs) and the standard deviation and range (box plot at top) of egg size in the native population is indicated. From Einum and Fleming (2000a).

size observed in the natural population. This result indicates that mean egg size in Atlantic salmon has evolved primarily to maximize maternal fitness, and acts via effects on offspring fitness. The close fit between predictions and observations demonstrates that the Smith–Fretwell model represents an important contribution toward understanding egg size and number evolution.

2.2. Optimal Egg Size and Post-Hatching Environment Quality: Experimental Tests

In our above treatment of the Smith–Fretwell model, we showed how it can be used to predict differences in egg size among populations if they experience different environments, and if the egg size/offspring fitness function depends on those environments (Figure 4.1b). Environment quality obviously differs among populations. For example, growth rates during the first year of life vary more than twofold among Norwegian populations of brown trout, primarily as a phenotypic response to differences in water temperature (Jensen et al. 2000). However, it is less obvious that the egg size/offspring fitness function depends on environment quality.

Two studies have tested for an effect of environment quality on the egg size/ offspring fitness function in salmonids. The first compared survival of juvenile brook charr originating from different egg sizes (Hutchings 1991). Twenty-seven different egg size groups, each consisting of 20 juveniles, were derived from females in three different populations and reared separately in aquaria at two different food levels (high and low). A comparison of regressions between egg size and juvenile survival for the high and low food levels showed that survival generally increased with increasing egg size, but that the effect depended on environment quality. Specifically, the slope relating egg size to survival was steeper under low than under high food levels, suggesting that juveniles from small eggs do just as well as those from large eggs when conditions are favorable. Thus, due to a tradeoff between egg size and egg number, the optimal egg size was smaller in the high-quality environment, as predicted from Section 1.2.

The second study compared growth of juvenile brown trout originating from small and large eggs (Einum and Fleming 1999). Growth rate has implications for later body size and is thus considered to be closely related to fitness in organisms with indeterminate growth. Egg size was manipulated as described in Section 2.1, and siblings from small and large eggs were reared together in seminatural enclosures, allowing competitive interactions. This design addressed two issues not considered in Hutchings' (1991) study: (1) it controlled for genetic correlations between egg size and other traits by comparing siblings from different-sized eggs, and (2) it included effects of competitive asymmetries (i.e., juveniles from small eggs competed with those from large eggs), which should be common in the wild. In this experiment, siblings from large eggs outgrew those from small eggs, but as predicted, the magnitude of this effect depended on environment quality, with the relative advantage of large juveniles decreasing with increasing environment quality (Figure 4.5). These results argue

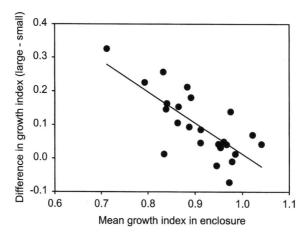

Figure 4.5. The relationship between mean growth index (growth rates relative to maximum growth for given size and temperature) of an enclosure, and the difference in growth index between sibling brown trout originating from large and small eggs and reared in competition ($y = 0.94 - 0.92x$, $r^2 = 0.56$). Each data point represents the mean value for one enclosure. The effect of egg size on relative growth decreases as environment quality increases. Modified from Einum and Fleming (1999).

that the optimal egg size will be smaller where the environment experienced by juveniles is of higher quality.

These two studies show that the egg size/offspring fitness function depends on environment quality. Thus, variation among populations in environment quality may promote divergent local adaptation. Is there any empirical evidence for such adaptation? Generally, demonstrating local adaptation is a laborious task (Rose and Lauder 1996), and no explicit tests of the above hypothesis have been performed for salmonids. However, correlative studies lend circumstantial support. Fleming and Gross (1990) found that among 17 populations of coho salmon, egg number increased and egg size decreased with increasing latitude. They suggested that this pattern might be explained by the existence of concurrent clines in competition for prey or size-selective predation (if competition or size-selective predation decrease with latitude). Alternatively, optimal egg size may be small at low temperatures (higher latitudes) because conversion efficiency (i.e., offspring mass/egg mass ratio) increases as incubation temperatures decrease (Heming 1982). In other words, the egg size required to produce a given optimal offspring size (i.e., maximizing maternal fitness) is smaller at lower temperatures.

Other sources of support for dependence of the egg size/offspring fitness function on environment quality come from within-population studies of Atlantic salmon (N. Jonsson et al. 1996), brown trout (Lobón-Cerviá et al. 1997) and white-spotted charr (Morita et al. 1999). In these studies, females that experienced high growth rates as juveniles produced smaller eggs as adults. This might represent an adaptive phenotypically plastic response to early growth conditions. Assuming that mothers can use the growth they experience as juveniles as a clue to the conditions their offspring will experience, they might adjust egg size according to the expected optimal value in that environment. Thus, females that experience higher growth as juveniles may benefit from producing larger numbers of small eggs. Similar correlations between growth rate and egg size have been found among populations of Japanese masu salmon (Tamate and Maekawa 2000).

Finally, two other recent studies are relevant but treated here in brief because they were published after the completion of this chapter. First, Heath et al. (2003) showed that smaller egg sizes evolve in hatchery salmon because the relationship between egg size and offspring survival is weaker than in the wild (but see Fleming et al. 2003 for an alternative explanation). Second, Koops et al. (2003) argued that more variable environments and smaller eggs are associated with increased variability in egg size.

2.3. Opposing Selection Stages: Pre-Hatching Mortality May Select for Small Eggs

The evolution of egg size is commonly thought to be driven by variation in post-hatching fitness. However, egg mortality may be caused by many factors (disease, predation, physical disturbance, temperature, dissolved oxygen, gravel size) and can be very high (Chapman 1988). For example, Bradford (1995) found that the average survival-to-emergence of pink salmon, chum salmon, and sockeye salmon was only 7%. Quinn et al. (1995) tested the hypothesis that selection during incubation may influence egg size by examining variation among 18 populations of sockeye salmon in two Alaskan lake systems. The populations all experience similar oceanic conditions, migrate at similar times and for similar distances, and have the same rearing environment for their offspring (large oligotrophic lakes). Thus, variation in egg size among populations was most likely attributable to variation in the incubation environment. One feature of the incubation environment that varies among these populations is the size of the spawning and incubation gravel (geometric mean gravel size varies 30-fold). Quinn et al. (1995) compared the average egg size of the populations to the average gravel size of their spawning sites.

As predicted, mean egg sizes, and egg sizes adjusted to a common female length, were positively correlated with geometric mean gravel size (Figure 4.6), and with other measures of gravel size (Quinn et al. 1995). It thus appears that pre-hatching mortality in sockeye salmon influences the evolution of egg size, and does so most strongly in the smallest gravels. One hypothesis for the positive correlation between egg size and gravel size was that smaller gravels reduce oxygen availability, which then selects against large eggs because they are less able to obtain the oxygen they need (Quinn et al. 1995). As noted above, however, a recent experiment suggested that large eggs are actually favored when challenged with low dissolved oxygen (Einum et al. 2002). Thus, a positive relationship between egg size and gravel size may arise for some other reason. Alternatives suggested by Quinn et al. (1995) are (1) size-selective predation on

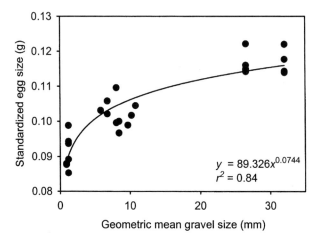

Figure 4.6. The relationship between standardized egg size (wet mass, adjusted to a common female length of 450 mm) and the size of incubation gravels for Alaskan populations of sockeye salmon. Multiple points for a given gravel size represent samples from the same population in different years. Data from Quinn et al. (1995).

eggs by gape-limited fishes (particularly sculpins) may favor larger eggs in larger gravels, where sculpins can gain access to incubating eggs, and (2) larger juveniles may become trapped when trying to emerge through smaller gravels.

2.4. Phenotype-Specific Optimal Egg Size

As we saw in Section 1.4, several existing models predict a correlation between optimal egg size and female phenotype within populations. Hendry et al. (2001b) suggested that their phenotype/habitat matching model was particularly well suited for salmonids, where large females are better competitors and obtain preferred nesting sites of higher quality (van den Berghe and Gross 1989; Foote 1990; Fleming and Gross 1994). According to their model, this should select for (1) a positive correlation between female size and egg size, (2) a greater proportional increase in egg number than in egg size with an increase in female size, and (3) a positive correlation between relative (i.e., adjusted to a common female size) gonad investment and relative egg number but not relative egg size. Hendry et al. (2001b) tested these predictions using empirical data from 101 populations of five species (43 populations of four species had data for both egg size and number). First, egg size was significantly positively correlated with body length within 47 of 61 populations having egg size data, supporting prediction 1. Second, the proportional increase in egg number was greater than the proportional increase in egg size within 37 of 43 populations having both types of data (Table 4.1), supporting prediction 2. Third, females with greater relative investment produced more but not larger eggs within all three collections having appropriate data, supporting prediction 3.

Table 4.1. Relationships between female length and egg size (mass) and egg number for studies of salmonid fishes.

Species	Egg size vs. body length				Egg number vs. body length				Paired t	
	Pops (fish)	Slope (S.D.)	r^2 (S.D.)	Pops sign.	Pops (fish)	Slope (S.D.)	r^2 (S.D.)	Pops sign.	Paired N ($S < N$)	t (P)
Sockeye salmon	14 (1755)	1.25 (0.18)	0.19 (0.15)	4	14 (4731)	1.86 (0.31)	0.31 (0.14)	14	14 (11)	2.64 (0.01)
Sockeye salmon	8 (202)	1.05 (0.31)	0.39 (0.16)	8	10 (160)	2.03 (0.30)	0.61 (0.16)	10	8 (8)	7.63 (<0.001)
Masu salmon	7 (153)	1.01 (0.93)	0.27 (0.30)	3	7 (149)	1.94 (0.90)	0.35 (0.23)	5	7 (5)	1.78 (0.04)
Coho salmon	22 (495)	1.15 (0.66)	0.32 (0.20)	2	17 (330)	2.28 (0.51)	0.68 (0.18)	16	14 (13)	5.75 (<0.001)

Columns indicate the number of discrete populations (Pops), with the total number of fish in parentheses, the slope coefficient averaged across populations (Slope), r^2 averaged across populations, and the number of populations with significant relationships (Pops sign.). The second last column gives the number of populations for which both egg size and egg number slopes were available (Paired N), with the number of populations in which the egg number slope was greater in parentheses. The last column provides the t statistic for paired sample t-tests of the difference in slope between egg size and egg number, both with respect to body length. Data from Hendry et al. (2001b).

How confident can we be that the proposed phenotype/habitat matching model is responsible for the observed correlation between female size and egg size? Both the sibling competition model (Parker and Begon 1986) and the predation model (McGinley 1989) are in some respects similar in their predictions. Both predict an increase in egg size with female size, and unless female size influences the egg size/offspring fitness function too strongly, they also predict that the proportional increase should be greater for egg number than for egg size. In one respect, however, these models differ from that of Hendry et al. (2001b); whereas the phenotype/habitat matching model assumes the effect of female body size on offspring fitness is caused by maternal behavior, the sibling competition and predation models assume an effect of egg number per se on offspring fitness. As a result, the sibling competition and predation models predict a positive correlation between relative gonad investment and relative egg size. That is, if a female produces large gonads relative to her body size, she should also produce large eggs relative to her body size. This prediction contrasts with the phenotype/habitat matching model, which predicts no such correlation, and conflicts with the available empirical data (Hendry et al. 2001b). The observed relationship between female size and egg size in salmonids therefore seems to fit the model of Hendry et al. (2001b).

However, the above results may be a case of getting the right answer for the wrong reason. The recent experimental evidence suggesting that large eggs actually have an advantage in low dissolved oxygen (Einum et al. 2002) counters one of the assumptions of Hendry et al. (2001b), and suggests that positive egg size/ female size correlations may arise for another reason. For example, larger females may actually provide incubation environments of lower quality because they lay larger clutches, which have greater total oxygen demand. Under these conditions, larger females should produce larger eggs if larger eggs are better able to cope with lower dissolved oxygen. Indeed, Hendry and Day (2003) have shown that this prediction follows when the results of Einum et al. (2002) are incorporated into the theoretical modeling approach of Hendry et al. (2001b). Furthermore, Einum et al. (2002) have shown that larger females should be favored to produce larger eggs because doing so reduces the oxygen demand per unit of clutch mass.

2.5. Non-Optimal States: Micro-Evolution in New Zealand Chinook Salmon

Trait values of new or disturbed populations will often not be at their equilibrium or optimal values. The rate at which traits in the population will then approach the new equilibrium (and potentially optimal) values will depend on the nature of selection, genetic variation, and gene flow. The time required to reach approximate equilibrium values for traits such as egg size and number is not known but significant deviations from equilibrium conditions, and hence strong selection, are likely for the newest of populations. For this reason, historical and experimental introductions provide useful systems for examining rates and patterns of evolution on contemporary time frames (reviews: Hendry and

Kinnison 1999; Kinnison and Hendry 2001; Reznick and Ghalambor 2001; Stockwell et al. 2003).

Chinook salmon introduced to New Zealand have recently been the subject of intensive research (e.g., Kinnison et al. 1998a, 2001; Quinn et al. 2001b; Unwin et al. 2003). All of the present day populations are descended from upper Sacramento River (California, USA) "fall-run" fish introduced into a single river system on the South Island of New Zealand between 1901 and 1907 (McDowall 1994). Thus, any genetic differences existing among the extant New Zealand populations have evolved on a time scale of around 30 or fewer generations. Kinnison et al. (2001) examined how the cost of freshwater migration by adults might influence the evolution of egg size and number in these new populations. In particular, they first predicted that an increasing cost of migration would result in a proximate cost to tissue energy reserves and gonad investment. To test this prediction, they released smolts from the same families from two sites with different migratory rigor, and then measured ovarian traits of maturing females returning to each site.

The experimental manipulation of migratory rigor (100 km and 430 m elevation vs. 17 km and 17 m elevation) resulted in a 17% reduction in metabolizable muscle mass and a 14% reduction in mean egg size and total ovarian mass (Kinnison et al. 2001; Figure 4.7). Altered costs of migration are thus one cause of immediate changes in ovarian investment (particularly egg size). The interesting question then becomes: are these changes non-equilibrium and non-optimal, and how do they influence the evolution of egg size? A comparison of egg size, number, and total ovarian mass for two New Zealand populations with different migratory rigor (100 km and 430 m elevation, vs. 60 km and 200 m elevation) confirmed the above proximate cost of migration: wild-caught females

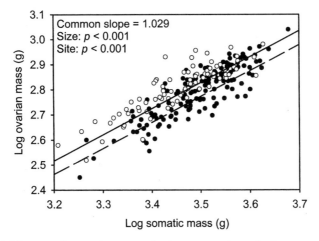

Figure 4.7. The cost of migration on ovarian investment shown by log ovary mass relative to log somatic mass for 3-year-old female chinook salmon from the same families returning after a long freshwater migration (filled circles—100 km, 430 m elevation) or a short freshwater migration (open circles—17 km, 17 m elevation). Dashed and solid lines represent the predicted ANCOVA relationships.

had a lower ovarian mass and smaller eggs in the longer migrating population (Kinnison et al. 1998a). When these same two populations were then reared under common-garden conditions, however, the farther migrating population had a 6.4% larger ovarian mass, suggesting evolutionary compensation for the proximate cost of their longer migration (Kinnison et al. 2001). Interestingly, the compensation came largely in the form of increased egg number rather than egg size.

Kinnison et al. (2001) suggested several possible reasons for why egg number evolved more quickly than egg size, despite the obvious proximate cost of migration on egg size and not egg number. First, evolutionary constraints, particularly those resulting from genetic correlations, may be important. In general, ovary mass shows higher genetic correlations with egg number than with egg size (Su et al. 1997; Kinnison et al. 2001), suggesting that an evolutionary increase in ovary mass should be accompanied by an initial increase in egg number rather than egg size. This effect would suggest that the smaller egg size in the longer migrating population is actually suboptimal, and that with enough time, it might increase to some new and larger equilibrium value. Second, optimal egg size may actually be smaller in longer migrating populations. For example, migration may diminish energy reserves available to females for the production of high-quality nest sites. If so, the model of Hendry et al. (2001b) (Section 1.3) would predict a decrease in optimal egg size as migratory costs increase. Another possibility is that there may be consistent correlations between distance from the ocean and environment variables that influence optimal egg size (Section 1.2).

Kinnison et al. (2001) also examined egg size and number variation among North American populations of Pacific salmon. They obtained estimates of mean egg size and number (most corrected for female size) for five geographic groups of populations in three species. Each group consisted of between four and 14 populations that enter fresh water at essentially the same location but that migrate different distances. For all five groups, the ratio of egg number to egg size increased with increasing migratory distance, driven by a strong tendency for egg size to decrease and egg number to remain relatively constant or increase (Figure 4.8). Other authors have also found that egg size decreases with increasing migratory difficulty (Beacham and Murray 1993; Healey 2001). This pattern parallels the costs of migration and evolutionary compensation for those costs in New Zealand. However, total egg production (ovary mass) tends to remain negatively correlated with migration distance (see also Fleming and Gross 1989; Beacham and Murray 1993; Healey 2001), suggesting that full compensation for migratory costs does not occur.

It remains unknown whether the observed patterns among indigenous populations reflect optimality or constraints. If they reflect constraints, then optimality may be very slow to evolve or may never be attained. If smaller egg size with more difficult migrations is actually adaptive, attention should be focused on determining how factors related to migration alter fitness functions. The quantitative precision of optimality theory in explaining the range of egg size variation among indigenous salmon populations awaits further empirical confirmation. Ultimately, deviations from optimality predictions may serve as a basis

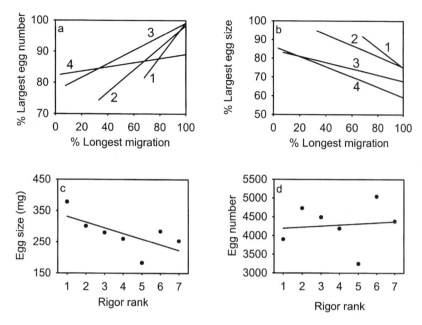

Figure 4.8. Relationships between migratory distance (rigor) and reproductive investment (egg size and egg number) for chinook, sockeye, and coho salmon from five geographic population groups. For (a) and (b), graphs represent regression lines through population values in which each population's mean egg size, mean egg number, and migration distance is converted to a percentage of the largest population mean value for that group. Trait values are standardized to a common body size in each population grouping. (1) Fraser River coho (a: $r = 0.68$, b: $r = -0.75$), (2) Puget Sound coho (a: $r = 0.71$, b: $r = -0.89$), (3) Fraser River sockeye (a: $r = 0.27$, b: $r = -0.82$) and (4) Fraser River chinook salmon (a: $r = 0.51$, b: $r = -0.34$). For (c) ($r_s = -0.71$) and (d) ($r_s = 0.14$), regressions for Columbia River chinook salmon are presented (with individual population points) relative to an *a priori* ranking of migratory rigor that takes into account migration distance and seasonality effects on time spent in the river. Data from Kinnison et al. (2001).

for determining when more complex evolutionary interactions are at play. Salmon adapting to hatchery environments provide another opportunity to examine what happens when selection on egg size or number is changed. Heath et al. (2003) showed that egg size decreased rapidly in a hatchery environment, presumably because selection for larger eggs was weaker than in the wild but selection for more eggs was not (but see Fleming et al. 2003).

3. How Do We Proceed from Micro- to Macro-Evolutionary Theory?

The Smith–Fretwell model and its extensions have been important in shaping our understanding of variation in egg size and number within and among populations. Some effort has also been devoted to understanding variation among higher phylogenetic groups (e.g., Ware 1975; Kölding and Fenchel 1981; Hutchings and Morris 1985; Sargent et al. 1987; Visman et al. 1996; Fleming 1998). This endeavor is particularly challenging because the life history of an organism (used here in the broad sense to include behavior, morphology, and other aspects of biology) likely influences the relationship between egg size and fitness. In this section, we posit factors that may be responsible for observed variation in egg size among species, pointing out aspects that may be particularly important in salmonids. This is not meant to be an exhaustive consideration, but rather a summary of several factors that may explain some of the variation. Our starting point is a consideration of constraints on egg size evolution. We then move to factors that might influence egg size evolution according to the Smith–Fretwell framework (including effects of egg size on survival before and after hatching). Finally, we consider how other life history traits may evolve to maximize fitness within the limits set by various constraints on egg size.

3.1. Mechanical Constraints

Adult body size will ultimately constrain egg size. In its simplest form, this constraint sets an upper limit on egg size because a female only has a given amount of resources (and space in her body cavity) available for egg production Thus, even if all resources are allocated to one offspring, it can still only be of a certain size. This inevitability was taken to absurdities in a tongue-in-cheek parody of the so-called adaptationist program: "Why are juveniles smaller than their parents?" (Ellstrand 1983). Such constraints would only be important for organisms producing small clutch sizes, and thus not salmonids. Additional constraints may arise because of design limits on the structures used for oviposition or parturition. For example, in most tetrapods, eggs or neonates must pass through the pelvic girdle, which faces competing selection for locomotory function. Similarly, many insects employ specialized ovipositors with design constraints associated with the egg-laying substrate. In theory, such constraints could also apply to species laying multiple eggs during a given reproductive event.

In salmonids and most other fishes, bony girdles do not completely enclose the oviposition canal and such constraints thus seem unlikely. This conclusion is supported by the observation that although non-anadromous salmonids are much smaller than their anadromous counterparts, their egg sizes are fairly similar (Hendry et al. 2003b—*this volume*). For example, although mature non-anadromous O. *nerka* (kokanee) may be one-third the length of anadromous O. *nerka*, their egg diameters differ by only about 4% (Wood and Foote 1996). If any mechanical constraints actually exist in female salmon they would most likely relate to overall ovary size. For example, larger ovaries will compete with organs for space in the body cavity and may reduce swimming ability and predator avoidance (see Miles et al. 2000 for locomotor constraints in lizards).

Sinervo and Licht (1991) argued that physical constraints exist in lizards based on experiments where they enlarged eggs by removing parts of the clutch during vitellogenesis, and found that such females often could not successfully oviposit. It is also interesting, however, that the optimal egg size expected in the absence of failed oviposition seemed to match the mean egg size in wild populations (Sinervo et al. 1992), suggesting that there is little if any post-oviposition selection for larger eggs. Thus, the question of whether egg size really is constrained by design limits on structures used for oviposition or parturition remains unanswered.

3.2. Genetic Constraints

Optimality theories assume that genetic architectures allow the traits under consideration to evolve to a value maximizing maternal fitness. Under certain circumstances this may not be realistic. The most obvious genetic constraint is a paucity of additive genetic variation, which would reduce evolutionary responses to phenotypic selection. However, heritabilities and additive genetic variance for life history traits in animals are generally extensive (Mousseau and Roff 1987; Houle 1992). Estimates of the heritability of egg size in salmonids vary widely but are often quite high (e.g., 0.60, Su et al. 1997; 0.78, Kinnison et al. 2001), suggesting that genetic variation will not appreciably limit the evolution of egg size.

A second genetic constraint is the presence of genetic correlations. Such correlations can be caused by linkage disequilibrium (i.e., genes determining two traits are closely located on the chromosome and certain allele combinations are more common than others), in which case they will likely be transient, or by pleiotropy (i.e., two traits are determined by the same genes), in which case they may be permanent. In this case, selection on one trait causes simultaneous changes in another trait, and they cannot evolve independently to their individual optima.

Given these potential complications, are the optimality approaches in this chapter likely to be fair approximations of what is really occurring in the wild? Ultimately, evolutionary constraints on egg size can best be demonstrated by showing that observed mean egg sizes do not maximize maternal fitness. However, as we discuss further at the end of this chapter, such demonstrations may be difficult to obtain if evolutionary changes in other traits "adapt" the organism to its constrained egg size, in which case the observed egg size may be optimal even where significant constraints are present.

3.3. Adaptations to Pre- and Post-Hatching Environments

Selective pressures potentially imposed by the pre-hatching environment were initially discussed in Sections 1.3 and 1.4. In those sections, we focused on the model of Hendry et al. (2001b) concerning oxygen limitations faced by eggs, and the relationship between sockeye salmon egg size and spawning gravel size (Quinn et al. 1995). Hendry et al. (2001b) also attempted to explain variation in the strength of relationships between female size and egg size among species. Using data from 181 freshwater fish species, they found that positive correlations between female size and egg size were more common in fish where female size was likely to influence the quality of the incubation environment (e.g., females construct nests). They argued that in these species, larger females could provide better oxygen conditions for their eggs, allowing the evolution of larger egg size. Again, however, this analysis rested on the assumption that large eggs suffer disproportionately in low dissolved oxygen, and subsequent work has found that, the true effect may actually be the opposite (Einum et al. 2002).

Thus, we may need to seek other explanations for this variation. In a comparative study of 119 fish species, Einum and Fleming (2002) found that intra-population variation in egg size was most pronounced in fish with demersal eggs and larvae, where the offspring environment is likely influenced by maternal phenotype, and least so in fish with pelagic eggs, which experience a relatively stochastic spatial distribution during incubation. These results indicate that there is selection favoring within-population variation in egg size in species where maternal phenotypes can influence the offspring environment. Furthermore, fish with demersal eggs and pelagic larvae did not differ from those with pelagic eggs, indicating that this selection may occur mainly post-hatching. However, at least for salmonids, oxygen limitations during incubation have the potential to select for variation within populations, although for a different reason than previously thought (Einum et al. 2002; Section 2.4). Pre-hatching environments may also pose many other forms of selection on egg size. For example, small eggs may be *favored* in some organisms to promote passive dispersal or reduce predation (e.g., Hammond and Brown 1995; Moegenburg 1996). Selection during this stage may therefore shift optimal egg size considerably from that expected based only on the post-hatching period.

Variation in post-hatching fitness functions are probably also important and here we list five potentially important factors. First, the abundance and size distribution of food may influence how starvation rates are related to egg size. If food abundance is low, it has been hypothesized that selection will favor large, developmentally advanced young (Itô 1980; Itô and Iwasa 1981). Some evidence from experimental work within salmonid populations supports this idea (Hutchings 1991; Einum and Fleming 1999). The size distribution of prey may also influence starvation rates relative to egg size. If small eggs produce larvae that are too small to exploit the available prey items, this may set a lower limit on egg size. If we assume that the types of food possibly exploited by offspring (e.g., carnivorous versus herbivorous diet) are under stronger phylogenetic constraints than egg size, organisms must adapt their egg size, feeding structures, or feeding style, to the available prey size and numbers.

Second, from a game theory perspective (Maynard Smith 1982), interference competition may select for large and developmentally advanced offspring if the outcome of competition is phenotype-dependent. The outcome of agonistic interactions often depends on body size (e.g., Johnsson et al. 1999; Cutts et al. 1999b), and in certain biological systems this may select for large eggs (e.g., species with territorial juveniles).

Third, if predation rates are size-dependent, this may shape the egg size/ offspring fitness function. Such size-selective predation appears common in many systems (e.g., Connell 1970; Caldwell et al. 1980; Palmer 1990; Moegenburg 1996). For a given intensity of size-selectivity, the potential for effects of predation on the evolution of egg size depends on the level of predation that juveniles commonly experience, which may vary among taxa.

Fourth, if the number of offspring strongly influences offspring fitness through effects on competition among siblings or predation (density dependence), this may favor larger eggs (Section 1.4, Parker and Begon 1986; McGinley 1989). We suggest that this effect may be particularly evident when comparing (1) organisms with different levels of dispersal during egg and larval stages, and (2) organisms with and without parental care. In the first scenario, organisms with high degrees of dispersal will have little opportunity for intra-clutch interactions in comparison with organisms where siblings show less dispersal. In the second comparison, particularly if parental care must be divided among offspring (e.g., feeding), producing too many eggs will be disadvantageous, and more resources should go into each egg. Evidence for such patterns is given by Sargent et al. (1987), who showed that fishes with parental care produce larger eggs (although their explanation was related to development time and egg mortality; see Section 1.4).

Finally, abiotic environment conditions may influence the egg size–offspring fitness function. For example, seasonal migrations may set a constraint on the minimum offspring size if juveniles must reach a certain size at a given time to be able to complete such migrations. Hydrodynamic stress may also select for large offspring to avoid dispersal if large juveniles are better able to withstand such stress. For example, juveniles from species of darters (Percidae) hatching from larger eggs drift in the water column less than those hatching from smaller eggs (Paine 1984). This may be the reason for the pattern observed in stream-living darters, where a high degree of gene flow among populations is associated with high egg numbers and small egg sizes (Turner and Trexler 1998).

3.4. Predictions and Perspectives

One may wish to speculate about causes of variation in egg size among salmonid species. In particular, the different North American Pacific salmon species provide a tempting target. These are similar in many aspects of their biology, such as patterns of adult migration, semelparity, spawning season, nest defense after spawning, and geographic distribution. Yet, they vary considerably in egg size and number (Table 4.2), and in some factors that may influence their evolution. Here we make some suggestions as to what selective factors may have played a role in this variation, if only to illustrate the complexity associated with such an endeavor.

As discussed in Sections 1.3 and 1.4, female size may have implications for the quality of the incubation environment and the intensity of sibling interactions. Thus, it is not surprising that chinook and chum salmon, being generally larger than the other species, also produce larger eggs (see also Fleming 1998). However, this can not be the whole story, because whereas sockeye salmon are larger than pink salmon, they produce substantially smaller eggs (Table 4.2). We

Table 4.2. Egg size (wet mass, means and ranges of population means), egg number (mean of population means), and general behavioral and life history traits of North American Pacific salmon.

Salmon species	Mean female body size (mm)	Post-emergent behavior	Egg size (mg)	Egg number
Sockeye	530–690	Lake rearing, shoaling	105 (69–132)	3500
Pink	460–560	Downstream migration to estuaries	162 (136–197)	1800
Coho	590–700	River rearing, initial aggregations, later territorial	188 (91–297)	3000
Chum	550–780	Downstream migration to estuaries	239 (164–282)	3200
Chinook	480–1030	Downstream migration to estuaries (ocean type) or river residence (stream type)	251 (138–378)	4300

Data from Groot and Margolis (1991) and Fleming (1998). Exceptions can be found to each of the generalizations about post-emergent behavior.

suggest two possible explanations for this pattern. First, whereas juvenile pink salmon disperse to estuaries at emergence, juvenile sockeye salmon rear in lakes. Levels of size-selective predation may thus be higher in pink salmon (although this has not been tested). Second, whereas pink salmon fry experience strong osmoregulatory stress as they reach the highly saline marine environment, sockeye salmon fry that rear in lakes do not. Effects of such stress are thought to depend on juvenile size because larger size generally confers better osmoregulatory ability (McCormick and Saunders 1987). These factors may be among those that place a premium on large eggs in pink salmon, relative to sockeye salmon.

It is also interesting that coho salmon produce much larger eggs than sockeye salmon despite their similar body sizes. Selection for large eggs in coho salmon may come about through effects on competitive abilities. Whereas juvenile sockeye salmon shoal in lakes, coho salmon are territorial in streams during their juvenile stage (Groot and Margolis 1991), and this potentially selects for large juvenile size.

The consideration of any one factor influencing egg size/offspring fitness relationships in isolation is clearly not sufficient to explain the observed variation in egg size among Pacific salmon. Abiotic, physiological, and ecological factors may all be necessary parts of any conceptual model predicting variation among higher phylogenetic groups. Although the number of factors influencing egg size evolution (including any constraints) and interactions among these factors makes such a research project immensely challenging, it also makes it interesting.

Considerations of the evolution of optimal egg size tend to be biased toward an expectation that egg size evolves to match other aspects of an organism's life history. An alternative view is that many features of an organism's life history actually evolve to suit a particular egg size that is otherwise determined by constraints. Perhaps the size of lizard eggs *was* constrained by pelvic girdle size during earlier evolutionary stages (Section 3.1), but other aspects of their life history have now evolved to suit that egg size. Although there is no doubt that egg size has evolved to different evolutionary stable states within taxonomic groups (e.g., Pacific salmon, see Table 4.2), simultaneous evolution in other traits has occurred. Focusing on the evolution of egg size alone biases us toward considering only part of what "optimality" may entail. In fact, instead of asking "which came first, the chicken or the egg?" (see Shykoff and Widmer 1998 for comparative method solution) perhaps we should be asking "which evolved most, the chicken or the egg?" Carefully designed experiments may be of aid in the search for possible evolutionary changes that could create coadaptations with egg size. Furthermore, comparative studies using a phylogenetic approach that simultaneously consider evolutionary changes in multiple traits (e.g., Crespi and Teo 2002) could possibly identify instances where constraints on egg size appear to have governed the evolution of other traits.

Norms of Reaction and Phenotypic Plasticity in Salmonid Life Histories

Jeffrey A. Hutchings

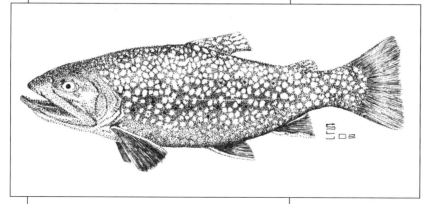

Mature male brook trout

5

Theory predicts that adaptive plasticity in life history strategies should be favored in organisms with widely dispersed offspring because of the increased likelihood of encountering spatial and temporal environmental heterogeneity. Phenotypic plasticity is the ability of a genotype to produce different phenotypes across an environmental gradient (Bradshaw 1965; Schlichting and Pigliucci 1998; Debat and David 2001). Plasticity does not represent genetic change, although the form of change (the trait may increase in value, decrease, or remain the same) may be a product of selection. Phenotypic plasticity can be heuristically and graphically described as a norm of reaction, a linear or nonlinear function that expresses how the phenotypic value of a trait for a given genotype changes with the environment (Schmalhausen 1949; Schlichting and Pigliucci 1998).

The primary purposes of this chapter are to explore the adaptive basis of plastic changes in salmonid life history traits, to examine how norms of reaction have been used to explain the persistence of life history variation among individuals within populations, and to hypothesize how selection, including that induced by anthropogenic activities such as fishing, might change the shapes of reaction norms, thereby influencing how individuals within populations respond to environmental change. Within the context of evolutionary biology, one of the themes that emerges from the study of reaction norms in salmonids is the need to determine how fitness changes as genotypes alter their phenotype along different reaction norms, a subject that has received little attention to date.

Life history strategies describe how genotypes vary their stage- or age-specific expenditures of reproductive effort in response to intrinsic and extrinsic factors that influence survival and fecundity (Schaffer 2003—*this volume*). As such, life histories reflect the expression of traits most closely related to fitness, such as age and size at maturity, offspring size and number, and longevity, as well as the timing of the expression of those traits throughout an individual's life. Implicit within any life history theory prediction is the assumption that natural selection favors those genotypes whose age(stage)-specific schedules of survival and fecundity generate the highest per capita rate of increase, that is, fitness, relative to other genotypes in the same population (Stearns 1992; Roff 2002). This genotypic rate of increase can be expressed as r, the intrinsic rate of natural increase, or as R_0, the net reproductive rate. When expressed at the individual level, r and R_0 represent fitness. When expressed at the population level, they represent rates of population growth, parameters that determine extinction probability, recovery rate, and sustainable levels of exploitation. From an evolutionary perspective, it is reasonable to focus on life history traits because of the direct link that exists between these characters and fitness.

When life history trait optima differ among environments inhabited largely at random with respect to genotype within and among generations, selection can be expected to act on the way in which a genotype alters its life history in response to environmental change. That is, selection will act on a genotype's norm of reaction (Schmalhausen 1949; Via and Lande 1985; Schlichting and Pigliucci 1998). Such adaptive phenotypic plasticity may underlie many life history responses by salmonid fish to environmental change, notably to non-genetic variation in individual growth rate, but also to differences in temperature, habitat quality, and food supply. Random distribution of salmonid genotypes among spatially heterogeneous environments may be most likely in early life, almost immediately after individuals begin to feed exogenously. Upon emergence from gravel egg nests, many salmonids are transported to areas of slow-moving water where they spend periods of time ranging from days to months of their first growing season (Northcote 1984; Hutchings 1996; Marschall et al. 1998). It is likely that individuals utilize these habitats at random with respect to genotype because of (1) the inability of recently emerged juveniles to swim against all but the weakest of water currents; (2) spatial and temporal variation in hydrography and food supply; and (3) significant discordance in emergence times among individuals from the same brood (e.g., Field-Dodgson 1988; Snucins et al. 1992).

Despite the voluminous life history literature on salmonids (Groot and Margolis 1991; Fleming 1996; Hutchings and Jones 1998; Marschall et al. 1998), only a handful of studies have focused on reaction norms for life history traits. Based on data for unexploited populations of brook charr in southeastern Newfoundland, Hutchings (1993a,b, 1996, 1997) constructed norms of reaction for age and reproductive effort at maturity, providing a basis for predicting how such plasticity might change under exploitation. Experimental studies have revealed how reaction norms for juvenile growth rate in brown trout (Einum and Fleming 1999) and juvenile survival in brook charr (Hutchings 1991) can be influenced by egg size, and there is corresponding evidence that growth rate in early life can negatively (N. Jonsson et al. 1996; Morita et al. 1999) and positively (Morita et al. 1999) influence egg size as an adult in Atlantic salmon and white-spotted charr, respectively. By modeling reaction norms as threshold traits, researchers have been able to account for the role of both environmental and genetic influences on age at maturity when explaining the maintenance of conditional alternative mating strategies in Atlantic salmon (Hutchings and Myers 1994) and coho salmon (Hazel et al. 1990), and within-population differences in time spent at sea prior to reproduction (Hutchings and Jones 1998). The most recent work on life history reaction norms in salmonids, described in a series of papers on European grayling, has provided some of the most compelling evidence to date for the hypothesis that population differences in reaction norms represent adaptive responses to local environments (Haugen 2000a,b,c; Haugen and Vøllestad 2000).

1. Theory: Reaction Norms and Adaptive Phenotypic Plasticity

1.1. Genotype × Environment Interactions

Reaction norms describe how individuals respond to environmental change. Strictly speaking, within an evolutionary context, norms of reaction pertain to responses by genotypes, the units of study when Woltereck (1909) first coined the term *Reacktionsnorm* in his work on *Daphnia*. Genotypic studies of reaction norms are not uncommon in plants, clonal organisms, and *Drosophila* (see Schlichting and Pigliucci (1998) for examples) because of the relative ease with which single genotypes can be generated and their responses to environmental change documented. However, for most sexually reproducing animals, the family level is the lowest level at which reaction norms can be studied.

Reaction norms inform us about the magnitude of trait plasticity, the presence of genotype × environment (G×E) interactions on the phenotypic expression of a given trait, and the extent to which the additive genetic variance (V_A) of a trait changes with the environment. Norms of reaction that run parallel to the environmental gradient axis reflect an absence of plasticity because the trait does not change its phenotypic value with changes in the environment (Figure 5.1A). Plasticity is reflected by reaction norms with non-zero slopes. In Figure 5.1B, the reaction norms have similar slopes, implying that the pattern of phenotypic response to environmental change is the same among genotypes, that is, there is no G×E interaction. In contrast, crossing reaction norms, indicative of G×E interaction, suggest the presence of genetic variation in plasticity (Figure 5.1C). If that genetic variation is additive, then selection can produce evolutionary changes to the shapes of reaction norms, thus making their study fundamentally important to evolutionary biology.

Genotype × environment interactions can affect the degree to which alleles are subject to selection. For the reaction norms in Figure 5.1C, the differences in phenotype among genotypes are greatest at the environmental extremes; if that genetic variation is additive, one can conclude that the V_A of the trait in question will be highest at the extremes of the environmental gradient. By contrast, V_A will be much lower near the center of the environmental gradient, where the reaction norms converge, because of the similarity in phenotypic values among the different genotypes. Thus, the additive genetic variation of the trait in question, and the rate at which it will respond to selection, will vary with the environment.

1.2. Inferring Individual Responses to Environmental Change

Several approaches can be used to infer the shapes of reaction norms. In a broad sense, this is a hierarchical process of obtaining data on how the phenotypic values of life history traits change with the environment. At the highest level, one focuses on population means of a given life history trait, such as age at maturity, that are plotted against an environmental variable, such as temperature. The line best fitting these data (a "species-level" reaction norm) has frequently been used to predict how individuals within a population might respond to environmental change. Examples in the salmonid literature of such population-level variation for life history traits abound (Marschall et al. 1998), for instance, egg size and fecundity versus latitude (a proxy for environmental variation) in coho salmon (Fleming and Gross 1990); Atlantic salmon smolt age versus metrics of growth rate (Power 1981; Metcalfe and Thorpe 1990;

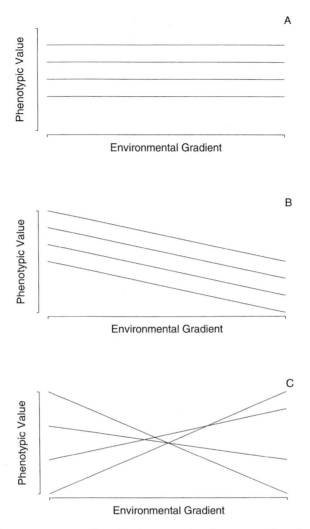

Figure 5.1. Reaction norms describe changes in the phenotypic value of a trait across an environmental gradient. Reaction norms can reflect an absence of plasticity (A), the presence of plasticity but an absence of genotype × environment (G × E) interactions (B), or the presence of both plasticity and G × E interactions (C).

Hutchings and Jones 1998); and incidence of the Atlantic salmon male parr maturity versus growth rate (Myers et al. 1986).

Population-level functions represent the second level of the hierarchical process of identifying reaction norms, wherein each graphical relationship could be described as a "population-level" norm of reaction (Figure 5.2A). This approach can be useful if the primary goal is to predict how individuals within a population will respond, on average, to specific changes in an environmental variable. But the use of population-level data to infer individual

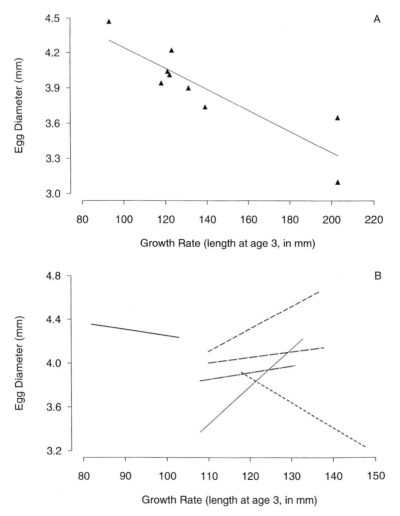

Figure 5.2. Bivariate plots of egg size against growth rate in brook charr reveal differences in the predicted associations between these traits when the correlations are calculated on the basis of population means within species (A) or among individuals within populations (B). The data presented in (A) are from Hutchings (1990); the data presented in (B) are from Hutchings (1990, 1996) and from Adams (1999).

responses to environmental change is subject to several statistical and biological caveats. For example, a plot of the average egg size against average individual growth rate (approximated by the mean length at age 3 years, a reasonable proxy for environmental variation) yields a significantly negative correlation among nine populations of brook charr ($r = -0.88$, $p = 0.0017$; Figure 5.2A; Hutchings 1990). However, when one plots the separate relationships between egg size and growth rate among individuals within populations, the correlation coefficients range from significantly negative to significantly positive (Figure

5.2B). The observation that populations respond differently to changes in the environment is consistent with the hypothesis that individuals within such populations are, to some degree, adapted to their local environments.

The study of co-gradient and counter-gradient variation provides an excellent empirical and theoretical framework for the study of population differences in the average response by individuals to their environment (reviewed by Conover and Schultz 1995). Co-gradient variation describes heritable traits that vary genetically and phenotypically in a manner predicted by the environment (e.g., the most rapid genetically based growth is found among those individuals inhabiting the environment most favorable to growth). In contrast, counter-gradient variation exists when genotypes are distributed such that genetic and environmental influences on a trait oppose one another (e.g., the population exhibiting the fastest genetically based growth inhabits the environment least favorable to growth). The observation that populations in apparently hostile environments (cold temperature, low food) express the most rapid rates of growth across a range in environmental variables provides the most widely cited example of counter-gradient variation in fish (Conover and Schultz 1995). In salmonids, Nicieza et al. (1994) found evidence of counter-gradient variation in Atlantic salmon digestion rate. Although support for counter-gradient variation in growth rate was equivocal in a recent study in the same species, B. Jonsson et al. (2001b) were able to document population-level norms of reaction for rates of growth and food consumption in Atlantic salmon.

However, if one wishes to estimate the genetic variation in reaction norms within a population, assess the degree to which reaction norms are under selection, or predict the rate at which the shapes of reaction norms might change in response to selection, a finer-scale approach needs to be adopted. This third level of the hierarchy involves the study of plasticity of individuals from the same family. This is usually the lowest level of analysis at which one can use empirical data to describe the shapes of reaction norms in salmonids (for an example in which intra-individual variation is examined in salmonids, see Einum and Fleming 1999).

Although data on family-level responses to environmental change are commonly gathered in aquaculture breeding programs, relatively few have been reported in the primary literature. Among those, Beacham and Murray (1985) describe interfamilial variation in phenotypic responses to environmental change for chum salmon from Nitinat River, British Columbia, Canada. Reaction norms between embryonic survival and incubation temperature reveal considerable genetic differences among families (Figure 5.3A). This $G \times E$ interaction is reflected by families for which survival during the egg stage increases, remains constant, or declines with increases in incubation temperature from 4 to 12°C. The most common norm of reaction is one in which offspring survival is highest at the intermediate temperature. If the genetic variation among families is primarily additive (and if other genetic sources of variation do not change with temperature), then one would conclude that V_A for egg survival is highest at high and low temperatures, and lowest at 8°C. By comparison, reaction norms between length at emergence and temperature exhibit comparatively little

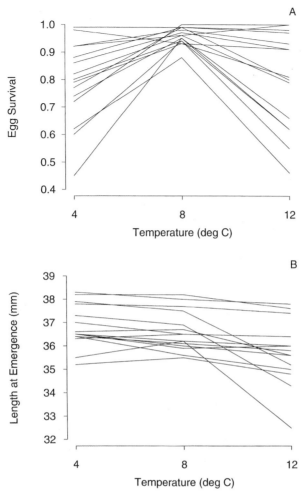

Figure 5.3. Differences in reaction norms for egg survival (A) and length at emergence (B), as functions of temperature, among families of chum salmon from the Nitinat River, British Columbia, Canada. Data were obtained from Beacham and Murray (1985).

genetic variation among families, as suggested by similarities in phenotypic response to environmental change (Figure 5.3B). This comparatively low level of G×E interaction occurs because reaction norms for alevin size fall primarily into only two types; size at emergence is either invariant with temperature or declines between 8° and 12°C.

On an explanatory note, attention should be drawn to the fact that studies of reaction norms in salmonid fish use growth rate as both a proxy for environmental change and as a trait in itself. When used in the former manner, growth rate is plotted on the abscissa and the trait whose reaction norm is being described is plotted on the ordinate. Although growth rate is generally a very good metric of environmental change in indeterminately growing organisms, one need always be cognizant of the implicit assumption that size-at-age is a reliable proxy for growth. Use of growth rate as a metric for environmental change is common in field-based research and in studies for which the precise source of environmental variation is unknown. In contrast, reaction norms for growth rate, such as those presented in Figure 5.3B, are generally restricted to laboratory work. Also, it is not uncommon to have reaction norms depicted as bivariate plots between two life history traits, such as age and size at maturity (e.g., Stearns and Koella 1986). Whether these can truly be called reaction norms is perhaps open to debate, given that environmental change is not made explicit in such plots. However, what such bivariate associations do illustrate is the fact that life history tradeoffs, reflected by negative genetic and phenotypic correlations between traits, can differ among individuals within the same population. In this sense, they are analogous to, if not the same as, norms of reaction.

It is also worth underscoring the point that reaction norms in themselves tell us little about the fitness associated with genotypic responses to environmental change. Temperature, for example, can affect growth rate, which can then affect age at maturity, which may be positively or negatively associated with size at maturity, all of which can have consequences to reproductive effort. Thus, although always presented in two dimensions, reaction norms may often be hierarchical or n-dimensional in nature, reflecting the general observation that neither traits nor their reaction norms are independent of one another. It is this interdependence, mediated in part through life history tradeoffs, that ultimately influences fitness.

2. Reaction Norms for Life History Traits in Salmonids

2.1. Continuous Norms of Reaction for Age at Maturity: Individual and Population Variability

The most widely reported association between a metric of environmental quality and a life history trait in fishes is that between individual growth rate and age at maturity. The observation that faster growers mature earlier in life has been documented repeatedly in the laboratory and in the field (Alm 1959; Hutchings 1993a; Roff 2002). However, it is not clear that this association is necessarily adaptive. Furthermore, changes in age at maturity probably have consequences to individual fitness and to population growth rate, both of which may influence the risk of extinction for the individual genotype and the population as a whole. Changes in r brought about by plastic changes in genotype could have important consequences to the conservation biology and sustainability of commercial and non-commercial species.

To evaluate the adaptive significance of phenotypic plasticity in age at maturity in brook charr, Hutchings (1996) examined how environmental variation in juvenile (pre-reproductive) growth rate can generate variation in fitness (r) across a range of potential ages at maturity. The empirical data were obtained from three populations located on Cape Race, southeastern Newfoundland, Canada (these populations have been described in detail elsewhere; Hutchings 1993a,b, 1994, 1997). In many respects, the Cape Race trout populations are ideally suited for ecological and evolutionary study. The populations are unexploited and have not been subject to anthropogenic disturbance. Brook charr are the only fish in the rivers and they are not subject to avian predation, negating the potentially confounding influences of interspecific competition and predation on life history. In addition, these non-anadromous populations, despite being in close geographic proximity to one another, are genetically distinct, with some allozyme loci having reached fixed differences (Ferguson et al. 1991). In addition to the relative physical and biological simplicity of these systems, extensive life history data have been collected for these trout, including information on age-specific survival and fecundity, egg size, growth rate, age and size at maturity, and longevity (Hutchings 1993a,b, 1994, 1996, 1997).

Individual variation in brook charr growth rate prior to maturity is associated with significant variation in overwinter survival and fecundity; smaller individuals experience higher overwinter mortality and lower fecundity than larger individuals (Hutchings 1994, 1996). (The implicit assumption that size-at-age is a reasonable approximation of growth rate for these trout is based on similarity in size-at-emergence and length of growing season among years; J. Hutchings unpublished.) The primary consequence of this variation in growth is a substantial range in predicted fitness among individuals who reproduce early in life (3 years of age for two populations, 4 years for the third study population). This raises the question of whether slower-growing individuals, because of the fitness costs associated with maturing at a small size, might be favored to delay maturity, continuing to allocate energy to somatic growth rather than gonadal development. Using optimality theory, Hutchings (1996) predicted the fitness consequences associated with different ages at maturity and different rates of juvenile growth (Figures 5.4A,B), permitting the construction of norms of reaction for age at maturity for each population (Figure 5.4C).

With respect to these reaction norms, two observations are of note. First, the fitness functions support the prediction that a reaction norm describing a negative association between growth rate and age at maturity can represent an adaptive plastic response to environmental change. In Watern Cove and Cripple Cove populations, the slowest growing females are predicted to maximize fitness by delaying maturity, while the fastest growing individuals maximize fitness by maturing early in life.

Second, reaction norms for age at maturity are predicted to differ among populations (Figure 5.4C). In contrast to females in Watern Cove and Cripple Cove populations, those in Freshwater River are favored to mature early in life, regardless of growth rate. This underscores the point that the fitness advantages of delaying maturity (increased fecundity and higher overwinter survival because of larger body size) are inevitably balanced by the probability of realizing those benefits. Compared to trout in the other two populations, Freshwater River adult females experience significantly higher overwinter mortality (87%, as compared with 64% and 60% in Watern Cove and Cripple Cove Rivers, respectively)

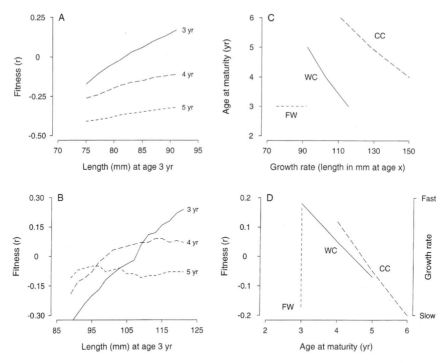

Figure 5.4. Fitness functions and reaction norms for age at maturity in populations of brook charr from Cape Race, Newfoundland, Canada. Associations between fitness at three ages at maturity are presented for females in Freshwater River (A) and Watern Cove River (B). Predicted reaction norms for age at maturity as a function of individual growth rate are presented in (C) for females in Freshwater River (FW), Watern Cove River (WC), and Cripple Cove River (CC). Predicted changes to fitness (r), as phenotypes change along the norms of reaction presented in (C), are given in (D). (Note that growth rate on the right ordinate is a qualitative representation of the lengths at age given in (C); one should not interpret growth rate as being perfectly correlated with fitness.)

and considerably slower growth in the wild (Hutchings 1993a, 1994). Thus, Freshwater River females appear to be favored to mature early in life, regardless of growth rate, because the survival costs of delaying maturity are too high relative to the apparently marginal benefits of increased fecundity.

Combining the information on fitness, age at maturity, and growth rate, one can assess the degree to which fitness (r) changes as individuals alter their phenotype (age at maturity) across an environmental gradient (reflected here by changes to individual growth) (Figure 5.4D). The primary point is that fitness can change as individuals alter their phenotypes along norms of reaction, an association all too rarely quantified in studies of plasticity. Thus, plastic changes in individual phenotypes can potentially have significant effects on the conservation biology of populations, affecting both their resilience and their ability to respond adaptively to environmental change.

Population differences in reaction norms for age, and size, at maturity have also been documented in grayling. Haugen (2000b) provided evidence of negative reaction norms between growth rate and age at maturity within four populations in Norway. Interestingly, however, he also found evidence of a non-plastic reaction norm for female grayling in one population, Øvre Mærrabottvatn, similar to that described above for Freshwater River brook charr. Regardless of growth rate, Øvre Mærrabottvatn grayling mature at 3 years of age, as compared to 4 to 8 years for grayling in the other populations. Haugen (2000b) attributed this invariant reaction norm in Øvre Mærrabottvatn grayling to the significantly higher mortality experienced by individuals in this population, during both the juvenile and adult stages, relative to that in the other populations. Given the recent common ancestry of the five populations (Haugen 2000a,b,c), this raises the possibility that the present invariant reaction norm observed for Øvre Mærrabottvatn grayling may represent a selective response to higher rates of mortality, a hypothesis discussed in more detail below.

2.2. Reaction Norms as Threshold Traits: Alternative Mating Strategies

Reaction norms need not vary continuously along an environmental gradient. This may be particularly true of those that underlie discontinuous variation in life history within populations. One example of discontinuous variation in salmonids is that of alternative maturation phenotypes within single populations (Gross 1985; Maekawa and Onozato 1986; Marschall et al. 1998). Male Atlantic salmon, for example, mature either as large (45–90 cm), relatively old (4–7 years) anadromous individuals, or as small (< 7–15 cm), young (1–4 years) mature parr (Jones 1959; Hutchings and Myers 1994; Metcalfe 1998; Fleming and Reynolds 2003—*this volume*). The former attain maturity following migration to sea (although some may have previously matured as parr), the latter mature without migrating. Prior to spawning, dominant anadromous males defend access to an anadromous female, while mature male parr establish what appears to be a size-based dominance hierarchy immediately downstream of the courting anadromous fish (Jones 1959; Myers and Hutchings 1987b). Mature parr compete with one another and with anadromous males for the opportunity to fertilize eggs. Since the first estimates of fertilization success in Atlantic salmon (Hutchings and Myers 1988), several studies have documented substantial reproductive success by parr as a group (e.g., Jordan and Youngson 1992) and as individuals (e.g., Thomaz et al. 1997), including recent estimates of the variance in individual fertilization success (Jones and Hutchings 2001, 2002).

The persistence of alternative mating "strategies" (or "tactics," depending on the interpretation, Gross 1996) has been explained as a product of negative frequency-dependent selection (Partridge 1988). As the frequency of a given strategy increases within a population, increased competition among individuals adopting that strategy will cause reduced average fitness. By contrast, the average fitness of individuals adopting the alternative strategy would increase because of reduced competition, resulting in a shift to the alternative strategy within the population.

There is good evidence that alternative mating strategies in salmonids have a genetic basis. Comparing the incidence of jacking (early maturity) among male progeny sired by jacks and by "hooknose" (late-maturing) chinook salmon, Heath et al. (1994) estimated the heritability of jacking to be about 0.4 (see also Heath et al. 2002c). Two studies have compared the incidence of male maturity as parr and as one-sea-winter Atlantic salmon, or grilse, among different families from several populations (Glebe and Saunders 1986; Herbinger 1987). Despite being reared in the same environment, the incidence of parr and grilse maturity differed significantly among families in both studies (Figure 5.5A),

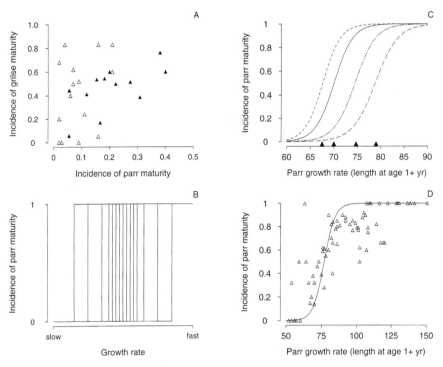

Figure 5.5. Genetic variation for alternative mating strategies in salmonids and the modeling of age at maturity reaction norms as threshold traits. (A) Genetic differences in the incidence of Atlantic salmon parr maturity are suggested by differences in age at maturity among families (represented by different points) reared under the same environmental conditions (solid triangles represent data from Herbinger (1987); open triangles are data from Glebe and Saunders (1986)). (B) Threshold reaction norms for age at maturity, redrawn from Hazel et al. (1990). (C) Hypothesized differences among populations in norms of reaction for Atlantic salmon parr maturity. Solid triangles represent the lengths at age at which 50% of the male parr in each population would be expected to adopt the parr reproductive strategy.(D) Incidence of male parr maturity plotted against individual growth rate for Atlantic salmon parr in the Little Codroy River, Newfoundland (redrawn from Myers et al. 1986).

providing strong evidence of a genetic basis for parr maturation, notwithstanding maternal effects (which could not account for the observed variation). Breeding experiments also suggest that the incidence of parr maturity is higher among the progeny of mature male parr than it is among those of anadromous males (Thorpe et al. 1983; Gjerde 1984a; Glebe and Saunders 1986).

Whatever the genetic basis to parr maturity, it is clear that environmental factors are also of significant importance. In particular, there is a significant correlation between growth rate and/or condition and the likelihood of maturing as parr (Myers et al. 1986; Thorpe 1986; Bohlin et al. 1994; Metcalfe 1998). Given that much of this variation in growth rate/condition is environmental in origin (e.g., broad-sense heritabilities for growth rate in the first year of life are typically between 0 and 0.3 for salmonids; Kinghorn 1983; Nilsson 1994; see also Garant et al. 2002, 2003a), there is a need to explain how genetically influenced alternative strategies can be evolutionarily stable within populations that also experience environmental heterogeneity largely at random with respect to genotype, and for which habitat quality significantly influences fitness.

Thus, it would seem that the most appropriate framework in which to address the evolutionary stability of alternative strategies in salmonids is the theoretical and empirical constructs encompassed by norms of reaction. In other words, are these alternative strategies maintained by adaptive phenotypic plasticity (Warner 1991)? Conceptually, this approach is no different than treating alternative mating strategies as conditional strategies, *sensu* Levins (1968). To incorporate both their environmental and genetic determinants, alternative strategies in salmonids have been modeled as threshold traits (Myers and Hutchings 1986; Hazel et al. 1990; Hutchings and Myers 1994; Hutchings 2002). In the quantitative genetic sense, threshold traits describe characters that are determined by alleles at multiple loci and that can be assigned to one of two or more distinct classes (Roff 1998). The loci affecting threshold traits are assumed to each have some small effect on a trait that varies continuously. For alternative strategies in salmonids, the continuously varying trait may be the concentration of a hormone, amount of lipid deposition, or metabolic efficiency (Thorpe 1986; Metcalfe 1998). Genotypes expressing less than the threshold value of this underlying trait will express one phenotype, while those exceeding the threshold will express the alternative phenotype. It has long been hypothesized that adoption of either the parr or the anadromous strategy in salmonids depends on whether an individual's growth rate in early life exceeds that specified by a growth-rate threshold, that is, that the strategies are conditional upon an individual's state (Leonardsson and Lundberg 1986; Thorpe 1986; Bohlin et al. 1990; Hutchings and Myers 1994). The hypothesis that alternative reproductive strategies may represent conditional strategies, originally developed in the study of salmonids, has since been extended to other groups of organisms (Gross 1996; Gross and Repka 1998).

Growth-rate thresholds can be modeled as norms of reaction for age at maturity. The existence of substantive differences in the incidence of parr maturity among families reared in a common environment (Figure 5.5A) suggests that differences in reaction norms for the probability of parr maturity exist among individuals in the same population. In their assessment of the utility of artificially selecting against parr maturity, Myers and Hutchings (1986) suggested that alternative mating strategies in Atlantic salmon could be modeled as quantitative genetic threshold traits, whereby the proportion p of individuals whose condition or status falls above the threshold adopts one strategy, and the remainder of the population $(1 - p)$ adopts the alternative strategy. As part of their study of alternative strategies in coho salmon, Hazel et al. (1990) adopted a reaction norm approach, depicting norms of reaction for age at maturity as a series of step functions varying among genotypes within a single population (Figure 5.5B).

Using age-specific survival data from the field (Myers 1984) and strategy-specific fertilization data from the laboratory (Hutchings and Myers 1988), Hutchings and Myers (1994) estimated the fitness associated with Atlantic salmon parr and anadromous male strategies as functions of mate competition and age at maturity. Although competition had been an integral part of previous treatments of this subject (e.g., Gross 1985; Myers 1986), Hutchings and Myers (1994) were also able to account for within-strategy differences in age at maturity, an inevitable consequence of environmental variation in growth rate and the influence that growth has on age at maturity. Given the existence of multiple age-specific sets of fitness functions for each strategy, they suggested that the fitnesses of alternative strategies were best represented as multidimensional surfaces. This is because phenotypic plasticity in age at maturity, caused by environmental heterogeneity in the prospects for individual growth, is associated with age-specific differences in the fitness consequences of adopting a given strategy. The points of interaction of these surfaces would then identify an evolutionarily stable continuum of strategy frequencies along which the fitnesses associated with each strategy are equal.

Two predictions from this model relevant to the study of alternative life histories are that: (1) the average fitness of individuals adopting the parr and anadromous strategies is not equal (Figures 2 and 3 in Hutchings and Myers 1994) and (2) selection maintains the frequencies of genes for anadromous male and mature male parr growth rate thresholds in an evolutionarily stable state, rather than the strategies themselves. Subsequent analyses of conditional strategies have drawn similar conclusions (Gross 1996; Gross and Repka 1998).

By adopting a reaction norm approach to the study of alternative life histories, Hazel et al. (1990) and Hutchings and Myers (1994) provide a mechanism by which alternative mating strategies, determined by both genetic and environmental factors, can evolve in variable environments (see also Thorpe et al. 1998). Based on this body of work, growth rate thresholds for parr maturity should differ among populations. This prediction can be represented graphically as a series of population-level reaction norms for each of several populations (Figure 5.5C), which reflect the high degree of variation in the incidence of parr maturity among populations (Myers et al. 1986). The only available data of this type (Little Codroy River, Newfoundland; Myers et al. 1986) do fit a normal cumulative density function (Figure 5.5D), providing the only empirical support to date of the existence of parr maturation threshold reaction norms within natural populations.

2.3. Natural Selection of Reaction Norms in the Wild

The existence of genetic variation in the shapes of reaction norms raises the possibility that the plastic responses of individuals to environmental change can be under selection. Evidence that the shapes of reaction norms are heritable, and can respond to selection, has been revealed by laboratory experiments on *Drosophila* and plants (Schlichting and Pigliucci 1998; Pigliucci 2001a,b). Nonetheless, evidence of such selection in the wild is limited at best. One exception to this may be recently completed research on plasticity in grayling. This work will be described in some detail as it provides some of the best and most compelling evidence for selection on reaction norms in wild populations. Furthermore, if natural selection can act on reaction norms, it is possible that selection induced by anthropogenic activities, e.g., fishing, also may be important. Evidence of plastic changes to life history induced by fishing, and resulting changes to the shapes of reaction norms, will also be discussed below.

The adaptive significance of interpopulation differences in plasticity in salmonids, and the possibility that selection is responsible for these differences, has recently been examined for five Norwegian populations of grayling (Haugen 2000a,b,c; Haugen and Vøllestad 2000). Since their introduction from Lesjaskogsvatn into Hårrtjønn and Øvre Mærrabottvatn in 1910, grayling have dispersed among several other lakes in south-central Norway, including Aursjøen and Osbumagasinet. Over a period of time ranging from 9 to 22 grayling generations, there has been considerable divergence in life history among these populations, notably with respect to age at maturity, size at maturity, and fecundity (Haugen 2000a,b,c).

Population differences in life history have also manifested themselves as differences in the shapes of reaction norms for age and size at maturity (Figure 5.6A; Haugen 2000b). Delayed age at maturity is associated with smaller size at maturity for grayling in four populations, with the population-level reaction norms crossing in state space. Interestingly, and as discussed earlier, the reaction norm for grayling in Øvre Mærrabottvatn expresses an invariance in age at maturity, all individuals maturing at age 3. Consistent with the explanation discussed earlier for population differences in reaction norms for age at maturity in Newfoundland brook charr (Figure 5.4C), there appears to be a direct association between the steepness of the grayling reaction norms and average adult mortality (Haugen 2000b). For individuals aged 4 through 8 years, the instantaneous rate of mortality, Z, was highest (0.77) for Øvre Mærrabottvatn grayling, those apparently favored to reproduce as early in life as possible. By contrast, the reaction norms with the shallowest slopes, encompassing the greatest ranges of ages, are those for Hårrtjønn and Aursjøen grayling, which have the lowest rates of mortality ($Z = 0.36$ for both).

To test the hypothesis that population differences in reaction norms are a result of selection, acting over a comparatively short period of time (9 to 22 generations), Haugen and Vøllestad (2000) undertook a common-garden experiment in which they reared grayling from three different populations under the same experimental conditions in the laboratory. Specifically, they measured

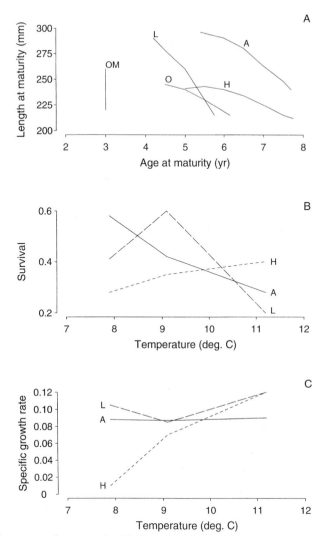

Figure 5.6. Norms of reaction for life history traits in grayling from five populations in south-central Norway. (A) Reaction norms for age and size at maturity for grayling in Øvre Mærrabottvatn (OM), Lesjaskogsvatn (L), Osbumagasinet (O), Hårrtjønn (H), and Aursjøen (A); redrawn from Haugen (2000b). Temperature-based reaction norms for grayling survival (B) and specific growth rate (C) during the first 180 degree-days of exogenous feeding; data are from Haugen and Vøllestad (2000).

survival and specific growth rate during the first 180 degree-days of exogenous feeding at three different temperatures. These temperatures corresponded to the average temperatures experienced by grayling in each of the three populations during this stage of life in the wild: 7.9°C (Aursjøen), 9.1°C (Lesjaskogsvatn), and 11.2°C (Hårrtjønn).

The common-garden experiments revealed significant genetic differences in reaction norms for survival and growth rate among populations. Survival declined with increasing temperature for the "cold" population (Aursjøen), peaked at the intermediate temperature for the "medium" population (Lesjaskogsvatn), and increased with increasing temperature for the "warm" population (Hårrtjønn) (Figure 5.6B). Thus, survival was highest at the temperatures typically experienced by these grayling in the wild. There were also clear differences in plasticity for growth rate among populations (Figure 5.6C), although the fastest growth rates were not always achieved at the temperatures experienced by grayling in the wild (given the existence of life history tradeoffs, this need not be unexpected).

Thus, life history research on grayling has revealed genetic differences in reaction norms for life history traits among populations, providing data consistent with the hypothesis that the shapes of at least some reaction norms (temperature-dependent survival in early life) are adaptive and are the product of natural selection (see also the work by Hendry et al. (1998) on sockeye salmon). The results are also in agreement with research indicating that evolution can take place over relatively short (10 to 20 generations) periods of time (Hendry and Kinnison 1999; Kinnison and Hendry 2001, 2003—*this volume*).

2.4. Anthropogenic Selection of Reaction Norms

The potential for fishing to cause significant evolutionary change within a population is no different than other forms of predator-induced mortality. Given that differential mortality among genotypes provides the basis for evolutionary change, fisheries can be accurately described as large-scale experiments on life history evolution (Rijnsdorp 1993).

While the potential for genetic change caused by fishing has been acknowledged by comparatively few fisheries scientists (for recent discussions, see Conover 2000; Hutchings 2000; Stokes and Law 2000), the idea that exploitation can cause significant phenotypic changes is widely accepted. Reductions in age at maturity, for example, are one of the best examples of phenotypic change associated with fishing (Policansky 1993). Life history changes often arise because of decreased competition for food and space, increased growth rate resulting from relaxed intraspecific competition, and the negative association that often exists between growth rate and age at maturity (Trippel 1995). Long-term changes in size at maturity have been interpreted as genetic responses to the size-selectivity of fishing gear, for example lake whitefish, *Coregonus clupeaformis* (Handford et al. 1977), pink and chinook salmon (Ricker 1981), Atlantic salmon (Bielak and Power 1986), and European grayling (Haugen and Vøllestad 2001). Recent selection experiments that mimic fisheries and theoretical analyses have also shown that genetic changes in life history are to be expected (Conover and Munch 2002; Hard 2003—*this volume*).

The question of whether fishing can change the shapes of reaction norms by selection has received comparatively little attention. Reznick (1993) hypothesized that the primary effect might be to change the elevation of the reaction norms, assuming that fishing would select against individuals genetically predisposed to mature at large body sizes. Considering how fishing might affect the slopes of reaction norms, Hutchings (1993b, 1997, 2002) used age-specific survival and fecundity data on brook charr populations to predict how reaction norms for age, size, and reproductive effort at maturity might change in response to increases in adult mortality. As fishing mortality increased, selection was predicted to favor a flattening of reaction norms, notably for age and effort at maturity, such that individuals would be favored to reproduce as early in life as possible and to expend the maximum amount of reproductive effort at that age, irrespective of growth rate (see also Schaffer 2003—*this volume*).

These hypothesized changes in plasticity resulting from fishing are consistent with the differences in life history reaction norms observed among salmonid populations, such as those described earlier for brook charr (Figure 5.4) and grayling (Figure 5.6). That is, a flattening of the reaction norm for age at maturity (Figures 5.4C, 5.6A) is expected to occur as the probability of realizing the fitness benefits of delayed maturity decline with increases in mortality due to fishing. Similarly, as longevity declines with increased fishing pressure, selection should favor increases in reproductive effort.

It seems reasonable to conclude that fishing can result in selective changes to reaction norms in heavily exploited populations (Hutchings 2002). These changes may involve both the slopes and the elevations of reaction norms for several life history traits. Detecting such changes, however, will be exceedingly difficult, given the near-absence of research on phenotypic plasticity and reaction norms on commercially exploited fishes.

3. Conclusions and Extensions

The evolution of adaptive phenotypic plasticity depends on the existence of additive genetic variation in the shapes of reaction norms (Schlichting and Pigliucci 1998), the sign, magnitude, and temporal constancy of genetic covariances among traits (Turelli 1988; Charlesworth 1990), and the persistence of environmental variation (Via and Lande 1985; Stearns and Koella 1986). Life history traits such as age and size at maturity, egg size, and fecundity are heritable in salmonid fishes (Gjedrem 1983; Thorpe et al. 1984; Gjerde 1984a, 1986; Robison and Luempert 1984; Kinnison et al. 2001), so it is not unreasonable to predict that heritable norms of reaction for life history traits exist in this family. Indeed, family-level differences in reaction norms (Beacham and Murray 1985; Haugen 2000c) and genetic differences in population-level reaction norms (Haugen and Vøllestad 2000; B. Jonsson et al. 2001b) suggest that selection on reaction norms in salmonids is inevitable.

To what degree has research on salmonid reaction norms informed us about the evolution of adaptive phenotypic plasticity? Studies of natural populations of brook charr and grayling reveal that the average life history response by individuals to environmental change can differ significantly from one population to the next. This variability will almost certainly affect the ability of populations, and of individuals within those populations, to adapt to environmental change. Population variation in the shapes of reaction norms for age and size at maturity also suggests conditions under which a form of trait canalization might take place. As adult mortality increases, plasticity is predicted to decline, leading to a relatively invariant phenotypic response across an environmental gradient. Models incorporating genetic and environmental determinants of alternative mating strategies in salmonids illustrate how threshold reaction norms can be applied to the study of conditional strategies influenced by both genetic and environmental factors. And common-garden experiments in salmonids provide evidence of adaptive phenotypic plasticity in the wild and of comparatively rapid responses in the shapes of reaction norms to selection, suggesting that genetic variation in plasticity can be both substantial and heritable in natural populations.

A comprehensive understanding of how genotypes respond to environmental change is fundamental to the study of evolutionary ecology. Recognition of this precept has underpinned the increasingly sophisticated research on reaction norms in the past two decades (Schlichting and Pigliucci 1998; Pigliucci 2001b). This work, embodied in several respects by the salmonid studies discussed here, has laid the foundation for future research initiatives and the questions that they might address:

1. How variable is plasticity among individuals within populations, among populations within species, and among species within and among clades?
2. To what extent do families within populations differ in their average response to environmental change? Is there genetic variation in the shapes of reaction norms within populations?
3. What is the additive component of this genetic variability? To what degree are the shapes of reaction norms heritable?
4. What constrains evolutionary changes in plasticity?
5. How does individual fitness and, by extension, rate of population growth change as phenotypes shift along norms of reaction?
6. How rapidly do reaction norms respond to natural and anthropogenic selection?

Among salmonids, most of the research on plasticity to date has been undertaken within the context of the first question, although relatively little work has been devoted to actually constructing population-level norms of reaction; this is particularly true of traits not directly related to fitness. From a general perspective, the fifth question, arguably the most important of all, has received remarkably little attention in studies of adaptive phenotypic plasticity for *any* organism. If one peruses the most recent reviews of reaction norms (Schlichting and Pigliucci 1998; Mazer and Damuth 2001; Pigliucci 2001a,b), for example, there is little or no discussion of how fitness changes along either individual or family-level norms of reaction.

These questions can be profitably addressed under controlled common-garden and quantitative genetic experiments in the laboratory. However, future research initiatives need also be mindful of the need to rectify the comparative dearth of plasticity studies that have been conducted under ecologically realistic conditions. Despite the difficulty in establishing appropriate experimental controls in the wild, particularly for many animals, the limited research that has been conducted on salmonids under such conditions suggests that the benefits of such research can be considerable.

Acknowledgments   The research was supported by a NSERC (Canada) Research Grant. I thank Andrew Hendry and Steve Stearns for the opportunity to undertake this work. Two anonymous referees provided very helpful comments on an earlier draft of the manuscript.

Ecological Theory of Adaptive Radiation
An Empirical Assessment from Coregonine Fishes (Salmoniformes)

Louis Bernatchez

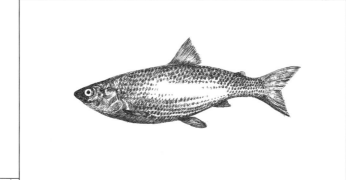

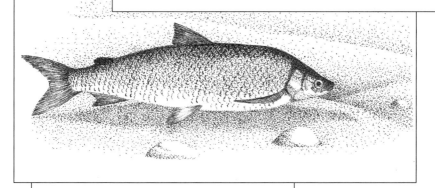

A zooplanktivorous lake herring (top) and an epibenthic lake whitefish (bottom)

6

Elucidating the causes of population divergence and species diversity is a central issue in evolutionary biology. As for any scientific discipline, progress in this field will be best achieved if studies are embedded into a strong, predictive, theoretical framework. In this view, perhaps the most comprehensive concept available to evolutionary biologists is the ecological theory of adaptive radiation. Its central elements were formalized in the first half of the twentieth century by founders of the evolutionary synthesis and others, namely Huxley (1942), Mayr (1942), Lack (1947b), Dobzhansky (1951), and Simpson (1953). The theory holds that adaptive radiation, including both phenotypic divergence and speciation, is ultimately the outcome of divergent natural selection stemming from environmental and resource heterogeneity, as well as competitive interactions. Schluter (2000) has recently re-evaluated and extended this theory in the light of studies that have been conducted since its formulation. Overall, this synthesis of knowledge has supported the great utility of the ecological theory of adaptive radiation, which indeed makes it one of the most successful theories of evolution ever advanced (Schluter 2000).

Progress in applying the concept of adaptive radiation and testing its processes is best achieved by the comparative study of evolutionarily young lineages, wherein processes of rapid phenotypic divergence and reproductive isolation are still active. Such lineages are most likely to be found in recently arisen areas of ecological opportunity, where the wealth of resources (heterogeneity in habitats and trophic resources) is underutilized by other taxa. These conditions have prevailed in northern temperate aquatic habitats that were cyclically affected by glacial expansion and retreats during the Pleistocene (Martinson et al. 1987). A multitude of empty habitats characterized by contrasting physicochemical conditions and trophic resources, such as freshwater versus marine environments, lakes versus streams and rivers, different habitat types (e.g., limnetic vs. epibenthic) and depths within lakes, were created following each glacial retreat. The limited number and duration of dispersal routes, and the small number of species that were physiologically tolerant to the low temperatures prevailing during early post-glacial aquatic environments (Lacasse and Magnan 1994) further contributed to the existence of empty niches.

Members of the north temperate fish fauna share several attributes that corroborate the predicted effect of post-glacial ecological opportunity in priming divergence, and make these taxa putative candidates for testing the ecological theory of adaptive radiation. Of particular interest is the occurrence of sympatric and parapatric morphs that predominate in salmonid fishes, but that also persist in other taxonomic families, including Gasterosteidae, Osmeridae, Centrarchidae, and Catostomidae (reviewed in Taylor 1999; Robinson and Schluter 2000). These morphs share similarities in general patterns (but not in the details) of phenotypic variation, suggesting that their diversification has been partly driven by the same selective processes. Molecular genetic studies generally confirm their post-glacial origin (10–15 ky before present) and reveal that the process of speciation has been initiated but is not complete in many cases (Taylor 1999).

In this chapter, I first summarize the basis of the ecological theory of adaptive radiation, largely referring to Schluter (2000). Secondly, I outline the principles of methods that can be used to evaluate processes implied by the theory. Comparative methods that do not rely on the experimental approach are emphasized. My main motivation in doing so is that an experimental approach in natural conditions may not be amenable to the study of numerous taxa due to logistical constraints. This explains why such studies have most often involved small, short-generation length and "easily handled" taxa (e.g., *Gasterosteus*). Alternative methods permitting indirect inference therefore represent an important complement to experimental studies. I then illustrate how the application of such methods to the study of coregonine fishes has contributed to the assessment of several important elements of the theory. I focus on this subfamily as it represents one the salmonid groups for which much progress has been achieved thus far. The other two important salmonid models, sockeye salmon and Arctic charr, have been recently reviewed elsewhere (Jonsson and Jonsson 2001; Skùlason et al. 1999; Taylor 1999).

1. Theoretical Basis of Adaptive Radiation

1.1. Definition

A major asset of the concept of adaptive radiation is that it can be formulated as an operationally explicit definition. Extending from the basic definition of Simpson (1953), Schluter (2000) defined adaptive radiation as "the evolution of ecological and phenotypic diversity within a rapidly multiplying lineage. It occurs when a single ancestor diverges into a host of species that use a variety of environments and that differ in morphological or physiological traits used to exploit those environments. The process includes both speciation and phenotypic adaptation to divergent environments."

1.2. Detecting an Adaptive Radiation

Identifying organisms that satisfy this definition is the first step in elucidating the processes of adaptive radiation. Putative candidates should satisfy four criteria:

1. Diverged populations or species must share a *common ancestry*. Evidence that common ancestry is recent is most desirable, as explained earlier. Common ancestry does not necessarily imply monophyly, that is, the theory does not predict that all descendants derived from a common ancestor will be included in the radiation.
2. Evidence for *rapid speciation* must be provided. Here, *rapid* is considered in relative terms, and refers to demonstrating that speciation has been episodic and resulted in a burst of species diversity compared to either

periods preceding or following an adaptive radiation, or compared to other lineages.

3. Populations or species included in an adaptive radiation must use a variety of environments or resources, and must differ in traits used to exploit those environments. That is, *phenotype-environment correlations* must be detected in order to fulfill the *adaptive* criterion, in the sense of current utility. The expression of such traits must also be genetically based rather than the mere outcome of phenotypic plasticity.

4. *Trait utility*, or evidence that particular phenotypic traits enhance performance in the environment with which they are associated, must be established. This also allows the identification of the proximal mechanisms responsible for phenotype-environment correlations and assists in ruling out "false" correlations.

1.3. Assessing the Role of Divergent Natural Selection in Adaptive Radiation

Once candidate organisms fulfill the criteria of an adaptive radiation, subsequent steps to test the theory share a common objective, which is to provide evidence that divergent natural selection is the main cause for the build-up of phenotypic differentiation and reproductive isolation. More specifically, the theory's foundation lies on three explicit processes that drive adaptive radiation by divergent natural selection.

The first process implies that *phenotypic differentiation between populations and species is caused directly by adaptation to the environments they inhabit and the resources they consume*. Thus, each environment subjects a species to unique selection pressures, owing to the advantages of particular combinations of traits for efficiently exploiting that environment. This process implies both that individuals from different environments experience divergent selection, and also that intermediate phenotypes are characterized by reduced fitness if no intermediate environment exists. The basic theory of this principle was first presented by Simpson (1953) as the concept of selection landscapes. Selection landscapes refer to surfaces that represent phenotypic traits and fitness. Fitness corresponds to the height of the surface where features of the environment shape its contours. Populations diverge when they are pulled toward different peaks (optimum phenotypes given the features of the environment) and away from the valleys of low fitness (Arnold et al. 2001). The number as well as the shape of the peaks and valleys are themselves generated by uneven fitness gains at different positions along gradients associated with the discreteness of environmental features. This framework explicitly predicts that in an adaptive radiation, phenotypic and environmental diversity will be correlated due to divergent natural selection, the latter being defined as natural selection that pulls the means of phenotypically distinct populations toward different adaptive peaks (Schluter 2000). The resulting phenotypic differentiation between populations is therefore caused directly by the inhabited environment and resources consumed.

The second process implies that *divergence in phenotype also results from competitive interactions, broadly defined to include ecological opportunity.* Competition, defined as the antagonistic interaction between phenotypes arising from depletion of shared resources, may drive sympatric populations or species to utilize new resources or environments where they will become subject to differential selective pressures. Competition is therefore an agent of divergent natural selection, just as differences between environments are: the competing populations or species being elements of the environmental landscape. In the same manner, ecological opportunity, which implicitly refers to a highly reduced potential for interspecific competitive interactions for utilizing new resources or habitat, can also drive rapid divergence until new available niches are filled. Ecological opportunity has been viewed as the major regulator of the rate and extent of phenotypic differentiation and speciation. Thus, Simpson (1953) explicitly recognized that for an adaptive radiation to occur, adaptive zones (a zone being a set of resources or environmental features necessary to complete a life cycle) must be occupied by organisms that are competitively inferior to the entering group or it must be empty. The novel selection pressures thus encountered drive adaptive radiation according to the first process (above).

The third and final process is *ecological speciation, by which reproductive isolation among lineages develops as a consequence of the two processes that drive phenotypic divergence.* Implicit to the theory, reproductive isolation within an adaptive radiation evolves within the same time frame and from the same processes that phenotypic and ecological divergence occur. Speciation is an essential step to the completion of the process of adaptive radiation since phenotypic and ecological differences that accumulate between diverging lineages would eventually disappear without reproductive isolation (should they become sympatric, or should the environmental gradients responsible for divergent selection vanish). Divergent natural selection may drive the speciation process in different ways. First, reproductive isolation may develop incidentally between populations that became differentially adapted to occupy distinct habitats or utilize different resources. This process is called *byproduct speciation* because reproductive isolation is not the direct target of selection itself (Rice 1987; Rice and Hostert 1993; Rundle and Whitlock 2001). For instance, post-mating isolation may develop as a consequence of genetic incompatibility between allelic combinations responsible for the expression of differentially adapted phenotypes. Reproductive isolation that develops as a consequence of divergent natural selection may also have a purely ecological basis without any intrinsic genetic incompatibilities. For instance, pre-mating isolation may incidentally build up if mating preferences are genetically correlated with phenotypic traits (e.g., adult body size) that are under the influence of divergent selection (Rundle et al. 2000). Byproduct speciation should be most effective when gene flow is highly reduced between populations by extrinsic barriers and/or when the environments differ most dramatically. The second process of ecological speciation implies that reproductive isolation itself will be favored by selection if intermediate genotypes experience lower fitness (Rosenzweig 1978). This process is referred to as *reinforcement* when intermediate genotypes are hybrids between previously allopatric populations and between which partial reproductive isolation has already developed. It corresponds to sympatric speciation if speciation is initiated by disruptive selection within a single population. Although once highly debated, there is now sound theoretical and empirical evidence that sympatric speciation, as well as reinforcement, may operate under various ecological situations (Noor 1999; Kirkpatrick 2000, 2001; Via 2001).

2. Empirical Tests of the Theory

2.1. Coregonine Fishes

Many biological attributes suggest that coregonine fishes are of particular interest for the study of adaptive radiation. This subfamily is composed of three genera, *Prosopium*, *Stenodus*, and *Coregonus*, and exhibits a broad circumpolar distribution in the Northern Hemisphere, with a southern limit that approximately matches that of the maximal extent of glacier advances during the Pleistocene (Bernatchez 1995). *Coregonus* is the most speciose genus among salmonids, with a conservative estimate of 28 taxonomically recognized species (Reshetnikov 1988). Depending on authorship, however, the number of biological species could vastly exceed this figure (e.g., Kottelat 1997). They have also evolved to occupy a wide array of habitats throughout the Palearctic and Nearctic regions, with their basic environmental requirement seeming to be the presence of cold, well-oxygenated waters (Scott and Crossman 1973). Some species, such as the Arctic (*C. autumnalis*) and Bering cisco (*C. laurettae*) are truly anadromous, and perform upstream migrations of many hundreds of kilometers to reproduce (McPhail and Lindsey 1970). Others, such as whitefish of the C. *clupeaformis/ lavaretus* complex, inhabit rivers and lacustrine habitats of all dimensions and water depths, and use reproductive habitats of all kinds, including small running streams, deep rocky shoals, and even pelagic areas.

Apart from the genus *Stenodus*, a highly distinctive large piscivore, morphological diversity in coregonine fishes can be simplified into two basic patterns; the "cisco" versus the "true whitefish" type (Lindsey 1981). Typically, ciscoes seem adapted for zooplankton feeding, being most often found in lacustrine pelagic zones, and characterized by a compressed body, large terminal mouth, large eyes and a silvery coloration (Scott and Crossman 1973). In contrast, the "true whitefishes" most generally, but not exclusively, occupy the epibenthic habitat at the adult stage, and are characterized by a subterminal mouth and a more robust body shape. However, the single most distinctive morphological trait that differentiates "ciscoes" from the "true whitefishes" is the number of gill-rakers (Lindsey 1981). Gill-rakers are cartilaginous projections located on the inner ridges of gill arches, and are used by fish to retain prey items when filter feeding. Although not strictly demonstrated experimentally, it is widely accepted that more numerous and longer gill-rakers increase the efficiency of small prey capture (e.g., zooplankton). Accordingly, ciscoes are on average characterized by a high number of longer gill-rakers relative to true whitefishes.

This phenotypic dichotomy is, however, highly complicated by the tremendous phenotypic variation among populations observed within many taxa labeled either as ciscoes or whitefish under a conservative taxonomic scenario. It is this level of complex phenotypic diversity that contributes to the notorious taxonomic confusion in coregonine fishes (Lindsey 1988). For example, several authors have claimed that all European populations of whitefish should be labeled as *Coregonus lavaretus* L. (Steinmann 1950; Reshetnikov 1988) whereas others consider that the differentiation observed among sympatric forms amply justifies their taxonomic distinction (e.g., Kottelat 1997; Svärdson 1998). As for the cisco/whitefish dichotomy observed at deeper phylogenetic levels, such morphs differ in traits that appear to be *a priori* adapted for pelagic versus epibenthic niche exploitation. Thus, the number of gill-rakers is the single most discriminant character between them. In the simplest cases, such as that observed for lake whitefish (*C. clupeaformis*) from North America, only two sympatric forms are found (e.g., Fenderson 1964). In other cases, such as for the European whitefish in several Fenno-Scandinavian and central Alpine lakes, up to five sympatric morphs are found (Svärdson 1979, 1998; Douglas and Brunner 2002). In the following section, I outline empirical results that test the four criteria that would define an adaptive radiation in coregonines.

2.2. Detecting an Adaptive Radiation

2.2.1. Common Ancestry

Molecular phylogenetic analysis of species relationships is the most straightforward way to test for common ancestry. The most complete coverage of phylogenetic relationships among coregonines was performed by Bernatchez et al. (1991) based on RFLP analysis of mitochondrial DNA variation (Figure 6.1). This study included 20 of the 28 commonly recognized taxa (e.g., Reshetnikov 1988), and a total of 310 character states (or approximately 1800 subsampled nucleotides) were analyzed using both distance and character-based methods. This analysis first confirmed that all *Coregonus* taxa share a common ancestor relative to the distantly related genus *Prosopium*. Secondly, cisco and true whitefish taxa do not compose reciprocal monophyletic clades, the ciscoes being a polyphyletic assemblage. Thirdly, members of the genus *Coregonus* share a paraphyletic relationship with the genus *Stenodus*. The extent of interspecific sequence divergence also suggests recent divergence from a common ancestor. Assuming a conventional molecular clock of 2% mtDNA sequence divergence per million years most commonly applied to estimate divergence time in northern fishes (discussed in Bernatchez 2001), the mean pairwise sequence divergence estimate of 3.00% (STD: ±1.35%; range: 0.31–5.62%) observed for the *Coregonus–Stenodus* clade suggests that most extant species originated during the Pleistocene over the last 2 million years (mean: 1.5 million years; range: 310,000 to 2.8 million years). Monophyly of *Coregonus* relative to *Prosopium* has also been confirmed by sequence analysis of the mtDNA control region and the ITS1 region of ribosomal DNA (Reist et al.

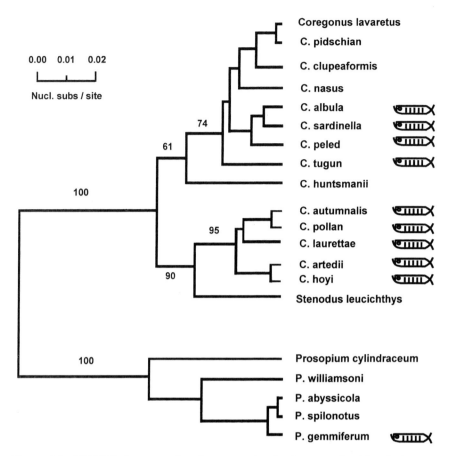

Figure 6.1. KITSCH phenogram based on a matrix of pairwise nucleotide substitutions per site estimated from mtDNA RFLP analysis among coregonine taxa. Values above the branches are bootstrap values > 50 (in percent) obtained in a character-based parsimony analysis. Cisco species are identified by fish symbols. Modified from Bernatchez et al. (1991).

1998; Sajdak and Phillips 1997), as well as allozyme variation (Bodaly et al. 1991). The polyphyletic nature of ciscoes within *Coregonus* was also supported in these studies. They also revealed low levels of sequence divergence among *Coregonus–Stenodus* taxa. MtDNA D-loop pairwise sequence divergence averaged 4.24% (STD: ±2.0%; range: 0.31–9.34%), and ribosomal DNA sequence divergence averaged 2.73% (STD: ±1.53; range: 0.15–4.61%). Nei's distance (*D*: Nei 1978) based on allozymes averaged 0.201 (STD: ±0.76; range: 0.038–0.36). Applying the commonly used molecular clock of 5×10^{-6} mutations per locus per generation for allozyme data (Nei 1987b) generates a time frame for the origins of coregonines comparable to that observed for mtDNA data (mean: 1 million years; range: 190,000 to 1.8 million years). In summary, molecular analyses of phylogenetic relationships are all supportive of the recent divergence (Pleistocene time frame) of most, if not all members of the *Coregonus–Stenodus* clade.

2.2.2. Rapid Speciation

The test of rapid speciation consists of establishing whether speciation rates are episodic and, if so, identifying the location of peak periods. Several criteria may be used for detecting speciation episodes, and these are generally based on analyses of branching rates of phylogenetic trees (Barraclough and Nee 2001). One such method was introduced by Wollenberg et al. (1996) and tests phylogenies estimated from data of extant species for departures from "null" trees that are generated by simulations under a Markovian process. The model is stochastic in the sense that equivalent probabilities of bifurcation (speciation event) and extinction are applied independently to each lineage.

I applied this statistical procedure to the data of Bernatchez et al. (1991) in order to test the null hypotheses that episodes of speciation (1) within the whole coregonine subfamily and (2) within the *Coregonus–Stenodus* clade have been random in time since the origin of extant coregonines. To test for bursts of speciation using the statistical approach of Wollenberg et al. (1996), the relative temporal placements of speciational nodes in a phylogeny must be estimated. This requires that the external nodes in a tree be "right justified" (contemporaneous) such that all extant species are at the same (present) position on the temporal axis of the tree. For this reason, I used a KITSCH tree that was generated using PHYLIP, version 3.57 (Felsenstein 1993) from the distance matrix of pairwise nucleotide divergence estimates. This tree generated primarily the same topology when compared to the character-based tree analyses (Bernatchez et al. 1991). A standardized temporal scale ranging from time zero (the earliest node in the tree) to time one (the present) was superimposed to identify the relative temporal placements of all internal nodes in the tree. Following Wollenberg et al. (1996), these nodal times were used to generate a cumulative distribution function (Sokal and Rohlf 1995) of normalized branching times for statistical comparison against the appropriate null expectations under the Markovian model in which speciation/extinction are stochastic with respect to time. The test used is the Kolmogorov–Smirnov (K-S) *D*-statistic calculated from the observed and expected cumulative distribution functions. Appendix II of Wollenberg et al. (1996) provided the cumulative frequency distributions of normalized branching times for various numbers of extant taxa, such that new simulations were not needed for generating the expected cumulative distribution functions.

The cumulative distribution functions for the putative coregonine phylogeny derived from mtDNA variation are presented in Figure 6.2. No evidence for bursts of rapid speciation events was detected when considering coregonines of all three genera (Figure 6.2A). Similarly, there was no evidence for distinct episodes of rapid speciation events when considering only the *Coregonus–Stenodus* clade (Figure 6.2B). However, the cumulative number of species consistently remained above random expectations, and this trend was nearly signifi-

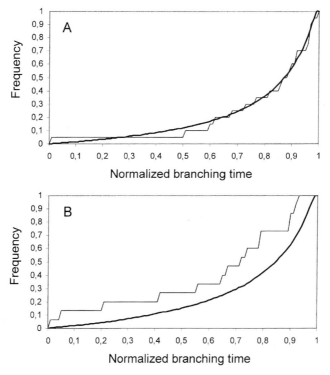

Figure 6.2. Cumulative distribution function for the putative coregonine phylogeny of Figure 6.1, and the appropriate null cumulative distribution functions built using the data of Wollenberg et al. (1996) for $n = 20$. (A) All coregonine taxa. (B) Species of the genus *Prosopium* are omitted.

cant (K-S, $D = 0.327$, $P = 0.10$). Because incomplete sampling of taxa biases the outcomes of the above method, these results must be interpreted cautiously. However, it is likely that adding the missing taxa would further support the conclusion of an excess in speciation rate relative to extinction rate within the *Coregonus–Stenodus* clade. The majority of missing taxa are North American ciscoes that diverged post-glacially (Turgeon et al. 1999; Turgeon and Bernatchez 2001a,b). In addition, the analysis did not include the sympatric morphs of either North American or European whitefish that also diverged post-glacially and fulfill the criteria of biological species (Kottelat 1997; Douglas et al. 1999; Lu and Bernatchez 1999; Douglas and Brunner 2002). The addition of these species would add internodes to the recent portion of the tree which would further increase the cumulative number of species above random expectation to the right of the cumulative distribution function.

   In summary, the analysis of cumulative distributions of species numbers through time does not provide evidence for distinct episodes of rapid speciation in coregonines. It suggests, however, that speciation rate has generally exceeded extinction rate within the *Coregonus–Stenodus* clade since its origin. Molecular evidence suggests that most members of this clade originated during the Pleistocene. It is also clear that the numerous Pleistocene glaciations repeatedly wiped out most of their current habitats, and have increased extinction rates at northern latitudes (Hocutt and Wiley 1986; Robinson and Schluter 2000). Consequently, a trend for the observed number of species to exceed that of random expectation is consistent with the hypothesis that conditions during interglacial periods were particularly favorable for increased speciation rate in the *Coregonus–Stenodus* clade.

2.2.3. Phenotype-Environment Correlation

A fit between the phenotypes of descendant species and their divergent environments fulfills the adaptive criterion. The statistical assessment that species differences in genetically based phenotypes are associated with use of different environmental resources is the principal means of depicting such phenotype-environment correlations. For coregonines, the number of gill-rakers is the main phenotypic trait for which such a correlation can be predicted. Species or morphs with numerous gill-rakers should be associated with the use of zooplankton as their primary diet, whereas the diet of morphs with low gill-raker numbers should mainly comprise larger prey.

Focusing on gill-raker numbers is also of particular interest because its genetic basis has clearly been demonstrated. Svärdson (1979) summarized the results of cross experiments that were conducted using two sympatric morphs of whitefish from Lake Locknesjön in Sweden. The "Storsik" is a large, sparsely rakered whitefish with a mean of 19 gill-rakers (range: 16–23). The "Planktonsik" is a densely rakered morph with a mean of 42 gill-rakers (range: 31–49). Pure and hybrid crosses of these morphs were artificially produced in eight consecutive years. Young of the year were reared in a common laboratory environment and then released into 2-hectare natural ponds. Samples were taken at the end of the growing season (September), and gill-raker numbers were compared with the parental sources. Figure 6.3 shows that the mean and variance of gill-raker counts of the pure progeny were very similar to those of their parents (Storsik: mean $= 19.5$ and 19.3, $t = -1.00$, $P = 0.32$; variance $= 2.29$ and 2.53, $F = 0.086$, $P = 0.77$; Planktonsik: mean $= 40.7$ and 41.9, $t = 4.27$, $P = 0.0002$; variance $= 7.12$ and 5.76, $F = 0.652$, $P = 0.42$). In contrast, the mean number of gill-rakers of the hybrid progeny was statistically different from that of either parental sources ($t = 45.4$ and -43.4, $P < 0.0001$), and almost perfectly intermediate between the two (29.1 vs. 30.1). The variance (3.24) of the hybrid progeny was also intermediate between the values observed for the parental sources (2.29 and 7.12). This pattern was repeated almost exactly over 8 years (see table 2, p. 10 in Svärdson 1979). More recently, Östbye (in preparation) estimated the broad sense heritability of gill-raker numbers using midparent–midprogeny relationships among families of several whitefish populations from Scandinavia. The gill-raker counts of midparent and mid-progeny were highly correlated ($r^2 = 0.89$, $P < 0.001$) and the h^2 value estimated from the regression slope was 0.79 (95% CI $= 0.63$–94). These

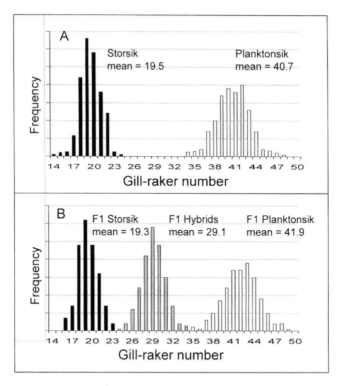

Figure 6.3. Frequency distribution of gill-raker numbers for (A) parental populations and (B) their F1 pure and hybrid progeny for the "Storsik" and "Planktonsik" whitefish forms from Lake Locknesjön, Sweden. Modified from Svärdson (1979).

observations are all indicative of a strong component of additive genetic variance for gill-raker counts (Lynch and Walsh 1998).

Svärdson (1979) also documented a phenotype–environment association between diet and differences in gill-raker counts between the sympatric morphs from Lake Locknesjön. The sparsely rakered morph typically fed on large epibenthic prey, whereas the diet of the densely rakered morph was almost exclusively zooplankton. The diet of the experimental progeny described above was also documented. The sparsely rakered progeny fed mainly on large prey, whereas the diet of the densely rakered progeny still concentrated on zooplankton. Interestingly, the diet of hybrids included both zooplantonic and epibenthic prey not heavily preyed upon by pure progeny. Age and seasonal variations in the extent of phenotype–prey association have also been documented for other sympatric morphs of the European whitefish (Bergstrand 1982; Sandlund et al. 1995).

In North America, the association between diet and number of gill-rakers was assessed for sympatric morphs found in four lakes of the Yukon (Bodaly 1979). Each lake harbors two sympatric morphs of whitefish referred to as the high-gill-raker (HGR, or Squanga whitefish, see Bodaly et al. 1988) and low-gill-raker (LGR) morphs. Using the data of Bodaly (1979), I calculated that the relative volume contribution of zooplankton was significantly higher in the HGR morph (mean = 0.92, range = 0.79–1.0) than in the LGR morph (mean = 0.37, range = 0.10–0.60) in all lakes (Mann–Whitney U-test; $Z = 2.31$, $P = 0.02$). Prey diversity index $(1 - \sum p_i^2$, where p_i = proportion of either benthic or zooplantonic prey) estimated using Bodaly's data also showed that the diet of the HGR morph was more specialized (mean prey diversity index = 0.13, range = 0–0.33) than that of the LGR morph (mean = 0.40, range = 0.18–0.50) (Mann–Whitney U-test; $Z = -2.02$, $P = 0.04$).

More recently, Bernatchez et al. (1999) compared the diet of sympatric "dwarf" and "normal" lake whitefish morphs found in several lakes of the St. John River drainage in northern Maine, USA, and southeastern Québec, Canada. Dwarf fish seldom exceed 20 cm in length and 100 g in weight, while normal fish do not mature until 30 cm in length, and commonly exceed 40 cm and 1000 g (Fenderson 1964; Chouinard et al. 1996). The dwarf whitefish are generally characterized by a higher number of gill-rakers but the overlap for this character between morphs varied among lakes (Lu and Bernatchez 1999). Under the hypothesis of phenotype–environment correlation, the extent of trophic niche differentiation should therefore also vary among lakes, and positively correlate with the differences in gill-raker counts between morphs. Bernatchez et al. (1999) tested this prediction by comparing the temporal variation of diet between sympatric dwarf and normal whitefish morphs from Cliff and East Lakes. Since the difference in gill-raker counts between morphs is more pronounced in Cliff Lake (Figure 6.4A), it was predicted that the extent of trophic niche partitioning should also be more pronounced in that lake. This study revealed that trophic niche overlap between dwarf and normal morphs varies among lakes, and is positively correlated with their differences in gill-raker counts. When estimated for both lakes and over all samples, the weighted importance of zooplanktonic prey was significantly higher in the dwarf (mean = 0.91, range = 0.70–0.99) than in the normal morph (mean = 0.28, range = 0–0.99) (Mann–Whitney U-test; $Z = 1.92$, $P = 0.05$). However, the extent of overlap in diet contrasted between both lakes (Figure 6.4B). Zoobenthos and fish prey predominated the diet of normal adult fish in Cliff Lake at the beginning (early June) and the end (late August) of the growing season, and zooplankton were essentially absent from their diet. In East Lake, the diet of normal adult fish was almost exclusively composed of benthic prey (large zoobenthos and molluscs) in June. In August, those prey were absent, and small zoobenthos represented approximately 50% of stomach content, the other half being composed of terrestrial insects and zooplankton, which composed 17% of the diet. Differences in trophic use among lakes were also observed between the dwarf ecotype and juvenile fish of the normal ecotype. In August, the diet of juvenile normals from Cliff Lake included almost exclusively small and large benthic prey, whereas that of dwarf fish mainly comprised zooplankton. This resulted in a low index of niche overlap $(D = 0.390)$ (Schoener 1970). In contrast, an almost complete overlap $(D = 0.968)$ was observed in East Lake in August.

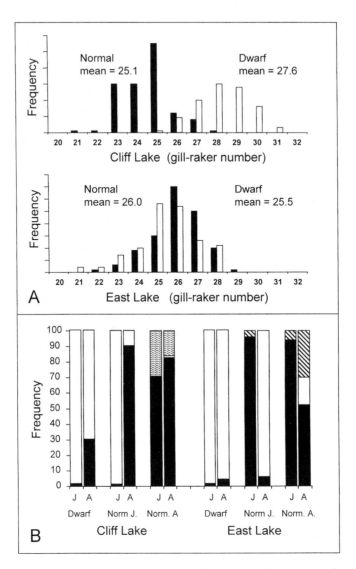

Figure 6.4. (A) Frequency distribution of gill-raker numbers between dwarf (white) and normal (black) forms of lake whitefish from Cliff and East Lakes. (B) Diet composition expressed as the weighted importance (in percent) of prey categories for the dwarf form, and the juvenile (Norm J.) and adult (Norm A.) stages of the normal whitefish form in East and Cliff Lakes for early (June: J) and late (August: A) in the growing season. White (zooplankton); black (zoobenthos); waves (fish); cross-hatched (terrestrial insects). Modified from Bernatchez et al. (1999).

Globally, the above results indicate that variation in gill-raker counts fulfills the adaptive criterion in sympatric whitefish morphs from Europe and North America. However, a recent study on North American Great Lakes ciscoes presented a contrasting exception to this rule and reminds us that phenotype–environment associations cannot be taken for granted without rigorous empirical appraisal. Turgeon et al. (1999) and Dupont (2000) compared morphological variation with trophic and spatial niches among Lake Nipigon (Ontario, Canada) ciscoes in two different years and at two different times (early July and late August). The ecomorphological data confirmed the existence of four distinct morphotypes that could be equated to the description of Koelz (1927) and Scott and Crossman (1973) for C. *artedii*, C. *zenethicus*, C. *nigripinnis*, and C. *hoyi*. Unexpectedly, morphotypes with similar gill-raker counts (C. *artedii* and C. *hoyi*, both with a modal count of 46) differed the most in their diet. In contrast, diet overlap was more important between C. *zenethicus* and C. *nigripinnis* (modal counts of 36 and 49, respectively). This pattern was observed for both time periods surveyed. Therefore, the null hypothesis of no phenotype–environment correlation between gill-raker counts and diet could not be rejected. Since their niche partitioning was largely based on vertical habitat segregation, it is possible that a phenotype–environment correlation in North American Great Lakes ciscoes may lie more in terms of physiological adaptation to occupy strikingly different water depths. This, however, remains to be evaluated.

2.2.4. Trait Utility

Trait utility is usually evidenced by testing whether particular phenotypic traits enhance performance in the environment with which individuals having different phenotypes are associated. This is generally done by reciprocal transplant experiments. For instance, by experimentally confining individuals of the limnetic and benthic morphs of the three-spine stickleback (*Gasterosteus* sp.) to their preferred and alternate habitats, Schluter (1995) showed that in its preferred habitat, each morph grows at about twice the rate it does in the alternate habitat, most likely due to difference in the efficiency of capturing and handling alternate prey types. Such experiments are not readily applicable to deep-water salmonids. Stress effects due to cage confinement would most certainly bias the experimental outcomes in such highly mobile organisms. Instead, trait utility could be tested without experimental manipulations by comparing energy budgets in natural populations using mass balance models (Rowan and Rasmussen 1997; Trudel et al. 2000). By quantifying the concentration of chemical elements such as mercury (Hg) or radioactive cesium (^{137}Cs), such models allow accurate quantification of food consumption rates and conversion efficiency of assimilated energy into growth. Trudel et al. (2000) have recently validated a mercury mass balance model for coregonines. Using this model, they showed that energy allocation to metabolism was 50–60% more important in dwarf lake whitefish relative to the normal morph, possibly due to increased swimming activity associated with capturing zooplanktonic prey. Interestingly, the energetic budget of the dwarf whitefish closely matched that of lake cisco (C. *artedii*), which occupies the same functional role and represents its most likely competitor (Trudel et al. 2001).

Trait utility remains to be tested in coregonines, but is currently being examined (L. Landry and L. Bernatchez unpublished) by documenting conversion efficiency using the mercury mass balance model of Trudel et al. (2000). The number of gill-rakers in the dwarf morph of lake whitefish is highly variable both among individuals within populations, and among populations from different lakes. Under the hypothesis that more gill-rakers increase the efficiency of feeding on zooplankton, we are testing the prediction that conversion efficiency of assimilated energy into growth should be more pronounced, both for individuals with higher gill-raker counts within populations, and in populations characterized by higher gill-raker counts on average. Of course, this assumes that different phenotypic variants of the dwarf morph feed on the same zooplanktonic prey, which we are also documenting.

2.3. Assessing the Process of Divergent Natural Selection in Adaptive Radiation

The next step in testing the theory is to provide evidence for the role of divergent natural selection in the three processes that drive adaptive radiation. The idea that divergent natural selection is the ultimate cause of phenotypic divergence and species origins is not new in the realm of researcher investigating coregonine fishes. Smith and Todd (1984) proposed almost 20 years ago that divergence and speciation of Great Lakes ciscoes may be driven by the three processes explicitly inferred from the theory. Firstly, they pointed out that lacustrine environments are characterized by steep ecological gradients that will submit members of a given lineage to distinct environments, and therefore, divergent selective pressures. Secondly, they proposed that competition may force divergence of subpopulations to different utilization of resources such as food and breeding sites. Consequently, Smith and Todd (1984) predicted that "... sister-species will differ primarily in characters that are functionally related to competition for such resources." Thirdly, they hypothesized that the use of distinct resources will also promote temporal and habitat segregation for spawning. That is, Smith and Todd (1984) envisioned that reproductive isolation may evolve as a byproduct of divergent natural selection. Since then, however, little progress has been made toward more rigorously establishing the role of divergent selection in driving phenotypic divergence and reproductive isolation in coregonines. In this section, I make use of available information from the literature to achieve this.

2.3.1. First Process: Phenotypic Differentiation between Populations and Species Is Caused Directly by the Environments They Inhabit and the Resources They Consume

Several approaches are available for testing the hypothesis that phenotype–environment associations are driven by divergent natural selection, including reciprocal transplant experiments, direct measurements of selection, and estimates of adaptive landscapes from resource distributions (Schluter 2000). The most readily applicable for species such as large salmonids, however, is the comparison of population differentiation with neutral expectations (the amount of differentiation expected by mutation and genetic drift only). To this end, the Q_{ST} method (Spitze 1993) has been the most frequently applied (Podolsky and Holtsford 1995; Lynch et al. 1999). Its theoretical principle is that divergence in quantitative traits should be similar to that of allele frequencies at nuclear loci, if they are evolving neutrally and have a quasi-pure additive genetic basis (Wright 1951). Under the influence of migration, mutation, and genetic drift, the among-population proportion of total genetic variance in phenotypic traits is expected to equal that of neutral molecular loci (Lande 1992). As an indirect method for the detection of natural selection, one can therefore compare the extent of population differentiation at quantitative traits (Q_{ST}) with that quantified at neutral molecular markers (F_{ST}). The prediction is that divergent selection will cause Q_{ST} to be larger than that expected on the basis of neutral loci.

Q$_{ST}$ is quantified as the proportion of additive genetic variance in quantitative traits attributable to differences among populations:

$$Q_{ST} = (\sigma^2_{GB}/\sigma^2_{GB} + 2\sigma^2_{GW})$$

where σ^2_{GB} and σ^2_{GW} are the additive components of genetic variance between and within populations, respectively (Spitze 1993). Strict use of the Q_{ST} method therefore implies that knowledge of genetic variation for quantitative traits must be obtained. However, the method has also been applied using phenotypic variance (e.g., Merilä 1997), under the assumption that it is an accurate surrogate for additive genetic variance. Merilä (1997) showed that only a very large environmental component of phenotypic variance would affect conclusions of Q_{ST} analyses. Many other investigators have also emphasized the similarity between genetic and phenotypic variance (e.g., Roff and Mousseau 1987; Cheverud 1988; Roff 1995, 1996a). Moreover, measures of phenotypic covariance have been shown to be nearly as successful as those of genetic covariance in predicting the direction of divergence between species (e.g., Schluter 1996). Consequently, results of Q_{ST} studies based on either phenotypic or genotypic variance do not differ in their general patterns of F_{ST}–Q_{ST} relationships (Lynch et al. 1999; Schluter 2000).

I performed a Q_{ST} analysis using the data from Lu and Bernatchez (1999) who documented phenotypic variation of eight meristic and 18 morphometric characters, as well as molecular genetic differentiation at six microsatellite loci among six sympatric pairs of dwarf/normal morphs of lake whitefish. Given the evidence for a phenotype–environment association of gill-raker counts with prey types, and that this character is the most discriminant one among coregonine species, the null hypothesis to be tested was that Q_{ST} estimated from gill-raker counts will not significantly differ from either F_{ST} derived from microsatellites or Q_{ST} values observed for other phenotypic traits. Meristic data were untransformed, whereas the univariate residual method was used to adjust each morphometric character for size heterogeneity among individuals (Fleming et al. 1994). That is, raw data were logged and used to establish regression lines describing the relationship between each character and length. Residuals from these regressions were used as variables for further analyses. Components of phenotypic variance were estimated by performing an analysis of variance. The phenotypic variance was equal to twice the observational component of variance for individuals within populations and used as a surrogate for $2\sigma_{GW}^2$. The phenotypic variance between populations was equated to the observational variance component for populations and used as a surrogate for σ_{GB}^2. Q_{ST} values were then contrasted with F_{ST} values estimated using the method of Weir and Cockerham (1984). Values were considered significantly different when their 95% confidence intervals did not overlap.

The first analysis compared Q_{ST} and F_{ST} estimates between dwarf and normal morphs from the same lake ($n = 6$ lakes) (Figure 6.5A, Table 6.1). Q_{ST} was calculated for each pair and then averaged across the six lakes. Mean F_{ST} and Q_{ST} values averaged over all traits and all lakes were very similar and not significantly different. When considering individual traits, the highest Q_{ST} was observed for gill-raker counts, which was also the only trait that was significantly higher than F_{ST}, exceeding its mean value by 0.27. This supports the hypothesis that differences in gill-raker counts between sympatric morphs are under the influence of divergent selection.

I then assessed the differential influence of divergent selection on dwarf and normal morphs. If selection is acting in the same way on both morphs, Q_{ST} estimates between either dwarf–dwarf comparisons and normal–normal comparisons should be similar, whereas they should be significantly different under differential selective regimes. Q_{ST} and F_{ST} estimates were compared in all pairwise comparisons of normal populations from different lakes (Figure 6.5B). Mean F_{ST} and Q_{ST} did not differ significantly (Table 6.1). Q_{ST} for gill-raker number did not differ from neutral expectation, indicating that this trait is not strongly influenced by divergent selection in normal populations. None of the other traits had a Q_{ST} significantly higher than F_{ST}, whereas two (interorbital width and anal fin length) were significantly lower. This suggests that these traits could be under the influence of balancing selection (Spitze 1993), although this

Figure 6.5. Comparison of F_{ST} and Q_{ST} values (with their 95 confidence intervals) for meristic and morphometric traits between (A) dwarf and normal forms of lake whitefish from the same lake (6 lakes), (B) normal populations from different lakes (15 pairwise comparisons), and (C) dwarf populations from different lakes (15 comparisons). Symbols from left to right in all panels refer to (1) mean F_{ST} values derived from microsatellite data, (2) mean Q_{ST} value over all traits, (3) scales above the lateral line, (4) suprapelvic scales, (5) lateral line scales, (6) dorsal ray counts, (7) anal ray counts, (8) pectoral ray counts, (9) pelvic ray counts, (10), gill-raker counts, (11) pre-orbital length, (12) orbital length, (13) post-orbital length, (14) trunk length, (15) dorsal length, (16) lumbar length, (17) anal fin length, (18) caudal peduncle length, (19) maxillary length (20) mandible length, (21) maxillary width, (22) pectoral length, (23) pelvic length, (24) body depth, (25) head depth, (26) caudal peduncle height, (27) adipose fin length, and (28) interorbital width. Horizontal dashed lines delineate 95% confidence intervals for F_{ST} measured from six microsatellite loci.

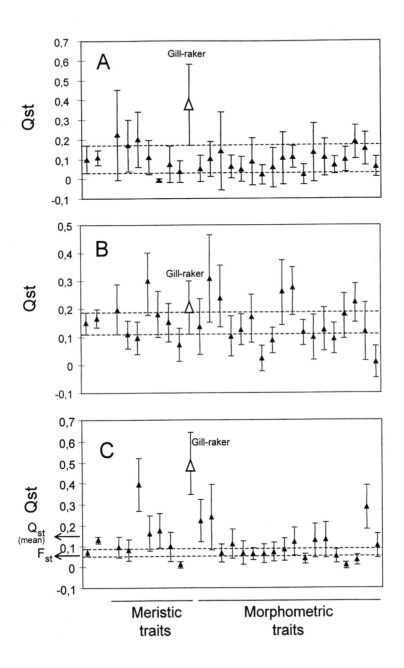

Table 6.1. Estimates of differentiation (means ± 1 S.E.) at six microsatellite loci (F_{ST}) and morphological data (Q_{ST}) between sympatric dwarf and normal forms of lake whitefish from the same lake, normal forms from different lakes, and dwarf forms from different lakes: East Lake, Témiscouata Lake, Crescent Pond, Webster Lake, Indian Pond, and Cliff Lake. Meristic and morphometric variables are listed in legend of Figure 6.5.

	Comparison		
	Dwarf–Normal	Normal–Normal	Dwarf–Dwarf
F_{ST}	0.100 ± 0.069	0.144 ± 0.037	0.068 ± 0.017
Q_{ST} (total)	0.107 ± 0.038	0.154 ± 0.167	0.130 ± 0.016
Q_{ST} (all meristics)	0.147 ± 0.092	0.163 ± 0.035	0.188 ± 0.026
Q_{ST} (all morphometrics)	0.089 ± 0.034	0.150 ± 0.023	0.104 ± 0.018
Q_{ST} (gill-raker number)	0.375 ± 0.204	0.198 ± 0.087	0.494 ± 0.147

remains speculative in the absence of a clear prediction of phenotype–environment associations for these characters.

If selection is indeed responsible for differential gill-raker counts between sympatric dwarf and normal morphs, and gill-raker counts in the normal morph are predominantly driven by neutral processes, it necessarily follows that divergent selection must be mainly acting on the dwarf morph. Q_{ST} for dwarf whitefish should therefore exceed that observed for the normal morph. This is clearly the case as the overall Q_{ST} estimate is significantly higher than F_{ST} (Table 6.1, Figure 6.5C). Gill-raker number is the trait most strongly deviating from the neutral expectation. Indeed, departures from neutral expectations were much more pronounced in dwarf–dwarf (0.42 above mean F_{ST} value) than in sympatric dwarf–normal (0.27 above mean F_{ST} value) comparisons. Q_{ST} for several additional traits was either significantly higher (lateral line scales, anal ray counts, preorbital length, adipose fin length) or lower (pelvic ray count, head depth) than neutral expectations. As a functional phenotype–environment correlation seems unlikely for at least some of these traits (e.g., lateral line scales, anal ray counts, or adipose fin length), these results may indicate the role of local, non-genetic environmental influences on phenotypic variance observed among populations. Therefore, the possibility that environmental influences could also affect the pattern observed for gill-raker counts cannot readily be dismissed, although evidence for its strong additive genetic basis makes it less likely for this trait than the others.

The results of the Q_{ST} analysis should be interpreted cautiously given that phenotypic rather than genetic variance was used. Yet, they leave little doubt that differences in gill-raker numbers are mainly driven by divergent natural selection that is predominantly acting on the dwarf morph. This is consistent with the fact that the diet of dwarf whitefish almost exclusively comprises zooplanktonic prey. Depth selection and swimming behaviors also seem to be under divergent selection between dwarf and normal whitefish, as revealed by F_{ST}/Q_{ST} comparisons (Rogers et al. 2002).

2.3.2. Second Process: Divergence in Phenotype also Results from Competitive Interactions and Ecological Opportunity

The role of interspecific resource competition in driving phenotypic divergence is best evidenced by the demonstration of ecological character displacement (Schluter 2001a). Such evidence may be predictive, experimental, and observational. The predictive approach uses optimality models to generate quantitative expectations of mean phenotypes under character displacement and compares these expectations with predictions of models not inferring competition. This approach has been rarely used, most likely because it requires *a priori* estimates of natural selection pressures stemming from resources. Field experiments on character displacement, whereby changes in competition, natural selection, and evolution are measured directly after addition and deletion of phenotypically distinct and closely related species, are also scarce. Schluter (1994) contrasted selection on a phenotypically intermediate species of three-spine stickleback raised alone with that when a zooplanktivorous stickleback species was added. As predicted by the character displacement hypothesis, addition of the zooplanktivore generated selection in favor of the more benthic phenotype within the intermediate species. See also Schluter (2003).

To a large extent, support for the role of divergent character displacement in adaptive radiation has been acquired by observational evidence (Schluter 2001a). The pattern most commonly inferred is "exaggerated divergence in sympatry," whereby phenotypic differences between evolutionary lineages are greater when found in sympatry than in allopatry. In coregonines, the best evidence for character displacement comes from the observations of shifts in gill-raker counts in North American whitefish populations when found in sympatry with closely related putative competitors. Lindsey (1981) compared mean gill-raker counts of allopatric populations (populations not occurring as members of sympatric pairs) of whitefish from the Yukon in the presence or absence of the Least cisco (*C. sardinella*), a densely rakered (averaging 45 gill-rakers) zooplanktivore. As predicted by the character displacement hypothesis, populations of whitefish found in sympatry with the cisco had significantly fewer gill-rakers (23.9 ± 0.83) than their congeners found in allopatry (26.0 ± 0.74) (t-test: $t = -5.99$, df $= 18$, $P < 0.001$), and were therefore phenotypically more divergent from *C. sardinella* than were the allopatric populations (Figure 6.6A).

Another example involves phenotypic variation in gill-raker numbers within the Acadian lineage of lake whitefish in Northeastern North America. This lineage, first identified by the phylogeographic analysis of mitochondrial DNA variation, evolved in geographic isolation from other whitefish lineages for approximately 150,000 years (Bernatchez and Dodson 1991). However, allopatric populations of this group and those from other glacial lineages are phenotypically indistinguishable, all harboring the normal phenotype (Lindsey et al. 1970). The St. John River basin is a zone of secondary contact between the Acadian and Atlantic lineages, and the occurrence of dwarf/normal sympatric whitefish pairs in lakes of this basin is typically associated with the occurrence of

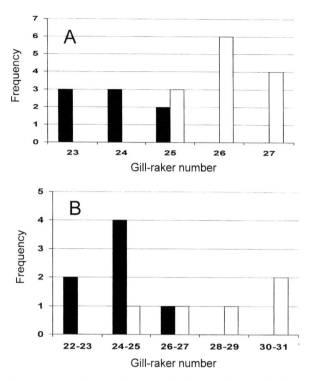

Figure 6.6. Frequency distribution of mean gill-raker numbers in (A) 21 whitefish populations from the Bering Sea drainage found in sympatry with the Least cisco (C. *sardinella*)
(black) and in allopatry (white), (B) 12 lake whitefish populations from the Acadian
lineage found in sympatry with lake whitefish from the Atlantic lineage (white) and in
allopatry (black). Data for (A) are from Lindsey (1981), and those for (B) are from Edge et
al. (1991) and Lu and Bernatchez (1999).

both lineages (Bernatchez and Dodson 1990; Pigeon et al. 1997). Although
introgressive hybridization has occurred between lineages, the dwarf populations
systematically remained genetically more similar to pure Acadian than to pure
Atlantic populations (Lu et al. 2001). This indicates that the dwarf phenotype
evolved post-glacially from ancestral Acadian populations of the normal phenotype, and following secondary contact with the Atlantic lineage. Thus, a higher
gill-raker count in the planktivorous dwarf whitefish relative to that observed in
other populations of pure Acadian origin would support the role of character
displacement in driving phenotypic divergence. This is indeed the case, as
revealed by significantly higher mean gill-raker counts (28.0 ± 2.55) in dwarf
populations (Lu and Bernatchez 1999) than in pure Acadian populations from
the Maritime provinces of Canada (Edge et al. 1991) (mean: 24.9 ± 2.48) (*t*-test:
$t = -2.18$, df $= 9$, $P = 0.05$) (Figure 6.6B).

    Strictly speaking, six criteria should be satisfied to conclusively infer char-
acter displacement (Robinson and Wilson 1994). Variation in gill-raker numbers
among whitefish populations satisfy at least four of them. (1) Phenotypic differ-

ences should have a genetic basis. As stated above, the genetic basis of gill-raker numbers in coregonines has been clearly established. (2) Chance should be ruled out as an explanation of the pattern. The significantly different distribution of gill-raker numbers among populations in presence or absence of putative competitors ruled out this possibility. (3) Differences must represent evolutionary shifts and not lineage sorting. This has been established by evidence from molecular phylogenetics that populations showing "displaced traits" are derived from ancestral populations that were characterized by the normal phenotype (Bernatchez et al. 1999). (4) Shifts in resource use should match phenotypic changes. This has also been demonstrated by the correlation between gill-raker numbers and prey size. (5) Environmental differences between sites of sympatry and allopatry must be controlled. The possibility of environmental differences among lakes that harbor whitefish populations with and without putative competitors cannot be ruled out. However, "displaced populations" are found in a wide variety of lake types that likely encompasses the diversity seen in lakes without putative coregonine competitors. (6) Independent evidence should be gained that similar phenotypes compete for resources. Direct evidence for competition is lacking, as in the majority of character displacement studies (Schluter 2001a). The only observational support for competition is provided by a generally slower growth rate observed among displaced populations (e.g., the dwarf morph of the Acadian lineage) relative to populations found in absence of putative competitors (Fenderson 1964).

Further evidence that competitive interactions influence phenotypic divergence is provided by associating an increased span of resource use (or increased variance in correlated phenotypic traits) with situations of ecological opportunity, such as the absence of unrelated competitors. The theory of character displacement predicts an increase in trait variance when a competitor is removed. An increase in variance may be manifested either as polymorphism within populations, as well as genetically distinct, sympatric populations (Robinson and Schluter 2000). In coregonine fishes, the generality of this phenomenon has been qualitatively documented by Lindsey (1981), who summarized data on the extent of phenotypic variance (largely in terms of gill-raker counts) in the presence or absence of potentially competing lineages (Figure 6.7). Lake Superior (center of the figure) harbors seven sympatric species, including three lineages typically associated with different trophic use. These include the more densely rakered ciscoes (then defined by Lindsey as the subgenus *Leucichthys*), the whitefish (then defined by Lindsey as the subgenus *Coregonus*) with an intermediate number of gill-rakers, and the genus *Prosopium*, typically characterized by very low gill-raker numbers. Such a differential pattern of gill-raker counts among lineages is repeated in thousands of North American lakes (Scott and Crossman 1973). On either side of Lake Superior, counts are shown for other lakes where one or two of the three lineages are lacking. In these lakes, the range of counts can be seen to be at least partially filled by a remaining group. In several lakes of the Yukon and Northern Maine, the absence of ciscoes have apparently favored the evolution of a high gill-raker morph of lake whitefish that occupy the zooplanktonic niche. That these morphs evolved in parallel in phylogenetically distinct

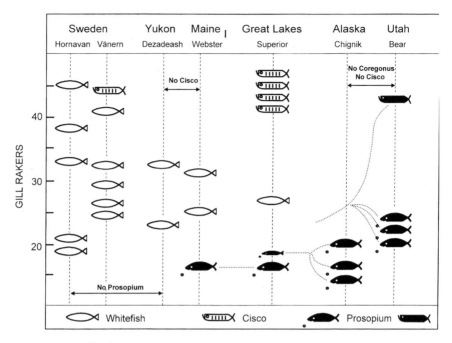

Figure 6.7. Gill-raker numbers of coregonine fishes from seven lakes. Each symbol represents the mean count of a given population or species in that lake. The figure shows that in the absence of one or the other members of the three basic coregonine groups (Whitefish, Cisco, *Prosopium*), the gill-raker "niche" of these groups tends to be filled by populations that diverge from the basic pattern of a given group. Modified from Lindsey (1981).

lineages further supports the view that their phenotypic divergence has been driven by natural selection (Bernatchez et al. 1996; Pigeon et al. 1997). Whitefish are more diversified in Europe, where there is only one species of cisco and *Prosopium* is absent. There, whitefish populations span the whole range of gill-raker numbers to include counts that typically characterize either the densely rakered ciscoes or the sparsely rakered *Prosopium*. Alternatively, Svärdson (1998) proposed that pronounced introgressive hybridization between phenotypically distinct ancestral whitefish lineages may account for increased phenotypic variance observed in these lakes. This hypothesis, however, remains to be rigorously tested by molecular phylogenetics.

Back to North America in the Bristol region of Alaska, pygmy whitefish (*P. coulteri*), typically characterized by low gill-raker counts, occupy lakes where both whitefish and cisco are absent, a very rare situation. Chignik Lake contains three sympatric morphs of *P. coulteri*, which differ in gill-raker counts and trophic use. The one with the highest number of gill-rakers is a plankton feeder, whereas the other two are shallow- and deep-water benthic feeders. An even more spectacular case of niche shift in the genus *Prosopium* has also been documented in Bear Lake, Utah, where both cisco and whitefish are also absent. Four species of *Prosopium* occur in the lake and are characterized by distinct gill-raker numbers. Most remarkable is *P. gemmiferum* which shows gill-raker counts typical of ciscoes. It is also externally more similar in appearance to ciscoes than any other *Prosopium* species. Molecular systematics, however, have confirmed that it recently evolved from a sparsely rakered ancestor (Bernatchez et al. 1991; Vuorinen et al. 1998). These qualitative observations support the view that in situations of ecological opportunity, divergent selection has driven the evolution of phenotypically distinct populations that occupy niches left empty by the absence of competitors.

More quantitative support for increased phenotypic variance in the absence of competitors comes from data compiled on lake whitefish in Northern Québec (Figure 6.8). Doyon et al. (1998) documented variation in size and age at maturity among 34 lakes of the La Grande Complex where lake cisco (C. *artedi*) is present (western part of the basin) or totally absent (eastern part) for biogeo-

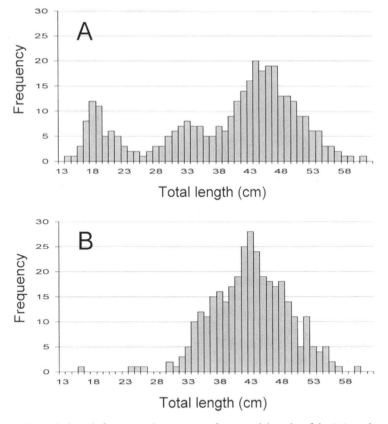

Figure 6.8. Body length frequency histograms of mature lake whitefish; (A) in the western part of the same drainage where cisco is absent (18 water bodies, $n = 623$), and (B) in the eastern part of the La Grande R. drainage where cisco (C. *artedi*) is abundant (16 water bodies, $n = 3149$). Modified from Doyon et al. (1998).

graphic reasons (Legendre and Legendre 1984). In lakes where cisco is present, lake whitefish populations are characterized by a unimodal distribution in size and age at sexual maturity, as typically observed in populations of the normal morph (Figure 6.8B). In contrast, whitefish populations from lakes where cisco is absent show increased variance in size at sexual maturity, and are characterized by the appearance of a second mode at approximately 200 mm (Figure 6.8A). Both the mean (Mann–Whitney U-test, $Z = -2.97$, $P = 0.003$) and variance (Levene test, $F = 142.93$, $P < 0.0001$) significantly differed between the two groups. It has also been shown that early maturing fish (dwarf morph by definition) compose populations that are genetically distinct from the sympatric late maturing fish (Bernatchez 1996). As the diet of both morphs differs (Doyon et al. 1998), these results support the view that an absence of cisco created the ecological opportunity necessary to promote the evolution of the recently derived dwarf whitefish populations in this river basin.

2.3.3. Third Process: Reproductive Isolation Develops as a Consequence of Divergent Natural Selection (Ecological Speciation)

The role of divergent selection in driving speciation has been the least investigated of the three processes inferred from the ecological theory of adaptive radiation. Only a handful of studies have empirically tested the hypothesis of ecological speciation in nature (Schluter 2001b). The most straightforward test is to compare the rate at which reproductive isolation evolves between regions differing in strength of natural selection. A typical example is the comparative analysis of the strength of reproductive isolation as a function of time of divergence between island and continental pairs of populations (Coyne and Orr 1997). An analogous test was recently performed for north temperate freshwater fishes, which included many coregonine taxa (Bernatchez and Wilson 1998). Pleistocene glaciations had a much more direct impact on fish habitats at northern than southern latitudes, particularly in North America. Cyclic glacial advances and retreats had two major effects. First, glacier advances have apparently been largely responsible for a steep decline in fish species diversity above 50° N (Robinson and Schluter 2000). Secondly, a multitude of pro-glacial lakes were formed following glacial retreats at northern latitudes, and access to these was partly hampered by the limited duration of dispersal routes (Hocutt and Wiley 1986). For aquatic species that were able to gain early access to them, these environments offered an ecological opportunity that may have promoted phenotypic divergence through divergent selection, as inferred from the second process of the theory. In contrast, fish species in nonglaciated regions would have had fewer opportunities for ecological release, due to the presumably greater temporal and environmental stability of communities and habitats. Under the hypothesis that this same process has also been responsible for the development of reproductive isolation, one would then predict that speciation events occurring in recent evolutionary times should be more prevalent at northern than southern latitudes. In order to test this prediction, Bernatchez and Wilson

(1998) quantified mitochondrial DNA sequence divergence between sister species as a function of their median latitude of distribution, under the assumption
that smaller divergence estimates reflect younger speciation events. A highly
significant negative relationship between sister-species divergence and latitudinal
distribution was observed, and a piecewise linear regression model explained
74% of the genetic divergence among species pairs (Figure 6.9). The breakpoint
between the two linear relationships was at 46°N, approximately coincident
with the median latitude of maximum Pleistocene glacial advance in North
America (44°N; Fulton and Andrews 1987). Five coregonine sister-species
pairs were included in this meta-analysis, and all ranked among the lowest
sequence divergence estimates observed, indicating that speciation events have
been particularly recent for this group as a whole. These results are consistent
with the hypothesis that ecological opportunity stemming from depauperate fish
diversity in new and favorable habitats has contributed to an elevated rate of
speciation in freshwater fishes at northern latitudes, and particularly in coregonines.

   A second test of ecological speciation in coregonine fishes compared the
strength of reproductive isolation among population pairs that evolved within
the same time frame but that differ in the extent of trophic specialization, under
the assumption that differentiation in traits involved in niche occupation reflects
the intensity of divergent natural selection in different environments. The

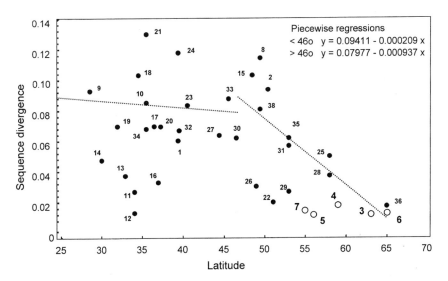

Figure 6.9. Relationships between sequence divergence estimates among pairwise comparisons of sister species as a function of median latitudinal distribution for Nearctic and
Palearctic freshwater and anadromous fish species. Numbers refer to species listed in
Appendix 2 of Bernatchez and Wilson (1998). Coregonine species pairs are indicated
by open circles. Piecewise regressions that best explained the total variance below and
above the breakpoint (46°) are given in the graph. Modified from Bernatchez and Wilson
(1998).

amount of gene flow occurring between sympatric morphs should decrease as the strength of their reproductive isolation increases. Thus, the extent of genetic divergence at neutral loci can be used as a surrogate for their reproductive isolation. If ecological processes are important in driving the reproductive isolation of sympatric morphs, gene flow should be more restricted between sympatric populations that are more specialized for distinct trophic niches. Lu and Bernatchez (1999) compared the extent of morphological differentiation between sympatric dwarf and normal whitefish morphs from six lakes with genetic differentiation at six microsatellite loci in order to test this prediction (Figure 6.10). Dwarf and normal morphs in each lake differed primarily in traits related to trophic specialization (particularly gill-raker counts), but the extent of differentiation varied among lakes. Variable extents of genetic divergence between morphs within lakes were also observed. The extent of gene flow between morphs within a lake and their morphological differentiation was negatively correlated ($r = 0.78$, $P = 0.06$), a result consistent with the prediction of ecological speciation. An alternative explanation is that the causal scenario is reversed, whereby the strength of divergent selection is similar everywhere, and morphological divergence is primarily driven by variable amounts of gene flow occurring in each lake due to factors independent of speciation processes (Hendry et al. 2001a). Available information, however, does not support this possibility. For instance, gene flow is most restricted in Cliff Lake, a relatively small lake where both morphs use the same spawning grounds located on midwater shoals (Chouinard and Bernatchez 1998). In contrast, gene flow was more important in Témiscouata Lake, a very large water body where both morphs use distinct spawning habitats located in the littoral zone and a small tributary

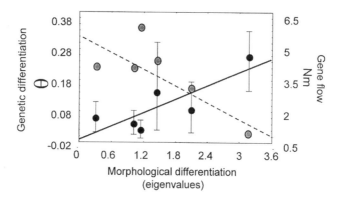

Figure 6.10. Relationships between morphological differentiation (eigenvalues of discriminant function analysis) and the extent of genetic differentiation (θ estimator of F_{ST} in black) or gene flow (Nm values in gray) estimated from the private-allele method between sympatric dwarf and normal forms of lake whitefish. Confidence intervals (means ± 1 S.E.) are provided for θ values. From left to right, symbols represent East Lake, Témiscouata Lake, Crescent Pond, Webster Lake, Indian Pond, and Cliff Lake, respectively. Modified from Lu and Bernatchez (1999).

(Lamoureux and Sylvain 1986). Secondly, measurements of the trophic resources available suggest that the strength of divergent selection differs among lakes (Bernatchez et al. 1999; L. Landry and L. Bernatchez unpublished). Consequently, the most parsimonious explanation at this time is that the extent of reproductive isolation reached between sympatric whitefish morphs evolved as a consequence of the intensity of local divergent selection. Interestingly, an analogous scenario involving a negative correlation between the extent of spatial niche partitioning and that of genetic divergence, independently of the potential for gene flow (geographic distance between spawning grounds), has recently been documented among four sympatric populations of landlocked Atlantic salmon (Potvin and Bernatchez 2001). A similar pattern has also recently been reported for the Arctic charr (Gíslason et al. 1999). These observations are consistent with the hypothesis that the extent of gene flow occurring among ecologically differentiated morphs of salmonids is more a consequence of selective processes that are driving the development of intrinsic reproductive isolation, than a primary cause for the observed patterns of phenotypic differences.

3. Future Directions

3.1. Coregonine Fishes: A Useful Model System for Further Investigation of Adaptive Radiation

Schluter's (2000) synthesis convincingly showed that our state of knowledge on the ecology of adaptive radiation has substantially progressed over the last few decades. Yet, the number of comprehensively documented cases of adaptive radiation are still relatively few, the vast majority of organisms having been investigated to test only a subset of the theory. In this chapter, I aimed at assessing the usefulness of coregonine fishes as a model system from which the ecological theory of adaptive radiation could be tested more broadly. Although our state of knowledge is still fragmentary on several issues, there is sufficiently detailed information to support the hypothesis that population phenotypic and ecological divergence, as well as reproductive isolation, has been driven by divergent natural selection. Several coregonine species complexes, particularly within the C. *clupeaformis–lavaretus* clade, have fulfilled the criteria tested to date that define an adaptive radiation. In contrast, results obtained for the Great Lakes ciscoes were more ambiguous.

Coregonine fishes should also be useful for studying several important ecological questions that have deserved less attention to date. For instance, because their divergence spans a time frame that includes both recent (several thousands of years) and more ancient (several millions of years) speciation events, they would be particularly relevant to investigate the rate of phenotypic divergence as an adaptive radiation progresses in time (e.g., Lynch 1990; Kinnison and Hendry 2001). The study of coregonine fishes should also move from correlative toward predictive approaches in inferring the link between phenotypic divergence and environmental features that generate alternate fitness peaks (e.g.,

Schluter and Grant 1984). Because lakes that harbor coregonine populations are often well circumscribed and relatively small ecosystems, it should be more feasible than for many other systems to systematically determine the environmental and resource features that vary among lakes and therefore, shape the multidimensional surface of the adaptive landscape. Available information to date suggests that the steepness of fitness hills and valleys differs from one lake to another and that the extent of both phenotypic divergence and reproductive isolation matches these expectations.

3.2. From Genetics to Ecology and Back

A major research area that will deserve particular attention in the following years, and to which the study of coregonine fishes can contribute, is the elucidation of the genetic basis of adaptive radiation. While a large amount of ecological information now supports the role of divergent natural selection as the main cause of population divergence and speciation, the genetic mechanisms underlying the process remain a black box. Because the genetic architecture of phenotypic traits determines how selection operates on them (Lynch and Walsh 1998), a refined understanding of the role of natural selection in an adaptive radiation requires detailed knowledge of the genetic basis of adaptation and speciation. The genetics of speciation has been a fruitful area of research for decades, however, researchers have previously been primarily interested in mechanisms of hybrid sterility and inviability, without much concern for genotype–environment interactions in natural conditions (Orr 2001). Given the state of progress on the ecological mechanisms of population divergence, and that technology now exists to search whole genomes for genotype–phenotype associations, it is time to investigate the genetic basis of phenotypic changes and reproductive isolation within the conceptual framework offered by the theory of adaptive radiation. In this context, the general question to be answered is: what "kinds" of genetic changes driven by divergent selection have actually led to phenotypic divergence and reproductive isolation during the course of an adaptive radiation? As for ecological mechanisms, the genetic basis of adaptive radiation should be studied in the early steps of the process, as genetic differences that accumulate after it is completed are less relevant. Elucidating the genetic basis of adaptive radiation will benefit from the synergic use of three complementary fields of genetics: "traditional" quantitative genetics, QTL (quantitative trait loci) mapping, and molecular/functional genomics. I outline below several important issues that can be addressed by each of these approaches.

3.3. The Genetic Basis of Adaptive Radiation

3.3.1 Traditional Quantitative Genetics

There are many reasons why the use of traditional quantitative genetics to document the extent and nature of genetic variance/covariance that constrains the evolution of phenotypic traits should be increasingly adopted in research programs on adaptive radiation. Most importantly, the magnitude and structure of genetic variation in characters, as well as the modes of inheritance (e.g., additivity vs. dominance), directly influence the rate and direction of both the response to natural selection and genetic drift (Lande 1979). Thus, the initial stages of divergence between some populations and species tend to accumulate in the direction of maximum genetic variance (G_{max}) within a population (e.g., Via and Lande 1985; Schluter 1996). It is also clear that phenotypic traits coevolve so as to optimize fitness in specific environments. Thus, morphological integration predicts that such functionally related traits will be genetically (and phenotypically) correlated and evolve as a cohesive unit. If true, the genetic variance/covariance matrix (G) for integrated traits should be subjected to stabilizing selection in order to maintain functional and developmental relationships (Lande 1980, Cheverud 1984). We therefore need to explicitly test the general hypothesis that G evolves to fit adaptive landscapes, and thus, G and the direction of adaptive divergence resemble each other because they respond to the same selective pressures. We also need to test whether the nature of genetic constraints to divergence varies depending on phenotypic traits (behavioral, morphological, life history). One possible way to assess these issues is to document genetic and phenotypic variance/covariance matrices in replicate systems in order to contrast the evolutionary direction of "selected" versus "neutral" traits as a function of their genetic variance (e.g., McGuigan 2001).

3.3.2. QTL Mapping

Understanding the genetic architecture of phenotypic variation and species differences involves estimating the number of loci that affect characters leading to different morphs. This also implies determining the magnitude and types of individual and joint effects of these loci in the expression of relevant phenotypic traits. Quantitative trait loci (QTL) mapping techniques offer a promising avenue to achieve this. These methods are not intended to pinpoint specific genes acting on phenotypic expression. Instead, they make use of linkage disequilibrium and recombination properties of experimental crosses to find statistical associations between genetic markers and a phenotype. This, in turn, allows the identification of QTL that correspond to chromosomal regions harboring one or more genes affecting a quantitative trait (reviewed in Tanksley 1993). A well-designed QTL study can determine the number, effects, interactions, and location of genetic factors contributing to reproductive isolation and phenotypic differences (e.g., Rieseberg et al. 1996; Bradshaw et al. 1998; Rieseberg et al. 1999). For traits controlled by small numbers of genes, it is also possible to resolve the epistatic effects of alleles at different loci (Kim and Rieseberg 2001).

At least four questions of high relevance to the study of adaptive radiation can be answered using QTL mapping. First, it can contribute to solving the long-standing conflict over the number and magnitude of genetic changes responsible for speciation (Fisher 1930; Barton 1998; Orr 1998; Schemske and Bradshaw 1999). Current empirical evidence is contradictory, with some studies implicating a large number of loci each with a relatively small effect, that is, infinitesimal

(Kim and Rieseberg 1999; Sawamura et al. 2000), whereas others have found evidence of a small number of loci each with a large phenotypic effect, that is, oligogenic (Tanksley 1993; Bradshaw et al. 1995, 1998). Secondly, QTL mapping can be used to infer the role of natural selection in shaping the genetic architecture of phenotypic traits and reproductive isolation, for instance, by performing a comparative QTL analysis between replicate sympatric morphs (e.g., dwarf and normal sympatric whitefish morphs) that evolved in parallel. Thus, evidence that phenotypically similar morphs that evolved independently differ by the same QTL of comparable effects and interactions would make a strong case in favor of divergent natural selection. Thirdly, the role of divergent natural selection in maintaining phenotypic divergence and reproductive isolation in the face of gene flow could be assessed by contrasting the extent of gene flow between QTL associated with adaptive phenotypic differences (e.g., gill-raker numbers in coregonine fishes) or genetic incompatibility (e.g., embryonic mortality rate; Lu and Bernatchez 1998) with other mapped "neutral" markers that are not in linkage disequilibrium with such traits. The prediction is that gene flow should be more constrained for QTL associated with adaptive phenotypic differences and that this pattern should be repeatable in different systems. Preliminary results illustrated the potential of combining the analysis of QTL mapping and introgressive hybridization in natural populations for testing future hypotheses that can detail the genetic basis of population divergence and reproductive isolation between whitefish morphs (Rogers et al. 2001). Finally, the integrated application of both traditional quantitative genetics and QTL mapping can be used to test the general hypothesis that genetic correlations (owing to pleiotropy or linkage disequilibrium) between traits involved in resource use and mate choice can facilitate both phenotypic/ecological specialization and speciation under divergent selection (Felsenstein 1981; Kondrashov and Mina 1986). For instance, Hawthorne and Via (2001) have recently proposed a model for the role of genetic correlations in specialization and speciation, and tested it by analyzing the genetic architecture of key traits in two specialized host races of the pea aphid (*Acyrthosiphon pisum pisum*; Hemiptera; Aphididae).

3.3.3. Molecular and Functional Genomics

Traditional quantitative genetics and QTL mapping will undoubtedly contribute to our understanding of the genetic architecture of traits involved in adaptive radiation. The major and ultimate challenge, however, will be to identify the actual genes involved, and what mutations (or sorting out of existing genetic variation) are responsible for generating genetic/phenotypic variance. Given the fact that phenotypes can rarely be explained by allelic variants at a single locus, elucidating the genetic basis of an adaptive radiation will be a daunting and lengthy task. Yet, progress accomplished to date in the field of molecular genomics is promising. Clearly, the most important development has been achieved in research on crops and farm animals (Andersson 2001; Mauricio 2001). For instance, a specific gene responsible for variance in fruit size has recently been found in tomato (Frary et al. 2000). The mutations in a gene affecting fertility in sheep have also been identified (Gallaway et al. 2000). Closer to speciation studies, Ting et al. (1998) have identified a fast-evolving gene (the homeobox-containing *Odysseus*) that contributes to sterility and inviability between *Drosophila* species. The possibility of performing similar searches in adaptive radiation studies will increase in the near future as molecular tools are rapidly being developed in many organisms, including salmonids (Danzmann and Gharbi 2001).

Evolutionists have now reached a consensus that many of the key genetic differences between organisms will not only be in the form of allelic variation at specific genes but will also be manifested as changes in gene expression during development (reviewed in Streelman and Kocher 2000). Accordingly, "post-genomics technologies" are being developed to characterize organismal "transcriptomes," that is, the set of genes expressed in a particular tissue at a specific time. The method that is now most commonly used to define transcriptomes is termed "DNA microarray" (Hughes and Shoemaker 2001). By comparing the differential pattern of gene expression between populations/species that are part of an adaptive radiation, we will then be able to identify what are the most significant shifts in gene expression involved in the origin of phenotypic and genetic diversity. Global patterns of transcription could also be investigated in order to test the hypothesis that certain chomosomal regions are more affected than others during the development of hybrids between populations/species involved in an adaptive radiation. This would also allow the identification of the genes that are misregulated during hybrid development and what molecular genetic networks are affected (White 2001). Ultimately, the full integration of functional genomics and evolutionary ecology will allow the prediction of how organisms with different genes and genotypes will respond to different environmental selective pressures.

There has never been a more exciting time to investigate the evolutionary processes of adaptive radiation.

Acknowledgments I am grateful to several colleagues for the thoughtful comments and criticisms on an earlier version of the chapter; D. Bodaly, A. Hendry, D. Schluter, S. Rogers, T. Smith, D. Véliz, and an anonymous reviewer. I am also indebted to K. Östbye for letting me use his unpublished data on heritability of gill-raker counts. My research program on the evolution and conservation of northern fishes is supported by grants from the Natural Sciences and Engineering Research Council (NSERC) of Canada, and the Canadian Research Chair for conservation genetics of aquatic resources.

From Macro- to Micro-Evolution
Tempo and Mode in Salmonid Evolution

Michael T. Kinnison

Andrew P. Hendry

Arctic grayling

7

What are the salient features of salmonid evolution? "Diversity" is perhaps the most obvious, as evidenced by many of the contributions to this book. Moreover, an appreciation of the prevalence and importance of salmonid diversity is reflected in the emphasis placed on protection and management at subspecific levels (Waples 1991a; Ford 2003—*this volume*). However, diversity without pattern can give the impression of chaotic evolution. The aim of this chapter is therefore to interpret the great diversity of salmonids in the context of the evolutionary patterns and processes through which it is generated. This is the domain of tempo (i.e., rate) and mode (i.e., pattern) in salmonid evolution.

G.G. Simpson's *Tempo and Mode in Evolution* (Simpson 1944) provided the inspiration for this chapter's title and subject. Although investigations of rate and pattern preceded Simpson's treatise, his integration of paleontological patterns with genetic principles was a landmark exploration of evolutionary patterns and their driving mechanisms. Simpson, as a paleontologist, focused heavily on apparent discontinuities at higher taxonomic levels (e.g., orders), which he referred to as mega-evolution. Most subsequent studies of tempo and mode have retained this focus on higher taxonomic levels and paleontological data (e.g., Fitch and Ayala 1995). This approach is not as tractable for salmon and their closest kin, owing to their close phylogenetic affinities (within the family Salmonidae or subfamily Salmoninae) and poor fossil record. Our chapter will therefore focus on lower phylogenetic scales (genera, species, and populations) and on extant evolutionary patterns and processes.

For the sake of expedience, we restrict much of our discussion to Atlantic and Pacific salmon, as well as their closest kin. Like Simpson, our approach is descriptive and logical, as opposed to formally analytical, although we are now afforded insights drawn from modern methods unavailable in Simpson's time. We first provide background information on some theoretical aspects of tempo and mode. We next consider a phylogenetic history of salmonids from the standpoint of tempo and mode. We then examine the evolution of intraspecific variation since the last major glaciation, particularly with regard to the diversity of life histories. Finally, we consider contemporary evolution and its relevance to larger phylogenetic and temporal scales.

1. Theory: Tempo and Mode in Evolution

How do differences among species, genera, or families, arise? Are they the result of gradual divergence among lineages accumulating micro-evolutionary changes, or are they the result of dramatic evolutionary innovations during periods of punctuated evolution? The former scenario is expected to reflect natural selection acting on mutations of small effect, and the latter, genetic innovations or dramatic changes in the selective environment. Although arguments in favor of one or the other viewpoint have received considerable attention (e.g., Gould and Eldredge 1977; Stanley 1979; Charlesworth et al. 1982), most evolutionary biologists would now concede elements of both. Adherents to the modern synthesis have built a compelling body of theory and empirical evidence linking the origin of species (i.e., macro-evolution), to underlying micro-evolutionary mechanisms (e.g., Charlesworth et al. 1982; Schluter 2000). However, a parallel body of literature has added credence to punctuational mechanisms through the evolution of developmental pathways, genes of major effect, and dramatic environmental events (e.g., Schwartz 1999; Jablonski 2000).

Are patterns of evolutionary diversification repeatable? In one extreme, similar selective regimes generate similar evolutionary outcomes (i.e., determinism). In the alternative, a rewinding and replaying of the "tape of life" (Gould 1989) would generate dramatically different outcomes owing to historical contingency. Given the obvious role of chance events in setting the evolutionary stage, a replay of life would certainly not produce an identical result. Still, there can be no doubt that determinism is also very important, as evidenced by numerous cases of parallel and convergent evolution (Losos et al. 1998; Taylor 1999; Schluter 2000; Taylor and McPhail 2000; Gilchrist et al. 2001). Arguments as to which scenario is more important are perhaps akin to arguments of whether the glass is half full or half empty. For example, even the classic examples of convergent and parallel evolution are likely contingent on historical events that exposed organisms with similar evolutionary potential to similar selective factors. In general, we expect that deterministic patterns should be stronger at shallower phylogenetic scales (e.g., populations, species, and genera), where genetic backgrounds and ecological niches are likely more similar. This prediction forms the rationale for our emphasis on convergence, parallelism, and continuity of evolutionary patterns in salmonids, a recently divergent and closely related set of taxa.

In recent years, interest has grown in the observation and experimental study of evolution over contemporary time scales (reviews: Hendry and Kinnison 1999; Kinnison and Hendry 2001; Bone and Farres 2001; Stockwell et al. 2003). Early observations indicated that evolutionary rates over such scales are much greater than those over longer time frames, particularly as revealed in the fossil record (e.g., Reznick et al. 1997). Such observations in turn fostered a renewed interest in the analytical basis for evolutionary rate inferences, and the significance of contemporary evolution to macro-evolution. Hendry and Kinnison (1999) reviewed the use of two evolutionary rate metrics, *darwins* and *haldanes*, for use in studies of contemporary evolution.

$$darwins = \frac{(\log_e \overline{X}_2 - \log_e \overline{X}_1)}{t}$$

where $\log_e \overline{X}$ is the natural logarithm of the mean trait value at time 1 and time 2 (in the case of a time series for one population, i.e., allochronic) or in population 1 and population 2 (in the case of divergence from a common ancestor, i.e., synchronic), and t is the time interval in millions of years (Haldane 1949).

$$haldanes = \frac{(\overline{X}_2/S_p) - (\overline{X}_1/S_p)}{g}$$

where $\overline{X}$ is the mean trait value at time 1 and time 2 (or population 1 and population 2), S_P is the pooled standard deviation, and g is the number of generations (Gingerich 1993; see also D'_H in Lynch 1990). When trait variation is expected to scale with the mean, raw data are $\log_e$ transformed to standardize variances.

Hendry and Kinnison (1999) concluded that haldanes are of greater utility to studies of contemporary evolution because: (1) they are dimension-independent, (2) they are applicable to a wider range of data types, (3) they express evolutionary change in generations (which facilitates comparisons among organisms with different reproductive schedules), and (4) they are equivalent to quantitative genetic predictions of evolutionary change. These benefits come at the cost of more parameters to estimate. Generation length can be particularly difficult to estimate but can at least be approximated in most extant organisms (Kinnison and Hendry 2001).

The significance of contemporary evolution can be assessed in the context of several broad questions, listed here in order of increasing complexity. First, do human-induced perturbations, such as translocations or selective harvest, cause substantial evolutionary change in wild populations? Second, are patterns of contemporary evolution consistent with variation among long-established indigenous populations, such as those established after the last glaciation? Third, in what fashion do rates and trends of contemporary evolution contribute to broader scales of diversification? We consider each of these three questions near the end of the following section.

2. Empirical Evaluations of the Theory

2.1. The Phylogeny of Salmon—Intergeneric Scale

The phylogeny of salmonid fishes has been extensively studied using morphological, allozyme, karyological, mitochondrial, and nuclear DNA characteristics, but nonetheless still engenders debate at both shallow (i.e., sister species) and deeper (i.e., sister genera) scales. This uncertainty reflects the group's rich evolutionary history and provides an interesting case study in elucidating pattern and process in taxonomic radiation. A brief review of the generic relations of salmon and their closest relatives (i.e., the subfamily Salmoninae within the family Salmonidae) provides some interesting insights into factors impacting the tempo and mode of evolution. A detailed history of salmonid (Salmonidae) and salmonine (Salmoninae) classification is provided by Stearley and Smith (1993) and Oakley and Phillips (1999), and a table of scientific and common names of the taxa discussed here is provided in Table 7.1. The trees presented in Figures 7.1 and 7.2 represent hypothesized phylogenies based largely on Phillips and Oakley (1997) and Oakley and Phillips (1999) but incorporate considerations from other studies of morphological (Stearley and Smith 1993), allozyme (Osinov and Lebedev 2000), mitochondrial (McVeigh and Davidson 1991), ribosomal (Phillips et al. 1992), and nuclear DNA (Murata et al. 1993) variation.

Table 7.1. Members of the Salmonidae discussed in this chapter, including scientific names, common names, geographic range, and notes on taxonomy and life history See Stearns and Hendry (2003—*this volume*) for a more extensive list.

Scientific name	Common name	Range	Notes[a]
Coregoninae	Whitefishes, ciscoes	N. temperate/polar	Mostly fresh water
Thymallinae	Graylings	Circumpolar	Mostly fresh water
Thymallus thymallus	European grayling	N. Europe and Asia	
Salmoninae	Salmon, trout, charr, and kin	Circumpolar	Most species capable of tolerating or using marine environment
Brachymystax	Lenoks	Russia	Fresh water only
Hucho	Danube salmon, Taimen, Huchen	E. Europe and Asia	Fresh water only; Huchen perhaps in own genus *Parahucho*
Salvelinus	Charrs	Circumpolar	Most species capable of diadromy
Salmo			
S. trutta	Brown trout	Europe	Often diadromous
S. salar	Atlantic salmon	N. Atlantic	Usually anadromous
Oncorhynchus			
O. clarki	Cutthroat trout	W. North America	Diadromous subspecies (*O. c. clarki*); other subspecies use inland rivers and lakes
O. mykiss	Rainbow trout, steelhead	N. Pacific	Steelhead anadromous; some freshwater forms considered distinct subspecies or species
O. masou	Masu, yamame, or amago salmon	N.W. Pacific	Anadromous individuals are semelparous
O. kisutch	Coho salmon	N. Pacific	Anadromous and semelparous
O. tshawytscha	Chinook salmon	N. Pacific	Anadromous and semelparous
O. nerka	Sockeye salmon	N. Pacific	Anadromous and semelparous, fry rear in lakes
O. keta	Chum salmon	N. Pacific and Arctic Ocean	Anadromous and semelparous, fry migrate immediately to sea
O. gorbuscha	Pink salmon	N. Pacific and Arctic	Anadromous and semelparous, fry migrate immediately to sea

[a] Species iteroparous unless otherwise stated.

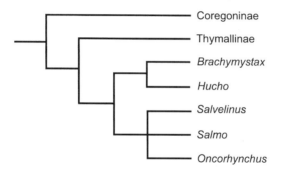

Figure 7.1. Hypothesized family and generic relations of the Salmonidae. Tree based primarily on Oakley and Phillips (1999), Osinov and Lebedev (2000), and Crespi and Fulton (see phylogeny in Stearns and Hendry 2003—*this volume*). The relationship between *Salmo*, *Oncorhynchus*, and *Salvelinus* is drawn as polytomy given that recent work does not support the traditional grouping of *Salmo* and *Oncorhynchus* as sister species (instead favoring *Oncorhynchus* and *Salvelinus* as sister taxa).

Total evidence DNA phylogenies are currently being developed (B. Crespi and M. Fulton unpublished), and these should be consulted for the most recent interpretations.

Although formal classification of Atlantic salmon (*S. salar*) and brown trout (*S. trutta*) to the genus *Salmo* dates from Linnaeus, Pacific salmon (*O. gorbuscha*, *O. nerka*, *O. keta*, *O. kisutch*, *O. tshawytscha* and *O. masou*) were only formally renamed to a separate genus (*Oncorhynchus*) in the mid-1800s (Stearley and Smith 1993). Pacific trouts (e.g., *O. mykiss*, *O. clarki*, and others) remained classified in *Salmo* until 1988 when they were reclassified into *Oncorhynchus* in accord with a review by Smith and Stearley (1989). This affinity previously had been recognized by a number of other investigators (e.g., Regan 1914; Rounsefell 1962; Behnke et al. 1962; Tsuyuki and Roberts 1966). Several taxa of uncertain standing, at times referred to as "archaic trouts," have been problematic and variously included in *Salmo* or separate genera (i.e., *Platysalmo*, *Salmothymus* and *Acantholingua*).

Affinities of the commonly regarded "salmon" clades, *Oncorhynchus* and *Salmo*, relative to other genera of the Salmoninae are also contentious. Until recently, most salmonid biologists would have aligned *Oncorhynchus* and *Salmo* as sister taxa, more distantly related in turn to *Salvelinus* (charr), *Hucho* (huchen and taimen) and *Brachymystax* (lenok). Indeed, a consensus view on the sibling status of *Salmo* and *Oncorhynchus* was likely greater than for any other generic relationship in the Salmoninae. However, this pairing, based initially on multiple morphological analyses, has not been universally upheld by recent molecular analyses. Parsimony analyses of growth hormone introns, ITS1, ITS2, 18s and cytochrome b sequences have provided comparable or greater support for phylogenies pairing *Salvelinus* with *Oncorhynchus* (McVeigh and Davidson 1991; Oakley and Phillips 1999; Crespi and Fulton unpublished, see phylogeny in Stearns and Hendry 2003—*this volume*).

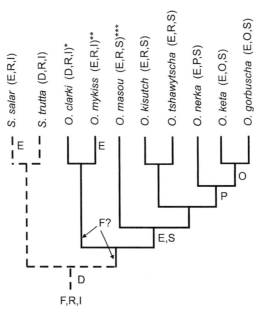

Figure 7.2. Phylogeny and life history of salmon and their congeners. The hypothesized phylogeny is based largely on relationships described in Stearley and Smith (1993), Oakley and Phillips (1999), and Osinov and Lebedev (2000). The dashed branch connecting *Salmo* and *Oncorhynchus* indicates questionable sister taxon status. Life history classifications shown in parentheses next to species names are for populations or subspecies with access to marine environments and are written in the following order: diadromy type, juvenile habitat, and parity. Abbreviations mapped onto branches indicate hypothesized evolution of life history synapomorphies. Abbreviations: F = freshwater resident, D = facultative diadromy or amphidromy, E = extensive anadromy (with long migration), R = riverine juveniles (territorial), P = pelagic shoaling, O = immediate ocean shoaling, I = iteroparity (potential in all individuals), S = semelparity (in diadromous individuals). Hypothesized life history of the salmonid ancestor is inferred from an outgroup consisting of other Salmoninae (*Brachymystax* and *Hucho*) along with Thymallinae and Coregoninae. Several Pacific trouts (*Oncorhynchus* spp.), regarded as morphologically plesiomorphic, are not shown but are all freshwater resident. Uncertainty about the phylogenetic position of these forms (basal to *O. mykiss*, basal to all Pacific trouts, basal to all *Oncorhynchus*?), and whether their freshwater dependence is vicariant or secondarily evolved, leads to uncertainty about a possible freshwater resident ancestry to all modern Pacific trouts or *Oncorhynchus* (reflected by "F?" on phylogeny). * The life history is for diadromous subspecies *O. c. clarki* (other subspecies restricted to inland freshwater habitats—diadromy in *O. c. clarki* thus possibly secondary). ** The life history is for the "steelhead" form. *** The life history is for diadromous individuals or populations (freshwater resident individuals or populations can be iteroparous).

Why have generic relations of the Salmoninae proven so difficult to resolve and what insights may be inferred regarding pattern and process at this scale? First, convergent evolution of form and life history, presumably a deterministic process, may occur even on this scale, thus resulting in a lack of concordance between some elements of phenotypic and molecular phylogenies. But why have even the molecular phylogenies been so variable and contentious? One possibility is that the Salmonidae (including Salmoninae) experienced a rapid radiation 25–100 million years ago (precise dating is difficult due a poor fossil record). Rapid radiations tend to result in star-shaped phylogenies and "hard" polytomies that are nearly insurmountable for phylogenetic reconstruction (Hoelzer and Melnick 1994). Such phylogenies are known for other taxa that have undergone rapid radiations (e.g., African cichlids—Kornfield and Smith 2000). If this scenario is true, the next logical question is, what evolutionary factors might favor a punctuated radiation of forms?

A tetraploid event (duplication of a diploid genome) that occurred in the ancestors of all Salmonidae has been hypothesized as a factor promoting their diversification (Allendorf and Thorgaard 1984). Tetraploidy is still detectable in the form of duplicate loci for many genes in salmonids, such as growth hormone introns. Under some conditions, duplication of a genome could theoretically allow for an increase in genetic variation and adaptive potential, opening the door for a rapid radiation (Otto and Whitton 2000). Polyploid events are common in plants and may also have contributed to the radiation of all vertebrates and of teleosts (Otto and Whitton 2000). On a scale more comparable to salmonids, polyploidy may underlie the diversity of catastomids (suckers and their relatives, Ferris 1984). However, the sizeable radiations within the subfamilies Salmoninae and Coregoninae (roughly 30 extant species each) have not been mirrored by a similar radiation in their tetraploid kin, the Thymallinae. This would suggest that tetraploidy is not the only factor contributing to rapid diversification.

Variability in phylogenetic signal among characters and of rates of evolution among lineages can further complicate phylogenetic reconstruction. The former problem is dramatically exemplified by alternate salmonine growth factor gene introns (GH1C and GH2C). These very closely related genes, likely derived from ancestral duplication of a common growth hormone sequence, are known to provide different phylogenetic signals (Oakley and Phillips 1999). The latter problem arises in the apparently faster evolution of allozymes in the genus *Oncorhynchus* than in the other genera, violating the assumption of a constant molecular clock. Accordingly, the use of allozymes to calculate genetic distances, assuming constant evolutionary rates, places *Oncorhynchus* at the root of the Salmoninae (Osinov and Lebedev 2000); whereas parsimony analyses of the same data place *Oncorhynchus* as a derived lineage (Osinov and Lebedev 2000).

The challenging intergeneric relations of the Salmoninae suggest potentially complex and intriguing patterns of tempo and mode in their history.

2.2. The Phylogeny of Salmon—Intrageneric Scale

Phylogenies at the intrageneric level of salmon and their closest kin have likewise been extensively debated. However, at this scale we can begin to focus on the evolution of an actual "salmon" life history and the insights it provides into tempo and mode.

Within the subgenus *Oncorhynchus*, *O. masou* have often been considered a sister group to the other Pacific salmon, within which *O. tshawytscha* and *O. kisutch* form one clade, and *O. nerka*, *O. gorbuscha*, and *O. keta* form another (Figure 7.2; Stearley and Smith 1993; Phillips and Oakley 1997). However, *O. masou* are usually neglected in molecular studies and when included occupy variable taxonomic positions (Shed'ko et al., 1996; Oohara et al., 1997; Osinov and Lebedev 2000; Crespi and Fulton unpublished, see phylogeny in Stearns and Hendry 2003—*this volume*). In the following, we adhere to original relationship supported by morphometric and allozyme studies (e.g., Stearley and Smith 1993; Osinov and Lebedev 2000).

The phylogenetic organization of the *O. nerka*, *O. gorbuscha*, and *O. keta* clade has been disputed extensively. Morphological, allozyme, and karyotypic (*O. nerka*, $2N = 56$; *O. gorbuscha*, $2N = 52$; *O. keta*, $2N = 74$) studies usually group *O. nerka* with *O. gorbuscha*, whereas mitochondrial, nuclear, and life history studies suggest an *O. keta* and *O. gorbuscha* sisterhood (Phillips and Oakley 1997; Crespi and Fulton unpublished, see Hendry and Stearns 2003—*this volume*). Observed hybridization between *O. gorbuscha* and *O. keta* led Stearley and Smith (1993) to hypothesize that their mitochondrial affinity may reflect introgressive capture of the mitochondrial genome of one species by the other. Subsequent analyses of additional sequence variation, however, have further supported an *O. gorbuscha*/*O. keta* clade. Although it may be more parsimonious to conclude that observed hybridization between these species simply reflects their close relatedness, it is worth noting that the introgression hypothesis is not definitively excluded by this outcome. Extensive hybridization could result in similarities in many parts of the genome. Hybridization has also been hypothesized to contribute to the difficulty of phylogenetic reconstruction among charrs of the genus *Salvelinus* (Phillips and Oakley 1997). More generally, hybridization significantly complicates the idealized view of evolution, wherein each species diverges ever further from its sister taxa (Taylor 2003—*this volume*).

Given the recent radiation of the Salmoninae and the subtlety of molecular differences that characterize species, the life history and morphological variation within genera is remarkable. As indicated above, Pacific salmon represent a monophyletic lineage more closely related to Pacific trouts and charrs, than to Atlantic salmon. This relationship provides interesting insights into the origins of life histories. Foremost, the appellations "salmon" and "trout" do not delineate natural phylogenetic groups, and the diversity of anadromous and freshwater life histories is clearly convergent within these lineages. Convergence in turn suggests an element of determinism, mediated most likely by a combination of similar genetic backgrounds and parallel selection. In the following, we consider the divergent and convergent evolution of several life history traits: anadromy versus non-anadromy and semelparity versus iteroparity.

Considerable debate has surrounded whether basal salmonids had a freshwater or marine origin. Accordingly, a number of authors have attempted to map diadromy (migrations between fresh water and the ocean) onto salmonid phylogenies (reviews: McDowall 1988; Stearley 1992; McDowall 2002).

Conclusions vary and McDowall (2002) has recently suggested that the ancestral salmonid was already anadromous rather than strictly fresh water or marine. This seems plausible because diadromy is known in the sister families to salmonids (Osmeridae and Galaxiidae). However, regardless of the life history of the ancestral salmonid, several evolutionary transitions between anadromy and non-anadromy, probably in both directions, have occurred within each lineage.

Given our central focus on "salmon", we will be concerned with the phylogenetic distribution of "extreme" anadromy, wherein salmon begin life in fresh water but then undergo extensive migrations in marine environments after true smoltification (distinct developmental changes in morphology, physiology, and behavior—Folmar and Dickhoff 1980). We use the term extreme anadromy to set this life history apart from facultative marine migrations that are short in duration and distance, and may not be accompanied by as distinct a smoltification process. An examination of salmonid phylogeny suggests that extreme anadromy has evolved independently within *Salmo* and *Oncorhynchus*. Supporting evidence includes the prevalence of less diadromous kin within both genera (Figure 7.2), a lack of similarly extreme anadromy in *Salvelinus*, and limited diadromy in the more distant sister taxa (e.g., *Brachymystax*, *Hucho*, Thymallinae and Coregoninae). The parallel evolution of extreme anadromy seems quite palpable given the prevalence of facultative marine migrations in the other species (Gross 1987; McDowall 1988). If the ancestors of Atlantic and Pacific salmon were already diadromous, then the existence of numerous non-anadromous populations within many species (Hendry et al. 2003b—*this volume*) represents further examples of parallel (within species) and convergent (between species) evolution (discussed below).

The evolution of anadromy/non-anadromy within salmonids appears linked with the evolution of other traits. For example, Stearley (1992) noted a phylogenetic association between the social and migratory behavior of juveniles, ranging from species that show juvenile territoriality and perhaps facultative marine migration after years in fresh water (e.g., coastal cutthroat trout) to others that shoal and migrate to the sea immediately after emergence (e.g., pink and chum salmon, Figure 7.2). Atlantic salmon are in many ways more analogous in these life history features to steelhead (anadromous O. *mykiss*) than to the generally regarded Pacific salmon. For example, both have a tendency to enter marine environments after multiple territorial years in fresh water and to then migrate extensive distances at sea (brown trout and most other Pacific trouts migrate only short distances in salt water). Most notably, however, they also share a propensity for iteroparity. We will now turn our discussion to parity and its ties to anadromy and other aspects of life history evolution in salmon.

Semelparity may sometimes seem a discrete genetic innovation but we agree with Stearley (1992) that "the reproductive difference between Pacific salmon species and other *Salmo* and *Oncorhynchus* is one of degree and not one of kind." Semelparity represents one end of a continuum that is bounded at the other end by substantial iteroparity. Semelparous salmon are predominantly anadromous and common ancestry from a semelparous/anadromous forebearer may contribute to this association (McDowall 2002; Figure 7.2). However, extensive marine migrations associated with anadromy may negatively impact survival and residual reproductive value, thus favoring increasing investment into current reproduction at the expense of post-reproductive survival (see also Schaffer 2003—*this volume*). Indeed, simultaneously considering parity and anadromy in the context of salmonid phylogeny suggests a general association between the traits. Pacific salmon, the most strictly semelparous group, are usually anadromous. One nominative "salmon" species, O. *masou*, occasionally has iteroparous males in the wild but those individuals also tend to be non-anadromous. Other species of *Salmo* and Pacific trouts, such as brown trout and cutthroat trout, are often less extensively anadromous and highly iteroparous. Application of paired contrasts between anadromous and non-anadromous populations within species, to control for phylogeny (Harvey and Pagel 1991), suggests that anadromy is indeed correlated with a lower probability of iteroparity (Fleming 1998).

Given that the evolution of parity may be viewed as a tradeoff between initial and future reproduction, it becomes logical that additional "salmon" traits tied to reproductive investment may be correlated with the evolution of parity. An independent contrasts analysis by Crespi and Teo (2002) confirmed that among iteroparous species the degree of iteroparity is negatively correlated with investment into ovarian mass, suggesting that selection for greater reproductive investment contributes to reduced iteroparity. However, semelparity was associated with a substantial increase in body size and egg size but not in total ovarian mass or egg number (Crespi and Teo 2002). A potential explanation for this result is that the additional reproductive investment associated with the transition from iteroparity to semelparity comes in the form of increased breeding competition (Fleming 1998; Fleming and Reynolds 2003—*this volume*). In support of this view, males of supposedly semelparous species can be functionally iteroparous if freed from some of the metabolic costs associated with breeding. For example, studies on captive chinook salmon have shown that males can mature repeatedly if they are not exposed to salt water, are freed from migration and breeding competition, and are raised on high nutrition diets (Unwin et al. 1999).

Another "salmon" trait potentially tied to semelparity is increased investment into secondary sexual development, particularly hump size and snout size at maturity. Mature individuals with larger humps and snouts generally have increased mating success (Fleming and Reynolds 2003—*this volume*) but may also expend more energy constructing the traits and may suffer increased predation or decreased swimming ability (Blair et al. 1993; Quinn and Kinnison 1999; Quinn et al. 2001a,c; Kinnison et al. in review). Additional trade offs may extend past the breeding season if post-reproductive individuals cannot revert to their original morphology (or if doing so is energetically costly) and if breeding morphology reduces the performance of post-reproductive individuals. Secondary sexual development does seem roughly correlated with parity. Most iteroparous Pacific trout species exhibit minimal secondary sexual trait development, whereas predominantly semelparous Pacific salmon display significant snout and hump development, with the latter reaching a peak in O. *nerka* and O. *gorbuscha*. Interestingly, these two species often experience some of the highest levels of breeding competition and are invariably constrained to semel-parity, even when freshwater-resident.

With only two recognized species, neither fully semelparous, it is more difficult to make phylogenetic comparisons within *Salmo*. Nevertheless, in comparison to Atlantic salmon, brown trout generally show less hump and snout development, and are more often freshwater resident and iteroparous. Furthermore, iteroparity and secondary sexual development covary within Atlantic salmon. Males that mature early in fresh water (mature parr) develop little if any secondary sexual characters and have a greater prob-ability of iteroparity (often later becoming anadromous) than do initially anadromous males (Fleming and Reynolds 2003—*this volume*). These mature parr do not attempt to compete with anadromous males for access to females but instead adopt a "sneaking" tactic (Fleming 1998; Fleming and Reynolds 2003—*this volume*).

Given the likely evolutionary parallels, it is interesting to consider some of the notable species differences. While enhanced secondary sexual trait develop-ment seems to be characteristic of both groups of "salmon," *Salmo* develop an extended lower jaw (mandible) and *Oncorhynchus* an extended upper jaw (max-illa and premaxilla). Furthermore, breeding coloration is a conspicuous element of all salmon mating systems but species color patterns differ extensively, even among Pacific salmon species. Do these variations on a common theme reflect contingency influencing otherwise deterministic patterns of parallel evolution (i.e., the evolution of secondary sexual traits), or adaptive divergence driven by different selective pressures? The former seems more likely, although diver-gent coloration might also result from selection to avoid hybridization, which now appears to be relatively rare among most Pacific salmon (Taylor 2003—*this volume*). The issue of color variation becomes even more intriguing when one considers the evidence for repeated evolution of a conserved color pattern within *O. nerka*, as discussed in the next section.

In summary, shared phylogenetic origins have probably contributed to the patterns of life history variation currently found in salmonines. However, par-allel and convergent evolution have probably also contributed strongly to the suites of traits found in different species. For example, both *Oncorhynchus* and *Salmo* have species that are predominantly anadromous and semelparous and species that are predominantly non-anadromous and iteroparous. Thus, no matter what their shared ancestral life history, a diversity of convergent life histories arose within each lineage. This diversity appears driven by selection for suites of traits compatible with optimal investment into initial versus future reproduction. This tradeoff seems to have led to a continuum of life history variation bound at one end by nominally "salmon" life histories exhibiting anadromy, semelparity (or low iteroparity), and exaggerated secondary sexual development. The covariance among these traits suggests a degree of determin-ism in their evolution but subtle variation among species suggests that con-tingency may also be important.

2.3. Post-Glacial Evolution

The existence of heritable interpopulation variation in salmonid phenotypic traits is incontrovertible. To review all such work is beyond the scope of this chapter and interested readers should refer to work by T. Beacham, C. Foote, B. Jonsson, T. Quinn, B. Riddell, E. Taylor, and others (e.g., Riddell et al. 1981; Beacham and Murray 1989; McIsaac and Quinn 1988; Taylor 1990b; Foote et al. 1992; Jonsson et al. 2001a,b). Furthermore, most salmonid biologists would agree that such differences generally reflect adaptations to local environments (Ricker 1972; Taylor 1991b; but see Adkison 1995). Here we consider a few examples of adaptive divergence that relate to one of our recurring themes: determinism versus contingency.

Most modern populations of salmon likely became established in the last 8000–15,000 years. Glaciation prior to this time restricted most salmonids to a limited number of isolated refugia (Hocutt and Wiley 1986; Wood 1995; Bernatchez and Wilson 1998). Following the recession of ice sheets, these refugial populations acted as sources for dispersal, leading to colonization of their current ranges in North America, Asia, and Europe. Although the refugial lineages may have undergone substantial evolutionary diversification prior to and during the most recent glaciation, a significant component of extant interpopulation variation has clearly arisen post-glacially. This scenario of multiple source populations colonizing multiple locations across a diversity of environments allows us to assess whether evolutionary diversification within species is best explained by contingency (e.g., vicariance events and colonization owing to glaciation) or determinism (i.e., natural selection driving parallel, convergent, and divergent evolution). We address this question by examining life history variation in O. nerka, O. tshawytscha, and S. salar.

For our first example, we return to a consideration of anadromy, or more appropriately, the secondary development of non-anadromy. Although Pacific and Atlantic salmon are primarily anadromous, many natural non-anadromous populations also exist and many others have been established by introductions. Even pink salmon, which possess one of the most marine-dependent life histories, have become established in fresh water following accidental introduction to the North American Great Lakes (Kwain and Lawrie 1981). However, only a few species of salmon frequently give rise to freshwater resident populations.

Many freshwater populations of O. nerka and S. salar have become naturally established following the last glaciation (Hendry et al. 2003b—this volume). These freshwater forms are sufficiently divergent from their anadromous counterparts for some taxonomists to consider them separate species or subspecies. Non-anadromous O. nerka, called "kokanee," live their entire lives within lakes in the northwestern United States, western Canada, and northeastern Asia. Natural populations of non-anadromous S. salar, often referred to as "landlocked salmon," are found in northeastern North America (including Lake Ontario) and in northern Europe (Berg 1985). Associated with their freshwater life history, non-anadromous salmon of both species are smaller at maturity than anadromous salmon, but the two forms are otherwise outwardly similar in general appearance (Hendry et al. 2003b—*this volume*). Kokanee populations may be found in systems without anadromous sockeye salmon but the two forms are also frequently found in sympatry, often breeding at the same times and places (Taylor et al. 1996; Taylor 1999). Anadromous and non-anadromous Atlantic salmon populations are occasionally found in sympatry but non-anadromous populations are more often "landlocked" above barriers to marine migration (Berg 1985; Hendry et al. 2003b—*this volume*).

Interestingly, analyses of presumed-neutral genetic markers indicate that freshwater populations are often, although not always, reproductively isolated from adjacent or even sympatric anadromous populations (Taylor et al. 1996; Primmer et al. 2000; Tessier and Bernatchez 2000; Hendry et al. 2003b—*this volume*). Furthermore, kokanee are more closely related to sympatric sockeye salmon than to allopatric kokanee (Taylor et al. 1996), a pattern suggesting multiple independent origins of the freshwater form. Multiple independent origins are also likely for different non-anadromous Atlantic salmon populations, which probably arose when migration routes to the ocean were lost through isostatic rebound (Berg 1985). With each independent origin of non-anadromous forms came the repeated evolution of a suite of traits appropriate for a wholly freshwater life history. Such trait changes include gill-raker morphology (associated with diet shift), growth rates, development rates, and other features (Wood and Foote 1996; Taylor 1999; Hendry et al. 2003b—*this volume*).

An excellent example of parallel, and probably deterministic, evolution is the ability of kokanee to acquire carotenoids from their carotenoid-poor freshwater diet, and to use those carotenoids in their characteristic red breeding coloration. Craig and Foote (2001) reared anadromous sockeye salmon and non-anadromous kokanee on similar diets in similar freshwater tanks. At maturity, the skin of the sockeye salmon was considerably less red than that of the kokanee, indicating that kokanee have evolved an improved ability to make use of carotenoids. Thus, genetic divergence in carotenoid uptake underlies phenotypic convergence of breeding coloration (both forms are red at maturity in the wild). This is thought to have occurred because red coloration is important in the breeding system of *O. nerka* (Craig and Foote 2001). Though different freshwater populations likely possess unique adaptations to their respective lake systems, it is noteworthy that, given a similar set of selective conditions and genetic architecture, evolution appears to follow parallel lines. Taylor (1999) reviewed how such multiple independent origins of ecologically similar forms generates an awkward taxonomic and management dilemma.

The parallel (and convergent) evolution of freshwater life histories in salmon raises a series of interesting questions. For example, although freshwater forms are outwardly similar (i.e., phenotypically), did they achieve this similarity in the same way or different ways (i.e., genetically)? Also, it remains to be determined how much variation among independently derived freshwater populations is due to recent adaptation versus historical contingency (e.g., different ancestral lineages). A fascinating test of parallel evolution would be to evaluate the outcome of a novel contact among independently derived freshwater populations. Would kokanee from different lake systems be more likely to select one another as mates than sockeye from their own lake systems? How would the fitness of hybrids between allopatric kokanee populations compare with the fitness of hybrids between sympatric kokanee and sockeye salmon? Questions such as these are fundamental to tests for the role of natural selection in the evolution of reproductive isolation, a process often called "ecological speciation" (Schluter 2000; Bernatchez 2003—this volume). Specifically, if adaptive divergence is driving the evolution of pre-zygotic reproductive isolation, mating should be more common between similar life history types (even from different systems) than between different life history types (even from the same system). Similarly, if adaptive divergence is driving the evolution of post-zygotic isolation, hybrids between allopatric populations of the same life history type might have higher fitness in nature than hybrids between sympatric populations of different life history types.

But what about contingency or phylogenetic constraints? As an obvious example, many freshwater populations would not have arisen without being isolated from the ocean by chance vicariance events. Another example may perhaps be found in the geographic distribution and phylogenetic relationships of alternative chinook salmon life histories. "Stream-type" populations of chinook salmon in North America migrate to the ocean after a year or more in fresh water, whereas ocean-type populations migrate in their first year of life (Healey 1983). Stream-type populations spawn in the upper reaches of the Fraser and Columbia River watersheds and in coastal rivers farther north (Taylor 1990a; Ford 2003—*this volume*). Ocean-type populations are found in the Sacramento River, the lower reaches of the Fraser, Snake, and Columbia River watersheds and in coastal rivers south of 56°N (Taylor 1990a; Ford 2003—*this volume*). Thus, large rivers like the Columbia and Fraser contain populations of both types. Common-garden experiments have revealed that juveniles from ocean-type populations have a genetic propensity to grow faster than juveniles from stream-type populations, and that the two forms have different ontogenetic patterns of aggression, rheotaxis (i.e., response to water flow), and smoltification (Taylor 1990b; Clarke et al. 1992).

We might anticipate that these alternative life histories evolved repeatedly following post-glacial colonization from a single source, in analogy to alternate forms of O. *nerka* and S. *salar*. However, analyses of allozyme variation suggest that the two forms may at times represent distinct colonizations. Stream-type chinook from the upper parts of different river systems in Washington (Myers et al. 1998; Ford 2003—*this volume*) and British Columbia (Teel et al. 2000; Figure 7.3), for example, are more closely related to one another than they are to ocean-type populations within their own drainages. It thus seems feasible that the two forms arose earlier, were preserved in separate glacial refugia, and recolonized overlapping ranges after glacial recession (Teel et al. 2000; Ford 2003—*this volume*). It is hence possible that contingency and phylogenctic history have played an important role in generating at least some of the life history variation in salmonids.

Even here, however, some deterministic elements may be at work. For example, the tendency for modern stream- and ocean-type populations to use

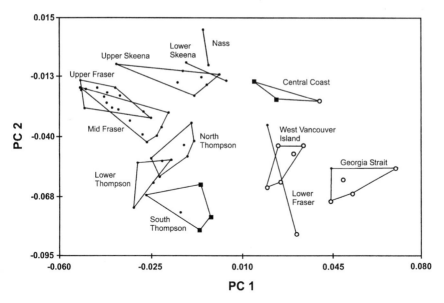

Figure 7.3. Genetic divergence of stream- and ocean-type lineages of chinook salmon as portrayed by mapping life history on to a scatter plot of the first two principal components of allozyme allele variation. Small filled circles = stream type; Large open circles = ocean type; Large filled squares = intermediate (mixed). Note that stream-type populations from the mid and upper Fraser River, and a major tributary of the Fraser (the Thompson) are genetically more similar to stream-type forms from other rivers (i.e., Skeena and Nass) than to Lower Fraser River salmon, which tend to group with nearby ocean-type coastal populations (Central Coast, Vancouver Island, Georgia Strait). A similar pattern is found for chinook salmon from U.S. rivers (see Ford 2003—*this volume*). Modified from Teel et al. (2000).

a characteristic set of habitats may bias the sites they colonize, thus favoring continued selection for the same form. It is also feasible that, had one form or the other not been preserved through the last glaciation, a similar life history would have re-arisen owing to parallel patterns of selection. Indeed, some stream-type populations in the Pacific Northwest appear to be derived from nearby ocean-type populations (Teel et al. 2000; Figure 7.3). Moreover, the presence of individuals with stream-type migratory timing within otherwise ocean-type populations suggests the potential for one form to evolve into the other. Finally, as a prelude to the next section, ocean-type chinook salmon transplanted to New Zealand have established populations with a range of juvenile life histories, including at least one with stream-type characteristics (Unwin et al. 2000).

There is tremendous diversity within all salmon species and many of their salmonine kin. One could likely make numerous further arguments for whether this variation is characterized by determinism or contingency. However, we suggest that such debates would be better supplanted by experiments aimed at using salmon and their kin to explicitly dissect the underlying genetic and selective bases of parallel evolution and their role in contributing to the cohesion or creation of species. In seeking this goal, investigators may not only find it useful to consider long extant cases of parallelism, but to also consider the early stages of population evolution and how common evolutionary patterns in salmon are initiated. In that spirit we now move on to a discussion of the tempo and mode of contemporary evolution in salmon and their kin.

2.4. Contemporary Evolution

Relative to rainbow trout, brown trout, and brook charr, introductions of Atlantic and Pacific salmon have been less frequent and less successful (Withler 1982; Lever 1996). The failure of many anadromous populations to become established or remain anadromous may be related to the additional complexity associated with an anadromous life history and a lower probability of obtaining a suitable initial match between genotype and environment (McDowall 1988; Utter 2000). Nonetheless, several notable cases of introduced anadromous populations can be found, and the examination of contemporary evolution in these populations allows a more precise examination of tempo and mode. In the following, we argue that it is at this evolutionary scale that determinism and continuity are most obvious, and that the trends observed in contemporary evolution contribute to long-term population divergence.

Our first example is from sockeye salmon that spawn on lake beaches versus in rivers (or streams). These two "ecotypes" are found within many natural lake systems and seemingly experience very different selective regimes (Wood 1995; Hendry 2001). For example, beach adults experience little water current and face little risk of predation or stranding (i.e., entering water too shallow to swim). Moreover, the nests of river females are much more likely to experience gravel scour that kills incubating eggs and embryos. Accordingly, beach and river populations show phenotypic differences that seem to reflect adaptation to these different environments: males on beaches are deeper bodied for their length, fish in small streams are often very small, and females in large rivers are often large (Blair et al. 1993; Quinn and Kinnison 1999; Quinn et al. 2001a,c). Hendry et al. (2000b) used these extant patterns to predict evolutionary divergence between sockeye salmon that colonized a beach and a river after their introduction to Lake Washington (Washington). In accord with natural patterns and expected selective regimes, beach males were deeper bodied than river males and river females were larger than beach females (Hendry et al. 2000b; Hendry 2001). The genetic basis for these adult phenotypic differences is not known but genetic differences have been confirmed for juvenile traits that reflect adaptation to incubation temperatures (Hendry et al. 1998; Hendry 2001). These adaptive differences appear to have arisen from a common ancestral source in about 13 generations.

Our second example is from chinook salmon introduced to New Zealand (NZ) in the early 1900s (McDowall 1994). From introductions to a single location, populations quickly became established in a number of locations that differ in migratory distance and rearing conditions. These environmental factors should be potent evolutionary forces, suggesting that adaptive divergence should be detectable among the present-day NZ populations. Within the finite energy budget available to maturing salmon, migratory costs may significantly decrease energy stores available for other aspects of reproductive allocation, including gametes, secondary sexual traits, and reproductive behavior (Fleming and Gross 1989; Hendry and Berg 1999). Kinnison et al. (2001; in review) showed experimentally that longer migrations result in a proximate reduction in ovarian investment, secondary sexual trait size, and remaining somatic energy stores. They therefore predicted that longer migrating populations should evolve compensatory patterns of energy allocation favoring investment into features most closely tied to fitness, such as ova production in females and energy for mate competition in males. Common-garden comparisons of short- and long-migrating NZ populations showed that long-migrating females evolved greater investment into ovarian production (Kinnison et al. 2001) and long-migrating males evolved smaller hump sizes that should be more efficient for migration (Kinnison et al. in review). Reviews of reproductive investment among North American populations of salmon suggest that analogous migratory costs have been important in shaping natural patterns of reproductive allocation (Kinnison et al. 2001; Crossin et al. in review).

New Zealand chinook salmon also appear to have evolved in response to different temperature regimes and productivity in fresh water. In North America, populations that spawn in warm water tend to mature and spawn later than those that spawn in cold water, a difference thought to synchronize larval development with the seasonal availability of food resources (Brannon 1987). Accordingly, stream-type chinook salmon juveniles from cooler inland streams stay longer in fresh water than ocean-type juveniles from coastal streams, whereas the former have a genetic tendency to grow more slowly (see above). Within NZ, Quinn et al. (2000) found that a population spawning in warmer water had a genetic tendency to return and mature later than a population spawning in colder water. Similarly, juveniles from most NZ populations migrate to the ocean in their first year (ocean-type), whereas those from a particularly cool and low-productivity site stay an additional year in fresh water (stream-type). Longer freshwater rearing in the latter was associated with a heritable tendency for significantly slower growth (Unwin et al. 2000), a result consistent with genetically based growth rates in North American "stream-type" populations.

How do new salmon populations, which form under straying and gene flow, diverge with such apparent speed? The answer is thought to lie in the interplay between adaptation and ecologically mediated reproductive isolation. We suggest the following scenario (see also Hendry et al. 2003a—*this volume*). (1) Natal imprinting rapidly causes some reproductive isolation between newly founded populations. (2) This restriction in gene flow allows some adaptive divergence in response to local environments. (3) This initial adaptive divergence reduces the fitness of individuals that stray between populations. (4) This reduction in the fitness of strays reduces gene flow and allows further adaptive divergence. Steps 3 and 4 then repeat until an equilibrium is reached. Some evidence for this process can be found in studies of introduced salmon. For Lake Washington sockeye salmon, Hendry et al. (2000b) showed that the rate of gene flow between the river and beach populations was much lower than the rate of physical straying by adults, implying the evolution of reproductive isolation as a result of adaptive divergence (see also Hendry 2001). Likewise, juvenile chinook salmon released in their natal stream survived and returned as adults at higher rates than juveniles from another NZ population released from the same site, implying that strays between sites would have lower fitness than locally adapted fish (Quinn et al. 2001b; Unwin et al. 2003). Quinn et al. (2000) argued that genetic divergence in spawning time may provide an additional factor contributing to reproductive isolation.

The above studies of introduced salmon suggest that evolution on contemporary time scales may be repeatable and may presage long-term patterns of population divergence. Such deterministic contemporary evolution may also be expected in other contexts (Reznick and Ghalambor 2001). For example, historical catch records have documented long-term changes in the size and age structure of many salmon populations (Figure 7.4). Some of these changes may be the result of phenotypic plasticity, as implied by their correlations with environmental factors such as sea surface temperature (e.g., Ricker 1981; Cox and Hinch 1997). However, environment/phenotype correlations do not negate the possibility of genetic changes. The most obvious way to determine whether genetic change has occurred would be to rear fish from different years under common conditions, an obviously daunting task. In lieu of this, a variety of evidence suggests that the opportunity for some genetic change exists. First, changes in size and age at maturity have been documented for a great diversity of exploited fish species (e.g., cod and other benthic fishes—Trippel 1995), and some of these changes remain even after correcting for correlated environmental factors (Rijnsdorp 1993). Second, changes in size and age often match those expected from the evolutionary effects of size-selective fisheries. For example, the size and age of European grayling in fresh water has changed in concert with documented patterns of selective fishing (Haugen and Vøllestad 2001). Third, experimental exploitation of Atlantic silversides (*Menidia menidia*) designed to mimic selective fishing resulted in heritable changes in growth-related traits consistent with fisheries trends (Conover and Munch 2002).

Genetic domestication (or "hatchery effects") represents another example of deterministic contemporary evolution. Artificial propagation in most salmonids uses broadly similar methodology and should therefore impose broadly similar selection, particularly on a genetically similar group of species, such as salmonines. Accordingly, general hatchery effects include changes in age at maturity, size at age, ovarian investment, morphology, behavior (agonistic, territorial, mating), predator avoidance, and fitness (Reisenbichler and Rubin 1999; Heath et al. 2003). Many of these changes in fully captive populations have long been known to have a genetic basis (Green 1952; Flick and Webster 1964). Genetic changes in supplementation programs, where adults are mated in captivity and juveniles released into the wild, are also now firmly established (Reisenbichler and Rubin 1999; Heath et al. 2003; Young 2003—*this volume*). Evidence of fishery and domestication effects is a potent reminder of our own

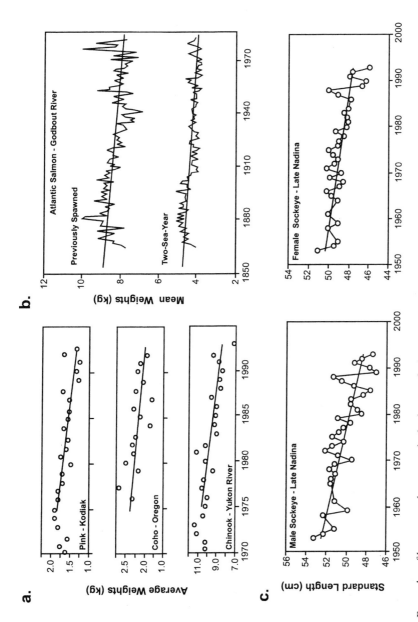

Figure 7.4. Examples of long-term changes in the body size of salmon potentially associated with environmental trends and size-selective fishing. Pink, coho, and chinook salmon data are for both sexes and all ages from commercial fishery catches (Bigler et al. 1996). Male and female sockeye salmon data are for adults measured on spawning grounds (Cox and Hinch 1997). Atlantic salmon data are from records of rod catches for fish that spent 2 years at sea before first spawning, and older individuals that had already spawned in previous years (Bielak and Power 1986).

role in the evolution of life, and indicates the importance of accounting for contemporary evolution in management strategies (Conover 2000; Conover and Munch 2002; Stockwell et al. 2003; Young 2003—*this volume*).

The above evidence clearly shows that contemporary evolution is observable in salmon and that it often has a deterministic element. Furthermore, contemporary evolution in wild populations is at least superficially similar to patterns found among indigenous populations. This suggests that contemporary evolution may reveal the seeds of long-term population divergence. In the following, we ask how rates of contemporary evolution in salmon compare with such rates in other "classic" examples of contemporary evolution, and whether such rates are consistent with increasing amounts of divergence over longer evolutionary scales.

Kinnison and Hendry (2001) computed evolutionary rates for salmon populations described in the above studies. Overall, rates of divergence and evolution for introduced and harvested salmon populations fall within the range of rates documented for many other taxa. However, evolutionary rates tend to scale negatively with time interval, largely because of a negative autocorrelation produced by plotting rates, which have time in the denominator, versus time itself. Still, we can draw some meaningful conclusions by examining residuals from a regression through the data. This approach reveals that rates of contemporary evolution in salmon are neither much greater nor much less than the range of rates for other organisms over similar time intervals (Figure 7.5). Salmon evolve just as rapidly as fruitflies (Gilchrist et al. 2001), Galápagos finches (Grant and Grant 1995), or Trinidadian guppies (Reznick et al. 1990), among others.

Although this conformity may imply that salmon have few additional insights to contribute to the field, this is far from the case. Most of the other organisms in which contemporary evolution has been measured have very short generation lengths, live their entire lives in confined habitats, and are often isolated from other conspecific populations. These features should favor particularly rapid adaptive divergence. In contrast, the complex life cycles of anadromous salmon, which span many different habitats, their long generation lengths, and the presence of ongoing gene flow (in Lake Washington and New Zealand) should limit the potential for adaptive divergence. The fact that substantial evolution has occurred, despite these potential impediments, suggests that salmon are actually exceptional in their conformity.

The evidence presented thus far suggests that contemporary evolution is consistent with patterns of diversification among indigenous conspecific populations. But does contemporary evolution actually contribute to macro-evolution (e.g., the origin of new species) and long-term evolutionary diversification? And is evolution punctuated or gradual? Kinnison and Hendry (2001) showed that evidence from salmon (particularly harvested populations) and other organisms suggests that longer time intervals are associated with increasing amounts of divergence. Moreover, the evidence reviewed above suggests that adaptive divergence in salmon contributes to the evolution of reproductive isolation. These results imply that contemporary evolution may contribute to larger changes— perhaps even macro-evolution.

At the same time, contemporary rates of evolution do not appear to persist unabated. Rather, Kinnison and Hendry (2001) found a tendency for the incre-

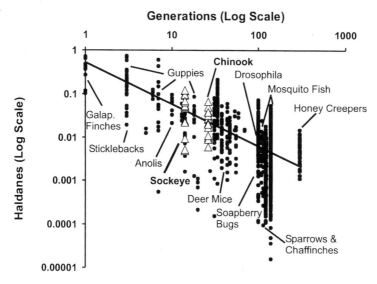

Figure 7.5. Evolutionary rates (haldanes) relative to time interval (generations) for studies of contemporary evolution. A 50% quantile regression line is provided as a standard for judging rates slower or faster than expected on a given time scale. Rates for Lake Washington sockeye salmon and New Zealand chinook salmon are labeled (large open triangles) along with rates for other organisms. Rate estimates include studies based on the measurement of wild individuals (i.e., including genetic and environmental effects on traits) and studies employing a controlled rearing design (i.e., genetic effects). Details of estimation methods and studies included can be found in Kinnison and Hendry (2001).

ment of maximum evolutionary change to decrease with increasing time interval. They suggested that large-scale divergence likely accrues from contemporary evolution in a micro-punctuational pattern of evolution "writ in fits and starts" (Kinnison and Hendry 2001). Unfortunately, few time series of contemporary evolution can be used to evaluate these questions, particularly because it is so hard to confirm that observed changes have a genetic basis. Time series of size and age in exploited salmon populations show little evidence that change is tapering off but the resolution of such time series is limited. Furthermore, harvest could be an unusual selective factor in that anthropogenic selection can remain strongly directional as the population evolves. For example, selective exploitation of the largest fish in a population may continue to target ever smaller fish as the mean size of fish in the population decreases. Harvest, therefore, may not be an apt representation of natural evolutionary processes.

3. Conclusions

We have considered the tempo and mode of salmonid evolution from the level of contemporary evolution within and among populations to the level of diversification among genera. From the standpoint of mode, we find ample evidence of parallel, convergent, and divergent patterns caused primarily by deterministic natural selection acting on similar genetic backgrounds. Picking apart the relative roles of selection versus genetic background would prove challenging and may be of questionable practical relevance. Most of evolution's experiments in parallel evolution, with salmon or other organisms, involve similar genotypes that often become established in similar habitats, which in turn may be subject to the same sorts of contingent events. By considering tempo and mode in a related group of fishes, we have constrained ourselves to witness a set of evolutionary replicates of only modest genetic or selective independence.

From the standpoint of tempo, we find strong evidence for continuity of evolutionary processes from the level of contemporary evolution through to post-glacial diversification. Specifically, patterns of population divergence on contemporary scales fit well with patterns among indigenous populations facing similar environmental conditions. It therefore seems likely that contemporary trends persist and develop to the scale of indigenous population diversity. The detailed pattern of that transition requires more research, but we argue that the pattern is unlikely to be one of complete linear gradualism. Indeed, if contemporary rates were to persist unabated (i.e., perfect gradualism) over even a thousand generations, the resulting phenotypic range would likely be outside the realm of current intraspecific variation. On the other hand, widespread presence of contemporary evolution argues against true "stasis" between large punctuated events. The most logical tempo is hence likely to be one integrating gradual and punctuated patterns. For example, at large time scales, evolution may seem driven by punctuated events that occur owing to dramatic shifts in selective regimes, such as climate change (e.g., glaciation). During the period of environmental change, however, net evolution might appear generally gradual. If one looks with even finer resolution, micro-punctuated events may again be evident within periods of apparently gradual change or stasis, reflecting variable and reversing selective conditions on generational scales.

A critical and yet tractable question of the tempo of population evolution remains to be addressed. How does the magnitude of contemporary divergence compare with the scale of indigenous population divergence measured over similar ranges of environmental conditions? Might divergence over a few decades approach 80% of the mean difference found among post-glacial populations or does it account for less than 20%? The implication of such work could be an appreciation of not only the nonlinear pattern of evolutionary rates, but also a more meaningful understanding of the relative significance of contemporary evolution in shaping modern population diversity.

Translating population-level variation to species-level variation is much more challenging. Although both intra- and interspecific variation in salmonids is extreme, some of the variation among species (e.g., breeding coloration) has seemingly little parallel to variation within those species. This is perhaps not surprising under an assumption that repeatability and determinism are less likely over larger phylogenetic scales. Nonetheless, some traits do seem to show parallel variation within and between species, such as along the iteroparity/semelparity continuum and the anadromy/non-anadromy continuum. Furthermore, we have presented evidence that processes that contribute to speciation, such as the evolution of reproductive isolation, can be seen acting in the divergence of populations within species.

In closing, we want to emphasize that appreciation of parallelism and continuity of scale in salmon and their kin by no means argues for the redundancy of populations, or undermines the value of preserving as much existing diversity as possible. The isolated traits we have emphasized vary overtly among populations in association with easily recognized environmental gradients. This vantage point skims over less overt variation of potentially significant adaptive value, (e.g., resistance to parasites) and over the potential integrated significance of adaptive trait complexes. Likewise, parallel phenotypic evolution does not necessarily dictate parallel genetic change, hence the potential for genetic change. Striking evolutionary convergence suggests that similarities among populations could at times belie critical genetic differences. Modern conservation of salmonids focuses on "evolutionarily significant" groups of populations that are likely to share common ancestry, adaptive features, and migrants (Waples 1991a; Ruckelshaus et al. 2002; Ford 2003—*this volume*; Young 2003—*this volume*). This focus fits well with maintaining the evolutionary processes that likely underlie evolutionary parallelism and continuity in salmonids.

Contemporary evolution in salmon, in particular, has at times been misrepresented as evidence against the need for preservation of indigenous populations. It is thus crucial to note that most attempts to introduce anadromous salmon have failed utterly, suggesting that contemporary evolution may often not be able to recapture lost diversity. Furthermore, the evolution observed in introduced populations, although short in evolutionary time, is still very long from the perspective of human interests, and we have little insight as to how its magnitude compares with post-glacial divergence. It would almost certainly be much more efficient to preserve existing variation than to attempt to re-create adapted populations *de novo*. A "fix it later" philosophy may further contribute to complicating factors associated with loss or degradation of habitat. The real implication of contemporary evolution in salmon is that we are indeed preserving an evolutionary "legacy" and not just a collection of interesting but stagnant populations. That legacy not only includes the evolutionary history of populations, but extends from today until extinction. Perhaps the most critical question surrounding tempo and mode is how our contemporary impacts will determine the future legacy of salmonid evolution.

Acknowledgments   Ian Fleming, Jerry Johnson, David Reznick, and Bill Schaffer provided constructive correspondence or reviews on earlier drafts. Bernie Crespi provided a prepublication copy of his salmonid phylogeny. M. Kinnison was supported in part by the Maine Agriculture and Forest Experiment Station during the writing of this chapter. A. Hendry was supported by the Natural Sciences and Engineering Research Council of Canada.

Evolution in Mixed Company

Eric B. Taylor

Evolutionary Inferences from Studies of Natural Hybridization in Salmonidae

Spawning bull trout

8

If the intellectual vigor of a field of study is reflected by the number and intensity of scientific controversies within that field, then evolutionary biology remains a vital discipline. Controversial issues have included, and in many cases continue to include: the nature and origin of species, the basis and mechanisms of inheritance, the extent and nature of genetic variation within populations, and the role of hybridization in evolution (Mayr 1982). In the latter case, I refer to successful mating between individuals from two populations that are genetically distinct in at least one measurable trait. This definition (cf. Harrison 1990; Arnold 1997) emphasizes that hybridization includes mating between conspecifics (from genetically distinct populations). Although Arnold's (1997) definition includes the additional point that at least some of the F_1 progeny are viable and fertile, I will consider "successful" hybridization as not necessarily requiring fertility because even infertile (but viable) hybrids can have evolutionary consequences. For instance, reinforcement, or selection for pre-mating isolation as the result of selection against hybrids, is an important model of speciation (Dobzhansky 1940). Introgressive hybridization, however, clearly requires that hybrids be both viable and fertile as it represents the movement of alleles from one (genetically distinct) population to another by hybridization and backcrossing. Within this context, a hybrid zone is a geographic area where genetically distinct populations overlap spatially and temporally such that hybridization between them occurs (cf. Arnold 1997).

Within the general field of hybridization, controversial issues have included: (1) the meaning of hybridization to the recognition of "good" species (i.e., can taxa that hybridize still be considered distinct species?), (2) the role of hybridization in creating evolutionary novelty and perhaps new species (i.e., is hybridization an evolutionary dead-end or a source of novel diversity?), and (3) the roles of endogenous (internal genetic mechanisms) versus exogenous (environmentally dependent) selection in structuring hybrid zones (Arnold 1997; Barton 2001). As Arnold (1997) has detailed, alliance to opposing sides of these issues tended to fall along taxonomic lines. In particular, zoologists, beginning at least with Darwin, tended to regard hybridization as a problem for recognizing hybridizing populations as distinct species. In addition, zoologists have tended to study hybridization with the view that it can reveal processes that influence gene flow and genetic divergence between species. By contrast, botanists, beginning with Linnaeus and the plant "hybridizers" regarded hybridization as an important creative process to generate evolutionary novelty including the emergence of "recombinant" species (Barton 2001). This taxonomic division regarding the role of hybridization in evolution has become increasingly blurred in recent years with the recognition that hybridization can act in both restrictive (but illuminating) and creative ways (Arnold 1997; Barton 2001). Despite this encouraging development, important uncertainties in the study of hybridization include the form of selection structuring hybrid zones, the origin (primary or secondary intergradation) of hybrid zones, and the role of hybridization in adaptation and speciation.

Within these theoretical contexts, this chapter reviews studies of hybridization in salmonid fishes and their contributions to resolving contentious issues in evolution. I begin by reviewing the taxonomic and geographic distribution of hybridization in salmonids, the environmental and ecological factors that appear to influence patterns of hybridization, and evolutionary inferences that have been made based on salmonids. I end by highlighting the major uncertainties that remain and the outstanding features of salmonid biology that could be used to address these uncertainties. Hybridization is increasingly viewed as important to biological conservation as one of the major causes of loss of biodiversity, along with overexploitation, habitat degradation, exclusion by exotics, and chains of extinction (Rhymer and Simberloff 1996). I do not, however, treat the significance of hybridization to salmonid conservation because this was recently reviewed by Allendorf et al. (2001). In addition, I do not review methods for the detection of hybridization because this subject is well covered in other reviews (e.g., Campton 1987; Scribner et al. 2000).

1. Taxonomic Distribution of Hybridization in Salmonids

Mayr (1963) discussed five categories of hybridization: (a) "occasional" hybridization between sympatric species where hybrids were inviable or sterile, (b) hybrids with at least some fertility such that occasional backcrossing occurs, (c) formation of zones of secondary contact where partial interbreeding takes place, (d) complete breakdown of reproductive isolation with the production of hybrid swarms, and (e) production of a "new specific entity" via hybridization and polyploidy (allopolyploidy). The taxonomic and geographic distribution of hybridization in salmonid fishes provides ample demonstration of classes (a)– (d), while the possibility of allopolyploidy as a speciation mechanism in salmonids is intriguing, but inconclusive (Mina 1992; Svärdson 1998).

Avise (1994) demonstrated that hybridization tends to be more common in the "lower" vertebrates relative to birds and mammals, and Hubbs (1955) provided the first summary to indicate that hybridization and introgression are widespread in fishes. In his review, Hubbs (1955) gave anecdotal accounts of hybridization between species in *Oncorhynchus*, *Salvelinus*, and *Coregonus*, and even an intergeneric mating in nature between brown trout and brook charr (see also Brown 1966). Aside from basic listings (see also Chevassus 1979; Schwartz 1982; Scribner et al. 2000), there has been little further discussion of the evolutionary implications of hybridization in salmonids. A summary of existing information compiled 42 years later (Arnold 1997) provided virtually no information on hybridization in salmonid fishes. Perhaps the most thorough work on hybridization in salmonid fishes was that of Svärdson (1998) who reviewed data on inherited variation in morphological and meristic traits used to investigate hybridization and introgression in Scandinavian *Coregonus*. Because of the limitations of morphological variation for studying hybridization (see Campton 1987), studies of hybridization in salmonids before the 1970s were limited to putative identification of hybrids in nature or to the morphological description of hybrids produced artificially in the laboratory. It was not until the development of protein electrophoresis in the mid-1960s and DNA-based technology in the late

1970s that the true taxonomic, temporal, and geographic extent of hybridization in salmonids became clear (Verspoor and Hammar 1991).

The first recorded case where biochemical markers were used to document interspecific hybridization in salmonids was that between Atlantic salmon and brown trout (Payne et al. 1972). The development of genetic markers for specific and hybrid diagnoses was clearly a crucial development for the study of hybridization. For example, despite being naturally sympatric in Europe and of intense fishery interest, these two species were not suspected or documented to produce hybrids in nature until the use of allozyme assays became common.

It is clear from a summary of documented cases of hybridization using diagnostic molecular markers, that hybridization is common to all major lineages of salmonids (Figure 8.1). Certainly, hybridization and introgression have been recorded in *Salmo, Salvelinus, Coregonus*, some members of *Oncorhynchus*, and even one of the most basal lineages (*Brachymystax*). Because of the cursory nature of most studies of hybridization in salmonids, identifying trends in the taxonomic distribution of hybridization is difficult. It does appear, however, that hybridization is much less common within the Pacific coast genus *Oncorhynchus* than within other salmonid genera. All but one incidence of natural hybridization within this genus are limited to hybridization between rainbow trout and various subspecies of cutthroat trout (Verspoor and Hammar 1991). What might be the reason(s) for this apparently reduced incidence of hybridization in *Oncorhynchus*?

First, the age of the species complex may be related to the propensity for hybridization. More distantly related species, by definition, show greater levels of genomic divergence, which would limit the *opportunities* for hybridization, owing to major ecological differences, as well as the outcomes of hybridization, owing to low hybrid viability (May et al. 1975; Verspoor and Hammar 1991; Avise 1994). For instance, estimates of average pairwise Nei's D in *Oncorhynchus* (0.55, Utter et al. 1973) are about twice that reported between Atlantic salmon and brown trout (about 0.3) and among various members of *Salvelinus* (about 0.2–0.3), and natural hybridization appears much more common in the latter two genera than in Pacific salmon. Within genera, however, there appears to be little clear correspondence between measures of genomic divergence and intrinsic isolation (Verspoor and Hammar 1991; see Section 4.3 below). Similarly, despite fewer records of natural hybridization among the five species of Pacific salmon, *Oncorhynchus*, they exhibit similar or lower divergences ($d = 1.68$) as measured by rDNA ITS sequences than do Atlantic salmon and brown trout ($d = 1.71$) or various species of charr ($d = 2.10$, Phillips et al. 1992).

A second argument that may explain variation in extent of hybridization is the stability of watersheds. Habitat disturbances from climate change, glaciation, sea level changes, and tectonic activity may promote secondary contact and the formation of hybrid zones between previously allopatric populations (e.g., Hewitt 2000). This hypothesis, however, probably does not explain the lower incidence of hybridization within *Oncorhynchus* because, if anything, the range of Pacific salmon has probably undergone more dramatic geological changes over the last 10 million years than the range of *Salmo* (Montgomery 2000).

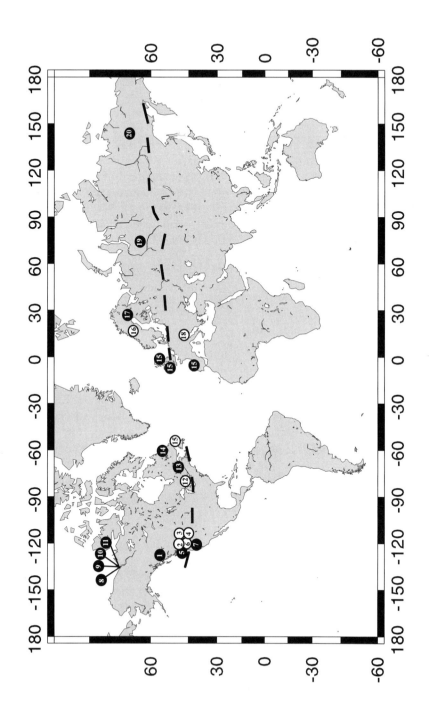

Figure 8.1. Geographic localities of hybridization in salmonids. Hybrids were detected using biochemical/molecular assays and, in some cases, morphological analyses. Closed symbols represent hybridization within the natural range of both species. Open circles represent examples of hybridization in nature, but involving one or both species introduced outside their native range(s). 1 = Dolly Varden (*Salvelinus malma*) and bull trout (*S. confluentus*), 2 = bull trout and brook trout (*S. fontinalis*), 3 = westslope cutthroat trout (*O. clarki lewisi*) and Yellowstone cutthroat trout (*O. clarki clarki*), 4 = rainbow trout (*Oncorhynchus mykiss*) and Yellowstone cutthroat trout, 5 = rainbow trout and coastal cutthroat trout (*O. clarki clarki*), 6 = rainbow trout and westslope cutthroat trout, 7 = chinook salmon (*O. tshawytscha*) and coho salmon (*O. kisutch*), 8 = inconnu (*Stenodus leucichthys*) and lake whitefish (*Coregonus clupeaformis*), 9 = lake whitefish and least cisco (*C. sardinella*), 10 = lake whitefish and Arctic cisco (*C. autumnalis*), 11 = Arctic cisco and least cisco (Reist et al. 1992, pp. 8–11), 12 = chinook salmon and pink salmon (*O. gorbuscha*), 13 = brook trout and Arctic charr (*S. alpinus*), 14 = Arctic charr and lake trout (*S. namaycush*), 15 = Atlantic salmon (*Salmo salar*) and brown trout (*Salmo trutta*), 16 = brown trout and brook trout, 17 = European whitefish (*Coregonus lavaretus*) and vendace (*C. albula*) (Pethon 1974), 18 = incipient species of brown trout (*Salmo marmoratus*) and (*Salmo trutta fario*) (Giuffra et al. 1996), 19 = European grayling (*Thymallus thymallus*) and Arctic grayling (*T. arcticus*) (Shubin and Zakharov 1984), 20 = ''sharp-'' and ''blunt-nosed'' forms of lenok (*Brachymystax lenok*). Example authorities for each case, except where noted, are reported in the text. The dashed line represents the approximate southward limit of ice sheets in the Northern Hemisphere during the most recent glacial advance. Localities are approximate and representative only; they do not necessarily show all areas of reported hybridization.

A third hypothesis is related to the idea that Pacific salmon are the most derived members of the family (Stearley and Smith 1993; Kinnison and Hendry 2003—*this volume*) and that their derived status and accompanying specialized life history traits may represent stronger selection against any hybrids that are produced, or stronger selection for reproductive isolation (via reinforcement) in sympatry. For instance, Utter (2000) reviewed evidence for anthropogenic introgression in two genera of salmonids, *Oncorhynchus* and *Salmo*, and demonstrated that hybridization and introgression were much more common in freshwater-resident populations within both genera than in anadromous populations. Other work by Campton and Utter (1985), Hawkins and Quinn (1996), and Young et al. (2001) on steelhead trout and coastal cutthroat trout, and Hansen et al. (2000b) on brown trout found fewer hybrids in adults (who have made migrations between fresh water and the ocean) than in juveniles sampled in fresh water. These results suggest that there may be strong selection against hybrids when one of the parental forms is anadromous. Consequently, it is possible that variation in the extent and timing of anadromous migrations that so characterize Pacific salmon (Hendry et al. 2003b—*this volume*), and their correlated effects on time and size at spawning, may more strongly limit hybridization in *Oncorhynchus* relative to *Salmo* or *Salvelinus*.

2. Geographic Distribution of Hybridization in Salmonids

The geographic distribution of natural hybridization in salmonids is strongly, but not exclusively, associated with portions of their range that were directly influenced by Pleistocene glaciation events (Figure 8.1). In large part, this association stems from fact that salmonids are north temperate in distribution and this was the major region of the Northern Hemisphere affected by glaciation (Pielou 1991). There are, however, some interesting subtleties within this general trend.

First, phylogenetic evidence has strongly indicated that hybridization between species occurred in areas where one of the two species involved no longer exists. For instance, Bernatchez et al. (1995) and Glémet et al. (1998) used combined analysis of nuclear and mitochondrial loci to demonstrate introgression of Arctic charr mtDNA into several allopatric populations of brook charr in eastern Québec. A similar phenomenon was demonstrated in an allopatric population of lake trout that was characterized by Arctic charr mtDNA (Wilson and Bernatchez 1998). Redenbach and Taylor (2002) observed Dolly Varden with bull trout mtDNA, and several Dolly Varden populations with introgressed bull trout mtDNA were located in areas (e.g., Vancouver Island) where bull trout are not currently found, clearly implying historical introgression. All of these instances of historical introgression in eastern and western North American charr indicate that the "footprint" of hybridization can be apparent in geographic localities where one of the participating species is no longer found, thus greatly expanding the geographic scale across which hybridization in salmonids has occurred.

Another subtlety in the geographic distribution of hybridization in salmonids is that hybridization between species may be higher in areas where salmonids have been introduced relative to where they are naturally sympatric. For instance, four species of *Oncorhynchus* have been widely introduced into the Great Lakes region of eastern North America (coho, chinook, and pink salmon, and steelhead trout). Rosenfield et al. (2000) demonstrated a low level of introgressive hybridization between pink and chinook salmon from the St. Marys River system (between lakes Superior and Huron), although hybrids between these two species have never been reported in their natural range. Further, rainbow trout and westslope cutthroat trout are largely allopatric or parapatric throughout their native ranges in western North America (Behnke 1992). Rainbow trout, however, have been widely introduced into the range of westslope cutthroat trout and introgressive hybridization between species typically results from such artificial secondary contact (Leary et al. 1984; Rubidge et al. 2001). McGowan and Davidson (1992a) reported hybridization between non-native brown trout and native Atlantic salmon in Newfoundland waters at a rate about twice that reported in Europe where both species are native. These cases of apparently greater hybridization following human-induced introductions may stem from fundamentally different causes. Hybridization between pink and chinook salmon in the Great Lakes is probably promoted, in part, by environmental differences between the Great Lakes and the Pacific basin. The two species have had a long history of sympatry in the Pacific basin where reproductive isolation is promoted, in part, by differences between species in age and size at maturity and spawning habitat distributions and preferences (e.g., Groot and Margolis 1991). Rosenfield et al. (2000) hypothesized that a lack of opportunities for spatial isolation in the St. Marys River system may promote hybridization between chinook and pink salmon in the Great Lakes. In contrast, the largely allopatric evolution of rainbow trout and westslope cutthroat trout raises the hypothesis that reinforcement may be important in salmonid speciation because pre-mating reproductive isolation between these species is incomplete, which may result from a lack of opportunities for historical interactions during mating in nature.

3. Temporal Distribution of Hybridization in Salmonids

Biochemical and molecular approaches to hybridization have been crucial for demonstrating that introgressive hybridization in salmonids is both a contemporary and an historical phenomenon. Leary et al. (1987) studied allozyme variation among seven subspecies of cutthroat trout and rainbow trout, found that westslope cutthroat trout were paraphyletic with respect to rainbow trout, and suggested that ancient hybridization explained the sharing of many alleles between these species that were not found in the other cutthroat trout subspecies. The finding of mtDNA introgression between *Salvelinus* species that are now allopatric (Bernatchez et al. 1995; Wilson and Hebert 1998) also strongly suggests historical hybridization in addition to contemporary hybridization between sympatric populations (Hammar et al. 1989; Wilson and Hebert 1993).

Discordance in phylogenetic relationships between different classes of genetic markers and variation in sequence divergence among haplotypes has been cited as evidence of historical hybridization in a number of animal and plant taxa (Arnold 1997). Dolly Varden and bull trout are parapatric and sympatric in western North America (Baxter et al. 1997). Redenbach and Taylor (2002) demonstrated discordance between mtDNA-based phylogenies and those based on two nuclear sequences (Figure 8.2). Although Dolly Varden and bull trout were clearly reciprocally monophyletic at two nuclear loci, the species were paraphyletic with respect to mtDNA. In particular, a "southern" clade of Dolly Varden mtDNA (from fish confirmed as Dolly Varden by morphological and nuclear DNA analyses) was more closely related to a "coastal" clade of bull trout mtDNA than to a "northern" Dolly Varden mtDNA clade (Figure 8.2). In addition, the geographic distribution of the paraphyletic Dolly Varden mtDNA clade was largely restricted to areas of current sympatry with bull trout. A notable exception, however, was the presence of paraphyletic Dolly Varden mtDNA on Vancouver Island, a region where bull trout do not currently occur. The presence of paraphyletic Dolly Varden mtDNA in otherwise "pure" Dolly Varden in currently allopatric Dolly Varden populations strongly suggests historical mtDNA introgression from bull trout to Dolly Varden (Arnold 1997; Redenbach and Taylor 2002).

The timing of such historical introgression events is difficult to determine, but some estimates are possible. For instance, the average allozyme divergence among cutthroat trout subspecies that are not suspected of historical introgression with rainbow trout is 0.217 (Leary et al. 1987). If this value is used as a baseline to date the separation between rainbow trout and cutthroat by using a standard rate of 1 unit of Nei's D for 5 million years (Nei 1987a), then the two species diverged from a common ancestor at least 1 million years ago (see also Behnke 1992). If the present Nei's D between rainbow and westslope/coastal cutthroat of about 0.11 (Utter et al. 1973; Leary et al. 1987) is then used as an estimate of time since an ancient introgression event, this suggests that hybridization first occurred on the order of 500,000 years ago. Behnke (1992), however, suggested that rainbow trout were likely absent from the upper reaches of the Columbia River until very late in the Pleistocene and likely did not come into contact with westslope cutthroat trout until between 32,000 and 50,000 years ago, suggesting a much more recent historical hybridization. Similarly, paraphyletic Dolly Varden mtDNA is found in several lake resident populations on Vancouver Island. Because bull trout are not found on Vancouver Island presently, the mtDNA introgression between the species must have occurred at least before the island became free of ice and accessible for post-glacial colonization some 12,000 years ago (McPhail and Lindsey 1986). Furthermore, these allopatric paraphyletic Dolly Varden mtDNAs are about 0.4–0.8% divergent from their closest bull trout mtDNAs which suggests (if mtDNA mutates at approximately 1% per million years, Smith 1992) that hybridization between these species could have occurred as early as 800,000 years ago. Such ancient

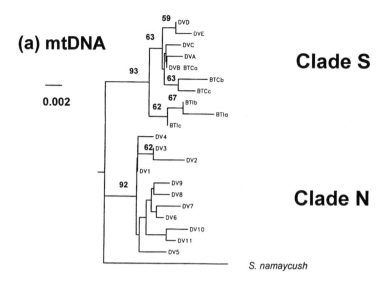

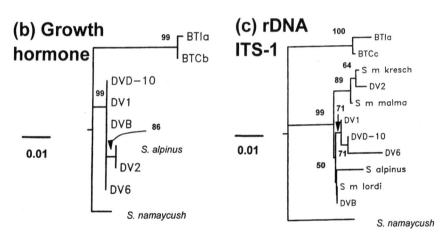

Figure 8.2. (a) Phylogenetic tree of *Salvelinus* mitochondrial DNA rooted with *S. namaycush* based on neighbor-joining analysis of Kimura 2-parameter distance inferred from 503 base pairs of the tRNA-Gln (36 bp), tRNA-Ile (73 bp), and NADH-1 (453 bp) genes. Bootstrap support levels from 1000 resamplings are reported (where over 50%) at branch points. (b) Phylogenetic tree of *Salvelinus* Growth Hormone 2 sequences rooted with *S. namaycush* sequence (from McKay et al. 1996), based on neighbor-joining analysis of Kimura 2-parameter distance inferred from 484 base pairs of intron C. (c) Phylogenetic tree of *Salvelinus* ribosomal DNA internal transcribed spacer region (rDNA ITS-1) sequences using *S. namaycush* as the outgroup. The tree is derived from neighbor-joining analysis of Kimura 2-parameter distance inferred from 410 base pairs of ITS-1. In all trees, "DV" are Dolly Varden (*Salvelinus malma*) haplotypes, "BTI" and "BTC" are interior and coastal, respectively, bull trout (*S. confluentus*) haplotypes. Bootstrap support levels (where over 50%) from 1000 resamplings are marked at branch points. All haplotypes are defined in Redenbach and Taylor (2002).

hybridization is not unreasonable. For instance, Smith (1992) reported a discordance between an mtDNA-based phylogeny of Pacific salmon that reported pink and chum salmon as sister species and one based on morphological/karyological/biochemical data where pink and sockeye salmon were sister species (see also Kinnison and Hendry 2003—*this volume*). He suggested that this difference in inferred relationships was caused by historical mtDNA introgression between pink and chum salmon that, from a consideration of the fossil record, occurred at least 2 million years ago.

4. Reproductive Biology, Isolation, and Hybridization in Salmonids

Aspects of reproductive biology are major factors that likely govern the incidence of interspecific hybridization, and may influence gene flow between hybrid and parental lineages. In this section, I briefly review conceptual issues of reproductive isolation and salmonid reproductive biology as it pertains to the incidence of natural hybridization. These topics bear directly on factors that may explain the genetic structure of hybrid zones in salmonids examined in the subsequent section.

4.1. Classification of Reproductive Isolation

The different aspects of genetic isolation (the prevention of hybrid formation) are most broadly categorized into extrinsic and intrinsic reproductive isolation. At this level, extrinsic isolation refers to the separation of potentially interbreeding populations by some geographic barrier such that they have no opportunity for mating. Intrinsic reproductive isolation refers to isolation imposed by properties of the organisms themselves (other than their geographic distribution) that prevent genetic exchange (see Dobzhansky 1937). Clearly, the lack of documented natural hybridization between some salmonid species results from extrinsic isolation (e.g., the masu salmon endemic to the western Pacific and cutthroat trout endemic to the eastern Pacific). The extent or breakdown of intrinsic reproductive isolation is, therefore, central to a discussion of hybridization and introgression.

Factors important to intrinsic reproductive isolation were originally classified by Dobzhansky (1937). Although he presented several different classes, they were most broadly differentiated into those that operate before mating (pre-mating isolation) and those that operate after mating (post-mating isolation). Pre-mating isolation typically involves aspects of behavior or reproductive physiology (e.g., time and place of reproduction, behavioral interactions) that restrict opportunities for interspecific pairings or actual mating (see Mayr 1963, p. 92; Palumbi 1998). Post-mating isolation encompasses phenomena that influence zygotic survival or fertility (e.g., developmental abnormalities, hybrid inviability, or infertility). More recently, it has been recognized that factors acting at the level of egg and sperm interactions may constitute important processes of reproductive isolation that act after mating (i.e., post-insemination, or in the case of salmonids, post-spawning), but before the formation of zygotes. Consequently, a more inclusive characterization recognizes pre-zygotic isolation as those processes acting prior to viable zygote formation and post-zygotic processes as those acting after viable zygote formation.

A second classification of processes has been defined that is relevant to the fate of hybrids produced when pre-zygotic reproductive isolation is incomplete. This nomenclature describes the role of selection and includes: (1) endogenous, or environmentally-independent processes and (2) exogenous, or environmentally dependent processes. Endogenous selection imposes reproductive isolation regardless of the environmental context via interspecies genomic or developmental incompatibilities that directly influence hybrid viability or fertility. The work by Dobzhansky (1933) on the genetics of hybrid male sterility, and its molecular dissection by Ting et al. (1998), in *Drosophila* are good examples of endogenous factors influencing selection against hybrids. Exogenous selection imposes reproductive isolation through the interaction between phenotype (as determined by genotype) and environmentally dependent natural or sexual selection against hybrid phenotypes (Rice and Hostert 1993; Rundle and Whitlock 2001; Bernatchez 2003—*this volume*). In this case, selection against (or for) hybrid genotypes is not "automatic" as in endogenous selection, but rather depends on the environmental conditions. The temporal variation in survival of hybrids between species of Darwin's finch (*Geospiza*) and its dependence on the distribution of seed sizes, which is influenced by rainfall patterns, exemplify the role of the environment on patterns of selection (see discussion summarized by Schluter 2000). Similarly, a close association between genotype survival and environmental variation within a hard-clam hybrid zone also points to the role of exogenous selection (Bert and Arnold 1995). Classification and discussion of reproductive isolation can be accomplished using a variety of nomenclature systems, but their real value lies in helping to focus research efforts on the processes involved in reproductive isolation and speciation (Berlocher 1998).

4.2. Pre-Zygotic Isolation in Salmonids

Some of the best studied aspects of salmonid reproductive biology (at least for salmon, trout, and charr) are habitat distribution and mating behavior. Several processes are involved in reproductive isolation between sympatric species including: macro- and micro-habitat utilization for spawning, secondary sexual traits, behavioral differences, and different sizes at maturity (which may influence habitat and mate choice). For instance, Pacific salmon mature at between 2 (pink salmon) and 8 (chinook salmon) years of age, and return to spawn in streams tributary to the northeast and northwest Pacific Ocean typically from late summer to late fall. During spawning, females construct and defend gravel nest sites and are attended by one or more males that compete for access to territorial females (see Groot and Margolis 1991, and Fleming and Reynolds 2003—*this volume*, for detailed descriptions). Notwithstanding these broad similarities among species, there are consistent differences that limit opportunities for hybridization (Fukushima and Smoker 1998). For instance, sockeye salmon exhibit perhaps the most specialized habitat choice in that they usually spawn in streams that are associated with lake systems (where the juveniles feed for a year or more before migrating to the ocean). In addition, sockeye salmon are the only *Oncorhynchus* that make extensive use of submerged gravel beaches for spawning in lakes (Burgner 1991; Taylor et al. 2000). In addition, pink salmon, the smallest Pacific salmon (typically 1.5–2.5 kg) usually do not make extensive upstream migrations for spawning and spawn in small coastal rivers or close to the sea in larger systems (Heard 1991). By contrast, chinook salmon, on average the largest of all Pacific salmon (up to 45 kg, Healey 1991), may make upstream migrations of 1000 km or more and tend to spawn in larger watersheds.

Pacific salmon are also well known for the development of exaggerated secondary sexual traits and coloration (see Fig. 2 of Hendry 2001), particularly in males of each species, which likely act in mate recognition in sympatric species. The variation in size at maturity, although also extensive within species, probably constrains interspecific hybridization owing to well-developed size assortative mating and size-dependent habitat use in salmonids (Foote 1988; Beacham et al. 1988b; Hagen and Taylor 2001). Indeed, the importance of these pre-mating isolation factors in limiting opportunities for hybridization in salmonids is suggested by the relatively high incidence of hybridization in systems where such isolating factors are absent or are less well developed. For instance, reproductive isolation between species may be weakened when there are constraints on spatial segregation in introduced populations (chinook and pink salmon; Rosenfield et al. 2000), greater similarity in size at maturity (e.g., rainbow trout and westslope cutthroat trout; Rubidge et al. 2001), or less extensive secondary sexual development (various *Coregonus*; Svärdson 1998).

Although I have argued that differences in size at maturity among species may act to constrain hybridization, there is at least one circumstance where such differences may actually promote hybridization; parasitic matings by smaller "sneaker" males. Such sneaking behavior is common in intraspecific matings in salmonid fishes. Sneaking occurs when a smaller male rushes into the nest of a larger mating pair of the same species and attempts to fertilize the eggs of the female (Svedäng 1992; Maekawa et al. 1993, 1994; Taborsky 1998; Fleming and Reynolds 2003—*this volume*). In cases where there are differences in size at maturity between species of salmonids, sneaking has been proposed as an explanation for interspecific hybridization (e.g., McGowan and Davidson 1992a; Kitano et al. 1994; Baxter et al. 1997; Redenbach and Taylor 2003). Consequently, although body size differences between species are usually sufficient to prevent interspecific *pairings*, they may, in some cases, actually promote interspecific *matings*. Under such conditions complete pre-zygotic reproductive isolation may be difficult to establish because sneaker males are employing an intraspecific parasitic reproductive trait that, presumably, is relatively successful in allopatric populations (e.g., Gross 1985; Maekawa et al. 1993, 1994; Fleming and Reynolds 2003—*this volume*). In addition, the selective penalty for interspecific sneaking by salmonid males may be reduced because sperm is relatively abundant and energetically "cheap," especially for iteroparous species where males have multiple opportunities to breed across years.

The extent of size differentiation between mature adults of sympatric species likely influences whether hybridization is uni- or bidirectional (Wirtz 1999). For instance, unidirectional hybridization has been reported in *Salvelinus* (e.g., Hammar et al. 1989; Wilson and Hebert 1993; Redenbach and Taylor 2003), *Oncorhynchus* (Dowling and Childs 1992), and *Salmo* (e.g., McGowan and Davidson 1992a). By contrast, hybridization may occur in both directions in *Oncorhynchus* (e.g., Rubidge et al. 2001). Several hypotheses, encompassing both pre-zygotic and post-zygotic processes have been proposed to explain different degrees of directionality to hybridization (reviewed by Wirtz 1999). One hypothesis involves sneak spawnings by males of the *smaller* species as outlined above. For instance, where Dolly Varden and bull trout are sympatric, males (and females) of the former mature at much smaller sizes than the latter (Figure 8.3) and hybridization, inferred from mtDNA analysis, is in the direction of male Dolly Varden × female bull trout (Figure 8.3a). This hypothesis predicts that

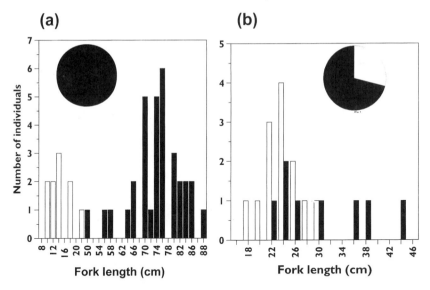

Figure 8.3. Size–frequency distribution of adult fish (bars) and mitochondrial DNA introgression (pies) in salmonid fishes: (a) Dolly Varden (*Salvelinus malma*, white bars) and bull trout (*S. confluentus*, black bars) and (b) rainbow trout (*Oncorhynchus mykiss*, white bars) and westslope cutthroat trout (*O. clarki lewisi*, black bars). The pies represent the percentage of hybrids with Dolly Varden (white shading) or bull trout (black shading) mtDNA ((a), N = 89 hybrids from five localities) or those with rainbow trout (white shading) or westslope cutthroat trout (black shading) mtDNA ((b), N = 16 hybrids from seven localities). Data for Dolly Varden and bull trout are from Hagen and Taylor (2001) and Redenbach and Taylor (2003). Data for westslope cutthroat trout and rainbow trout are from Rubidge et al. (2001) and Rubidge and E. Taylor (unpublished).

where sympatric species are the same or more similar in body size, hybridization should be bidirectional. Data collected from hybrid zones between sympatric rainbow trout and westslope cutthroat trout are consistent with the hypothesis; these species show greater overlap in size at maturity and hybridization is bidirectional (Figure 8.3b; Rubidge et al. 2001).

One aspect of pre-zygotic isolation that has received comparatively little attention in salmonids is gamete recognition. Considerable recent work on marine invertebrates (with external fertilization) has demonstrated the potential for physical and molecular bases of sperm–egg interactions and their influence on the probability of natural hybridization (Palumbi 1998; Eady 2001). Only a limited amount of work has been conducted on comparative analyses of gamete morphology and structure in salmonids (e.g., Groot and Alderdice 1985) and these data appear to indicate conservation of the basic structure of egg membranes. Subtle differences, however, in relative egg membrane thickness may be important in constraining hybridization, particularly under competitive conditions (Arnold 1997). This general area bears further investigation in salmonids, as experimental hybrid matings in salmonids have not been attempted under competitive conditions.

4.3. Post-Zygotic Isolation in Salmonids

Post-zygotic isolation in salmonid fishes, like other organisms, may take the form of endogenous, or environmentally independent, factors resulting from genomic or developmental incompatibilities between species that result in hybrid inviability or infertility. Alternatively, post-mating isolation can result from exogenous factors, or environmentally dependent selection against hybrids. Inferences about the importance of endogenous factors come largely from experimental crosses between species. Such inferences are limited by different procedures among experiments as well as by *intra*specific variation in the response to hybridization between species (i.e., population source used in the crosses) (Chevassus 1979). A review of salmonid experimental hybridization studies (Chevassus 1979) indicated that viable and fertile hybrids were routinely produced in intrageneric crosses in *Salmo*, *Salvelinus*, and *Oncorhynchus*. Hybrid viability and fertility was markedly reduced in crosses between genera, but at least some crosses produced fertile hybrids (e.g., *Oncorhynchus mykiss* × *Salmo trutta*). Some quantitative data exist for percent survival to hatching for experimental matings within *Oncorhynchus* (R. Devlin unpublished; Alverson and Ruggerone 1997) and these data, when associated with genomic measures of similarity (either Nei's genetic distance from allozymes, or ITS rDNA sequence divergences) show a positive, but not significant relationship (Figure 8.4).

In addition, although chromosome numbers vary from 52 to 84 among *Salmo*, *Oncorhynchus*, and *Salvelinus* (Allendorf and Thorgaard 1984), chromosome arm numbers are relatively conserved at about 100. Most chromosome number changes appear to be the result of centric fusions and fissions (Allendorf and Thorgaard 1984) suggesting few intrinsic chromosomal challenges to hybridization. Biases in the survival of reciprocal crosses in experimental matings

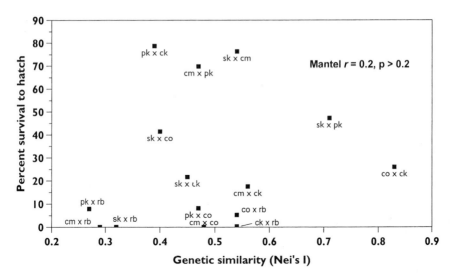

Figure 8.4. The relationship between a measure of nuclear genomic similarity (Nei's *I*) and survival to hatching in hybrid crosses within *Oncorhynchus*. Each point represents the mean of both reciprocal crosses between species. The hybridization experiments were conducted by R. Devlin (unpublished; Alverson and Ruggerone 1997) and involved two adults of each sex of each species. For each cross, approximately 500 eggs were fertilized from two females of each species. Values of Nei's genetic identity were obtained from Utter et al. (1973) and represent variation across 18 allozyme loci. Species acronyms are: sockeye salmon (sk), chum salmon (cm), pink salmon (pk), coho salmon (co), chinook salmon (ck), and rainbow trout (rb).

(e.g., McGowan and Davidson 1992b; R. Devlin unpublished; Alverson and Ruggerone 1997) and the general lack of extensive nuclear introgression in most salmonid systems where hybridization has been reported, however, suggest that differences in chromosome number may constrain introgression in some species through reduced fertility of hybrids. In summary, although results can be highly variable, it appears that endogenous factors operate strongly, but not necessarily completely at the intergeneric level and in many intrageneric crosses, but there is considerable potential for intrageneric gene flow. This conclusion is consistent with reports of natural introgression within genera in at least four genera: *Oncorhynchus* (Rosenfield et al. 2000), *Salvelinus* (Bernatchez et al. 1995; Baxter et al. 1997), *Salmo* (Verspoor and Hammar 1991), and *Coregonus* (Svärdson 1998).

Nevertheless, gene flow between salmonid species is likely constrained to some extent by endogenous selection against hybrids. For instance, there has been some experimental work on the performance of hybrid crosses in traits that would likely impact on their survival in nature. For instance, Lu and Bernatchez (1998) demonstrated three to four times higher embryonic mortality in hybrids between "dwarf" and "normal" lake whitefish (*Coregonus clupeaformis*) relative to parental crosses. The ecotypes of whitefish were from different glacial phylogroups that diverged on the order of 150,000 years ago, suggesting that genetic differences driving intrinsic developmental incompatibilities can evolve within this time frame. Leary et al. (1995) reported that experimental hybrids between rainbow trout and westslope cutthroat trout had equal or higher survival to hatching than did pure-type fish, but experienced slower growth and survival to 112 days post-fertilization under laboratory conditions. Leary et al. (1993) suggested that introgression between native bull trout and brook charr introduced into parts of western North America was constrained by F_1 sterility. Experimental matings between divergent lineages of brown trout also suggest some disruptions during meiosis in hybrids (Poteaux et al. 2000), an observation that is associated with low levels of introgression where the two lineages come into artificial contact in nature (Largiadèr and Scholl 1996). Hybrids between "even-year" and "odd-year" populations of pink salmon from the same river showed reduced survival to maturity that were suggested to arise from the disruption of co-adapted gene complexes, one mechanism of outbreeding depression (Gharrett and Smoker 1991; Gharrett et al. 1999). Finally, it is often observed when making reciprocal hybrid matings that one type of cross displays higher survival than the reciprocal cross. For instance, R. Devlin (unpublished, summarized in Alverson and Ruggerone 1997) reported 5% survival to hatching in progeny of female Atlantic salmon × male pink salmon crosses, but 0% survival in the reciprocal cross. Similarly, female sockeye salmon × male coho salmon crosses had 82.9% survival, but male sockeye salmon × female coho salmon crosses experienced 0% survival. Although the mechanism remains obscure, some of this variation may be explained by differences in egg size. It has been noted previously that, for instance, the cross involving females with the smaller egg size usually suffers higher mortality (e.g., Wilson and Hebert 1993; Lu and Bernatchez 1998). There may also be problems in terms of the developmental interaction between egg and sperm nuclei after fertilization that influence viability of matings in certain directions (R. Devlin, pers. comm.). It has also been suggested that in some instances, interspecific hybridization can result in skewed sex ratios in progeny (Forbes and Allendorf 1991) that can influence the extent of subsequent introgression.

One aspect of post-zygotic isolation that bears investigation in salmonids is Haldane's rule; i.e., the tendency for sex-biased mortality or infertility of hybrids where the sex that shows the lowered viability or fertility is the heterogametic one. For instance, Coyne and Orr (1989) summarized data that supported Haldane's rule in mammals, birds, and insects. Haldane's rule results in incomplete post-zygotic isolation (because one sex develops normally) and this would predict higher mortality/infertility in male salmonid hybrids (the heterogametic sex). In fact, Forbes and Allendorf (1991) observed a sex ratio in favor of females in some populations of hybrids between westslope cutthroat trout and Yellowstone cutthroat trout and discussed similar results in other fishes. As stressed by Forbes and Allendorf (1991), however, the effects on hybridization on sex-dependent fertility or survival has received little detailed study in salmonids. The relatively late age of maturity, difficulty of determining sex at early ages (but see Devlin et al. 1991), and the generally limited applicability of

Haldane's rule within the context of most situations were hybrids are produced (e.g., production facilities) probably explain the lack of attention. The relative ease of crossing and raising salmonid fishes and the broad range of divergence times of species that can be crossed (e.g., between "species pairs" to between genera) suggests that some insights into the role of sex-linked versus autosomal genes in reproductive isolation, for instance, could be gained by investigation of Haldane's rule in salmonids.

Exogenous selection against hybrid progeny has been less well studied in salmonid systems, but indirect evidence suggests considerable potential for this process to constrain gene flow in nature. Experimental matings between anadromous (sockeye salmon) and freshwater-resident (kokanee) *Oncorhynchus nerka* produced hybrids that were phenotypically intermediate in traits such as developmental timing and success, salinity tolerance, and swimming performance, which may compromise the survival of hybrids in nature (reviewed by Taylor 1999). Other indirect evidence for exogenous selection against salmonid hybrids comes from genetic surveys of natural populations. In at least two hybrid systems (cutthroat trout × rainbow trout and Dolly Varden × bull trout), hybrids tend to predominate in the younger age classes and adult hybrids are relatively rare (Campton and Utter 1985; Rubidge et al. 2001; Young et al. 2001; Redenbach and Taylor 2003), which suggests that hybrid survival to maturity is rare. In these cases, the hybridizing parental species show marked differences in life history (one is migratory and the other is not) and the complexities and rigors of migration may be an important source of selection against hybrids (cf. Hawkins and Quinn 1996; Utter 2000). By contrast, Forbes and Allendorf (1991) studied the genetic structure of different hybrid swarms between Yellowstone cutthroat trout and westslope cutthroat trout from lakes in Montana. Although taxonomically classified as subspecies, these fishes show levels of allozyme ($D = 0.3$) and mtDNA divergence (4–5%) comparable to some full taxonomic species of *Oncorhynchus* (e.g., Leary et al. 1987; Thomas and Beckenbach 1989). Forbes and Allendorf (1991) used estimates of linkage disequilibrium to test for selection against hybrids, and despite the high levels of structural gene divergence between subspecies, the authors observed stable hybrid swarms with little evidence of selection eliminating hybrids. As summarized by Utter (2000), it is perhaps the non-migratory or "simple" migratory (tributary streams to lake) life histories of these populations that provides limited opportunities for ecological selection against hybrids.

Sympatric and hybridizing populations of Dolly Varden and bull trout have also been studied to test the idea that parental species exploit alternative niches to which phenotypically intermediate hybrids may be poorly adapted. Stream-resident sympatric juvenile salmonids typically demonstrate well-marked differences in habitat use (e.g., Hartman 1965; Taylor 1991b). Juvenile Dolly Varden and bull trout coexisting in a northcentral British Columbia watershed, however, demonstrated little habitat and diet partitioning, suggesting limited potential for habitat or trophic-based ecological selection against hybrid progeny. By contrast, the two species showed marked habitat and diet partitioning when the entire life cycle was considered. Dolly Varden remain as stream-resident specialists on stream invertebrates and mature at 15–20 cm in length, whereas bull trout migrate to an adjacent lake where they become fish-eating specialists and mature at sizes ranging from 60 to 100 cm. Ecological selection against hybrid progeny (which are intermediate in size at maturity—see Hagen and Taylor 2001) may act either through disrupted migratory behavior or via size-dependent mate or habitat selection upon maturation.

Exogenous selection against hybrids may also be sexual in nature. For instance, mate selection in salmonids is often size-dependent, with fish tending to choose mates of their own size or larger (Foote 1988). In addition, for salmonids that reproduce in streams, habitat choice is also usually size-dependent with larger species using faster, deeper water with larger substrates (Wood and Foote 1996; Hagen and Taylor 2001). Consequently, if mature hybrids are intermediate in size to the parental species (e.g., Hagen and Taylor 2001), then they may be at a disadvantage during reproduction. For instance, Vamosi and Schluter (1999) demonstrated sexual selection against hybrids between benthic and limnetic species of three-spine stickleback (*Gasterosteus aculeatus*) in nature. The well-developed secondary sexual characteristics and behavioral repertoires of salmonids during mating suggest that such processes may also select against hybrids in salmonids.

5. Models of Hybrid Zone Structure and Evolution

One of the fundamental issues in hybridization research focuses on the evolutionary processes that determine the genetic structure of hybrid zones (Arnold 1997). All of these models incorporate dispersal and selection against hybrids, but they differ in the relative importance of these processes and in the role of the environment in selection against hybrids. These models of hybrid zone evolution were reviewed by Arnold (1997). In this section, I provide brief overviews of these models and how data on salmonids have, or could, be used to resolve remaining uncertainties.

5.1. The Tension Zone Model

The first and "simplest" model is the tension zone model (see Barton and Hewitt 1985), which posits that hybrid zones reflect a balance between dispersal of parental genotypes into the hybrid zone and endogenous (i.e., environment-independent) selection against hybrids that generates genetic clines of various widths. The mathematical theory and expectations of tension zone models described by Barton and Hewitt (1985) strongly influenced empirical studies of hybrid zones to explain the basis for variability in cline widths. Tests of tension zone models are perhaps best conducted in situations where hybridizing populations are broadly parapatric and meet as opposing "phalanxes" of genotypes. The studies of Szymura and Barton (1986, 1991) on *Bombina* toads and Hewitt (1993) on *Chorthippus* grasshoppers meeting along broad areas of secondary contact in Europe are longstanding examples of hybrid zones interpreted within the context of tension zones.

5.2. Bounded Hybrid Superiority Model

The bounded hybrid superiority model is one of the two major environment-dependent models of hybrid zones (Moore 1977) and posits that hybrids show superior fitness, relative to parental genotypes, in certain environments, and hence, that ecological factors determine the fitness of genotypes. The habitats in which hybrids show superior fitness are typically identified as transitional or intermediate ecotones between habitats in which the parental genotypes are most fit (Arnold 1997). The best examples of such hybrid zones are found in instances of "disturbed" habitats (Hubbs 1955) or where environmental change results in altered habitat characteristics through time (Arnold 1997). A potentially powerful example of such hybrid zones involves habitats that were disturbed during the many glacial and interglacial periods of the Pleistocene. The massive environmental disturbances in terms of water temperature, clarity, and drainage patterns would have provided abundant opportunities for secondary contact between divergent lineages, as well as hybridization, particularly if one species was less abundant than the other (Wilson and Bernatchez 1998; Svärdson 1998; Redenbach and Taylor 2002). The dynamic environmental changes may have favored some hybrid genotypes and perhaps allowed these lineages to persist (e.g., mtDNA introgressed charr, Glémet et al. 1998).

Bounded superiority hybrid zones are typically suggested to be ephemeral in that they persist only as long as the intermediate environments persist either temporally or spatially (Arnold 1997). Such intermediate or disturbed habitats may also be anthropogenically derived and lead to increased rates of hybridization or increased persistence of hybrid lineages (Pethon 1974; Bartley et al. 1990: Scribner et al. 2000). The apparently long-term persistence of some salmonid hybrid zones, i.e., those that appear to have persisted since deglaciation (e.g., Baxter et al. 1997) argues that they may be stable through time and are not solely dependent on transitional environments.

5.3. Mosaic Hybrid Zone Model

A second kind of environment-dependent hybrid zone model is the mosaic model. This model, however, differs from the bounded hybrid superiority model in that rather than representing a smooth transition of genotypes through an environmental gradient or ecotone, the distribution of parental and hybrid genotypes is "patchy" and results from adaptation of parental and hybrid genotypes to alternative habitats that are themselves patchy in distribution (Howard 1986; Harrison 1990). Differential adaptations to soil types were implicated in a mosaic genetic structure in hybridizing species of crickets, *Gryllus* (Rand and Harrison 1989). Although in its original form, mosaic hybrid zones posited that hybrid genotypes were less fit than parentals, Arnold and Hodges (1995) and Arnold (1997) presented a review of evidence that mosaic hybrid zones may include habitats where hybrid genotypes are more fit than one or both parental genotypes.

5.4. Evolutionary Novelty Model

Finally, Arnold (1997) proposed a new model of hybrid zones known as the "evolutionary novelty model" as a new theoretical framework to accommodate some of the empirical inconsistencies observed when adopting either environment-independent or environment-dependent models in isolation. The model essentially allows for both endogenous and exogenous selection *and* for the possibility that hybrid genotypes may demonstrate higher fitness than parentals and, consequently, establish stable evolutionarily distinct lineages (Arnold 1997).

5.5. Salmonids and Hybrid Zone Models

In general, there have been few empirical treatments of hybridization in salmonids within the context of evaluating different models of hybrid zone evolution and structure. The shortage of such studies is likely due to the dendritic nature of stream habitats or the isolated nature of many lakes and, therefore, the low probability of forming broadly parapatric contact zones across large geographic areas. The nature of the aquatic habitats of salmonids means that hybrid zones may be limited by historical contingencies that may be quite local in effect (e.g., watershed exchanges; Duvernell and Aspinwall 1995). This leads to the hypothesis that most salmonid hybrid zones probably follow the mosaic model in structure where broadly allopatrically or parapatrically distributed taxa only come into contact and hybridize where local contingencies permitted connection of specific watersheds. This idea also suggests the importance of the spatial scale of sampling when characterizing hybrid zones. For example, hybrid zones could be mosaic in structure within individual, isolated watersheds that contain two species, but where local sympatry depends on distribution of particular habitats. Alternatively, hybrid zones could be mosaic on much broader geographic scales where sympatry may be established in some watersheds, but not others, owing to historical patterns of dispersal.

A good example of such mosaic structure involves areas of hybridization between Dolly Varden and bull trout. These two species of charr are broadly parapatric in northwestern North America; Dolly Varden are coastal in distribution from western Alaska to the Olympic Peninsula, whereas bull trout are largely interior in distribution from northern California to the Yukon and Northwest Territories; that is, their ranges meet broadly along the Coastal/Cascade Mountains that separate coastal from interior drainages. Taylor et al. (2001) and Redenbach and Taylor (2002) reported areas of sympatry and hybridization between the species in at least seven independent watersheds. These areas were found in or near tributaries of large river systems that cut through the mountain crest (e.g., the lower Fraser River) or in the headwaters of coastal (or interior) watersheds that interdigitate with interior (or coastal) watersheds.

Historical and fish distribution data suggest that watershed exchanges between these coastal and interior watersheds have taken place post-glacially (Baxter et al. 1997). Consequently, the areas of hybridization form a mosaic pattern highly dependent on historical or contemporary opportunities for exchange between coastal and interior drainages. A further example is the mosaic distribution of hybridization between dwarf and normal lake whitefish among lakes in eastern North America (see Lu et al. 2001). On a broader geographic and temporal scale, however, hybrid zones between salmonid lineages may take the form of extensive clinal variation. For instance, Turgeon and Bernatchez (2001b) described a clinal pattern of microsatellite DNA variation along an east–west gradient in temperate North America in the cisco, *Coregonus artedi*. The clinal variation in allele frequencies in samples from populations that are currently largely isolated in distinct watersheds was apparent on a continent-wide scale. In addition, the cline was probably a signature of historical isolation between two glacial lineages of cisco that came into secondary contact via post-glacial dispersal through pro-glacial lakes (Turgeon and Bernatchez 2001b). More generally, such watershed exchanges may contribute to the formation of aquatic "suture zones"; that is, bands of geographic overlap between biotic assemblages (Remington 1968) that in turn may explain the origin of hybrid zones between previously isolated taxa.

The well-developed habitat partitioning that is often observed between salmonid species within drainages probably also promotes mosaic-like structure of salmonid hybrid zones at smaller spatial scales. For instance, Hartman and Gill (1968) studied the local distribution of coastal cutthroat trout and steelhead trout in southwestern British Columbia. Their analysis showed that cutthroat trout were found in smaller streams with lower gradient, and steelhead in larger, higher gradient streams. Where they occurred in the same streams, cutthroat tended to be found in the upper tributary reaches, whereas rainbow tended to be found in the lower reaches of the mainstem (Hartman and Gill 1968). Henderson et al. (2000) showed similar within-tributary broad-scale habitat partitioning between Yellowstone cutthroat trout and introduced rainbow trout in Idaho. Finally, Hagen and Taylor (2001) described differences in spawning habitat utilization between Dolly Varden and bull trout that were broadly sympatric within a single tributary stream. Bull trout tend to spawn in lower mainstem reaches, whereas Dolly Varden, which mature at a much smaller size than bull trout, concentrated their spawning in upper and much smaller tributary channels with groundwater seepage. These local habitat preferences should promote a mosaic pattern of hybridization within drainages, that is, hybridization will be most likely only where the distinct parental habitats are in close proximity. For example, Redenbach and Taylor (2002) found that most 0+ (i.e., those younger than 1 year of age) Dolly Varden × bull trout hybrids were found where tributary seepage channels (Dolly Varden spawning habitats) were close to mainstem habitats (areas utilized by spawning bull trout).

Notwithstanding the constraints imposed by aquatic habitats to the structure of hybrid zones in salmonid fishes, mosaic hybrid zones could also be structured by processes inherent to tension zones (e.g., high intrinsic mortality of hybrids; see, e.g., Lu and Bernatchez 1998). In general, however, given the evidence for relatively high survival and fertility of hybrids in the most common natural hybrid zones (i.e., those within genera), and the unique aspects of aquatic habitats it seems unlikely that salmonid hybrid zones would fall into the category of classic tension zones. Rather, intrinsic viability and fertility of many natural salmonid hybrids, and the tight linkage between salmonids and the distribution and nature of aquatic habitats, argue that environment-dependent processes generate salmonid hybrid zones that are largely mosaic-like in structure, at least at intermediate (watershed-level) geographic scales. In addition, hybrid zones within particular pairs of species may have different structures (i.e., hybrid swarms vs. relatively low numbers of hybrids relative to parentals) in different localities even within single streams (e.g., Campton and Utter 1985). This variability in structure of hybrid zones argues that local environmental conditions, either through their effects on assortative mating or selection against hybrids, are important in structuring salmonid hybrid zones.

Some recent molecular insights into salmonid hybrid zones suggest that the evolutionary novelty model may be applicable in some instances. Bernatchez et al. (1995), Glémet et al. (1998), and Wilson and Bernatchez (1998) have reported the complete introgression of mtDNA from Arctic charr into populations of brook charr and lake trout, respectively. In both instances, the authors argued that this introgression may have been promoted by positive natural selection for Arctic charr mtDNA. The mechanism of selection could have been the greater physiological performance of brook charr having mtDNA from Arctic charr, which has a more northerly latitudinal distribution in North America and may have given hybrid charr a selective advantage during cooler periglacial times (Glémet et al. 1998). In another charr hybrid system, Dolly Varden and bull trout, Redenbach and Taylor (2002) argued that asymmetrical introgression of bull trout mtDNA into Dolly Varden was a result of the mating system, and not natural selection. In sympatry, Dolly Varden are much smaller at maturity than bull trout and the Dolly Varden males probably act as "sneak" spawners on larger bull trout pairs (see also Wirtz 1999). These same authors, however, reported introgression of type 2 growth hormone from bull trout into Dolly Varden that was significantly (three times) greater than introgression from Dolly Varden to bull trout (Redenbach and Taylor 2003). The biased introgression of growth hormone was in marked contrast to introgression of three other nuclear loci that were roughly equal between species (i.e., bidirectional) and among themselves (Figure 8.5). Such biased introgression of a biparentally inherited locus could be a result of positive selection for bull trout growth hormone allele in Dolly Varden. Although the specific mechanism of selection is unknown, it is possible that Dolly Varden (the smaller of the two species in sympatry) obtain some advantage in terms of growth, metabolism, or even aggression (Devlin et al. 1994) from introgression of growth hormone from bull trout. Verspoor and Hammar (1991) also reported several possible cases of positive selection for allozyme alleles in brown trout and Arctic charr that resulted from introgression from Atlantic salmon and brook charr, respectively;

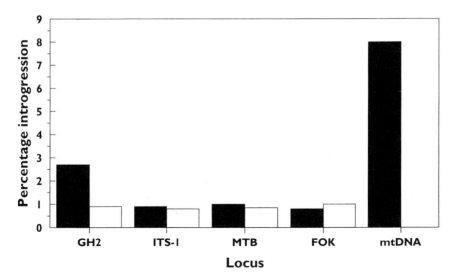

Figure 8.5. Introgression of species-specific alleles in Dolly Varden (*Salvelinus malma*) and bull trout (*S. confluentus*) at four nuclear loci and one mitochondrial locus. Percent introgression refers to the number of alien alleles present divided by the total number of alleles (i.e., $2n$ for nuclear, n for mitochondrial). Black bars represent percentage introgression of bull trout alleles into Dolly Varden, white bars represent percentage introgression of Dolly Varden alleles into bull trout.

foreign alleles that may be maintained in the recipient species by thermal selection.

6. Hybridization and Evolution: the Window Is Still Open

6.1. Hybrid Zone Structure and Speciation

Barton and Hewitt (1985) characterized hybrid zones as "natural laboratories," while Harrison (1990) described them as "windows" through which to study the process of speciation. In essence, the observation of natural hybridization signals incomplete reproductive isolation and, depending on the level of hybridization, different stages in speciation (Jiggins and Mallet 2000). The production of hybrids, therefore, provides the opportunity to examine their evolutionary fate. If hybrids are completely viable through maturity, are fertile, and exhibit equal or greater reproductive success to parental genotypes, hybrid swarms will likely result and few insights into speciation will be obtained. If the performance of hybrids, however, is inferior to that of parentals such that the genetic integrity of the latter is maintained in the face of gene flow, two key inferences are possible concerning the ecology and genetics of speciation. First, the ecological and genetic processes that constrain gene flow between hybridizing species may signal the same processes that initiate speciation. Second, if selection against hybrids results in enhanced assortative mating and pre-mating isolation, then this would constitute evidence for reinforcing selection being an agent of speciation (Butlin 1989). Below, I outline some ways in which more nascent fields of speciation research may profit from examination of salmonid hybrid zones.

Recently, Harrison and Bogdanowicz (1997) classified hybrid zones into three basic kinds based on the frequency distribution of genotypes. Hybrid zones consisting largely of a hybrid swarm of F_1, F_n, and early generation back-crosses are "unimodal," those consisting largely of genotypes resembling parental types are "bimodal," while those consisting of an intermediate distribution of genotypes are known as "flat" hybrid zones. Jiggins and Mallet (2000) suggested that bimodal hybrid zones were characterized by strong pre-zygotic barriers to hybridization, particularly in terms of assortative mating or fertilization. Unimodal hybrid zones, by contrast, are typified by little assortative mating (Jiggins and Mallet 2000). Salmonid hybrid zones that have been extensively studied support these generalizations. For instance, Redenbach and Taylor (2003) reported a strongly bimodal hybrid zone between Dolly Varden and bull trout in a northcentral BC watershed (Figure 8.6a) using variation across four nuclear markers to describe the frequency distribution of genotypes. In this system, the two species differ strongly in size at maturity and size-dependent spawning habitat use (Hagen and Taylor 2001) and likely mate choice as well (e.g., Maekawa et al. 1994). First-generation hybrids made up only 3–5% of all genotypes present in the watershed, with parentals and advanced generation backcrosses or hybrids predominating (see also Bartley et al. 1990). By contrast, Campton and Utter (1985) reported a nearly unimodal distribution of genotypes in a hybrid swarm between rainbow trout and coastal cutthroat trout from a creek in western Washington (Figure 8.6b). These same authors, however, also reported other sites from within the same stream that were either characterized by genotypes of one species or the other, or both with little evidence of hybridization, indicating the importance of scale of sampling when characterizing hybrid zones. Nevertheless, hybrid zones within western trout are often unimodal hybrid swarms (Forbes and Allendorf 1991; Allendorf et al. 2001) as are many hybrid zones reported between species of *Coregonus* in Scandinavian lakes (e.g., Vuorinen 1988). In these later cases, the breakdown of pre-mating isolation may be an important cause of introgression and may be a result of greater similarity in body size between species (e.g., Rubidge et al. 2001), environmental changes (Seehausen et al. 1997), or recent contact in environments with limited opportunities for spawning habitat segregation (Vuorinen 1988). The implied importance of pre-mating isolation in structuring hybrid zones in salmonids is consistent with the general lack of evidence of major endogenous genetic incompatibilities between species forming natural hybrids and with the general proposition that pre-mating isolation evolves first and/or more easily than endogenous post-mating isolation (i.e., hybrid inviability or sterility, Coyne and Orr 1997).

A major uncertainty in the process of speciation concerns the relative importance of deterministic factors, such as sexual or natural selection, or random

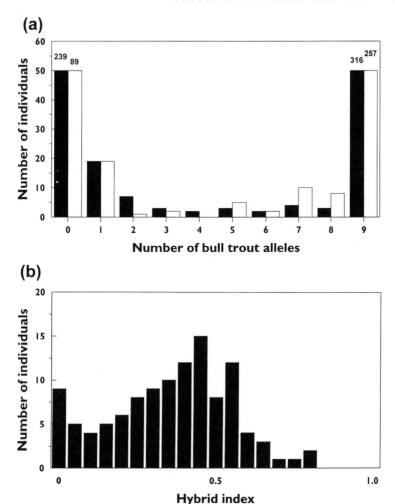

Figure 8.6. Genetic structure in two salmonid hybrid zones. (a) Bimodal hybrid zone between Dolly Varden (*Salvelinus malma*) and bull trout (*S. confluentus*) in Kemess Creek (black bars) and from five other tributaries of the Thutade Lake watershed (white bars), British Columbia. Shown is the frequency distribution of individual juvenile charr ($N =$ 990 fish) with different numbers of bull trout alleles ranging from 0 (equivalent to homozygous for Dolly Varden alleles at all for nuclear loci and exhibiting Dolly Varden mtDNA) to 9 (equivalent to homozygous for bull trout alleles and exhibiting bull trout mtDNA). Owing to small numbers of hybrid individuals, the heights of bars at 0 and 9 have been set to a maximum of 50 and actual numbers of individuals fixed for Dolly Varden or bull trout, respectively, alleles/haplotype are given above each bar. (b) Unimodal hybrid zone between coastal cutthroat trout (*Oncorhynchus clarki clarki*) and rainbow trout (*O. mykiss*) in a Washington State stream. Genotypes are designated by a hybrid score (see Campton and Utter 1985) that represents the multilocus probability that an individual fish arose from random mating within each of the two parental taxa. A score of 0 represents a high probability of cutthroat trout and 1 represents a high probability of rainbow trout. Redrawn from Campton (1987).

factors, such as genomic reorganization via random allopatric divergence or founder events, in driving the evolution of reproductive isolation (Howard and Berlocher 1998). Hybrid zones can provide insights into this question under the assumption that factors influencing the fitness and frequency of genotypes in species with incomplete reproductive isolation are also important in initiating speciation (Barton and Hewitt 1985; Jiggins and Mallet 2000; Barton 2001). This is an area where detailed study of salmonid hybrid zones can make an important contribution. In the best-studied hybrid zones in charr and western trout, there is little evidence that endogenous selection structures these hybrid zones. Rather, and as argued above, ecological factors appear to be important in constraining gene flow between species, either in terms of reproductive (e.g., Hagen and Taylor 2001; Redenbach and Taylor 2003) or migratory ecology (Campton and Utter 1985; Young et al. 2001). Further, studies of "species pairs" (*sensu* Taylor 1999) of sympatric salmonids, such as "dwarf" and "normal" lake whitefish (Lu and Bernatchez 1999; Lu et al. 2001), suggest that variation in opportunities for trophic differentiation within single lakes may explain differing levels of introgression between species (see also Bernatchez 2003—*this volume*). This hypothesis suggests that differences in the trophic selective environments of ancestral populations may have initiated divergence between species that have come into contact secondarily. The multifarious aspects of natural selection that act against hybrids between sympatric anadromous and non-anadromous sockeye salmon, and maintain genetic isolation between them in the absence of endogenous selection (reviewed by Taylor 1999), are also consistent with the importance of exogenous selection in initiating speciation in salmonids. The importance of exogenous selection in salmonid divergence is also suggested by experimental work on conspecific populations. An excellent example involves the experiment by Bams (1976) who demonstrated reduced homing ability of hybrids between two populations of pink salmon. The key inference from this study was that local adaptation to specific streams was important to the evolution and maintenance of spatial reproductive isolation between conspecific populations.

Taken together, studies of hybrids in salmonids across different scales of genetic divergence (from taxonomically distinct species to "species pairs" to conspecific populations) offer rich opportunities to study the importance and kinds of exogenous (and endogenous) selection in initiating divergence and speciation. The observation that many of these sympatric populations, hybrid zones, and genetically divergent populations have formed post-glacially (i.e., over the last 15,000 years) argues that such selection may operate rapidly (see also Hendry et al. 2000b; Hendry 2001; Kinnison and Hendry 2003—*this volume*).

Another issue relevant to speciation is the role of reinforcement, that is, when natural selection acts directly on mate choice to reduce the production of unfit hybrids (e.g., Noor 1995; Liou and Price 1994). Liou and Price (1994) and Jiggins and Mallet (2000) presented arguments that reinforcement should be most likely when there is a high degree of assortative mating that has already evolved, that is, in bimodal hybrid zones. Further, mosaic hybrid zones that seem to be common in salmonids should also promote reinforcement owing to a high initial level of assortative mating via differences in habitat use (Jiggins and Mallet 2000). Consequently, a comparative analysis of the degree of reproductive isolation between sympatric and allopatric populations of naturally hybridizing salmonids would appear to provide excellent models to test for reproductive character displacement predicted under reinforcement. A possible example of such a process may be found in the Dolly Varden–bull trout system. The two species show greater differences in size at maturity when sympatric than when allopatric; sympatric Dolly Varden are much smaller than bull trout (e.g., Hagen and Taylor 2001), but they are similar in size at maturity when allopatric (Baxter 1997; Underwood et al. 1996; McPhail and Baxter 1996; E. Taylor unpublished). Selection against hybrids in sympatry may drive the evolution of different sizes at maturity, with correlated effects on size-dependent mate choice and habitat utilization, increasing reproductive isolation in sympatry. A complication for this hypothesis, and for the more general idea that reinforcement may be more likely in bimodal hybrid zones, however, is the "sneaking" mating behavior of small-sized male salmonids. As discussed above, sneaking of smaller male Dolly Varden on larger bull trout mating pairs may promote hybridization and compromise selection for size-based assortative mating.

6.2. Hybrid Speciation

Hybrid speciation encompasses the idea that hybridization events may produce recombinant lineages that are more fit than the parental genotypes, either in parental or novel environments, and become the basis for new evolutionary lineages. This idea is broadly accepted by botanists and many plant species may have arisen by hybridization (e.g., Arnold 1997; Rieseberg 1998). In addition, genome duplications stemming from hybridization and subsequent divergent resolution of duplicate genes is thought to have been important in speciation within the Catostomidae (North American suckers) (Ferris et al. 1979). Furthermore, there is good evidence that the cyprinid fish *Gila seminuda* has arisen by homoploid (non-polyploid) speciation stemming from hybridization between *G. robusta* and *G. elegans* (DeMarais et al. 1992).

Perhaps the best hypothetical example of hybrid speciation in salmonids is the work of Svärdson (1979, 1998) who summarized an extensive dataset describing hybridization and introgression in Scandinavian *Coregonus*. Kosswig (1963) and Svärdson (1998) suggested that secondary contact between two to five formally allopatric species had resulted in hybridization and introgression with the production of novel recombinant genotypes that may have fostered adaptation to distinct environments resulting in up to five sympatric species (Svärdson 1998). Given the importance of trophic characters in resource partitioning in whitefishes and its possible role in reproductive isolation (e.g., Bernatchez 2003—*this volume*), as well as the possible novel trophic morphologies that may result from hybridization (McElroy and Kornfield 1993), hybrid speciation in these fish remains an interesting possibility. It will require rather careful phylogenetic analyses (e.g., DeMarais et al. 1992) to test this idea rigorously. Similarly, Mina (1992) suggested that a stable hybrid lineage of lenok

(*Brachymystax*) had been established in eastern Russia via hybridization between formerly allopatric "blunt-nosed" and "sharp-nosed" forms of *B. lenok*. Another interesting case of hybridization that may have promoted diversification and speciation in salmonids are charrs of the Pacific basin, in particular, charrs found in waters of the Russian far east. These fish consist of a complex of forms that at one time had been described as consisting of up to 13 distinct species and a large number of undescribed sympatric forms (Behnke 1972, 1980; Savvaitova 1980). Reassessment of the systematic relationships of these *Salvelinus* suggests that these forms are part of a relatively recent (10,000–20,000 year) divergence of the *S. alpinus–S. malma* (Arctic charr and Dolly Varden) complex resulting from vicariant events surrounding the high Arctic and Pacific coast of North America (Savvaitova 1980; Brunner et al. 2001). Given evidence both for historical and contemporary hybridization and introgression among Pacific basin charr (McPhail 1961; Baxter et al. 1997; Redenbach and Taylor 2003), some of the extant diversity in the genus, including sympatric forms, may stem from evolutionary novelties originating from hybridization (Behnke 1980).

6.3. Hybrid Zones and Molecular Aspects of Speciation

Finally, both natural and experimental hybridization and introgression can be exploited to resolve the genetic bases of ecological divergence and speciation. There are three basic approaches that can be adopted to address this fundamental unknown in evolutionary biology. First, high-resolution genetic markers such as tandem-repeat DNA loci that are so useful for individual parentage determinations (e.g., Bentzen et al. 2001; Taggart et al. 2001) could be used to study the fitness of genotypes in nature. The documentation of genotype-based fitness across multiple generations within a population consisting of parental and hybrid matings and their degree of association with habitat offer the possibility of directly assessing competing hybrid zone models (e.g., tension vs. mosaic models, endogenous vs. exogenous selection against hybrids; Rieseberg 1998). A second approach is to exploit natural variation in extent of hybridization/introgression among localities across a natural zone of contact. Harrison (1990) suggested that genes contributing to reproductive isolation could be identified by their association with linked genetic markers that introgress at lower than expected rates. Genetic mapping of these linked markers would then make it possible to determine the general number and genomic locations of "speciation genes." Rieseberg (1998) reviewed the application of this approach to a natural hybrid zone in *Helianthus* sunflowers. The integration of this approach with surveys in hybrid zones with different levels of hybridization and introgression and the possible environmental correlates of such variation (Lu and Bernatchez 1999; Lu et al. 2001; Redenbach and Taylor 2003) could reveal which ecological factors influence introgression and the genomic regions where such variable introgression is concentrated. Furthermore, a chief advantage of using salmonid hybrid zones for resolving the genetic architecture of divergence and speciation is the rich literature on ecological differentiation between species, including sympatric species pairs (e.g., Taylor 1999), that hybridize in nature. In addition, a comparably rich literature and understanding exists for traits that are of importance in sexual isolation (Craig and Foote 2001; Fleming and Reynolds 2003—*this volume*). This knowledge, coupled with the increasing base of mapped molecular markers (Danzmann and Gharbi 2001), means that it will be possible to map the genomic distribution of quantitative trait loci (body size, growth rate, trophic characters, secondary sexual characters) that contribute to reproductive isolation in natural hybrid zones. Although work on plant species (e.g., reviewed by Rieseberg 1998) has exploited this approach the best to date, recent work on whitefish *Coregonus* species pairs exemplifies the steps and possibilities applied to salmonids. Rogers et al. (2001) described the production of a preliminary linkage map in experimentally produced hybrids between "dwarf" and "normal" populations of lake whitefish. Segregation across 998 amplified fragment length polymorphism (AFLP) loci was used to infer the number of linkage groups and genomic locations of genes that may be responsible for ecological and reproductive divergence in sympatric ecotypes of whitefish.

The large number of genome mapping projects in salmonids (Danzmann and Gharbi 2001) means that this approach offers great promise, although a general constraint may be the relatively lengthy generation time of salmonid fishes. Hybrid zones in salmonids, however, may be particularly useful to explore these issues because specific systems are in various stages of evolutionary divergence and it may be possible, therefore, to examine the genetic basis of divergence at different stages in speciation. For example, it would be interesting to know if the same genomic regions are involved in divergence between recently derived sympatric "species pairs" (e.g., dwarf and normal lake whitefish) compared to older divergences. Furthermore, the independent evolution of species pairs that still hybridize in different lakes or streams offers the opportunity to test whether or not similar genomic regions have been involved in replicate divergences and, if so, would support the argument that natural selection is central to evolutionary diversification. Replicate species pairs of dwarf and normal whitefish are discussed by Bernatchez (2003—*this volume*) as an excellent model system to examine this question. Other replicate species pairs, such as anadromous and nonanadromous sockeye salmon (Taylor 1999) offer similar opportunities.

Hybrid zones between more divergent lineages (e.g., Dolly Varden and bull trout; rainbow and cutthroat trout) that have formed independently (i.e., through independent episodes of secondary contact) could also be examined in terms of levels of introgression of loci associated with quantitative ecological or genetic differences between species relative to that for neutral loci. As Rieseberg (1998) detailed, the idea that ecological differences contribute to reproductive isolation predicts that introgression of such traits should be constrained relative to independently segregating neutral traits. Such genetic mapping of traits under conditions of differential introgression within hybrid zones can help identify the traits important to isolation, but the comparative analysis of introgression of such traits across hybrid zones and their correlation with environmental differences could help identify the crucial ecological factors initiating speciation (cf. Lu et al., 2001; Redenbach and Taylor 2003).

A further contribution that salmonid hybrid zones could make to speciation involves their use in identifying specific genes important to reproductive isolation. In particular, comparative analysis of hybridization between pairs of species within genera that show differing levels of post-zygotic isolation may help resolve genetic regions important in, for instance, hybrid sterility. Comparative sequencing projects have, for instance, identified a gene important in hybrid sterility and inviability among species of *Drosophila* (Ting et al. 1998). Although the genomes of *Drosophila* are comparatively simple compared to salmonids (i.e., they have only four chromosomes), in principle, comparative sequence analysis of charr species that produce fertile offspring (e.g., Dolly Varden and bull trout) and those that produce sterile hybrids (e.g., bull trout and brook charr) may resolve the gene (or genes) contributing to sterility.

Finally, molecular genomics offers the opportunity to study the role of variable levels of gene expression in phenotypic divergence and speciation. Molecular genomics involves, in part, the identification and characterization of particular sets of genes that are expressed differentially during particular stages of development (Schulte 2001). By exploiting natural and experimental hybrids in salmonids and developing gene expression technology it will be possible, for instance, to characterize genes that are differentially expressed in different phenotypes (e.g., closely related species), including the resultant phenotypic novelties in hybrids. For instance, genes that contribute to endogenous selection against hybrids may have dysfunctional gene expression during development. Further, the incorporation of measures of differential expression in different phenotypes and the fitness of such phenotypes in different environments could provide insights into the potential adaptive nature of differential gene expression and how this contributes to exogenous selection against hybrids.

7. Conclusions

Given their importance to humans in a variety of ways and their conspicuous presence in natural ecosystems, salmonids have been well studied ecologically and genetically relative to most other fish. Such a scientifically high profile has resulted in numerous reports of natural hybridization between species across the Holarctic range of salmonids, as well as the study of numerous experimental hybrid populations. Nevertheless, salmonid hybrid zones remain relatively poorly characterized in terms of their structure and ecological and genetic factors that influence that structure. In large part, the underexploitation of salmonid hybrid zones results from the difficulty in working in aquatic ecosystems that are often geographically remote. Consequently, most salmonid hybrid zone studies have been opportunistic in nature (e.g., Baxter et al. 1997; Wilson and Hebert 1998). More recent and focused studies of salmonid hybrid zones (e.g., Campton and Utter 1985; Glémet et al. 1998; Lu et al. 2001; Redenbach and Taylor 2002, 2003), however, have helped to better characterize the genetic structure of hybrid zones, which, in general, appear to form mosaics owing to the tight linkage between salmonids and their aquatic ecosystems, and the role of watershed interconnectedness in promoting secondary contact between species (e.g., Redenbach and Taylor 2002). These studies, aided by the development of molecular approaches to hybrid studies, have identified: (1) both historical and contemporary components to hybridization (Bernatchez et al. 1995; Redenbach and Taylor 2002, 2003), (2) the assessment of endogenous (Lu and Bernatchez 1998) and exogenous selection (Lu and Bernatchez 1998; Hagen and Taylor 2001) in structuring hybrid zones, and (3) the possible adaptive consequences of introgression (Glémet et al. 1998; Redenbach and Taylor 2003). These contributions have helped to integrate salmonids into the broader context of hybrid zones and their significance to evolution. Encouragingly, salmonid hybrid zones will probably increase in their profile and utility for understanding the genetic and ecological factors important to reproductive isolation and speciation because (1) they offer a broad range of divergences (involving hybridization between post-glacial species pairs to divergences between lineages that are on the order of millions of years old), (2) the ecological and reproductive factors that differentiate species are well characterized, and (3) there is an increasing supply of genetic markers to probe the genomes for the loci that control these differences.

Acknowledgments I thank the editors, A.P. Hendry and S. Stearns, for the invitation to write this chapter. R.H. Devlin and F.M. Utter provided useful information or discussion for portions of this chapter. The comments on this chapter by L. Bernatchez, A. Hendry, and J. Wenburg are much appreciated. The illustration for the frontispiece of this chapter is based on a photograph by E.R. Keeley. The research of EBT is supported by NSERC (Canada) and the Province of BC.

Salmonid Breeding Systems

Ian A. Fleming

John D. Reynolds

Anadromous male Atlantic salmon with two mature male parr in the nest of a female Atlantic salmon

9

Sexual reproduction often involves substantial amounts of time and resources devoted to competition for mates, courtship, and mate choice. These behaviors may be underpinned by elaborate physical traits, such as acute sensory systems, bright color patterns, gawdy ornaments, and formidable weapons. The evolution of these characteristics has attracted a great deal of attention from behavioral ecologists over the past two decades, with a particular interest in tradeoffs with other aspects of life histories, as well as benefits to either sex. This research has enhanced our understanding of many aspects of the evolution of breeding systems.

In this chapter, we will explore the diversity of sexual selection and breeding systems exhibited by various species and populations of salmonids. The word "diversity" appears often in this book, thanks to the tendency of salmonids to be divided into small, genetically isolated populations, thereby evolving rapidly in response to local selection pressures. As reproductive behavior is intimately linked to life histories, we should not be surprised to find that the diversity of salmonid life histories is matched by a diversity of breeding systems. By "breeding system", we are referring to parental care and sexual selection, including the numbers of mates obtained by both sexes and the manner in which they are obtained through competition for mates and resources, courtship, and mate choice (Emlen and Oring 1977, Reynolds 1996, Reynolds and Gage 2002).

We begin by reviewing the general theory that seeks to explain the evolution of breeding systems, including sex differences in sexual selection and parental care. This includes subsections devoted to specific components of the theory, with predictions for salmonids. We then test these ideas with salmonid fishes, concentrating primarily on species in the genera *Salvelinus*, *Salmo*, and *Oncorhynchus*. Throughout, we emphasize the tremendous diversity in salmonid breeding behavior at scales ranging from differences between individuals within populations to variation among species. Along the way, we hope to accomplish a two-way goal: to see what salmonids can tell us about the theory, and to see what the theory can tell us about salmonids.

1. Theory of Breeding Systems

1.1. Reproductive Investment

Before we begin our review of breeding system theory, it is helpful to consider overall patterns of reproductive investment in relation to other aspects of life history because these set the stage for the breeding system. This is particularly important for salmonid species because there is extreme variation both within and among populations in the total reproductive efforts made by individuals. Optimal allocations toward reproduction at any given time will depend on trade-offs between benefits in terms of numbers of offspring produced, and costs to survival and future breeding attempts (Williams 1966a). If animals are faced with a low probability of being able to breed in the future, they should "put all their eggs into one basket" and breed at a maximum rate (Roff 1992, Stearns

1992). This leads to semelparity, or "big-bang" reproduction. Conversely, if adult survival is high and there are benefits from breeding more than once, this can select for repeated breeding, or iteroparity. Benefits of breeding repeatedly include age-related gains in fecundity and selection for bet-hedging in environments in which juvenile survival is highly variable.

This basic theory of life history tradeoffs leads to a number of predictions about reproductive investment in salmonid species. For example, the percent of energy reserves that fish allocate toward reproduction should be inversely related to the extent of repeat breeding. Of course, there is an issue of cause and effect in such correlations, because high costs of reproduction will compromise survival. We should also expect greater reproductive investment in migratory populations than in landlocked populations, due to costs of migration, which reduce the likelihood of being able to reproduce again. The same reasoning predicts that long and arduous migrations should be correlated with high reproductive investment. For additional discussions at reproductive investment in relation to migration, see Hendry et al. (2003b—*this volume*).

1.2. Differences Between the Sexes

Males and females are defined according to differences in the sizes of their gametes. To understand the evolution of such differences, game theory models have been used, whereby the fitness payoffs to individuals from adopting a particular tactic depend on the tactics adopted by others in the population. Anisogamy evolves when gametes are selected for either large size in response to gains in offspring survival, or toward a small size, due to benefits in fertilization success (Parker et al. 1972). As long as there are large gains available from investment into large gametes, ova will evolve toward the size that maximizes the number of surviving offspring, whereas sperm will evolve toward a size that maximizes their ability to survive prior to fertilization and compete with other sperm (Parker 1984). Thus, sexual selection, that is, selection for traits that enhance mating success, is implicated right at the start of the evolution of two sexes.

The stage is thus set for sexual selection to have profound impacts on animals throughout their lives. Picture millions of organisms swimming a marathon in order to reproduce, many dying before they reach their destination, and the survivors facing fierce competition from rivals when they get there. This is what male salmon do, and it is what their sperm do. Thus, when these differences in gametic allocation between the sexes are scaled up to the individual, we have the general (and greatly oversimplified) paradigm that female breeding traits tend to be shaped principally by natural selection for offspring production and survival, and those of males by sexual selection for access to mates. The stereotypical male competes more for mates than does the stereotypical female, which is more selective of its mates (Andersson 1994). Of course this is a simplistic caricature of the differences between the sexes, and below we go into more detail.

1.2.1. Parental Care

As in the case of anisogamy and sexual selection, the evolution of parental care involves tradeoffs within individuals and conflicts between individuals. Again, game theory models have provided the appropriate framework for understanding the conflicts that occur between males and females over the amounts of care to be provided to the young (Székely et al. 1996). The key benefit of care is enhanced survival of the young, and the key costs are reductions in growth, survival, and mating success of the parents. We therefore expect parental care by either or both sexes to evolve when the offspring benefit greatly from protection or provisioning, especially if the costs to future reproduction are low. Parental care exists in 21% of families of bony fishes, with female-only care being the rarest form (7% of families with parental care; Gross and Sargent 1985). The rarity of uniparental male care in the animal kingdom is attributed to strong tradeoffs for males in their ability to obtain multiple mates while caring for young (reviewed by Gross and Sargent 1985; Clutton-Brock 1991; Reynolds et al. 2002).

The mating costs of care provide a key link between the evolution of parental care and sexual selection, thereby cementing the two main components of breeding systems. But can the sexual selection tradeoff hypothesis explain the absence of male care in salmonids? Not necessarily, because many other species of teleost fishes show male care (Blumer 1982; Gross and Sargent 1985; Reynolds et al. 2002). For example, exclusive care by males occurs in stickle-backs (*Gasterosteus* spp.), damselfishes, many gobies, and freshwater sunfish species. Yet males are still the most competitive sex. This is probably because in these species males can care for more than one clutch of eggs at a time, due to the lower demands of time and energy from eggs and larvae in fish than in species such as birds. Parental care in many fish species therefore presents males with less of a tradeoff against multiple mating than in most other taxa. Furthermore, in many fishes the costs of care may be offset or even overthrown by *benefits* for sexual selection, as females in a number of species have been shown to prefer to spawn with males that already have eggs in their nests (reviewed by Jamieson 1995, Reynolds and Jones 1999). The lack of male care in salmonids, and presence of female care in many species, therefore demands an explanation, which we will try to provide in Section 2.2.1.

1.3. Operational Sex Ratio

The interplay between parental care and sexual selection in males and females can be scaled up to the level of the population by considering the operational sex ratio (OSR; Emlen and Oring 1977). This was originally defined as the ratio of sexually active females to males, although later authors have sometimes reversed the ratio, or defined it as available males versus the sum of available males plus females (Kvarnemo and Ahnesjö 1996). Whichever formulation is used, if a population has an equal adult sex ratio, the OSR will be determined by differences between the sexes in their "time out" while providing care and performing ancillary activities such as making nests and recovering between periods of reproductive activity (Clutton-Brock and Parker 1992). In salmonids, as with nearly all animal species, the OSR during the breeding season is usually male-biased with males having very little time out between periods of reproductive activity, compared with females, which are limited by egg production, nest production, and guarding of the embryos in some species. Benefits to males (especially high-quality individuals) from access to multiple mates reinforces selection for intense male–male competition, and provides greater scope for female choice among males.

Figure 9.1 illustrates relationships between parental inputs by each sex, the impacts of these on potential rates of reproduction, and hence the OSR and sexual selection. The reproductive efforts by each sex are divided between parental and gametic input, competing for mates and mate choice. Stereotypical males provide little care per offspring compared with the inputs from females. This is due to lower confidence of paternity (hence lower benefits of care) and higher benefits for the subset of males that are successful in mating competition (reviewed by Kokko and Jennions 2003). This biases the OSR toward excess males. Females, conversely, have less to gain from compet-

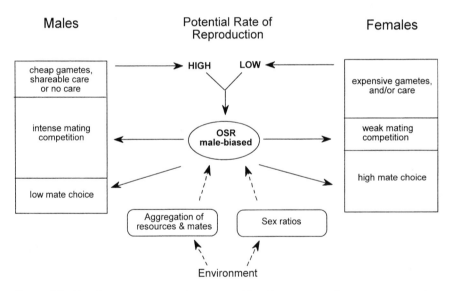

Figure 9.1. Breeding systems are characterized by the amount of time or energy that males and females devote to offspring, through gametic production and parental care, as well as the two forms of sexual selection: mating competition and mate choice (from Reynolds and Gage 2002). These allocations are shown in three boxes for each sex, indicating that in most species, males devote less to care and gametic production per brood than do females, and are more competitive for mates. These processes reinforce one another, with higher potential rates of reproduction by males causing a male-biased operational sex ratio (OSR : numbers of available females : males). This allows females to be more selective of their mates. The extent of aggregation of resources and mates in space and time will also affect the OSR, as will the adult sex ratio.

ing for males, as these are not in short supply, and they can afford to be more selective of their mates.

At any given time and place, we know that adult sex ratios are often unequal. For example, sex differences in timing of arrival at the breeding grounds can have strong impacts on the sex ratio, and hence OSR, in many species. Thus, in many salmonid populations, males typically arrive on the spawning grounds a few days earlier than most females (Morbey 2000). There may also be differential mortality of the sexes. These differences in adult sex ratio may amplify or reduce differences in the operational sex ratio (Figure 9.1). However, it is probably generally the case that in salmonids the disparity between males and females in potential rates of reproduction far outweighs disparities in adult sex ratios. This leads to the prediction that OSRs should be strongly male biased (Section 2.3).

1.4. Relationship between Sexual Selection and Ecology and Behavior

The spatial distribution of resources and the temporal distribution of mates determine the "environmental potential for polygamy," that is, the potential number of mates that animals can obtain (Emlen and Oring 1977). For example, when territories and other resources needed for reproduction are economically defendable, then variation in the quality of these resources will determine the benefits to males of competing for them. Similarly, if females arrive asynchronously at breeding sites, a single male may be able to mate with each one as it arrives, thereby monopolizing all of the matings. Conversely, high female synchrony should reduce the ability of individual males to dominate their rivals (Emlen and Oring 1977). Such aspects of the environment and ecology therefore affect the OSR, and hence the extent of sexual selection (Figure 9.1).

1.4.1. Breeding Time and Spatial Distribution

The timing of male arrival at the breeding grounds relative to that of females is an important component of a breeding system, as it affects the operational sex ratio. Earlier arrival of males is known as "protandry," not to be confused with another use of this term referring to patterns of sex change in which individuals begin life as males (Forsgren et al. 2002). Protandry is a common pattern of sex-biased timing in many animal taxa (reviewed by Morbey and Ydenberg 2001). There are two classes of hypotheses for protandry. The first involves hypotheses where *relative* arrival time itself has no direct fitness consequences for males or females, but rather the sex bias is an indirect result of selection on a correlated character, such as intrasexual competition for nesting territories that favors early arrival (e.g., Ketterson and Nolan 1976). The second class involves hypotheses where selection acts directly on *relative* arrival time because of the fitness consequences for males and females, such as mating opportunities (e.g., Wiklund and Fagerström 1977; Reynolds et al. 1986).

1.4.2. Secondary Sexual Characters

We expect greater intensity of sexual selection to lead to the evolution of more pronounced secondary sexual characters. In the case of salmonids, these include adult color patterns, hooked jaws (kypes), and humped backs which may serve as weapons or shields (see quote from Charles Darwin in the Preface). This can be tested with comparisons among populations, though one must also consider whether the costs of such characters also vary among populations. If secondary sexual traits make individuals susceptible to predation, we can predict a negative relationship among populations between elaboration of secondary sexual characters and predation risk.

1.4.3. Alternative Reproductive Behaviors

Intense breeding competition often leads to the evolution of alternative reproductive behaviors, whereby small males may forgo fights with larger, more aggressive males, and sneak into spawnings (Gross 1996; Taborsky 1998). Sneakers may be less successful per mating, but they may also have more mating attempts, if they reach maturity at an earlier age. Game theory models have been used to understand the evolution of such alternative behaviors (e.g., Gross 1984; Hazel et al. 1990; Hutchings and Myers 1994; Gross and Repka 1998). As salmon species have figured prominently in the development of such models, we will save a detailed commentary on the theory until the section on empirical tests (see 2.4.3). Briefly, alternative behaviors may be either genetically based, or they may depend on an individual's status and condition (Gross 1996). The maintenance of two alternative behaviors in the population is generally ascribed to negative frequency-dependent selection, whereby each behavior is most successful when it is rare. This should lead to evolutionary stability.

1.4.4. Sperm Competition

Sperm competition occurs whenever the ejaculates of two or more males compete for the ability to fertilize a female's ova (reviewed in Birkhead and Møller 1998). The intensity of sperm competition is known to influence gonadal investment in a wide variety of taxa. For example, Stockley et al. (1997) found that species of fish with high risk of sperm competition have relatively larger testes and greater sperm output than those with low risk. This effect also extends to comparisons within species, where phenotypes with a higher incidence of sperm competition (e.g., sneakers) have relatively larger testes and milt volume for their size than phenotypes with a lower incidence (e.g., fighters) (Gage et al. 1995; Taborsky 1998; Vladić and Järvi 2001). Thus, variation in intensity of sperm competition makes salmon species excellent test cases for such patterns.

1.4.5. Female Competitive Tactics

A shortcoming of the early development of breeding system theory was the strong emphasis on males, which ignored the fact that females often should (and do) display an equally rich array of morphological and behavioral adaptations for winning the breeding games that animals play (Ahnesjö et al. 1993; Reynolds 1996; Amundsen 2000). One can expect females to compete for high-quality males if such males are in short supply, for example because of long "time outs" while breeding (Section 1.3). More often, natural selection can lead to female–female competition, for example due to shortages of suitable nest sites. In the case of salmonid species, we can therefore expect to see more fighting and greater development of aggressive traits such as kypes in females from populations where suitable nesting sites are in short supply.

1.4.6. Mate Choice

Females are usually more selective of their mates than are males, as predicted from the disparity between the sexes in reproductive inputs (Andersson 1994; Reynolds 1996). This does not mean that males are totally unselective of their mates. Indeed, there may be some choosiness exhibited by both sexes in most species, depending on their differences in potential rates of reproduction and variation in quality as partners (Johnstone et al. 1996). Female choice may be based on direct benefits afforded by the male, such as territory quality and parental care (reviewed by Reynolds and Gross 1990; Andersson 1994). Alternatively, males may be chosen for the genetic (indirect) benefits they confer to the female's offspring (e.g., "good genes," "runaway" coevolution, or sensory bias). For example, female guppies from a population in Trinidad prefer larger-bodied and longer-tailed males, and these males sire faster-growing daughters which are more fecund (Reynolds and Gross 1992). One cost of mate choice is the loss of willing partners. This is usually less of a problem for females than for males, due to male-biased operational sex ratios (Section 1.3). Other costs include search time and energy and predation risk. In some taxa female choice is prevented by male–male aggression. Thus, the general prediction from breeding system theory when applied to salmonid species is that females should be choosier than males, unless males prevent them physically from being so.

2. Empirical Tests of the Theory

2.1. Reproductive Investment

Several members of the genus *Oncorhynchus* (Pacific salmon) are semelparous or "big-bang" breeders, investing all their resources in a single breeding episode before dying. Other salmonids are iteroparous, conserving resources and breeding multiple times. Individuals within some of these species may breed ten or more times over their lifetime (reviewed in Hutchings and Morris 1985). Several species also undertake oceanic feeding migrations (in some cases over thousands of kilometers) where they rapidly accumulate resources that will be used for breeding (e.g., increased fecundity, body size, and energetic resources for competition) when they return to their natal freshwater environments (anadromy). Others remain resident within fresh water for their entire lives. Moreover, populations within species and even individuals within populations may show differing propensities to become anadromous or remain resident (reviewed in Jonsson and Jonsson 1993; Hendry et al. 2003b—*this volume*).

Reproductive investments by salmonids involve far more than gamete production, and include time and energetic expenditures on breeding migrations, elaborate secondary sexual traits, competition for nesting resources and mates, courtship, and mate choice. Energetic expenditures during a single reproductive bout range from as little as 13% of total body energy for resident male brown trout to nearly 80% for some populations of semelparous Pacific salmon (*Oncorhynchus*) (Figure 9.2; Hendry and Berg 1999). The high reproductive investment that epitomizes semelparous salmon reflects their inability to breed repeatedly (for rare exceptions, see Tsiger et al. 1994; Unwin et al. 1999). By contrast, iteroparous salmonids generally show lower levels of energy depletion during reproduction (Figure 9.2). These findings match the general theory of life history tradeoffs (Section 1.1). Atlantic salmon, however, appear to be an exception to this, with males and females in some populations expending as much as 60–70% of their energy during migration and reproduction (N. Jonsson et al. 1997). Not unexpectedly, however, Atlantic salmon generally show low levels of iteroparity, with less than 10% of spawners typically returning to breed again (Figure 9.3; but see Ducharme 1969). Moreover, N. Jonsson et al. (1997) found post-spawning survival rates among populations to decrease with increasing energy expenditures during spawning (estimated). Of the iteroparous Salmoninae, only the anadromous steelhead trout, the nearest ecological parallel to the Atlantic salmon in the Pacific, show similarly low levels of iteroparity (Figure 9.3).

When viewed across iteroparous species, the level of repeat breeding is correlated negatively with female gonadal investment (i.e., gonadosomatic index) (Crespi and Teo 2002). This is consistent with predictions that lower survivorship is associated with higher reproductive effort (Section 1.1). Despite this, semelparous females do not invest more heavily into gonads than iteroparous females (Figure 9.4; Crespi and Teo 2002). This suggests that semelparous females are investing relatively more energy into other activities, such as migration, competition for nest sites, and defense of eggs, than are iteroparous females (reviewed in Fleming 1998). As a result, gonadal investment is likely to underestimate reproductive effort more in semelparous than iteroparous salmonids (Vøllestad and L'Abée-Lund 1994; Crespi and Teo 2002).

There is an apparent increase in reproductive investment/cost associated with anadromy. Paired contrasts of anadromous and resident forms of the same species indicate a significantly ($P < 0.01$) reduced probability of repeat breeding for anadromous forms (Fleming 1998; Figure 9.3). This probably reflects the additional costs of long migrations, which are traded off against

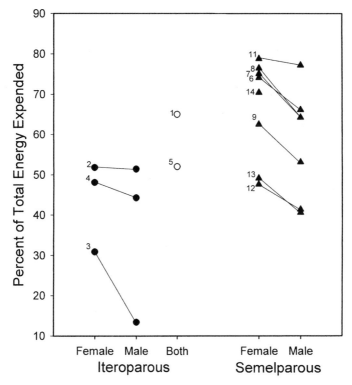

Figure 9.2. Estimated female and male energy costs of reproduction for different populations of iteroparous and semelparous salmonids. 1–2 Atlantic salmon (*Salmo salar*), 3–4 brown trout (*S. trutta*); 5 Arctic charr (*Salvelinus alpinus*); 6–9 sockeye salmon (*Oncorhynchus nerka*); 11–12 chum salmon (*O. keta*); 13 pink salmon (*O. gorbuscha*); and 14 chinook salmon (*O. tshawytscha*). In two cases, energy costs were not separated by sex. Data are derived from Table A1 of Hendry and Berg (1999), with the exception of those for chinook salmon, and the numbers in the figure correspond to those in Table A1 of Hendry and Berg (1999). Chinook salmon data are derived from Greene (1926; Tables 4 and 6), incorporating mass-specific energy loss (methods of Hendry and Berg 1999) and energy loss through gonads.

the benefits of rapid growth and increased body size afforded by feeding in nutrient-rich temperate seas (Northcote 1978; Gross et al. 1988; Hendry et al. 2003b—*this volume*). The exception to this is the resident form of sockeye salmon, "kokanee," which appear phylogenetically constrained to being semelparous like their anadromous conspecifics. Interestingly, the degree of iteroparity of anadromous and resident forms is correlated, suggesting species-specific effects of breeding systems (Fleming 1998). Anadromy also affects the pattern of energy use. For example, anadromous brown trout store a greater proportion of their energy as fat than do resident trout (Jonsson and Jonsson 1997). Similarly, the proportion of energy stored as fat increases with the distance of the freshwater spawning migration among sockeye salmon populations (Gilhousen 1980; Hendry and Berg 1999). Fat provides a more concentrated

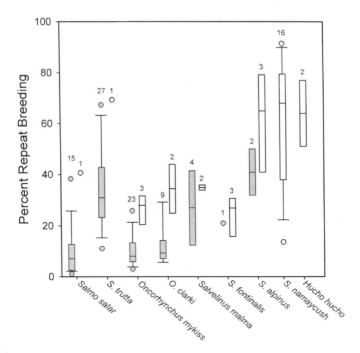

Figure 9.3. Percent of spawners that breed more than once (repeat breed) in anadromous (gray) and resident (open) populations of iteroparous salmonids. Median, sample size (number of populations), and 5th, 25th, 75th, and 95th percentiles as vertical boxes with error bars are shown. Circles are individual data points, and for species having sample sizes greater than nine, they represent data points lying outside the 5th or 95th percentiles. Data are derived from Fleming (1998) and literature sources contained therein.

and readily mobilized energy source for migration than does protein (Brett 1995). Furthermore, females in anadromous populations tend to invest more in gonads, having greater gonadosomatic indices (gonadal wet weight/total body weight × 100) than those in resident populations, a pattern that is found both among and within species (Fleming 1998; Figure 9.4). The reduced probability of repeat breeding of anadromous forms should favor increased present versus future reproduction (Section 1.1; Williams 1966a; Stearns 1976; Roff 1992). The two forces, however, are linked causally because increased present reproductive investment will simultaneously decrease the likelihood of repeat breeding.

The large body size afforded by an anadromous life history provides larger total energy reserves for reproduction than those available to smaller salmon (Hendry et al. 2003b—*this volume*). Large size, however, is also associated with a decrease in the likelihood of breeding more than once (Dutil 1986; N. Jonsson et al. 1991a; Jonsson and L'Abée-Lund 1993). This reflects the general increase in the proportion of body energy expended during reproduction by larger fish (N. Jonsson et al. 1997; Hendry and Berg 1999; Jonsson and

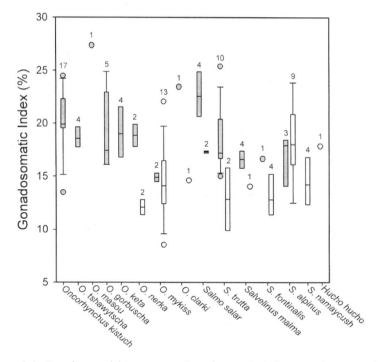

Figure 9.4. Female gonadal investment (gonadosomatic index: gonad wet weight/total body weight × 100) in anadromous (gray) and resident (open) salmonid populations. Median, sample size (number of populations), and 5th, 25th, 75th, and 95th percentiles as vertical boxes with error bars are shown. Circles are individual data points, and for species having sample sizes greater than nine, they represent data points lying outside the 5th or 95th percentiles. Data are derived from Fleming (1998) and literature sources contained therein.

Jonsson 2003). Moreover, large fish appear to have greater difficulty restoring the energy lost during reproduction (Ware 1978; Dutil 1986). Crespi and Teo (2002) argue that these patterns provide strong links among large size, anadromy, and breeding once, and are consistent with their finding that the origin of semelparity in salmonids involved an increase in body size. It is interesting to note, however, that among extant species, body size and parity do not correlate well, suggesting that subsequent diversification was not constrained by such linkages.

2.2. Differences Between the Sexes

Females, on average, expend more energy in total during reproduction than do males across populations and species (Figure 9.2). Females invest 8–27% of their body weight in gamete production (Figure 9.4) or about five times that by males (range: 1.5–10%; Fleming 1998). When considered in energetic terms, this difference is even more pronounced because the mass-specific energy content of ovaries is 1.4 to 2.3 times that of male testes (e.g., Jonsson and Jonsson 1997; N. Jonsson et al. 1997; Hendry and Berg 1999). For example, in female Atlantic salmon this represents about half of the energy expended during reproduction (N. Jonsson et al. 1991b, 1997). As expected from breeding system theory (Section 1.1), male salmonids appear to invest more heavily in breeding competition, including development of secondary sexual characters, as well as fighting and mate-searching behavior (e.g., N. Jonsson et al. 1991b; Brett 1995; Hendry et al. 2000a). Despite the greater energetic investment by females into gametes during reproduction, males often have a lower likelihood of surviving the breeding season (reviewed in Fleming 1996) and breeding again in another year (e.g., Atlantic salmon, Shearer 1992; brown trout, Evans 1994; rainbow trout, Busby et al. 1996; Dolly Varden, Armstrong and Morrow 1980; brook charr, Hutchings 1993a). This suggests that reproductive costs other than gamete production are involved. Intense male–male competition for mating opportunities may be partly responsible, because males frequently experience higher levels of wounding and predation risk during breeding than do females (e.g., B. Jonsson et al. 1990; Fleming et al. 1997; Quinn et al. 2001a,c). The differences in reproductive allocation between the sexes therefore support the expectation from breeding system theory that female traits are likely to have been shaped principally by natural selection for offspring production and survival, and those of males by sexual selection for access to mates.

2.2.1. Parental Care

Females of the genera *Salvelinus* (excluding *Salvelinus namaycush*), *Salmo*, and *Oncorhynchus* are unusual among the bony fishes (Osteichthyes) in providing uniparental care in the form of egg burial within a nest for protection during incubation (e.g., against predation, gravel shifts caused by freshets and digging by other females, desiccation, and freezing). Females use a series of tail-beating sequences to construct a depression in gravel substrate within which they deposit a portion of their eggs and then cover them as they proceed to dig their next nest (e.g., Jones 1959; Tautz and Groot 1975; Fleming 1996). Males do not participate in parental care, neither helping to construct nor defend nest sites.

Gross and Sargent (1985) proposed that female-only care evolved in salmonids, unlike many other fishes, because the cost for females is small because no potential matings (egg releases) are forfeited and the energy expended (i.e., lost fecundity) is outweighed by the benefits of nest construction (i.e., offspring survival). Moreover, salmonids are capital breeders, relying on energy reserves for reproduction, and thus females need not forage between egg releases within the reproductive season to accumulate energy for further egg production. Costs for males, however, would be large if they remained to help construct and/or defend nest sites rather than seeking additional mating opportunities. While females of all egg-burying salmonids defend their nests from reuse by other females during their nesting period, only the semelparous Pacific salmon provide aggressive defense until death. Nest defense is energetically expensive and risky (e.g., injury and predation) and for iteroparous species, which typically breed at lower densities and have the opportunity to breed again another year, the benefits may not exceed the costs.

2.3. Operational Sex Ratio

We have seen that breeding system theory predicts that investment differences between the sexes will be reflected in the ratio of sexually receptive females to males, the operational sex ratio (OSR) (Section 1.3). This should play a fundamental role in determining the level of competition for mates and hence the direction and strength of sexual selection (Figure 9.1). In salmonids OSR is indeed usually male biased (e.g., Schroder 1982; Fleming and Gross 1994; Hendry et al. 1995; Quinn et al. 1996; Blanchfield and Ridgway 1997), even when the sex ratio of returning adults is female biased (Fleming 1998). This reflects the short time window (a few days) within which females must deposit their eggs once ovulated (because of overripening and infertility, de Gaudemar and Beall 1998), and the asynchronous nature of spawning among females, which may occur over many weeks (reviewed in Groot and Margolis 1991; Fleming 1996). By contrast, males may remain sexually active for considerably longer periods, spawning repeatedly and frequently, and thus bias the OSR. Females also pay a cost of "time out" from reproductive activity (*sensu* Clutton-Brock and Parker 1992) to prepare and guard their nests, which males do not incur. The resultant biased OSR generates intense male–male competition for mating opportunities and female choice among males.

In most salmonids, a male's breeding success is largely determined by his mating opportunities. A secondary effect is through potential male choice among females, because of relationships between female traits and fecundity, egg quality, spawning site quality, and parental care. Intense aggression among males usually leads to local dominance hierarchies in the vicinity of nesting females, with one or more males occupying secondary or "satellite" positions further from the female than the dominant, consort male (e.g., Schroder 1981; Gross 1985; Maekawa and Onozato 1986; Fleming 1996; Blanchfield and Ridgway 1997). At oviposition, the males rush in beside the female and release sperm, attempting to attain priority and/or proximity to the female and thus increase paternity (Schroder 1981; Chebanov et al. 1984; Maekawa and Onozato 1986; Foote et al. 1997; Mjlønerød et al. 1998). As males do not participate in nest construction or guarding, they frequently depart shortly after spawning in search of more mating opportunities (but see Morbey 2002). The intense nature of male–male competition, including physical aggression, sperm competition, and high mortality in several salmonids, has meant that they have been a focus of a number of studies examining the role of sexual selection, particularly intrasexual competition, in breeding system evolution.

2.4. Relationships between Sexual Selection and Ecology and Behavior

Both sexes of salmonids show high variance in reproductive success (Table 9.1; cf. Clutton-Brock 1988a), reflecting intense breeding competition. It is this variance that generates significant opportunity for selection to shape the behavior, morphology, and life history of salmonids. Moreover, the sources of variation in reproductive success differ between the sexes, with that of females deriving mainly from variation in offspring production and survival, and that of males from variation in access to mates. This, coupled with the somewhat greater variance in reproductive success among males than females (Table 9.1), means that selection is likely to target traits with differing intensities in the sexes, giving rise to sexual dimorphism (Section 2.5.3).

2.4.1. Breeding Time and Spatial Distribution

2.4.1.1. Females Among salmonids, females have the greatest influence on embryo and early juvenile survival, as they choose the breeding time and location, construct the nest, deposit the nutrient-rich eggs, and in some species, guard the eggs. They respond to the local environment, which as we have seen, is expected to set the ground rules for the breeding system, through the environmental potential for polygamy (Section 1.4). Differences among populations in breeding time correlate with the water temperature during incubation, which probably ensures appropriate timing of hatching and initial feeding for the offspring (Brannon 1987; Heggberget 1988). However, female options may be constrained by water-flow conditions and river freeze-up, which limit access to the breeding grounds. While these restrictions might be expected to exert stabilizing or directional selection on breeding time, within-population variability can be significant (Groot and Margolis 1991; Fleming 1996). Competition among females for nesting sites (e.g., Schroder 1981; Fleming and Gross 1993; Elliott 1994) and among their emergent offspring for territories (e.g., Chandler and Bjornn 1988; Brännäs 1995; Einum and Fleming 2000b) generates a frequency- and condition-dependent evolutionary game among females. It is frequency-dependent in that success gained by breeding early, for example, is influenced by the breeding time of other members of the population. Early breeding females often have ready access to the highest quality nest sites and their offspring emerge early, establishing territories, feeding and growing before late emergers. These advantages, however, are traded off against susceptibility to nest destruction by the digging activity of later breeding females (e.g., McNeil 1969; van den Berghe and Gross 1989; Kitano 1996; Blanchfield and Ridgway 1997; Taggart et al. 2001; Morbey and Ydenderg 2003) and the probability of unfavorable environmental conditions early in the season. The tradeoff is also condition-dependent in that a female's phenotype (e.g., size, energy stores, physical condition) relative to that of other females in the population almost certainly influences the payoffs for early versus late breeding, and thus the decision as to when to spawn. Larger females dig deeper nests than smaller females,

Table 9.1. Estimates of the variance in breeding success (standard variation: σ^2/mean^2), a measure of the opportunity for selection during breeding, for coho salmon and Atlantic salmon. Data are means, with ranges in parentheses.

Species	Environment	Fish origin	Stage measured	Standardized variation		Reference
				Female	Male	
Coho salmon	Spawning channel	Wild and hatchery	Egg deposition	0.15 (0.08–0.19)[a]	1.30 (0.65–2.11)[a]	Fleming and Gross (1994)
Atlantic salmon	Spawning arena	Wild	Eyed embryo	0.60 (0.30–0.90)	0.86 (0.30–1.35)	Fleming et al. (1997)
Atlantic salmon	Spawning arena	Hatchery	Eyed embryo	0.79 (0.31–1.23)	1.43 (0.14–2.82)	Fleming et al. (1997)
Atlantic salmon	Spawning arena	Combined	Eyed embryo	0.88 (0.51–1.23)	1.6 (0.36–2.82)	Fleming et al. (1997)
Atlantic salmon	River stretch	Transplant	Fry	0.53	0.59	Garant et al. (2001)

[a] Excludes variation in egg survival other than that estimated to result from nest destruction.

reducing their susceptibility to nest destruction (e.g., van den Berghe and Gross 1984; Crisp and Carling 1989; Steen and Quinn 1999). Moreover, female egg size affects offspring size at emergence (Einum et al. 2003—*this volume*) and subsequently influences growth, social status, susceptibility to starvation, and predation (reviewed in Fleming 1996). Thus larger females might be predicted to spawn before smaller females, as evidenced in sockeye salmon (Hendry et al. 1999). While many components of the tradeoff have been identified, the conditional nature of the strategy remains unresolved.

Hendry et al. (1999) proposed an alternative view for differences among females within populations in breeding time. They suggested that heritable differences in maturation schedules might allow adaptation to specific breeding times. Maturation date is indeed heritable in salmonids and closely linked to the onset of breeding (e.g., Siitonen and Gall 1989; Gharrett and Smoker 1993; Quinn et al. 2000). The adaptation-by-time hypothesis maintains that breeding time is a *cause* of variation in life history traits rather than an *effect* of such variation. While condition dependence and adaptation-by-time involve different mechanisms, their effects are not mutually exclusive and both may operate simultaneously within populations. In support of adaptation-by-time, Hendry et al. (1999) found that reproductive lifespan and patterns of energy allocation by sockeye salmon varied directly with breeding time. Early-breeding fish lived longer than late breeding fish because they invested less into gamete production (females) and retained more energy for breeding activity, such as nest defense. While the findings are correlative, they do support the existence of evolved differences in allocation strategies by early and late breeders. The maintenance of such differences depends on temporal variation in selection being sufficiently strong to counteract gene flow among fish breeding at different times. Thus, adaptation-by-time is likely to be most relevant where reproductive lifespans are short and breeding seasons extended, such as breeding systems characterized by semelparity and capital breeding (e.g., Pacific salmon; Hendry et al. 1999). In these situations, the relative importance of adaptation-by-time versus condition dependence is likely to reflect the degree to which adaptive phenotypic plasticity (i.e., the ability of phenotype to respond adaptively to environmental conditions; Hutchings 2003—*this volume*) can act as a mechanism to successfully modulate among alternative phenotypes.

The female's choice of nesting site dictates not only the survival of her embryos during incubation (reviewed in Chapman 1988), but also the environment her offspring will experience following hatching, and thus their growth and survival (reviewed in Gibson 1993). While breeding sites in many rivers may appear non-limiting, females do show strong preferences for particular sites, often clumping nests in such areas (Heggberget et al. 1988, Blanchfield and Ridgway 1997, Essington et al. 1998). Clumping of nest sites may reflect continuous rather than discrete (threshold) female preference criteria and may explain why nest superimposition is common even among iteroparous salmonids that spawn at low densities relative to semelparous salmonids (Kitano 1996; Blanchfield and Ridgway 1997; Taggart et al. 2001). Similarities in preference criteria lead to competition for nest sites, influencing female distribution and resulting in occupation of less-preferred sites (Quinn and Foote 1994; Blanchfield and Ridgway 1997; Hendry et al. 2001c). Female choice of nest site may balance intrinsic habitat quality against the cost of competition (Fretwell and Lucas 1970) or the risk of predation (Sih 1994). Moreover, choice is likely to be influenced by individual status and its effects on competitive ability (Hendry et al. 2001c). In this context, prior residence at a particular nest location appears to be among the most important determinants (Schroder 1982; Foote 1990), reflecting the unequal payoffs to residents versus intruders; for instance, residents may have more invested in the site (e.g., acquisition, nest construction, and eggs deposited) and more information about its intrinsic value.

2.4.1.2. Males   As expected from breeding system theory, the timing of arrival at the breeding grounds is commonly protandrous in salmonids, with males arriving before females. For example, in a study of seven Pacific salmon populations (pink, coho, sockeye, and chinook), Morbey (2000) found significant protandry in 90% of the years and in all populations (see also Pritchard 1937; Groot and Margolis 1991). Despite the consistency, the degree of protandry averaged only 1–5 days. Protandry in salmonids appears to be consistent with a male strategy to maximize opportunities to mate, with selection acting directly on male arrival time *relative* to that of females (Quinn et al. 1996; Morbey 2000, 2003). In polygynous breeding systems, such as that of salmonids, where males compete intensely for mates and reproduce for a longer period of time than females, initiating breeding early may afford males access to more mates over their reproductive life span. The idea is that the fitness-maximizing degree of protandry is a tradeoff between female availability and the reduction in mating opportunities caused by the presence of competitors or death (e.g., Wiklund and Fagerström 1977; Bulmer 1983). Female fecundity may also be linked to protandry if, for example, early arriving females have higher fecundity (Kleckner et al. 1995). There is evidence that this may be the case in brown trout, where early arriving females are often larger than those arriving later, and are thus likely more fecund (Elliott 1994; see also Hendry et al. 1999). Males may also benefit from mating with virgins (Wiklund and Fagerström 1977). In the case of salmonids, mating with virgins is advantageous because the number of eggs a female spawns declines as she proceeds through her breeding cycle from her first to last nest (e.g., Schroder 1981; Fleming 1996; de Gaudemar et al. 2000a). Protandry may also benefit females by minimizing waiting costs and facilitating mate choice through increased male–male competition during the formation of dominance hierarchies that singles out the higher-quality males (Morbey 2000).

Arrival date may also affect the selective advantage of male body size in accessing reproductively active females. In a study of pink salmon in a southeast Alaskan creek, Dickerson et al. (2002) found that benefits of large size for males decreased through the breeding season and suggested that an increasingly male-biased operational sex ratio was responsible. The ability of males to dominate access to breeding females appears to decline with more male-biased operational sex ratios in Pacific salmon (Fleming and Gross 1994, Quinn et al. 1996).

Males respond not only to the temporal distribution of receptive females (i.e., protandry), but also to their spatial distribution. At a broad scale, Quinn et al. (1996) found that the abundance of male sockeye salmon on the spawning grounds was sensitive to the availability of receptive females. However, the operational sex ratio (OSR) varied considerably at smaller spatial scales within the spawning grounds, with particular sites having more male-biased OSRs than others. These sites tended to have larger females than other sites and were the sites preferred by females (i.e., they were first settled) and thus may have been of higher quality (Quinn et al. 1996). Many males showed limited movement (see also Hendry et al. 1995), which may reflect the importance of prior residence in the outcome of male–male competition (Foote 1990) and the costs of establishing dominance relations at a new site.

2.4.2. Secondary Sexual Characters

Some salmonids develop among the most elaborate secondary sexual characteristics seen in breeding fishes. Darwin (1871, p. 676) noted this in his classic book that introduced the concept of sexual selection. Foerster (1968, p. 105) was led to remark that "changes which take place in salmonoid fishes and in particular the Pacific salmon, genus *Oncorhynchus*, are so great that were one not cognizant of the transformation which takes place as the fish mature, one would not consider them the same species." Among the most pronounced of the changes is the reshaping of the skull during maturity, as "feeding" teeth are shed and canine-like "breeding" teeth are grown, fusing gradually to elongating jaw bones that may nearly double in length (Figure 9.5; Davidson 1935; Tchernavin 1938; Vladykov 1962). These changes create a hooked snout (Figure 9.5d) or "kype" (Figure 9.5b), specialized weapons for fighting (Darwin 1871; Gross 1984), perhaps equivalent to horns, antlers, or tusks. Breeding competition has also been responsible for other changes exhibited by salmonids, such as the development of a dorsal hump, formed by a bar of cartilage under the skin. This may serve as a shield against attack (Schroder 1981), block the access of competitors to ovipositing females (Fleming and Gross 1994), and/or serve as a display of status (i.e., body size and condition; Quinn and Foote 1994). Similarly, thickening of the skin and embedding of scales within it may serve to minimize damage incurred on the spawning grounds, particularly from fighting. Changes in skin coloration from the silvery marine state to bright breeding coloration (Craig and Foote 2001) and enlarged adipose fins (Järvi 1990; Petersson et al. 1999) may be important status signals during intrasexual contests and mate choice.

The expression of secondary sexual characters appears to follow differences in intensity of breeding competition, with the semelparous Pacific salmon having the most exaggerated characters among salmonids (Davidson 1935; Tchernavin 1938; Vladykov 1954, 1962), reflecting their total investment in a single reproductive bout. Furthermore, among the iteroparous salmonids, the degree of expression appears to reflect the degree of iteroparity, with Atlantic salmon typically having greater expression than the more iteroparous species (e.g.,

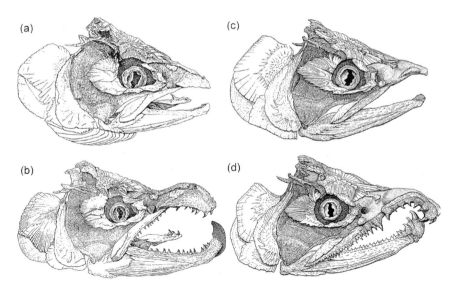

Figure 9.5. Changes in the skull morphology of male Atlantic (*Salmo salar*; a and b) and chum salmon (*Oncorhynchus keta*; c and d) from the "clean" state as the fish move coastwards to spawn (a and c) to the breeding state (b and d) on the spawning grounds (from Tchernavin 1938). In the "clean" state, the "feeding" teeth have been shed and new immature "breeding" teeth lie in connective tissue, not yet fastened to the bones. By the breeding state, the jaw bones have lengthened considerably and the large, canine-like "breeding" teeth are fully developed. In Atlantic salmon, a "kype" (hook) of conjunctive tissue has developed on the lower jaw, while in Pacific salmon it is the upper jaw that shows the most considerable curving to form the "hooked snout."

Salvelinus spp.; Vladykov, 1954). No study, however, has tested this in a phylogenetically controlled manner. Differences in the intensity of breeding competition, generated by intra- and intersexual selection, are also likely responsible for the sexually dimorphic expression of these traits within salmonids (Fleming and Gross 1994; Quinn and Foote 1994). In an experimental study quantifying selection during breeding in coho salmon, Fleming and Gross (1994) found that competition alone can generate a 52-fold increase in the opportunity for selection among males and a 6-fold increase among females. They concluded that the nearly ninefold higher opportunity for selection in males than females was in part responsible for the sexually dimorphic expression of the secondary sexual traits (see also Quinn and Foote 1994). Although overall body size, which is highly correlated with the expression of secondary sexual traits, is often the principal target of such sexual selection, selection also acts directly (i.e., controlling for body size and other morphological characters) on secondary sexual traits. For example, Fleming and Gross (1994) demonstrated direct selection for the male hooked snout, which provides males with increased access to females, participation in spawning events and, ultimately, breeding success. Similarly, in a study of sockeye salmon, Quinn and Foote (1994) found that variation in male breeding success was more strongly related to the species' secondary sexual traits, particularly the relative size of the dorsal hump (i.e., controlling for body size), than it was in females. Taken together, these findings indicate that breeding competition is driving the elaborate and sexually dimorphic expression of secondary sexual characters in salmonids.

Differences in the intensity of competition in relation to other selection pressures may also help explain among-population variation in the expression of secondary sexual characters (Fleming and Gross 1989; see also Blair et al. 1993; Quinn et al. 2001a). For example, we have seen that the intensity of competition will be affected by the sex ratio of returning adults, which can range from around 0.1 to over four anadromous males per female among Atlantic salmon populations (Fleming 1998), and also by spawner density (Fleming and Gross 1994; Quinn et al. 1996). Natural selection, on the other hand, opposes the exaggeration of secondary sexual characters that are otherwise favored by sexual selection (reviewed in Andersson 1994). The development of secondary sexual characters is presumably energetically costly and must be traded off against other energetic expenses, such as immuno-competence to fight off potential pathogens and parasites (Skarstein and Folstad 1996) and migratory costs. Furthermore, sexual selection for more exaggerated breeding traits may involve faster or longer growth and higher nutritional requirements, and thus more aggressive foraging tactics that may expose individuals to higher risks of mortality before maturity (Holtby and Healey 1990). Secondary sexual characters may also compromise survival on the breeding grounds, increasing susceptibility to predation and stranding, though evidence for such direct selection has been elusive (Quinn et al. 2001c). However, Quinn et al. (2001a) found that body and dorsal hump size of sockeye salmon in southwestern Alaska varied across populations in relation to the water depth of their natal river, a measure of the susceptibility to bear predation and stranding.

Within-population variability in the expression of secondary sexual characters can also be considerable, with some individuals having relatively larger characters for their size than others (e.g., Tchernavin 1944; Fleming et al. 1994; Quinn and Foote 1994). The pattern of expression appears to fit the idea of a tradeoff between potential benefits (e.g., more matings) and costs, with individuals varying their investment in relation to their ability to bear the costs of producing and/or maintaining the trait(s) (cf. Zahavi 1975; Andersson 1986; Jennions et al. 2001). That is, the secondary sexual characters are condition-dependent in their expression. In sockeye salmon, for example, individuals with high levels of stored somatic energy also have large secondary sexual characters and high gonadal investments (Hendry et al. 2000a), suggesting that secondary sexual characters are honest signals of individual condition/quality. Such variation in the ability to bear the costs may be purely environmental in origin, or reflect underlying differences in individual genetic quality.

Research on salmonids has revealed that counter-gradient variation can play a role in the evolution of condition-dependent sexual traits, promoting convergence in their expression. Craig and Foote (2001) discovered such counter-gradient variation in the use of the pigments (carotenoids) necessary to create the bright red breeding coloration of *Oncorhynchus nerka*. The non-anadromous kokanee morph, which occupies carotenoid-poor lakes, is approximately three times more efficient at sequestering carotenoids in their flesh than are the anadromous sockeye salmon morph, with whom they breed sympatrically in freshwater streams. A preference for red mates was apparently maintained in non-anadromous populations during their evolution and drove the re-evolution of the red phenotype in kokanee through more efficient use of dietary pigments. Such genetic divergence in carotenoid use, hidden by phenotypic similarity, may be present in other systems given the widespread use of carotenoid-based signals in animal breeding systems (reviewed in Houde 1997). Moreover, countergradient variation may be a mechanism by which sexual selection can promote the genetic differentiation of sympatric populations (Craig and Foote 2001).

2.4.3. Alternative Reproductive Phenotypes

Salmonids have played a key role in our understanding of the evolution of alternative reproductive phenotypes (Gross 1984, 1985). For example, in coho salmon, large "hooknose" males mature after at least 18 months in the ocean and, displaying well-developed secondary sexual characters, breed as fighters. They form size-structured dominance hierarchies next to nesting females. In contrast, "jack" males mature after only 6 months and, at about one-fifth the body weight of hooknose males, breed as sneakers. The younger age at maturity provides a survival advantage and the small size an aid for sneaking close to females, allowing them to avoid aggression from hooknose males and dart in to fertilize eggs at oviposition (Gross 1985, 1991). In sockeye salmon, such jack males may achieve nearly comparable paternity to that of hooknose males when competing one-on-one for fertilization success at a spawning event (Table 9.2), although little is known about their relative lifetime breeding success. While hooknose and jack males can employ either behavioral tactic, the life history decision to mature early or late will make the individual predominantly a sneaker or fighter, respectively.

Early male maturation reaches its greatest expression in terms of frequency and magnitude of the mature male size difference in Atlantic salmon (Fleming 1998). In some populations, up to 100% of males have been estimated to mature early as parr during their life history (Bohlin et al. 1986), with some later becoming anadromous. Unlike coho jacks, they mature before ever having migrated to sea, and at a body weight that is often two orders of magnitude smaller than that of the anadromous fighter males with whom they compete (see chapter frontispiece). Such early maturing male parr, as a group, may fertilize a considerable proportion of the eggs within populations (11–65%; Table 9.2). However, as individuals, their relative fertilization success may frequently be more than an order of magnitude less than that of anadromous males (Table 9.2). Yet, the iteroparous life history of Atlantic salmon means that mature male parr have the possibility of breeding again as parr in another year and/or subsequently migrating to sea (Bohlin et al. 1986) and returning to breed as fighters (i.e., adopting the early maturity life history tactic does not preclude the anadromous tactic).

Table 9.2. Relative fertilization success (paternity) of alternative male reproductive phenotypes: fighters and sneakers.

Species, study	Scale	Group paternity (sneakers)	Individual paternity		Reference
			Fighters	Sneakers	
Oncorynchus nerka					
Experiment	Nest	42.2 ± 8.1%	57.8 ± 8.1%	42.2 ± 8.1%	Foote et al. (1997)
Salmo salar					
Experiment	Redd	16 ± 4%	84 ± 4%	3 ± 1%	Hutchings and Myers (1988)
River[a]	Redd and population	10.8 ± 5.2%	—	—	Jordan and Youngson (1992)
Experiment	Redd	51.2 ± 19.6%	48.8 ± 19.6%[b]	6.3 ± 1.7%[b]	Morán et al. (1996)
Experiment	Redd	30 ± 3%	70 ± 3%	14 ± 5%	Thomaz et al. (1997)
River[a]	Redd	65.1 ± 9.5%	8.7 ± 3.7%[b]	3.1 ± 0.3%[b]	Martinez et al. (2000)
Experiment	Redd and population	65% (42–86%)[c]	15.8%	13.2%	Garcia-Vazquez et al. (2001)
Experiment	Redd	42.0 ± 8.8%	54.2%	3.9%	Jones and Hutchings (2001)
River	Population	42.3 ± 3.9%	—	—	Taggart et al. (2001)
River[a]	Nest	11.0 ± 2.6%[b]	67.0 ± 20.8%[b]	4.7 ± 1.1%[b]	Garant et al. (2002)
Experiment	Nest and population	30.1 ± 5.8%	17.5 ± 1.5%	1.6 ± 0.3%	Jones and Hutchings (2002)
Salvelinus malma miyabe					
Experiment	Nest	16.9 ± 9.4%	83.1 ± 9.4%	16.9 ± 9.4%	Maekawa and Onozato (1986)

[a] Controlled river section.
[b] Based only on males that had some fertilization success.
[c] Range among redds.

The alternative phenotypes involve jack and hooknose males for *Oncorhynchus nerka*, anadromous and parr males for *Salmo salar*, and lake-run and stream-resident males for *Salvelinus malma miyabe*. "Study" refers to whether the work was conducted in experimental or natural river environments and "scale" refers to level of analysis, i.e. nest (individual spawning events), redd (groups of nests of a single female), or population. The data are means ± standard errors.

Early models to explain the coexistence of the two male phenotypes in salmonids proposed that they represented genetically distinct strategies (Gross 1985; Bohlin et al. 1986; Myers 1986). Support for this came from evidence of a weak, but heritable basis to early maturity (e.g., Iwamoto et al. 1984; Glebe and Saunders 1986; Heath et al. 1994, 2002e). It quickly became evident, however, that there was a much stronger environmental component and that a genetically determined threshold related to growth or state (e.g., lipid, weight) existed, which once achieved results in early maturity (Myers et al. 1986; Thorpe 1986; Bohlin et al. 1990; Gross 1991; Metcalfe 1998). Thus, the coexistence of two male phenotypes in salmonids probably represents alternative tactics within a single conditional strategy (Bohlin et al. 1990; Gross 1991; Hutchings and Myers 1994), the equilibrium of which is determined by the combined effects of frequency- and condition-dependent selection (Gross 1996). It is frequency-dependent in that the success gained from either fighting or sneaking is affected by the tactics that other members of the population use. Neither tactic is in itself evolutionarily stable, but rather a mixture of the two tactics. In addition, the relative success gained by using a tactic probably depends on an individual's competitive ability or "state" relative to others in the population, and is thus also condition-dependent. Although the tactics may have unequal average fitnesses, the switch-point between them (i.e., their frequencies) will be evolutionarily stable even when the status cue (e.g., body size, growth rate) influencing the switching decision shows some heritability (Hazel et al. 1990; Gross and Repka 1998). In this sense, the alternative tactics entail a norm of reaction to environmental conditions (Hutchings 2003—*this volume*).

An interesting outcome of this research has been to question the presumption that animals adopting a sneaker tactic have lower average fitness than individuals that fight. In the case of salmonids, the opposite may well be the case as the sneaker life history tactic is chosen by faster-growing/larger juveniles (e.g., Thorpe 1986; Gross 1991; Heath et al. 1996), and thus likely the highest-status individuals in the population (Gross 1996; see also Garant et al. 2002). Only the largest or most dominant males may be able to effectively adopt the early maturation tactic, while other males are better off delaying maturity. In this vein, the propensity for early maturation among populations will be sensitive to environmental conditions, such as growth opportunities (L'Abée-Lund et al. 1990), migration costs (Young 1999), and habitat structure (Gross 1991; but see Koseki et al. 2002), as these factors affect costs and benefits. The game theoretical framework of conditional strategies thus provides a powerful means of interpreting patterns of early male maturity among salmonids.

The OSR has almost certainly contributed to the evolution of conditional strategies in male salmonids by affecting the shape of selection during breeding. In a study involving hooknose male coho salmon, Fleming and Gross (1994) found that the ability of males to control access to females decreased with increasing OSR (see also Schroder 1982; Quinn et al. 1996), leading to disruptive selection on male body size, with small and large males having higher breeding success than intermediate-sized males. Under such conditions, small size and crypsis probably afford males an advantage in sneaking access to ovipositing females, which likely selects for early maturity. Accordingly, Young (1999) found that the proportion of jack males among coho populations in two Oregon river basins varied directly with breeding density, which correlates positively with the extent of male bias in the OSR (Schroder 1982; Fleming and Gross 1994).

2.4.4. Sperm Competition

Sperm competition is a prominent feature of salmonid spawning behavior. Male fertilization success at a spawning event depends largely on the number of competitors, with sperm competition ensuing when more than one male spawns simultaneously with the female. Under such circumstances, male proximity to the female allows for closer and/or quicker access to the eggs during oviposition (Gross 1985; Foote et al. 1997; Mjølnerød et al. 1998) and thus sperm precedence for increased fertilization success (Schroder 1982; Chebanov et al. 1984; Maekawa and Onozato 1986; Mjølnerød et al. 1998; but see Foote et al. 1997). This may be a common pattern in animals with external fertilization (cf. Yund and McCartney 1994; Brockmann et al. 2000). In salmonids, large body size among fighter males facilitates proximity through its effects on dominance and thus increases sperm precedence. Moreover, larger males are likely to release more spermatozoa because of their larger ejaculate volumes (Kazakov 1981; Gjerde 1984b; Foote et al. 1997). Large, and thus dominant males, however, may suffer decreased fertilization success in competition over time because of high levels of spawning activity (Mjølnerød et al. 1998), which negatively affects ejaculate volume, and spermatozoa concentration and activity (Kazakov 1981; Gjerde 1984b).

Despite their small size, early maturing males can attain significant fertilization success (Table 9.2). They can use their size to advantage by either rapidly sneaking next to the female's vent at spawning, as in sockeye salmon (Foote et al. 1997), or remaining within the female's nest until oviposition, as in Atlantic salmon (e.g., Jones 1959; Thomaz et al. 1997; Jones and Hutchings 2002). It has been estimated by genetic analyses that mature male Atlantic salmon parr may fertilize from 11 to 50% of all eggs in a population (Jordan and Youngson 1992; Taggart et al. 2001). At individual spawning events in the wild, parr success may range from 17 to 86%, with an individual male parr fathering up to 47% of the eggs (Martinez et al. 2000). In sockeye salmon, the fertilization success of jacks has been observed to range from 3 to 96% under experimental conditions (Foote et al. 1997).

Sperm competition should influence gonad investment (Section 1.4.4). Within salmon species, the phenotypes that face the highest risk of sperm competition (e.g., sneakers) have relatively larger testes and milt volume for their size than phenotypes with lower risk (e.g., fighters) (Gage et al. 1995; Taborsky 1998; Vladić and Järvi 2001). The pattern holds when comparing alternative male reproductive phenotypes of other salmonids (Figure 9.6; paired contrasts $t_4 = 4.306$, $P = 0.013$). Investment in sperm competition may also involve producing higher ''quality'' sperm (e.g., size, longevity, motility). Evidence from Atlantic salmon indicates that mature male parr have greater spermatozoa con-

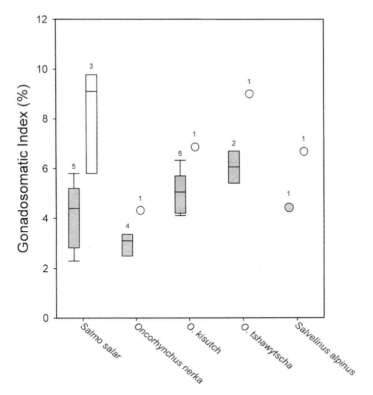

Figure 9.6. Comparison of the gonad investment (gonadosomatic index: gonad wet weight/total body weight × 100) by alternative male phenotypes within different salmonid species (fighters = gray, sneakers = white). Median, sample size (number of populations), and 5th, 25th, 75th and 95th percentiles as vertical boxes with error bars are shown. Circles are individual data points. Data are derived from Fleming (1998) and literature sources contained therein.

centrations and motility, and longer spermatozoa life span than anadromous males (Daye and Glebe 1984; Gage et al. 1995; Vladić and Järvi 2001). Similarly, Hoysak and Liley (2001) found significantly greater spermatozoa concentrations and motility in sockeye salmon jacks than hooknose males but, however, found no relation between motility and fertilization success during sperm competition trials.

2.4.5. Female Competitive Tactics

We saw in Section 1.4.5 that researchers have traditionally focused on male–male competition. Yet female–female competition is difficult to ignore in salmonids, and various species have helped to clarify the relative importance of sexual versus natural selection in the evolution of traits used for competition. Breeding density, rather than OSR, should be the principal determinant of the intensity of female–female competition in salmonids, because females are the limiting sex, and should thus rarely face difficulty in obtaining mates. Access to breeding resources, specifically nesting territories, should be the driving force of female–female competition during breeding and this should be functionally related to breeding density (Fleming and Gross 1989). Accordingly, Fleming and Gross (1994) found that mean breeding success of coho salmon females declined and variance in success increased when breeding density was increased experimentally (see also Schroder 1982). Competition over nesting sites can result in delays in breeding, female displacement and nest destruction by super-imposition (e.g., Schroder 1982; van den Berghe and Gross 1989; Essington et al. 1998), all of which will reduce reproductive success.

Female breeding success in salmonids often shows a positive relation to body size partly due to increased fecundity (e.g., van den Berghe and Gross 1989; Fleming and Gross 1994; Fleming et al. 1996, 1997; Fleming 1998; but see Garant et al. 2001). Large size also constitutes an advantage during breeding competition, particularly in semelparous species, through possible relationships with duration of nest guarding (van den Berghe and Gross 1989; but see Hendry et al. 1999) and nest depth (e.g., van den Berghe and Gross 1984; Crisp and Carling 1989; Steen and Quinn 1999), which reduces susceptibility to nest site reuse by other females and gravel scour. Furthermore, body size may enhance competitive ability (Fleming et al. 1997), affecting a female's ability to establish a territory within a preferred area (Foote 1990) and avoid delays in breeding (Fleming and Gross 1994). Investment in fighting morphology may also enhance competitive ability, at least indirectly through the effects of overall body size (Fleming and Gross 1994). Among coho salmon populations, the expression of the hooked snout (Figure 9.7) and bright breeding coloration in females corre-lates with breeding density (Fleming and Gross 1989). Moreover, the decoupled development of such traits between the sexes (i.e., females develop some sec-ondary sexual traits seen in males but not others) suggests that some traits are directly responsive to selection (Hendry and Berg 1999) and not the product of a genetic correlation between male and female characters (*sensu* Lande 1987).

2.4.6. Mate Choice

Females are predicted to be choosier than males when they have a lower poten-tial rate of reproduction, and the population's OSR is therefore skewed toward males (Section 1.3). This appears to be the case for salmonids (Section 2.3). Females appear to express mate choice through aggression toward males (e.g., Keenleyside and Dupuis 1988; Fleming et al. 1997; Petersson and Järvi 1997), delays in breeding (e.g., Schroder 1981; Foote 1989; Berejikian et al. 2000; de Gaudemar et al. 2000b), and possibly tricking males to join spawnings through "false orgasm" (Petersson and Järvi 2001). Females may also manipulate the number of eggs they lay (cf. Côté and Hunte 1989; van den Berghe et al. 1989) and bias paternity by directing eggs toward a preferred male during multi-ple male spawnings, though the latter has never been examined. The extent of female choice in salmonids, however, appears to be constrained or circumvented

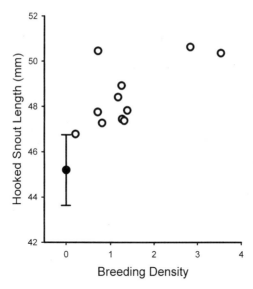

Figure 9.7. The correlation between female hooked snout length (size adjusted) and estimated breeding density (number of females per capacity) among coho salmon populations (open circles) ($r = 0.673$, $P = 0.023$). Also shown is the mean ± one standard error of hooked snout length among five hatchery populations (closed circle), where artificial breeding has relaxed selection for secondary sexual traits (Spearman rank $r = 0.672$, $P = 0.004$ including hatchery and wild populations). Data are from Fleming and Gross (1989).

by male–male competition, because dominant males can monopolize access to females (e.g., Schroder 1981; Fleming et al. 1997; Petersson et al. 1999). Female incitation of male–male competition (e.g., through delays in breeding, moving away from a suitor, or false orgasms), however, may be viewed as a means of "passive" choice (cf. Cox and Le Boeuf 1977), although its role in salmonids is unclear. There is also the possibility of cryptic female choice of male sperm, for example, if eggs discriminate among sperm on the basis of genotype at a surface protein locus (Palumbi 1999; see also Stockley 1999). In these cases, thousands of sperm cells may attach anywhere on the egg surface and undergo acrosome reaction before one achieves success. However, in teleost fishes, such a process of discrimination is unlikely because fertilization occurs via a single small opening (micropyle) in the egg membrane and there is no acrosome reaction. If egg discrimination among sperm occurs in salmon, it will probably involve some influence on the sperms' ability to locate the micropyle or on the swimming speed of the sperm (Hoysak 2001).

There are no direct, material benefits (e.g., territories, nutrition) available to female salmonids from mate choice, although there can be direct, non-material benefits, such as safety from disruption and injury during spawning, reduced risk of infection and assurance of fertilization. In brook charr, for example, female choice of large males significantly reduces brood cannibalism by satellite males (Blanchfield and Ridgway 1999; see also Maekawa and Hino 1990). The only evidence in salmonids for choice based on genetic criteria comes from a study on Atlantic salmon reporting a significant, non-random increase in heterozygosity at the major histocompatibility complex (MHC) in progeny relative to parents (Landry et al. 2001). This increase occurred independently of the rest of the genome, leading Landry et al. (2001) to conclude that Atlantic salmon must be choosing mates to increase the heterozygosity of their offspring at the MHC, presumably to provide better defence against parasites and pathogens. Diversity at the MHC may be important in disease and pathogen resistance in other organisms (e.g., Penn et al. 2002; Bernatchez and Landry 2003).

Although female mate choice has received more attention, there is good reason to believe that male choice of females also occurs. The strongest determinant of mate choice by males appears to be female proximity to oviposition (Schroder 1981; Hamon et al. 1999), which may increase a male's mating opportunities over the breeding season by reducing the time between breeding events. Males may also be choosy with regard to female size (Sargent et al. 1986; Foote 1988; de Gaudemar et al. 2000b), because larger females are more fecund, have larger eggs, and offer better parental care (reviewed in Fleming 1996, 1998). Consequently, the number of males attending a female tends to increase with female size (e.g., Hanson and Smith 1967; Sargent et al. 1986; Blanchfield and Ridgway 1999). Female nesting stage may be another choice criterion of males (de Gaudemar and Beall 1999), because the number of eggs a female deposits in a nest decreases strongly as she proceeds through her nesting cycle (Fleming 1996; de Gaudemar et al. 2000a). The expression of male mate choice, however, will be affected by intrasexual competition and should be condition-dependent. This may explain the common observation of size-assortative mating in salmonid (e.g., Hanson and Smith, 1967; Schroder, 1981; Jonsson and Hindar 1982; Foote 1988; Maekawa et al. 1994; Blanchfield and Ridgway 1999, but see Dickerson et al. 2003). When all individuals of one sex have the same preference (e.g., males for large females), but only some of them are able to achieve it (e.g., large, most dominant males), assortative mating will likely ensue (Burley 1983).

3. Conclusions and Extensions

Salmonid species have figured prominently in studies of breeding systems, particularly in research concerning alternative reproductive behaviors, secondary sexual characters, female–female competition, and sperm competition. Not only has salmonid research informed the development of these fields, but it has also been interesting to note how interpretations of salmonid reproductive behaviors have changed as the fields have advanced. Thus, tiny males on the spawning grounds have gone from being maladaptive aberrations to genetically superior beings! Elaborate secondary sexual characters were once considered to be atavistic traits with no purpose (Chernavin 1918, 1921), outcomes of wounding (Cunningham 1900) or pathological developments due to dysfunction of excretory organs (Barrett-Hamilton 1900). Now they are adaptations for intense breeding competition, particularly among males. Moreover, female reddish coloration and enlarged jaws have gone from being genetic spillovers of male adaptations to adaptations by females for fighting with other females. And fertilization success has gone from being a lottery based on numbers and proximity to females to an outcome of adaptive changes in ejaculate volume, motility, and energy reserves.

Where do we go from here? Many refinements to our understanding of breeding system theory come to mind, for which salmonids are well suited. For example, we can take advantage of our ability to rear salmonids in hatcheries to test for genetic benefits of mate choice, including the role of immunocompetence and genetic compatibility. After a two-decade obsession with the benefits of female choice in mating systems in which males have little to offer, the jury is still out on the relative importance of genes for viability and male attractiveness. Sexual conflict is still a hot topic in breeding systems (Chapman et al. 2003), and such conflicts abound in salmonids, for example involving mate choice by both sexes, and involving sneakers and hooknose males. Should females prefer to mate with sneakers or avoid them? What tactics do they use to bend the mating rules in their favor? The evolution and maintenance of alternative reproductive tactics remains an unresolved issue. What forces led to, and maintained such tactics in some salmonids and not others? Salmonids lend themselves well to manipulations in artificial streams, which should help provide the answers.

We would also like to see reproductive allocation and breeding systems embedded more firmly into an understanding of population dynamics (Einum et al. In press). Each of these subjects is usually studied in isolation, yet a well-integrated model of links between breeding behavior and population ecology could be very useful for studies of conservation (e.g., Sutherland 1996). Most of the elements of such a synthesis are already available, but few people have put them together. For example, we know how breeding competition by both males and females affects reproductive success of individuals. We are also beginning to realize that non-genetic contributions from females (e.g., egg size, breeding location and time; Einum and Fleming 2000b, Einum et al. 2003—*this volume*) and males (e.g., egg swelling; Pakkasmaa et al. 2001) can have significant implications for offspring success. Furthermore, we know a lot about juvenile growth, mortality, and competition, as well as salmon population dynamics. These can be combined into a model that translates breeding competition into production of fry, and then into growth and survival of juveniles, and then into the numbers of fish that can occupy a given habitat. We need to put these components together both to provide a more holistic understanding of behavioral and population ecology, and also to inform conservation management.

For example, habitat destruction has been a major cause of the extinction of salmon populations, and we know how competition and reproductive success of individuals is linked to the quality and spatial distribution of habitats. An understanding of the population consequences of this aspect of breeding behavior should therefore feed into management advice concerning habitat preservation and restoration. If a certain percentage of the habitat of a given breeding population is destroyed, we should be able to predict the impact on the population size according to how individuals compete with each other for the available spawning resources. Similarly, exploitation by fisheries is size-selective (Hard 2003—*this volume*). We have seen that body size is a key determinant of many features of salmonid breeding systems, including alternative reproductive behaviors. We should therefore be able to use this information to predict the consequences of management options for breeding systems, and therefore for productivity. This will be important both for short-term predictions, and also for evolutionary impacts (Young 2003—*this volume*). One management option often applied is the artificial culture of salmonids in hatcheries, oceanic net-pens, and gene banks. We should recognize that this disrupts the natural breeding system and alters fitness-related traits (e.g., Gross 1998; Fleming and Petersson 2001). Moreover, the implications, both genetic and ecological, of the intentional and unintentional release of these fish for wild populations will be largely dependent on what occurs during breeding and the subsequent performance of the offspring (Fleming et al. 2000; Garant et al. 2003b). Effective management of cultured fishes will need to recognize the role that the breeding system plays here. Finally, genetically modified salmon and trout have now been developed for use in aquaculture. What impacts will these large, fast-growing fish have on wild populations when they escape? Models that integrate competitive interactions with productivity should be able to tell us the answer. There is a pressing need to address all of these issues, and the answers will depend in part on our understanding of breeding systems.

Salmonid Insights into Effective Population Size

Robin S. Waples

Spawning sockeye salmon

10

Conservation biology can be defined as "the science of scarcity and diversity" (Soulé 1986). The discipline is concerned with conservation of important components of biological diversity that are at risk because of their scarcity. The intersection of evolutionary biology and conservation biology, therefore, comprises the study of the consequences of scarcity and diversity for the persistence and evolution of species and populations.

The genetic consequences of scarcity and diversity all depend, either directly or indirectly, on the concept of effective population size (N_e). Wright (1931 and later) recognized that populations of the same census size (N) but with different demographic parameters do not necessarily respond to evolutionary forces in the same way, and he developed the concept of effective population size to provide a means of characterizing the evolutionary behavior of populations with different census sizes and demographic traits. Thus, N_e (rather than N) determines the rate at which random genetic drift acts in a population. Furthermore, because drift may swamp the effects of migration and selection in small populations, N_e also determines the relative importance of these evolutionary forces. As a consequence, estimating N_e or the ratio N_e/N in natural populations has attracted a great deal of interest among both evolutionary and conservation biologists. The unusual life history of Pacific salmon—they reproduce only once but have variable age structure—presents some analytical challenges but also provides an opportunity for novel evolutionary insights into effective population size. In this chapter I consider some of the insights inspired by these challenges.

1. Evolutionary Theory

1.1. Definition of N_e

Effective size is an artificial construct that allows us to characterize the evolutionary behavior of populations. A population of census size N behaves (in an evolutionary sense) as if it were of size N_e; hence N_e is the "effective" size of the population. To make this concept more concrete and unambiguous, Wright defined effective population size in terms of an "ideal" population in which $N_e = N$. The effective size of a population of N individuals can be thought of as the census size of an ideal population of N_e individuals. Put another way, N_e is the size of an ideal population that would have the same rate of genetic change as the actual population under consideration.

An "ideal" population has specific demographic features: random mating, equal sex ratio, discrete generations, and constant population size. In addition, in an ideal population reproductive success follows a Poisson distribution—that is, the variance among individuals in progeny produced equals the mean number of progeny produced. This would occur, for example, if all adults contributed equally to a large gamete pool, and progeny for the next generation were formed by random union of gametes from this pool. This "Wright–Fisher" process can be modeled in a computer simulation by forming each individual in the next generation by randomly choosing one male and one female parent (with replacement) from the pool of potential breeders. In what follows, therefore, I will refer to this scenario as "random" reproductive success. In real populations, departures from ideal conditions generally cause the variance in reproductive success to be greater than would occur under random conditions, with the result that N_e is less (and sometimes much less) than N.

N_e determines the rate at which random genetic processes (genetic drift, increase in inbreeding, and loss of neutral alleles) occur in a population. Separate effective sizes have been proposed to describe each of these processes (variance, inbreeding, and eigenvalue effective sizes, respectively; Crow and Kimura 1970; Ewens 1979), but the differences among them are small unless the population changes substantially in size. Here I will consider primarily the variance effective size. A general expression for the variance effective population size is (Crow and Kimura 1970; Kimura and Ohta 1971):

$$Variance\ N_e \approx \frac{\bar{k}(N-1)}{1+(V_k/\bar{k})} \tag{1}$$

where N is the number of parents, k is the number of progeny in the next generation produced by a parent, and $\bar{k}$ and V_k are the mean and variance of k among individuals. This formula assumes separate sexes and an equal progeny distribution for each sex, but it provides a good approximation for most species and a variety of mating systems.

Most species do not have discrete generations, as assumed in Wright's formulation of effective population size. Felsenstein (1971), Hill (1979), and others have shown that discrete generation models for N_e are fairly robust for organisms with overlapping generations, provided that the populations are demographically stable. Stable age structure is more important than constant population size in determining whether the discrete generation model is appropriate (Crow and Denniston 1988). If population size changes over time, the multigenerational N_e is approximately the harmonic mean $(\tilde{N}_e)$ of the effective sizes in the individual generations (Wright 1938). This result is important for conservation and evolutionary biology because the smaller terms dominate in a harmonic mean.

1.2. The N_e/N ratio

Because of the difficulties in computing or estimating N_e for natural populations, the ratio N_e/N is of considerable interest to evolutionary biologists and conservation biologists. If rules-of-thumb can be developed for predicting this ratio, then estimates of N_e can be generated from estimates of N, which are available for a much wider range of species. Whether it is reasonable to expect that rules-of-thumb can be developed that have broad applicability is a topic of active debate in the scientific literature. Nunney (1993, 1996) developed theoretical expectations for effective size under a variety of mating systems, including the extreme case of harem polygamy in which the effective sex ratio is very skewed, and concluded that in most species single-generation N_e/N values should be approximately 0.5 or higher. This theoretical prediction was tested empirically by Frankham (1995), who reviewed published estimates for over 100 species and found that the mean of single-generation N_e/N estimates was 0.35. Although this mean value was significantly lower than Nunney's prediction of 0.5, collectively the studies suggest that a rule-of-thumb that N_e is approximately $\frac{1}{3}$ to $\frac{1}{2}$ of N might be widely applicable.

However, others believe that N_e/N can be much lower than 0.5 in some organisms. For example, Husband and Barrett (1992), using both demographic and genetic methods, found a mean N_e/N ratio of only about 0.1 in 10 populations of an annual plant, while Nei and Tajima (1981) suggested that N_e/N might be less than 0.1 in small organisms with large population sizes. Several studies using genetic methods have yielded estimates of N_e in both aquatic and terrestrial species that are several orders of magnitude less than N (Hedgecock et al. 1992; Turner et al. 1999; Scribner et al. 1997; Hauser et al. 2002).

These differences in viewpoint can be attributed to two major issues. First, some confusion has arisen as a result of the comparison of single-generation N_e/N estimates to long-term estimates over a number of generations. One common way of computing long-term N_e/N (e.g., Frankham 1995; Vucetich et al. 1997) is to divide the harmonic mean N_e by the arithmetic mean N over the time period in question. However, as pointed out by Waples (2002a), with variable N this approach will lead to low multigeneration N_e/N ratios even if the single-generation N_e/N is constant. Many of the studies mentioned above that discussed N_e/N ratios included data for multiple generations or a mix of single- and multigeneration estimates. Kalinowski and Waples (2002) discussed various approaches for dealing with this issue. To avoid these complications, in this chapter I will primarily consider single-generation estimates of N_e/N.

Second, most of the claims for extremely low N_e/N ratios are for species with Type III survivorship curves (i.e., those with high fecundity and high mortality in early life stages). Many marine species have Type III survivorship, as do salmon and many insects and plants. Hedgecock (1994) suggested that low N_e/N ratios in marine species can be explained by a scenario involving "sweepstakes" survival of entire family groups during critical life stages. Under this hypothesis, the vast majority of families may produce no offspring that survive to adult because their eggs are not fertilized or larvae do not find suitable conditions for survival during critical period. In contrast, under a random (Poisson) distribution of reproductive success in a population of constant size ($\bar{k} = 2$), family sizes of 1, 2, and 3 are expected to occur more frequently than family sizes of 0.

Evaluating N_e/N ratios directly in natural populations requires detailed information on the mean and variance in reproductive success across families. This can be demonstrated by rearranging Equation (1) to provide an approximate expression for the ratio N_e/N (after Crow 1954; Hedrick 2000):

$$\frac{N_e}{N} \approx \frac{\bar{k}}{1 + (V_k/\bar{k})} \tag{2}$$

This formulation makes it apparent that the ratio of effective to census size depends heavily on the term $V_k/\bar{k}$, which Crow and Morton (1955) called the "index of variability." I will follow Geiger et al. (1997) and Waples (2002b) and use $R = V_k/\bar{k}$ in the formulae that follow, with R being subscripted as necessary to indicate a particular life stage. R is similar in form to the Index of Total Selection ($I = V_k/\bar{k}^{-2}$; Crow 1958), which has been used in studies of reproductive success in a wide variety of organisms (e.g., Arnold and Wade 1984a; Clutton-Brock 1988b). I can be interpreted as a standardized variance in fitness and mathematically is the square of the coefficient of variation in family size.

In estimating N_e from demographic data, parents and their offspring should be counted at the same life stage, and if so the estimate of N_e will apply to a full generation. Since families are typically defined for adults, family size should be evaluated at the adult stage as well. Accomplishing this, however, is very difficult for species with Type III survivor curves. In such species, if data on family-specific survival can be collected at all, they typically are available only at an early life stage, at which point each family might be represented by hundreds or thousands (in some cases millions) of individuals. Because the variance effective size primarily reflects the number in the progeny generation (and hence is strongly influenced by $\bar{k}$; see Equations (1) and (2), using early life history data can lead to substantial bias in estimating N_e or N_e/N.

Crow and Morton (1955) were the first to consider this problem in detail. Assuming that family size data were collected at an early (e.g., juvenile) life stage, they considered how two extreme models of post-enumeration survival might affect N_e/N at a later (e.g., adult) life stage:

Model 1 (random survival): every individual in the population has an equal probability of surviving to the later stage.
Model 2 (correlated survival): each family either survives or does not as a unit, and which families survive is random.

Under these two models of survival, R and $\bar{k}$ following an episode of mortality (subscript 2) can be expressed as a function of R and $\bar{k}$ at an earlier life stage (subscript 1) (Crow and Morton 1955):

$$\text{Random:} \qquad E\left[\frac{R_2 - 1}{\bar{k}_2}\right] = \frac{(R_1 - 1)}{\bar{k}_1}$$

$$\text{Correlated:} \quad E\left[R_2 + \bar{k}_2\right] = R_1 + \bar{k}_1$$

where E indicates the expectation. These formula can be rearranged as follows (Waples 2002b):

Model 1: random survival

$$E[R_2] \approx 1 + \bar{k}_2 \frac{(R_1 - 1)}{\bar{k}_1} \qquad\qquad (3)$$

Model 2: correlated survival

$$E[R_2] \approx R_1 + \bar{k}_1 - \bar{k}_2 \tag{4}$$

These formulae provide a means for scaling (adjusting) demographic data collected early in life history for Type III species to values expected at the adult stage. In a stable population, the long-term mean family size must be close to 2. Crow and Morton (1955) suggested, therefore, that an estimate of N_e can be obtained for Type III species by adjusting juvenile estimates of $\bar{k}$ and V_k to their expected values at $\bar{k}_2 = 2$. For example, if one is interested in how much N_e or N_e/N has been reduced by mortality that has occurred through the juvenile stage, juvenile estimates of $\bar{k}$ and V_k can be scaled to their expected values at $\bar{k}_2 = 2$ using Equation (3). This equation assumes random mortality between juvenile and adult phases, so after scaling to constant population size any reduction in N_e or N_e/N can be attributed to non-random processes that occurred prior to the time of enumeration. If, however, one is interested in evaluating how extreme a reduction in N_e/N might be expected from an episode of family-correlated survival, Equation (4) can be used to adjust early life history data for survival during the next phase. Waples (2002b) showed that if multiple episodes of random and family correlated survival occur during a complete life cycle, the order in which they occur does not matter. Further, the composite effect on effective size of all family-correlated survival is to reduce N_e/N to approximately S, the overall survival rate through all family correlated phases.

In the next section stage-specific family size data for salmon will be used to gain insights into N_e and N_e/N.

2. Insights Regarding Effective Size Gained from Salmon

2.1. Definition of N_e in Salmon

In discrete generation or standard overlapping generation models, N_e represents the effective size of a population over a generation. Salmon life history does not conform to either of these paradigms, although it shares features of both. Pacific salmon experience 100% turnover in the breeding population each year, but because of variable age structure they do not have the same simple Markov-chain dynamics as do organisms with discrete generations. These unusual life history features—semelparity with variable age structure—make the concept of effective population size difficult to apply directly to Pacific salmon. In particular, although effective population size is defined with respect to a generation, the typical sampling unit for a salmon population is a single year of spawners (or their juvenile progeny).

Waples (1990a) considered this problem and developed a method to relate the effective number of breeders per year (N_b) to the effective size per generation (N_e) in Pacific salmon. Using computer simulations, he modeled populations with Pacific salmon life history and monitored the rate of change in allele frequency over time, measured as the mean squared difference (variance) between allele frequencies (P) at times 0 and t. Comparison of empirical results with theoretical expectations allowed him to relate the Pacific salmon model to the rich body of population genetics theory for organisms with discrete or over-lapping generations. Figure 10.1A shows results of a simulation with generation length (average age at spawning) of $g = 4$ years and $N_b = 50$ each year. (A more precise definition of generation length is the average age of parents that produced the current generation. If age-specific fecundity and fertility data are available, this information could be used to weight the age at maturity data to provide a better estimate of g.) The slope (rate of genetic change) of the observed data points agrees with the slope expected for $N_e = 4N_b = 200$—that is, the modeled Pacific salmon population with 50 effective breeders per year experienced genetic change at a rate equivalent to a population with effective size of 200 per generation. Simply using N_b instead of N_e in standard population genetics models would greatly overestimate the rate of change in salmon populations (Figure 10.1A; theoretical slope for $N_e = N_b = 50$ is much steeper than was observed in the simulations). More generally, Waples (1990a) showed that in the simplest case where N_b is constant, the rate of genetic change over time in a Pacific salmon population with N_b effective spawners per year agrees with the-oretical expectations for a population with an effective size of $N_e = gN_b$. This result reflects the fact that an individual salmon's reproductive output in a single spawning year also represents its lifetime contribution to the next genera-tion—hence N_e in salmon is an additive function of the N_b values over a period of a generation.

If, however, a salmon population is measured at any two points in time (e.g., by sampling spawners or their juvenile progeny in two different years), the observed *magnitude* of genetic change will be higher than expected assuming $N_e = gN_b$ (e.g., observed variance in allele frequency in Figure 10.1A is consis-tently higher than expected for $N_e = 4N_b$, even though the slopes are the same). This effect arises because a single year of spawners or their progeny (the typical sampling unit) represents only part of a salmon generation, and this additional source of error in sampling from the generation as a whole inflates the observed variance. Waples (1990a) showed that this effect can be adjusted for in Pacific salmon by adding a correction factor, $C = P_0(1 - P_0)/N_b$, to the expected var-iance (Figure 10.1B). With this correction, the model correctly predicts both the rate and magnitude of genetic change over time in Pacific salmon populations. This sampling effect is most apparent in semelparous species such as salmon, but a similar difficulty (the problem of obtaining a random sample from an entire generation) applies to many other species. Waples (2002c) discussed this issue and some ways to reduce sampling bias for iteroparous species with overlapping generations.

Although some Atlantic salmon and steelhead (anadromous *O. mykiss*) spawn more than once, the incidence of repeat spawning by adults appears to be low in most populations. Therefore, the Pacific salmon model for defining N_e should hold reasonably well for these species as well (unless repeat spawning by male Atlantic salmon as parr and adults proves to be a significant factor).

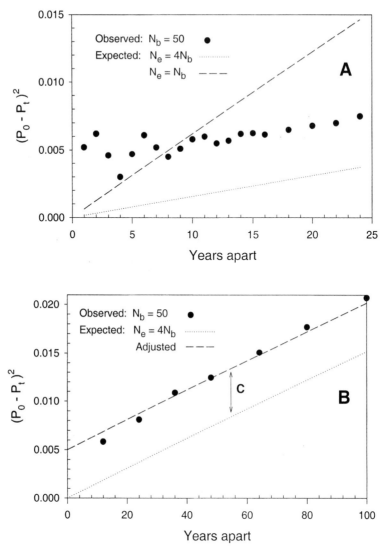

Figure 10.1. (A) Variance of allele frequency $(P_0 - P_t)^2$ observed (filled circles) in a simulated Pacific salmon population with $N_b = 50$ effective spawners per year and generation length of 4 years (50% of adults mature at age 4 and 25% each mature at ages 3 and 5). Dotted and broken lines show theoretical variance assuming $N_e = 4N_b$ and $N_e = N_b$, respectively. Initial allele frequency was $P_0 = 0.5$; P_t is the frequency at time t. (B). As in A, extending time period to 100 years. Rate of genetic change (slope of the line) in the modeled salmon population agrees with theoretical expectation for a population with $N_e = 4N_b = 200$, but the magnitude of variance measured over any given time period is higher than expected. Adjusting expected variance by adding $C = P_0(1 - P_0)/N_b = 0.25/50 = 0.005$ accounts for the fact that the spawners in a single year represent only a portion of a generation. Modified from Waples (1990a).

However, other trout and charr (e.g., coastal cutthroat trout, brown trout, and Dolly Varden) more closely conform to the iteroparous/overlapping generation model. Jorde and Ryman (1995, 1996) presented a method for estimating N_e in species such as this based on demographic and genetic data. This approach was later modified and applied by Laikre et al. (1998 and 2002) and Turner et al. (1999) for use with data for mtDNA and applied to several fish species.

2.2. Ryman–Laikre Effect

Salmon (both Atlantic and Pacific) are unusual in that they are the subject of serious conservation concern throughout much of their native range (Ford 2003—*this volume*), yet hatchery and harvest programs for these species are conducted on a truly gigantic scale (Mahnken et al. 1998). Both types of programs can have profound genetic effects on natural populations (Hindar et al. 1991b; Hard 2003—*this volume*). Ryman and Laikre (1991) examined the consequences for effective population size of hatchery supplementation, which they termed "supportive breeding." They showed that the effective size of a combined captive–wild population is a function of the effective sizes of the captive and wild breeding phases (N_{ec} and N_{ew}, respectively) and the proportion of the next generation that is produced by the captive breeders (X; $0 < X < 1$) and the wild breeders $(1 - X)$:

$$N_e = \frac{1}{\dfrac{X^2}{N_{ec}} + \dfrac{(1 - X)^2}{N_{ew}}} \qquad (5)$$

Ryman and Laikre's simple model demonstrated for the first time that the overall effective size of a combined system could be reduced by selectively enhancing only part of the gene pool (i.e., spawners taken from the wild into a hatchery). As an example, consider a hatchery–wild system with the following characteristics: 100 spawners return to a stream, and 20 of these are taken into a hatchery to spawn. N_e/N is 0.5 in both environments, so $N_{ec} = 0.5 \times 20 = 10$ and $N_{ew} = 0.5(100 - 20) = 40$. The 80 remaining wild fish produce 80 spawners the next generation, and the 20 hatchery spawners also produce 80 returning adults, so $X = 1 - X = 0.5$. N_e of the combined system is thus $1/[(0.25/10) + (0.25/40)] = 32$ compared to $N_e = 50$ if the 20 fish had remained in the wild instead of being taken into the hatchery. This population has grown in size as a result of supplementation but also has a higher overall level of inbreeding because half of the population can now be traced to just 20 parents.

As originally developed, Ryman and Laikre's model had the following features: it assumed discrete generations; it applied to the inbreeding effective size; it described only a single episode (generation) of enhancement; and it assumed the total population size would not increase as a result of supplementation (because the excess fish would be harvested). Subsequently, Ryman et al. (1995) extended the analysis to consider variance effective size and showed that, because the population size changes with supportive breeding, the effects on variance and inbreeding effective size are out of phase. Wang and Ryman (2001) and Duchesne and Bernatchez (2002) further extended the model to consider multiple generations of supportive breeding. The latter authors also showed that, not surprisingly, increases in inbreeding due to the Ryman–Laikre effect are reduced if the target population is actually part of a meta-population, in which case the genetic effects are diluted by gene flow with unaffected subpopulations. These results are applicable to a wide range of species with captive propagation programs.

Waples and Do (1994) adapted the Ryman–Laikre model to salmon life history and relaxed the other restrictive assumptions mentioned above. They used computer simulations to model the rate of change in overall levels of inbreeding and found that: (a) N_e of the composite (captive + wild) system depends more on the number of spawners used in the hatchery than the proportion of the population taken for broodstock; (b) age structure can profoundly affect the rate of change in inbreeding and reduction in N_e, particularly if only part of a generation is subject to enhancement; (c) if N_e/N is high in the captive phase relative to the wild phase (e.g., due to equalization of progeny number per family), this factor can at least partially offset increases in inbreeding due to the Ryman–Laikre effect; (d) successive generations of enhancement will lead to continued erosion of genetic diversity. Marking the progeny of hatchery spawners so that they can be avoided in subsequent broodstock collection can delay the rate of increase in inbreeding but not avoid it entirely; (e) Whether the population remains large after supplementation can be the biggest factor determining the long-term genetic effects of selective enhancement. If the excess fish are harvested randomly, or if the population increases during supplementation but crashes back to its former size after supplementation ends (e.g., because the root causes of decline have not been addressed), the net effect of selective enhancement will be a reduction in N_e compared to an unsupplemented population. However, if the population grows substantially as a result of supplementation and remains large after the program is terminated, the short-term increases in inbreeding due to enhancing only part of the gene pool can be quickly offset by a lower rate of genetic change in the larger population. Collectively, these results emphasize that, to minimize deleterious inbreeding effects associated with supportive breeding, programs should be appropriately scaled and implemented in conjunction with efforts to address factors that are limiting natural recovery.

Hansen et al. (2000a) used several analytical approaches based on molecular markers to evaluate effects of supportive breeding in three brown trout populations and found strong evidence for reduced effective population size. Hedrick et al. (2000a) presented data that allowed an empirical evaluation of the consequences for N_e of a supplementation program for Sacramento winter-run chinook salmon, which is listed as an endangered species under the U.S. Endangered Species Act. Broodstock collected from the wild were spawned in a hatchery, and their progeny released as juveniles in the hope that they would help boost population size when they returned as adults. The exact magnitude of the Ryman–Laikre effect could not be measured because an estimate of N_b was not available for the natural component of the population. However, based on estimates of the ratio of N_b to N_i (number of spawners in a single year) for other salmon populations, Hedrick et al. (2000a) concluded that the captive program probably did not result in a significant reduction in effective size and may even have increased it. This result can be attributed to two major factors: (1) Efforts were made to equalize reproductive success among families in the hatchery, and these efforts resulted in fairly high (about 0.6–0.8) estimates of N_b/N_i in the captive phase (hence a relatively high N_{ec} from Equation (5); and (2) the hatchery fish contributed relatively few adults to the population (X in Equation (5)).

As discussed below, the basic Ryman–Laikre approach can be extended to apply more generally to evolutionary issues relating to effective population size.

2.3. N_e and Fluctuating Salmon Populations

The results shown in Figure 10.1 were derived for a population of constant size. Based on a limited number of computer simulations, Waples (1990a) also concluded that if N and N_b fluctuate over time, N_e in a salmon population can be expressed as

$$N_e \approx g\tilde{N}_b \quad \text{(Harmonic mean method)} \tag{6}$$

where $\tilde{N}_b$ is the harmonic mean of the N_b values over a generation. This result is analogous to the well-known property that multigeneration N_e is approximately the harmonic mean of the single-generation N_e values (Wright 1938 but see Iizuka 2001). It turns out, however, that the result in Equation (6) depends on a rather restrictive assumption about population dynamics—specifically, that each spawning population contributes equally to the next generation regardless of the number of spawners. Waples (2002a) showed that an alternative assumption—that each spawning population contributes to the next generation exactly in proportion to the number of spawners—leads to the result that N_e is an additive function of N_b even when N_b varies:

$$N_e \approx \sum N_{bi} = g\bar{N}_b \quad \text{(Additive method)} \tag{7}$$

where $\bar{N}_b$ is the arithmetic mean N_b over the period of a generation. The sensitivity of these genetic results to assumptions about population dynamics demonstrates that a more generalized model relating N_e to N_b in fluctuating salmon populations must consider not only family size variation within years (which affects N_b) but also variation in mean reproductive success among families from different years. This latter source of variability can be quantified from cohort-specific spawner-recruit information, which is routinely collected for many salmon populations. In this context, recruits are spawners in the next generation produced by spawners in a particular year or cohort. Because of variable age at maturity in all salmon except pink salmon, recruits must be summed over two or more years to get the total contribution of a particular cohort to the next

generation. Spawner-recruit data for salmon indicate that adult–adult replacement rates in salmon can vary among years by several orders of magnitude (Cass
and Riddell 1999; Beamesderfer et al. 1998; Waples 2002a), and these population dynamic processes can greatly increase the variance in mean reproductive
success among families in different cohorts but within the same generation.

Waples (2002a) showed that the genetic consequences of this phenomenon
can be evaluated using a generalized form of Equation (5) (Ryman and Laikre
1991):

$$N_e = \frac{1}{\sum_i (X_i^2 / N_{bi})} \qquad (8)$$

In this notation, N_{bi} is the effective number of breeders in year i, X_i is the
proportional contribution of spawners in year i to the next generation, and the
summation is over all years in a generation. X_i can be computed from spawner-
recruit or run-reconstruction data as $X_i = \partial_i / \partial_T$, where ∂_i = recruits produced
by spawners in year i and $\partial_T = \sum \partial_i$ = total recruits produced by all the spawners in a generation. Application of this model to empirical data for chinook
salmon (Waples 2002a, unpublished) found that, in most cases, variation in
replacement rate across years caused N_e per generation to be even less than
predicted from the harmonic mean method (Equation (6)). Simulations using
survival rates that varied randomly across years led to similar results.
Implications of this result for the N_e/N ratio are discussed below.

2.4. N_b/N_i and N_e/N in Salmon

Pacific salmon share some of the life history features of the highly fecund species
that have been the subject of speculation about very low N_e/N ratios. A female
salmon typically will lay several thousand eggs (Groot and Margolis 1991; Einum
et al. 2003—*this volume*), so less than one in a thousand progeny needs to survive
for the population to remain stable. This high mortality allows for the possibility
of large differences in reproductive success among families and, consequently,
greatly reduced N_e/N. In this section I examine empirical data and theoretical
considerations that provide insight into this topic.

2.4.1 Demographic Estimates

No comprehensive demographic estimates of N_e/N exist for Pacific salmon, but
two approaches can shed some light on this issue. First, it is interesting to consider what N_b/N_i would be for species with salmon life history if survival over at
least some stage was as extreme as in Crow and Morton's Model 2 (family
correlated survival). To evaluate this, I considered a hypothetical salmon population with mean fecundity of 5000 eggs/female, with a coefficient of variation of
0.25 (realistic values for chinook salmon; J. Hard unpublished). I assumed that
males have the same variance in fertilization success, and that mean survival is
2% from both the egg-to-smolt and smolt-to-adult stages, resulting in an average

of $\bar{k} = 2$ progeny per parent that survive to adult. Although the 0.25 coefficient of variation in fecundity leads to an index of variability that is quite large at egg deposition ($R > 300$; Table 10.1), random survival throughout the rest of the life cycle would lead to N_b/N_i only slightly less than one. This indicates that this variation in fecundity would not, by itself, result in significant reduction in N_b or N_b/N_i. In contrast, completely correlated survival in either life stage would reduce the ratio to only about 0.02, and correlated survival throughout the life cycle would reduce N_b/N_i by several orders of magnitude. Note that if the life history includes episodes of both random and correlated survival, the reduction in N_b and N_b/N_i is not affected by the order of the different survival episodes. Table 10.1 also illustrates the close agreement predicted by Waples (2002b) between N_b/N_i adjusted to $\bar{k} = 2$ and the total survival rate through all stages of family-correlated survival (S_T). The results in this table demonstrate that it is theoretically possible for species with salmon life history to have N_b/N_i ratios as low as has been proposed for some marine species, but this would occur only under rather extreme conditions.

Second, a few published studies have examined the demographic parameters necessary to compute N_b and N_b/N_i for at least part of the salmon life cycle or part of a generation (Table 10.2). Each of these studies (Simon et al. 1986 with coho salmon; Geiger et al. 1997 with pink salmon; Hedrick et al. 1995, 2000b with chinook salmon) involved captively rearing fish from individual families and marking them before release as smolts. Adult family size was then monitored when the surviving fish returned to spawn. Waples (2002b) used the modified Crow and Morton approach to evaluate stage-specific effects on N_b and N_b/N_i in these experimental populations. Table 10.2 summarizes these results. The subscript "*J*" denotes juvenile (smolt) estimates and "*A*" denotes adult estimates. At the smolt stage, $\bar{k}_J$ was approximately 2000 for chinook and coho salmon, leading to $R_J \approx 100-600$. In pink salmon $\bar{k}_J$ and R_J were both somewhat smaller. At the adult stage, $\bar{k}_A$ was also greater than 2 in all but two cohorts (1984e pink salmon and 1995 chinook salmon). Equation (3) was used to scale juvenile and adult indices of variability to their expected values at $\bar{k} = 2$, and these scaled indices of variability (R_J^*) were used in Equation (2) to obtain a scaled estimate of N_b/N_i for each life stage. As an example, for the 1974 cohort of coho salmon, $R_J > 200$ for the smolt stage (Table 10.1), much higher than the value ($R_J = 1$) that would be expected if survival to smolt had been completely random with respect to family. However, random survival from that point on to $\bar{k} = 2$ at the adult stage would reduce the index of variability to only $R_J^* = 1.23$. This result shows that non-random survival that occurred prior to the smolt stage would lead to only a small reduction in N_b/N_i (from 1.0 to 0.9, or 10%). A similar pattern is seen in the remaining cohorts: at the smolt stage R_J was large but after adjusting to $\bar{k} = 2$, R_J^* was not substantially larger than 1. As a result, the scaled N_b/N_i estimates all were fairly high (0.8–0.95). This is not surprising and does not provide any real insight into what effects juvenile survival in the wild might have on N_b/N_i, as the hatchery represents a benign environment intended to maximize survival during the egg–smolt stage. Furthermore, varying degrees of effort were made to equalize family size at release in these experimental populations.

Table 10.1. Demographic parameters for a hypothetical salmon population.

Survival by stage	Egg			Smolt			Adult			N_b/N_i	S_T
	$\bar{k}$	V_k	R	$\bar{k}$	V_k	R	$\bar{k}$	V_k	R		
Ran/Ran	5000	1,562,500	312.5	100	723	7.23	2	2.24	1.12	0.094	1
Ran/Cor	5000	1,562,500	312.5	100	723	7.23	2	210.4	105.2	0.019	0.02
Cor/Ran	5000	1,562,500	312.5	100	521,250	5212.5	2	210.4	105.2	0.019	0.02
Cor/Cor	5000	1,562,500	312.5	100	521,250	5212.5	2	10,621	5310.5	4×10^{-4}	4×10^{-4}

At spawning, mean fecundity is 5000 eggs/female, with CV = 0.25. Subsequently, mean survival is 2% to smolt stage and 2% from smolt to adult, resulting in a constant population size. V_k is calculated for smolt and adult stages based on Equations (3) and (4), assuming either random (Ran) or correlated (Cor) survival by stage, respectively. $R = V_k/\bar{k}$ (the index of variability). N_i is the effective number of breeders in a cohort. N_b/N_i is calculated from Equation (2) using mean and variance of family size at the adult stage. S_T is the survival rate through all family correlated live stages.

Table 10.2. Variation in female family size in three studies using Pacific salmon marked and released from hatcheries as juveniles (smolts).

	Juvenile (smolt)			Scaled		Adult			Scaled		Percent N_b/N_i reduction in marine phase
	$\bar{k}_J$	V_{kJ}	R_J	R_J^*	N_b/N_i	$\bar{k}_A$	V_{kA}	R_A	R_A^*	N_b/N_i	
Coho salmon (Simon et al. 1986)[a]											
1967	?	?	?	—	0.90[b]	15.5	58.7	3.79	1.35	0.85	6
1968	2309	273,816	118.6	1.10	0.95	8.4	13.2	1.57	1.13	0.94	1
1971	1245	206,847	166.2	1.27	0.88	2.7	6.8	2.52	2.13	0.64	27
1974	1812	373,785	206.3	1.23	0.90	2.2	5.5	2.50	2.39	0.59	34
1977	1756	447,821	255.0	1.29	0.87	11.2	42.8	3.82	1.50	0.80	8
Pink salmon (Geiger et al. 1997)[a]											
1982e[c]	942	152,604	162.0	1.34	0.85	2.8	5.0	1.77	1.56	0.78	8
1983e	773	75,193	97.3	1.25	0.89	9.3	46.2	4.97	1.86	0.70	21
1983l	706	78,154	110.7	1.31	0.87	13.2	53.2	4.03	1.47	0.81	7
1984e	630	65,916	104.7	1.33	0.86	0.7	1.7	2.45	5.14	0.33	62
1984l	838	85,027	101.5	1.24	0.89	2.0	3.2	1.58	1.56	0.78	12
Chinook salmon (Hedrick et al. 2000b)											
1994	2709	340,601	125.7	1.09	0.96	5.8	7.2	1.23	1.08	0.96	0
1995	2387	1,386,797	580.9	1.49	0.80	1.1	2.1	2.04	2.98	0.50	38

[a] Unpublished smolt data were provided by J. McIntyre (coho) and H. Geiger (pink).

[b] Estimated as the mean for other four cohorts.

[c] e = early run; l = late run.

Scaled values of the index of variability ($R^* = V_k/\bar{k}$) and estimates of N_b/N_i were calculated using Equations 3 and 2, respectively, assuming random survival to a final adult population with $k = 2$. Percent reductions in N_b/N_i were computed as $[100(\lambda_J - \lambda_A)/\lambda_J]$, where λ_J is the scaled N_b/N_i at the juvenile (smolt) stage and λ_A is the scaled N_b/N_i at the adult stage. Modified from Waples (2002b).

Of more interest is the comparison of scaled estimates at the juvenile and adult stages, which permits an evaluation of the effects of marine survival on the N_b/N_i ratio. In most cohorts R_A was larger than R_J^*, indicating that smolt–adult survival was not entirely random, but again the departure was not extreme enough to substantially reduce effective population size. The scaled adult estimate of N_b/N_i was > 0.7 for most cohorts. Overall, the estimated percent reduction in N_b/N_i during the marine phase ranged from zero (no evidence for non-random smolt–adult survival in the 1994 cohort of chinook salmon) to 62% (1984e cohort of pink salmon). Means across species in percent reductions in N_b/N_i during the marine phase were correlated with the proportion of time spent in the marine environment (highest for pink salmon; lowest for coho salmon; Waples 2002b).

Several caveats must be kept in mind in interpreting these data. First, computing V_k for one sex only (females in this case) will tend to underestimate the overall variance in k because an uneven sex ratio will lead to a higher $\bar{k}$ in the less numerous sex. Furthermore, males might be expected to have a higher variance in reproductive success than females. Second, the N_b/N_i estimates were based on adult returns before spawning. Effective population size is also affected by differential spawning success, which may have a genetic (and hence family) component. Finally, correlated mortality of entire families due to environmental factors or chance events is much more likely to occur in the freshwater phase in nature, when family groups are more likely to coexist in close proximity. For these reasons, N_b/N_i ratios for a full salmon life cycle in nature would be expected to be lower than those shown in Table 10.2. Nevertheless, it is interesting that the high mortality in the marine phase (>99% in most or all years in each species) does not appear to have depressed the N_b/N_i ratio below the range typically found in other species.

Although these data do not prove that salmon have N_b/N_i ratios in the same range as most other organisms, they do suggest that if the ratios are substantially lower, the explanation must be sought in either (1) effects in the freshwater stage of the life cycle, or (2) variance in mean family size across years within a generation. Of course, these data also cannot rule out the possibility that high family size variance occurs in the marine environment for at least some populations in some years.

As noted above, it is easy to see how essentially random events might lead to strong family-correlated survival in the freshwater phase because family groups tend to co-occur in space and time. It is not so easy to see how this might occur in the marine phase of the life cycle. Furthermore, the half-sib experimental design in the pink salmon study allowed Geiger et al. (1997) to conclude that the higher-than-random variance in family size in their study cannot be wholly attributed to environmentally correlated survival—it also has a genetic component (see also Unwin et al. 2003). If so, however, one might reasonably ask why substantial heritability for marine survival should exist, since natural selection presumably would rapidly push the population toward the optimal phenotype. Geiger et al. (1997) considered this conundrum and concluded that their results could be explained if selective pressures in the marine environment fluctuate with changing environmental conditions, so that different genotypes are favored in different years. Hard et al. (2000) also demonstrated a genetic component for resistance to toxic effects of a marine algal bloom in chinook salmon—a trait that could be expected to be important only in certain years. If this hypothesis is true it emphasizes the importance of maintaining a large enough N_e in salmon to support a rich store of adaptive variation.

2.4.2. Genetic Estimates of N_b

Genetic estimates of effective population size have the advantage that they encompass all demographic features throughout the entire life cycle (and even over multiple generations); however, they may have limited precision and can be subject to various sorts of biases (reviewed by Waples 1991b; Schwarz et al. 1998). This approach has been used in only a few published studies of salmon. Bartley et al. (1992c) used gametic disequilibrium analysis (Hill 1981) to obtain an N_b/N_i estimate of only 0.05 in a sample of Sacramento winter-run chinook salmon. However, this estimate is difficult to interpret because the sample of juveniles could have included individuals from three other populations that also spawn in the river, and genetic admixture would create linkage disequilibrium that would downwardly bias the estimate of effective population size (Waples and Smouse 1990). Waples (1990b) modified the temporal method (Nei and Tajima 1981; Waples 1989) for estimating effective population size to accommodate salmon life history, and used this method as well as gametic disequilibrium analysis to estimate N_b and N_b/N_i in 15 Snake River chinook salmon populations sampled over a 3–6 year period (preliminary estimates based on unpublished data reported by Waples 2002c). The median N_b/N_i estimate was 0.25, and two-thirds of the estimates fell in the range 0.1–0.4.

2.4.3. Estimating N_b/N_i Directly

Anderson (2001) developed an analytical method for estimating the effective: census size ratio directly, rather than estimating each component separately. This method uses both demographic and genetic data in a Bayesian framework. Initial results are promising, although the algorithm is computationally intensive and adopted a number of simplifying assumptions.

All of the above studies provided estimates of N_b or N_b/N_i, not N_e or N_e/N. Genetic estimates of N_e/N are discussed below.

2.4.4. Variable N_b, Variable Replacement Rate, and N_e/N in Salmon

Equations (5) and (6) provide complementary approaches to estimating N_e in salmon based on yearly N_b values; they have important implications for the N_e/N ratio in salmon because they account for demographic processes that occur across as well as within years. If year-specific replacement rate data are available, Equation (5) can be used to estimate N_e and hence N_e/N. Empirical results for chinook salmon (Waples 2002a; R. Waples and M. McClure unpublished) show that temporal variation in replacement rate typically reduces N_e/N in this species by approximately 25–50%, and this reduction is in addition to factors discussed above that reduce N_b/N within years.

If adequate demographic data are not available but it is possible to estimate N_b (e.g., using genetic data), then Equation (6) may provide a reasonable approximation for N_e in salmon. Inspection of Equation (6) makes it apparent that, unless N_b is constant, N_e/N per generation will be less than the mean N_b/N_i per year. This is because N_e in salmon is related to the harmonic mean N_b, while N per generation is simply the sum of the spawners in the individual years (and hence is a function of the arithmetic mean N_i). Figure 10.2 illustrates this effect, using the estimates of N_b/N_i for Snake River chinook salmon described above. I used Equation (6) to estimate N_e and N_e/N in these populations and compared this with the annual estimates of N_b/N_i. The results demonstrate that estimates of annual N_b/N_i provide an upwardly biased picture of generational N_e/N in salmon. In every population, the estimate of N_e/N was $\leq 80\%$ of the mean of the N_b/N_i estimates, and several were below 50%. This agrees with the demographic data discussed above in which Waples (2002a) showed that inter-annual variation in productivity could depress N_e/N by an average of about

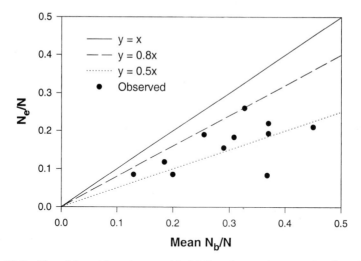

Figure 10.2. Plot of the arithmetic mean N_b/N_i (based on preliminary data from Waples et al. 1993; Waples 2002c; R. Waples unpublished) and estimated N_e/N for 11 Snake River chinook salmon populations sampled in 3–6 years. In each population each year, N_b was estimated from allozyme data using both temporal and disequilibrium methods and N_i was estimated from expansion of counts of spawners or redds (nests). N_e/N was calculated as follows using all available data for each population: $N_e = g\hat{N}_b$ (Equation (6); $g = 4.5$ years); N = total number of spawners in a generation $= \sum N_i = g\overline{N}$. Two populations with a yearly N_b estimate of ∞ were omitted from the analysis. Each filled circle represents data for one population; the solid, broken, and dotted lines indicate N_e/N = 100%, 80%, and 50% of mean N_b/N_i, respectively.

25–50% in chinook salmon, in addition to factors that reduce N_b/N_i within years. Combining the empirical genetic and demographic information (majority of N_b/N_i estimates in the range 0.1–0.4; N_e/N typically reduced by an additional 25–50%), it seems reasonable to conclude, at least tentatively, that N_e/N for a generation of chinook salmon can be expected to fall in the range of approximately 0.05 to 0.3 (0.5×0.1 to 0.75×0.4).

As a cautionary note, it should be pointed out that the demographic estimates obtained by Waples (2002a) could be biased downwards (by an unknown amount) due to measurement error in counting and aging spawners. However, the N_e/N range proposed above is similar to that reported by Heath et al. (2002b), who used the temporal method to analyze steelhead (anadromous O. *mykiss*) populations over a period of 20–40 years and obtained N_e/N estimates ranging from 0.06 to 0.29.

3. Summary

Variable age structure complicates the interpretation and estimation of effective population size in salmon, but semelparity compensates for this complication in two ways. First, because an individual's contribution to a cohort is also its lifetime reproductive output, it is much easier to evaluate variance in lifetime family size in salmon than in iteroparous organisms. Second, semelparity sharpens the focus on the link between demographic and genetic processes in the population. For example, application of the Ryman and Laikre (1991) model provides a means of partitioning the effects on N_e/N of within- and between-cohort processes. Empirical data and modeling suggest that both factors can be important in reducing N_e/N in salmon.

Although salmon are renowned for the dramatic features of their life history (esp. long migrations and a strong homing tendency), they are not unique in being semelparous with variable age at maturity. This combination of life history traits can be found in a wide variety of organisms, from freshwater crustaceans with diapausing eggs to annual plants with seed banks (Hairston and Caceres 1996; Leck et al. 1989). Thus, insights regarding N_e and N_e/N gained from salmon should be more broadly applicable (e.g., see Nunney 2002).

Genetic estimates suggest that yearly N_b/N_i ratios in natural salmon populations typically are less than (and perhaps only about half of) Nunney's theoretical prediction that $N_e/N \geq 0.5$. This presumably is because semelparity does not provide the same buffer against reproductive failure in one breeding season that is found in organisms with overlapping generations. In addition, variation in mean reproductive success across cohorts within a generation further reduces single-generation N_e/N in salmon, perhaps by another 25–50%. Empirical estimates for chinook salmon suggest that the generational N_e/N ratio may typically fall in the range of approximately 0.05–0.3. Although these values are still well above the tiny N_e/N ratios that have been proposed for some marine species, they suggest that the implications of relatively low N_b/N_i and N_e/N ratios must be carefully considered in evolutionary investigations and conservation applications for Pacific salmon.

The next decade promises to be an exciting one for empirical research into effective population size of salmon as well as other organisms. Increasing availability of a large number of highly polymorphic molecular markers, together with improved analytical and computational methods, makes it feasible to use genetic methods to obtain a detailed pedigree of natural populations. As a result, it is now possible to collect the detailed demographic information necessary to directly estimate N_e and N_e/N with high precision. Some key research questions include analysis of reproductive success of alternative life history strategies (e.g., precocious maturation); changes in V_k over time and life history stage; differential survival of families based on life history or other traits; and relative reproductive success of captively reared fish in the natural environment. The first wave of publications reporting research of this type in salmonids has recently appeared (e.g., Bentzen et al. 2001; Berejikian et al. 2001; Garant et al. 2001; Garcia-Vazquez et al. 2001; Jones and Hutchings 2001, 2002; Letcher and King 2001; Taggart et al. 2001), and as this book goes to press many more studies are in progress or being planned.

Acknowledgments   Jack McIntyre and Hal Geiger kindly provided unpublished data on family size at smolting for use in Table 10.2. This chapter benefited from comments by Mike Ford, Jeff Hard, Andrew Hendry, and two anonymous reviewers. Some of the data used in this chapter were developed as part of contract DE-A179-89BP0091 funded by the Bonneville Power Administration.

Evolution of Chinook Salmon Life History under Size-Selective Harvest

Jeffrey J. Hard

Chinook salmon returning to spawn

11

The consequences of selection for phenotypic diversity have important implications for conserving life history variation, especially in exploited species like Pacific salmon. Life histories reflect evolutionary processes that affect variation in fitness within populations and subsequently result in local adaptations that can lead to divergence among populations (Roff 1992, 2002; Stearns 1992). Williams' (1966a) description of fitness as "effective design for reproductive survival" may be construed as a reference to the consequences of variation in life history. The evolutionary processes that mold life history involve many traits, some of which are likely to interact in their expression. A comprehensive analysis of life history evolution therefore requires recognition that evolution involves multiple traits and that multiple traits contribute to adaptation. Genetic correlations among life history traits, due to pleiotropic gene effects or to linkage, may constrain the genetic repertoire available to populations for adaptive evolution, at least in the short term. If so, these correlations would restrict the ability of selection to simultaneously optimize all traits with respect to fitness. Dickerson (1955) referred to such constraints as "genetic slippage," a phenomenon Dingle (1984) called the "cost of correlation." On the other hand, genetic correlations that reinforce selection on a particular trait may accelerate evolutionary responses. Any evolutionary constraints imposed by correlations among traits may be short-lived and thereby not preclude the independent evolution of correlated traits (Via and Lande 1985; Zeng 1988; Hard et al. 1993). However, to what extent evolutionary constraints imposed by genetic correlations among traits limit the evolutionary pathways that populations have available to them has received little attention empirically (but see Etterson and Shaw 2001).

In this chapter, I evaluate some potential effects of size-selective harvest of chinook salmon on life history evolution in this species to determine how correlations among traits might affect evolutionary responses to harvest. I apply available genetic data to models that predict short-term responses of correlated traits to selection, and show how directional and disruptive selection on a single trait affect life history patterns when stabilizing selection on size is sufficiently strong. These analyses rely on a quantitative genetic approach to emphasize the importance of considering relationships among traits in evaluating response to selection on life history.

1. Evolution of Life History

1.1. Quantitative Genetic Inference

Most life history traits have some key features in common. First, in general the expression of these traits is influenced by both environmental and genetic factors. A large body of quantitative genetic research (Falconer and Mackay 1996; Roff 1997; Lynch and Walsh 1998) has demonstrated that a trait's genetic component—or, more precisely, the magnitude of its variance relative to that of the composite environmental component—determines the response of that trait to selection. Because quantitative genetics does not focus on individual genes but rather on composite genetic and environmental components of phenotypic expression, the power of this approach to resolve the details of genetic architecture is limited. Nevertheless, quantitative genetics is useful in investigating life history evolution because it can resolve how heredity and environment interact broadly to produce patterns of adaptation, something that analysis of variation at neutral genetic markers cannot do (Hard 1995; Lynch 1996; Storfer 1996; Crandall et al. 2000). Quantitative genetic approaches are therefore more synoptic (and clearly less mechanistic) than molecular genetic approaches. Moreover, quantitative genetic approaches are prospective rather than retrospective (Ewens 1979): inference in quantitative genetics focuses on the potential evolutionary consequences of particular genetic states rather than on past evolutionary processes that have produced them. Thus, quantitative genetics provides a useful framework for characterizing the adaptive consequences of genetic change, at least over the short term.

Second, these traits typically do not fall into discrete phenotypic classes but instead exhibit continuous phenotypic distributions (threshold traits such as survival, disease resistance or migratory tendency are thought to have an underlying distribution of genetic effects that is approximately continuous; Falconer and Mackay 1996). Even for discrete traits, however, analytical methods are available to dissect genetic and environmental components of variation (Crittenden 1961; Falconer 1965; Reich et al. 1972; Lynch and Walsh 1998), making them amenable to quantitative genetic inference.

Third, these traits appear to be under the control of several—perhaps very many—genes of varying effect on the trait (Wright 1968; Lande 1981). Quantitative genetic analyses use the statistical properties of these traits to describe the collective behavior of the underlying genes. From the phenotypic resemblance of individuals of known relationship, these analyses permit inferences about the inheritance and evolution of the traits, processes that are controlled by hidden variation in gene frequencies and their effects (Falconer and Mackay 1996; Lynch and Walsh 1998). In this way, observed patterns of means, variances, and covariances in a population, when combined with appropriate breeding designs and statistical techniques, can be used to estimate the genetic parameters that determine response to selection.

However, although ascertaining whether a trait simply has a genetic basis is relatively straightforward (Lawrence 1984; Crow 1986), estimation and interpretation of genetic parameters can be difficult and may require several important assumptions. Most parameter estimates are specific to the population and environment in which they are measured. The estimates are sensitive to the influence of migration, mutation, selection, non-random mating, level of inbreeding, genotype by environment correlation or interaction, and ecological factors (Barker and Thomas 1987). In addition, quantitative genetic estimates assume that the underlying genes are unlinked. Despite these limitations, quantitative genetic methods have proven to be effective in measuring the inheritance of quantitative traits and their responses to evolutionary forces (Barton and Turelli 1989; Roff 1997; Lynch and Walsh 1998). Careful studies of natural populations have amply demonstrated the power of quantitative genetic approaches in characterizing genetic variation and evolutionary response in the wild (e.g., Gustafsson 1986; Grant and Grant 1995; Kruuk et al. 2000).

1.2. Selection on Correlated Traits

Selection requires phenotypic variation, a concept recognized since Darwin proposed natural selection as the primary mechanism for observed evolution, a process he called "descent with modification." Estimating selection on traits and their responses is important for two main reasons. First, it allows at least some consequences of variation to be identified, illuminating potential evolutionary constraints and opportunities in a population. This approach can help to uncover evolutionary ramifications of phenotypic variation without explicit knowledge of its underlying genetic mechanics. Second, if the phenotypic response to selection can be measured, it reflects the inheritance of phenotypic variation and therefore permits prediction of prominent aspects of a population's actual evolutionary pathway. The distinction between variation resulting from selection and the evolutionary response to selection, pointed out by Fisher (1930) and Haldane (1954), is a simple but fundamental issue, and is now firmly entrenched in both plant and animal breeding (Mayo 1980; Falconer and Mackay 1996) and evolutionary biology (Arnold and Wade 1984a). Analysis of phenotypic variation does not require knowledge of a trait's inheritance, but analysis of its evolutionary response to selection does.

1.2.1. Analytical Approach

The theoretical foundation for measuring selection was pioneered by Pearson (1903) and extended by Crow (1958), Van Valen (1965), and O'Donald (1970). In the intervening decades, empirical work has supported a growing recognition that selection in nature seldom influences only one trait. Current approaches are based on an assumption of polygenic control of more or less continuously distributed phenotypic traits. Despite the fact that variation in some single-gene traits can have fitness consequences (e.g., Frankel 1983; Mitton and Grant 1984; Zouros and Foltz 1987), the correlation between allelic diversity and fitness or performance is generally weak (Nei 1987b; Lewontin 1991). Milkman (1982) and Lynch (1984) argued that when a continuous phenotypic trait is exposed to weak stabilizing selection, its constituent allele frequencies are generally affected more by random genetic drift than by selection. The clear implication is that allelic variation and fitness are poor correlates of one another (Hedrick and Miller 1992; Wang et al. 2002), so that variation in single-locus traits is often unsuitable for detecting selection.

Important contributions that extended this work to multiple traits have been made by Lande (1976, 1979), Lande and Arnold (1983), Arnold and Wade (1984a,b), and Schluter (1988), with reviews of these approaches by Manly (1985), Endler (1986), and Lynch and Walsh (1998). In two seminal papers, Lande (1979) and Lande and Arnold (1983) provided a framework for analyzing selection on correlated traits. This framework has proven to be a powerful one because it provides a means of estimating the actual selection gradients, that is, the phenotypic changes occurring within a generation in response to selection that account for indirect selection due to correlations among traits. The method does not identify or require knowledge of the source of selection nor does it address modes of trait inheritance. Empirical measurement of selection and its response in natural populations were reviewed recently by Hoekstra et al. (2001) and Kingsolver et al. (2001). These reviews were instrumental in specifying some key selection parameters for the models applied in this chapter.

The models are based on the breeders' equation describing the response to selection for a single trait, $R = h^2 S$, where R is the single-generation response to selection, h^2 is the trait heritability, and S is the selection differential (the difference between the phenotypic mean before vs. after selection; Falconer and Mackay 1996). The multivariate, discrete-generation form of this equation is (Magee 1965; Yamada 1977; Lande 1979):

$$\Delta z = GP^{-1}s \tag{1}$$

where Δz is a vector of changes in phenotypic means, composed of elements Δz_i (the change in mean of the ith trait); G is the additive genetic covariance matrix, composed of elements G_{ij} (the covariances of breeding values of the ith and jth traits within an individual); P is the phenotypic covariance matrix, composed of elements P_{ij} (the covariances of phenotypes of the ith and jth traits), with inverse P^{-1}; and s is the vector of selection differentials (Lande 1979). The matrix product of P^{-1} and s is the vector β, the multivariate selection gradient. Under the assumption that the traits are multivariate normal, this vector describes directional selection acting on the set of traits identified in P (Lande and Arnold 1983). Because the selection differential on a trait is equal to the covariance between the trait's value and individual relative fitness (the Robertson–Price identity; Lynch and Walsh 1998), each element in β is equivalent to the gradient of the relative fitness surface for each trait, scaled by the trait's variance and its covariance with the other traits in P. The vector of directional selection gradients can also be expressed as

$$\beta = P^{-1}\sigma[w, z_i] \tag{2}$$

because the vector of selection differentials, s, is equivalent to the vector of the covariances of the individual traits with individual relative fitness ($\sigma[w, z_i]$ for directional selection).

In the following section, I describe a genetic analysis of life history variation in salmon that illustrates the importance of considering both genetic and phenotypic trait correlations in evaluating selection response. To my knowledge this is the first application of empirical genetic data from salmon to a multivariate selection model (see also Hard In press). This example demonstrates that response of life history variation in salmon to selection imposed by harvest can provide some insights into the short-term consequences of selection in exploited populations.

2. Size-Selective Harvest of Chinook Salmon as an Evolutionary Model

The extensive marine migrations of anadromous salmon expose many populations to several fisheries, most of which are selective for size (e.g., Healey 1986; Hamon et al. 2000) and often migration timing. Current salmon harvest management is based on stock-recruitment theory (Ricker 1954) and focuses almost entirely on the demographic consequences of fishing. Managers strive to achieve a constant escapement (total spawners, adult females, or an estimate of potential egg deposition) that is thought to generate the maximum sustainable yield (MSY) to fisheries (Ricker 1958; Larkin 1978; Hilborn 1985). Except for designating particular size restrictions and time and area closures to limit catch of the largest fish and increase potential aggregate escapements, harvest management does not take into account selective effects in regulating fisheries (Ludwig et al. 1993; Law 2000; Essington 2001; Ratner and Lande 2001).

In recent decades, concerns have arisen over the erosion of genetic variability within populations owing to human exploitation (Larkin 1981; Walters 1983; Nelson and Soulé 1987; Smith et al. 1991; Altukhov 1994; Laikre and Ryman 1996; Levin and Schiewe 2001). Furthermore, evidence is accumulating that harvest can alter patterns of adaptive phenotypic variation (especially size and age) that affect viability for both Atlantic salmon (Bielak and Power 1986; Jensen et al. 1999) and Pacific salmon (Ricker 1981, 1995; Hankin and Healey 1986). The possible effects of harvest on life histories of fish, including salmonids, are widely acknowledged (Ricker 1981, 1995; Hankin and Healey 1986; Healey 1986; Riddell 1986; Law and Grey 1989; McAllister et al. 1992; Rijnsdorp 1993; Stokes et al. 1993; Altukhov 1994; Trippel 1995). Simulation studies have demonstrated that size-selective harvest is likely to alter size and growth, age at maturity, and fecundity of exploited populations, with potentially serious consequences for population productivity and yield (Kaitala and Getz 1995; Heino 1998; Martínez-Garmendia 1998; Ratner and Lande 2001).

Healey (1986) described the likely effects of size-selective fisheries on variation in size, growth, and age of salmon populations. He argued that these effects will often be difficult to quantify due to poor quality of catch data, the influence of large-scale environmental variability on salmon size and age, the implications of age structure for selection on size, phenotypic and genetic relationships among traits, and natural selection on body size through reproductive success in the wild. Hankin and Healey (1986) showed that the effects of exploitation on age-structured populations like chinook salmon depend heavily on the population's inherent productivity. They argued that populations that mature at older ages are likely to respond to selection for earlier age at reproduction and smaller size at maturity, with a greater potential for collapse from overfishing.

Although several examples of changes in salmon life history under fishery exploitation exist in the published literature, a direct causal relationship between
harvest selection and life history variation remains difficult to establish because
of the high sensitivity of this variation to environmental factors (Riddell 1986).

2.1. Evolutionary Responses to Harvest Selection on Size

Most empirical studies have not been able to evaluate genetic and phenotypic
consequences of selective harvest on salmon populations, because few relevant
data are available regarding harvest selectivity and genetic underpinnings of
variability in size and age (Law 2000). In 1994, I initiated a breeding study to
determine the consequences of inbreeding in a population of chinook salmon
(Hard et al. 1999 outline the breeding design): This study permitted estimation
of genetic and phenotypic variance components of size, age, and other life history traits that may respond to harvest selection. In this chapter, I apply these
empirically derived estimates to multivariate models of selection response to
ascertain how size-selective harvests alone might affect size and other traits.
An underlying objective was to determine whether size-selective harvests of
various kinds could produce trends in size similar to those that have been
observed for some chinook salmon populations (Ricker 1981, 1995; Bigler et
al. 1996).

I apply estimates of genetic and phenotypic variance for several salmon traits
to two forms of a multivariate selection model; Figure 11.1 depicts prominent
features of these forms. The ocean-type chinook salmon population originated
from Grovers Creek Hatchery (Puget Sound, Washington, USA), where adults
for this study returned to spawn between 1996 and 1999. I first derived phenotypic and genetic covariance matrices for length expressed at different ages to an
age-structured model to predict direct response to selection on length in fisheries. I then derived analogous matrices for several correlated traits expressed in
adults to a discrete-generation model to estimate correlated responses to selection on length. The correlated traits were maturation age, adult body length,
adult body weight, growth rate in length from entry into sea water until adult
return, and spawn timing. The fish evaluated here were progeny of adults mated
in 1994 in a hierarchical design (North Carolina Design 1, with 30 males each
mated to an average of 3.3 females); I identified the progeny to individual full-sib
families with coded-wire tags implanted before release (257,093 tagged fish in
95 full-sib families nested within 30 half-sib families) in May 1995 (Hard et al.
1999). Fish in this population mature between ages 1 and 5, but predominantly
at ages 3 and 4. After transforming data where necessary to reduce their departures from normality, I estimated the genetic and phenotypic variance components underlying these traits with restricted maximum likelihood (REML)
methods. Maximum likelihood methods are useful for estimating variance components when family structure is unbalanced and the resulting covariance
matrices are sparse (Shaw 1987). When I could not estimate genetic covariances
directly from the covariance matrices (e.g., for traits expressed at specific ages), I
computed the genetic covariance between traits x and y as $h_x h_y \sigma_x \sigma_y$, where h is

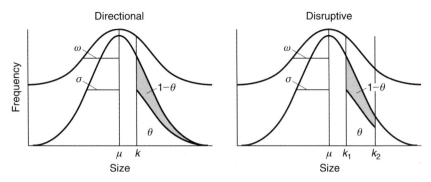

Figure 11.1. Size-selective harvest under a threshold can be simulated with five parameters for directional and six for disruptive selection. This simple case depicts selection on size for a single age class (or for a species with no age structure). The left panel depicts directional selection mediated by harvest of fish above a fixed threshold, k. The panel on the right depicts disruptive selection mediated by harvest of fish between two thresholds, k_1 and k_2. The mean size of the population before selection is μ. In each case, the proportion of fish larger than k that survive harvest is θ; the proportion harvested is $1 - \theta$. Size is assumed to be normally distributed with standard deviation σ. ω defines the width of the fitness function corresponding to size. In the models explored here, ω varies from 1 to 4 phenotypic standard deviations (σ) to encompass stabilizing selection intensities that range from very strong (1σ) to relatively weak (4σ). For an age-structured species, a separate fitness function can be derived for each age-specific size distribution. See text for discussion.

the square root of the heritability estimated by REML and σ is the phenotypic standard deviation. Where an REML estimate of heritability could not be computed for each age-specific trait, I set h^2 equal to 0.2 (a value often observed for life history traits; Falconer and Mackay 1996; Mousseau and Roff 1987; Roff 1997) and the phenotypic correlation r_p between traits to 0.01 (therefore assuming independent traits with the provision that $r_p \neq 0$).

The variance component estimates first parameterized an age-structured model determining direct responses to selection in length. This model integrates non-equilibrium age-structured demography and multivariate response to selection in predicting the responses of m lengths in a population of maximum age n over t years. The model, a derivative of Equation (1), determines responses in age-specific length according to the formula (Law 1991)

$$X_t + 1 = (X_t + PP_0^{-1}S_t)T_tC + A_{t+1}C' \tag{3}$$

where X_t and X_{t+1} are matrices of order m by $n + 1$ of phenotypes corresponding to years t and $t + 1$, P is the phenotypic covariance matrix for the lengths expressed at different ages, P_0 is a modification of P that has non-zero elements on the diagonal corresponding to traits expressed at the same age (P_0^{-1} its inverse), S_t is the matrix of selection differentials, T_t is a transition matrix that increases the age of present cohorts, C is a diagonal matrix of order $n + 1$ by $n +$

1 with elements equal to unity except for the first which equals zero, $\mathbf{C}'$ is a diagonal matrix of equivalent size but with the first element equal to unity and zeros elsewhere, and $\mathbf{A}_{t+1}$ is the matrix of breeding values (see also Charlesworth 1994). For a given trait, an individual's breeding value is a measure of the additive genetic component of the phenotype and reflects its average effect on the population mean as measured by the mean phenotype of its progeny (Falconer and MacKay 1996). Computing the breeding values each year requires estimating the intensity of selection on size expressed at each age and combining this with estimates of age-specific reproduction and genetic variance in length. Law (1991) describes the method in considerable detail and provides some simple numerical examples of predicted response to selection. For each cohort, I assumed the starting number each year to be 5,000 and estimated age-specific annual survival rates from the values in Ricker (1976) and Bradford (1995); I derived reproductive rates from age structure and sex ratio data for the Grovers Creek population. After Law (1991), I computed $\mathbf{A}_{t+1}$ as

$$\mathbf{A}_{t+1} = (\mathbf{A}_t + \mathbf{GP}_0^{-1}\mathbf{S}_t)\mathbf{T}_t \tag{4}$$

where $\mathbf{G}$ is the genetic covariance matrix (see Table 11.1).

The parameters that describe the response to size-selective harvest under a size (length) threshold are illustrated in Figure 11.1. The left panel depicts directional selection mediated by harvest of fish above a fixed threshold, k. The panel on the right depicts disruptive selection mediated by harvest of fish between two thresholds, k_1 and k_2. The mean size in the population is μ. In each case, the proportion of fish larger than k that survive harvest is θ; the proportion removed by harvest is $1 - \theta$. Size is assumed to be normally distributed with

Table 11.1. Phenotypic (above diagonal) and genetic (below diagonal) covariance matrices estimated by restricted maximum likelihood for the traits used to evaluate correlated responses to selection (J. Hard, unpublished).

	Adult age (yr^3)	Fork length (mm)	Adult weight $(\sqrt{g})$	Spawn date (ln (Julian d))	Growth rate $(mm\ [\ln(d)]^{-1})$
Adult age (yr^3)	719.7 (251.4)	3278.9	375.4	0.069	549.7
Fork length (mm)	1705.7	18,650.7 (6258.3)	2420.3	0.524	2510.5
Adult weight $(\sqrt{g})$	143.1	688.0	108,267.0 (1082.7)	0.039	353.2
Spawn date (ln (Julian d))	0.046	0.524	0.041	0.008 (0.002)	0.027
Growth rate $(mm\ [\ln(d)]^{-1})$	575.0	874.8	88.4	0.009	334.1 (102.7)

Phenotypic variances are on the diagonal (genetic variances in parentheses on the diagonal). The traits were transformed as indicated to reduce deviation from normality.

standard deviation σ. The parameter ω defines the width of the fitness function corresponding to size, expressed as the deviation of the inflection point from the optimum in the current generation before harvest.

The mean phenotype for length i in individuals of age j after selection in year t is used to estimate the selection differentials for length according to

$$x_{ijt}^* = \overline{W}_{jt}^{-1} \int \cdots \int x_{ijt} W(\mathbf{x}_{jt}) f(\mathbf{x}_{jt}) d\mathbf{x}_{jt} \tag{5}$$

In this case $\overline{W}_{jt}$ is the mean fitness of an individual of age j in year t, x_{ijt} is the length of that individual for trait i, $f(\mathbf{x}_{jt})$ is the joint probability distribution in year t for all traits expressed at age j and $W(\mathbf{x}_{jt})$ is the corresponding survival probability; integration is over all lengths expressed at age j (Law 1991). Mean fitness for trait j in year t is

$$\overline{W}_{jt} = \int \cdots \int W(\mathbf{x}_{jt}) f(\mathbf{x}_{jt}) d\mathbf{x}_{jt} \tag{6}$$

In each year, the mean fitness for trait j in year t under directional selection with a minimum length threshold can be computed from (Law 1991)

$$\hat{W}_j = \int_{-\infty}^{k_j} \frac{1}{\sqrt{2\pi(\sigma^2 + \omega^2)}} \exp\left(-\frac{y_j^2}{2(1+\omega^2)}\right) dy_j + \theta \int_{k_j}^{\infty} \frac{1}{\sqrt{2\pi(\sigma^2 + \omega^2)}} \tag{7}$$
$$\exp\left(-\frac{y_j^2}{2(1+\omega^2)}\right) dy_j$$

where y_j is the length scaled as a deviation from the mean value in standard deviation units (σ), k_j is the threshold length of fish exposed to selection (harvest), ω is the width of the fitness function in units of σ, and θ is the probability of fish escaping harvest (Figure 11.1). The selection differential on length j is

$$\hat{W}_j^{-1} \left\{ -\frac{1-\theta}{\sqrt{2\pi(\sigma^2 + \omega^2)}} \exp\left(-\frac{k_j^2}{2(1+\omega^2)}\right) \right\} \sigma_j \tag{8}$$

where σ_j is the phenotypic standard deviation (Law 1991).

For disruptive selection as might be imposed by harvest with a "slot limit" (Figure 11.1), I computed the mean fitness for trait j in year t with a length threshold from

$$\hat{W}_j = \int_{-\infty}^{k_{1j}} \frac{1}{\sqrt{2\pi(\sigma^2 + \omega^2)}} \exp\left(-\frac{y_j^2}{2(1+\omega^2)}\right) dy_j + \theta \int_{k_{1j}}^{k_{2j}} \frac{1}{\sqrt{2\pi(\sigma^2 + \omega^2)}}$$

$$\exp\left(-\frac{y_j^2}{2(1+\omega^2)}\right) dy_j + \int_{k_{2j}}^{\infty} \frac{1}{\sqrt{2\pi(\sigma^2 + \omega^2)}} \exp\left(-\frac{y_j^2}{2(1+\omega^2)}\right) dy_j \qquad (9)$$

where k_1 and k_2 are the lower and upper length thresholds, and with the other parameters equivalent to those in Equation (7). I computed the corresponding selection differential on length j with the equation

$$\hat{W}_j^{-1} \left\{ -\frac{1-\theta}{\sqrt{2\pi(\sigma^2 + \omega^2)}} \exp\left(-\frac{k_{1j}^2}{2(1+\omega^2)}\right) + \frac{1-\theta}{\sqrt{2\pi(\sigma^2 + \omega^2)}} \exp\left(-\frac{k_{2j}^2}{2(1+\omega^2)}\right) \right\} \sigma_j$$

$$(10)$$

The first part of the equation reflects the selection on length due to harvest of fish larger than k_1, and the second part reflects the selection due to harvest of fish smaller than k_2; the combination of these parts simulates size-selective fishing operating between the two thresholds. In the simulations I used 450 and 770 mm or 711 and 813 mm for k_1 and k_2, respectively, to represent realistic lower harvest thresholds with fish allowed to escape harvest if sufficiently large (at least 1 σ longer than k_1).

In the models I consider here, the width of the fitness function ω varies from 1 to 4 phenotypic standard deviations (σ) to encompass stabilizing selection intensities that range from very strong (1 σ) to weak (4 σ) but represent possibilities that extend from those supported by the empirical evidence (3–4 σ) to those that *might* be possible in some intense fisheries (1–2 σ). In recent reviews of empirical studies, Hoekstra et al. (2001) and Kingsolver et al. (2001) concluded that evidence from natural populations of a variety of taxa indicates that natural selection (directional and stabilizing) is seldom likely to be very strong; the strength of stabilizing/disruptive selection (measured as the quadratic selection gradient, γ) is typically weak, with a mean near 0, suggesting that stabilizing selection is not more common or stronger in nature than disruptive selection. Assuming that $1 + \omega^2/E = 20$, where E is the environmental variance (assumed here to be 0.5, Turelli 1984), then ω is 3.1, toward the upper end of the range modeled in this chapter. The width of the fitness function ω is related to the quadratic selection gradient γ approximately by $\gamma = -1/w^2$ (Arnold et al. 2001), so that a value of $\omega = 4$ corresponds to a value of γ of about 0.06, close to the median value reported by Kingsolver et al. (2001). Thus, a range of ω from 1 to 4 σ should encompass most selection scenarios.

The primary assumptions of the selection models considered in this chapter are that **P** and **G** are constant over time, that phenotypes are determined simply by a breeding value and an environmental deviation (i.e., additive gene expression and no genotype–environment interaction), that harvest intensity and environmental influences on salmon productivity do not vary over time, and that the distributions of breeding values, environmental deviations, and phenotypes are multivariate normal, even under exploitation. All of these assumptions are likely to be violated to varying degrees. The effects of such violations are difficult to predict with any precision, but they may yield in some cases response estimates considerably higher than those presented here. The first and last assumptions may be especially problematic; violating them might produce inflated short-term response estimates. The simulated responses to selection described here should therefore be viewed with considerable caution.

To determine how correlated traits such as mean adult age, weight, growth rate, and spawn timing would respond to harvest selection on mean length (adult length should be highly correlated with length at harvest, especially for older fish), I used a discrete-generation analog of the age-structured model (equivalent to Equation (1)) to evaluate correlated responses. This model does not take into account age structure, which is a factor likely to influence selection response (Hankin and Healey 1986; Hankin et al. 1993), but the absence of estimates for genetic covariances among these traits expressed at each age rendered the use of the age-structured model impractical for evaluating correlated responses.

Under the discrete-generation model, $\mathbf{G}$, $\mathbf{P}$, and $\mathbf{s}_t$ are assumed to be constant over the number of generations simulated (five, or approximately 25 years for chinook salmon). To estimate the selection differentials on these traits, traits are again assumed to be under stabilizing selection with the mean trait value representing the selective optimum and fitness decaying with deviation of the mean from this value. For each trait I computed fitnesses and corresponding selection differentials on mean adult length, S_{lg}, over all ages in the population from Equations (5)–(10) under the assumption that mean adult length in a cohort represented the mean length of its members susceptible to fishing. I computed the correlated selection differentials on the means of the other adult traits, S_j', from the relation

$$S_j' = \frac{\mathrm{cov}_{\mathrm{lg}\,y}}{\mathrm{var}_{\mathrm{lg}}} S_{\mathrm{lg}} \tag{11}$$

where $\mathrm{cov}_{\mathrm{lg}\,y}$ is the phenotypic covariance between each pair of traits and $\mathrm{var}_{\mathrm{lg}}$ the phenotypic variance of length (Falconer and Mackay 1996).

2.2. Direct Responses to Selection

2.2.1. Directional Selection

When the genetic and phenotypic data summarized in Table 11.1 are applied to the age-structured model, it predicts that harvest imposing directional selection on size would lead to modest reductions in age-specific size of these fish over 25 years (approximately five generations). In these scenarios, θ varies from 0.2 to 0.7, which encompasses a range of harvest rates experienced by chinook salmon

($\sim 25 - 75\%$), based on unpublished coded-wire tag recoveries in the Pacific States Marine Fisheries Commission database. Under a directional harvest threshold of 711 mm (28 in., the threshold used for this species in troll fisheries in much of the western U.S.), the responses are greatest in 3- and 4-year-old fish. (In all simulation runs, responses of fish younger than age 2 are weak over the time frame considered here, and only age-specific responses at age 4—the modal age—are shown in the figures.) For example, for a probability of harvest for fish susceptible to catch equal to 0.6 (corresponding to a harvest rate of approximately—but somewhat less than—60% for fish ≤ 711 mm) and strong stabilizing selection on size, predicted declines in mean length of fish ages 2 to 5 over 25 years vary from 11 mm for 2-year-old fish (2.3%) to 18 mm for 4-year-old fish (2.2%) (results for age-4 fish shown in Figure 11.2A). Under this directional harvest threshold, declines in length vary from 6–9 mm (0.8–1.6 %) at the lowest harvest rate ($\theta = 0.7$) to 10–18 mm ($1.2 - 2.5\%$) at the highest harvest rate ($\theta = 0.2$).

Stabilizing natural selection on size has a substantial effect on the response to directional harvest selection, and this effect clearly depends on the harvest rate. When $\omega = 2$, predicted responses in length under a directional harvest

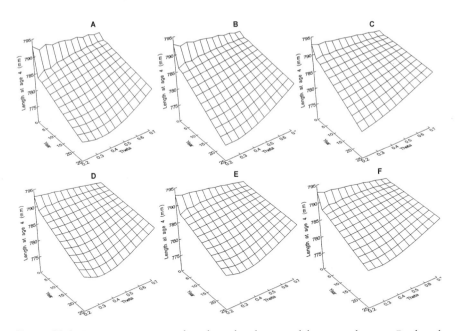

Figure 11.2. Direct responses in length under directional harvest selection. Predicted trends in length of chinook salmon at age 4 under directional selection over 25 years (5 generations) for harvest size thresholds of 711 mm (A–C) or 450 mm (D–F) and three different scenarios of stabilizing selection on length at age with an optimum at the current mean (left column: $\omega = 1\sigma$, central column: $\omega = 2\sigma$, right column: $\omega = 4\sigma$). The results are from the age-structured model (Equation (3)). Theta (θ) is the probability that fish larger than the size threshold escape the fishery, and represents a fixed proportion of the fish above the threshold length k.

threshold of 711 mm are diminished by 10–30% at all but the highest harvest rate, where they increase slightly (results for age-4 fish shown in Figure 11.2B). Responses are weaker at ages 3–5 for all but the highest harvest rates when $\omega = 4$ (results for age-4 fish shown in Figure 11.2C); the length of 2-year-old fish actually increases modestly at high harvest rates under this scenario.

In fisheries that impose a shorter harvest threshold (e.g., 450 mm, implemented in some Canadian fisheries), predicted responses to harvest selection are somewhat diminished for this population because selection differentials are smaller (the threshold is closer to the phenotypic mean). The age-specific responses in length to directional harvest selection are largely shifted to younger fish. Predicted declines over 25 years in mean length of fish ages 2 to 5 vary from 25 mm for 2-year-old fish (5.2%) to 17 mm for 5-year-old fish (1.5%) under strong stabilizing selection on length at age when $\theta = 0.2$ (results for age-4 fish shown in Figure 11.2D). Predicted responses when $\omega = 2$ are reduced at high harvest rates by about 25% for 2-year-old fish but not at all for older fish; responses to low harvest rates are lower by about half for 2-year-olds and by less than 15% for older fish (results for age-4 fish shown in Figure 11.2E). Responses are diminished at all ages for all harvest rates when $\omega = 4$ (results for age-4 fish shown in Figure 11.2F). Under directional selection, moving the harvest threshold appears to elicit the largest response in fish with age-specific lengths nearest the threshold.

2.2.2. Disruptive Selection

Disruptive selection, as might be imposed by slot limits in some net fisheries, tends to sharply reduce responses to selection on length over 25 years. In most cases, the directional response is predicted to switch direction at high harvest rates. When the largest fish (fish $>\sim 1$ S.D. above the population mean) are allowed to escape harvest (slot limits of 711–813 mm), length at age 3 declines by only 2 mm ($< 1\%$) at all harvest rates regardless of ω, while lengths at ages 2, 4, and 5 decline by less than half that amount (results for age-4 fish shown in Figure 11.3A–C). The model predicts that age-specific lengths actually increase modestly under a left-shifted slot with limits of 450-770 mm, if harvest rate is high and stabilizing selection on length is strong ($\omega = 1$) (results for age-4 fish shown in Figure 11.3D–F). For example, under the highest harvest rate ($\theta = 0.2$), length at age 4 is expected to increase by 4.1% over 25 years under disruptive selection with these slot limits (Figure 11.3D). Weaker stabilizing selection ($\omega > 2$) flattens the response in length to this type of selection under either harvest scenario (Figure 11.3B–C, E–F).

2.3. Correlated Responses to Selection

2.3.1. Directional Selection

The discrete-generation form of the model predicts that harvest imposing directional selection on size would lead to modest reductions in mean size of adult

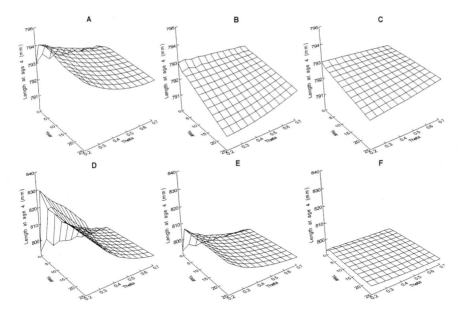

Figure 11.3. Direct responses in length under disruptive harvest selection. Predicted trends in length of chinook salmon at age 4 under disruptive selection over 25 years (5 generations) for harvest size limits of 711–813 mm (A–C) or 450–770 mm (D–F) and three different scenarios of stabilizing selection on length at age with an optimum at the current mean (left column: $\omega = 1\sigma$, central column: $\omega = 2\sigma$, right column: $\omega = 4\sigma$). The results are from the age-structured model (Equation (3)). Theta (θ) is the probability that fish between the size limits escape the fishery, and represents a fixed proportion of the fish between the lower and upper threshold lengths k_1 and k_2.

fish over five generations (approximately 25 years). Under a directional harvest threshold of 711 mm and strong stabilizing selection on mean adult length, the direct decline in mean length ranges from 23 mm (3.2%) at the lowest harvest rate to 87 mm (12.1%) at the highest harvest rate (Figure 11.4A). Proportionate reductions are predicted for mean body weight (7.4–26.9%; Figure 11.4D–F). Reduced stabilizing selection on adult length and weight ($\omega = 2$ and $\omega = 4$) tends to diminish the response in both traits on the order of 15–20% at low harvest rates but by more than 50% at high harvest rates (Figure 11.4).

Selection on size is predicted to yield detectable changes in means for growth rate, adult age, and spawning date. With strong stabilizing selection on growth rate ($\omega = 1$), mean growth rate from smolt to adult declines nearly linearly each generation at the highest harvest rate, showing a decline of nearly 15% over five generations; at the lowest harvest rate, the decline is only 3.6% (Figure 11.4G). The reduction in growth rate is only about half as large when $\omega = 2$ and is eliminated almost completely when $\omega = 4$ (Figure 11.4H–I). Directional harvest selection combined with strong stabilizing selection on size results in a shift in mean adult age from 4 to 3 years within three generations at

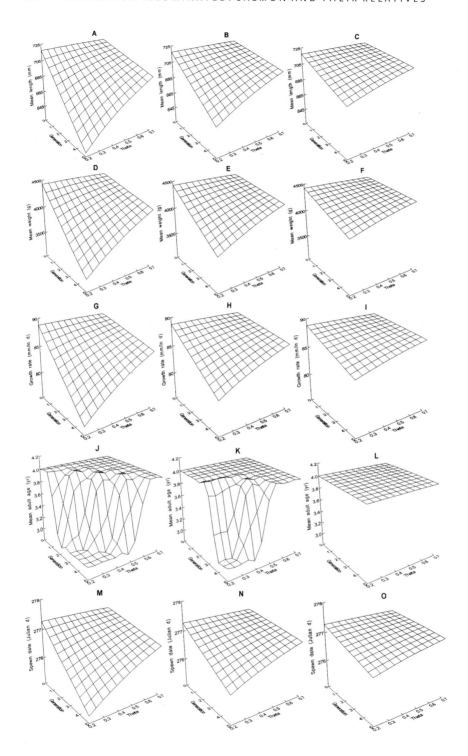

the highest harvest rate and within five generations at moderate harvest rates (Figure 11.4J). Lower harvest rates or weaker stabilizing selection on age yield no correlated response in age (Figure 11.4J–L). The response in spawning date depends on the harvest rate and intensity of stabilizing selection but is modest under all harvest scenarios (Figure 11.4M–O).

Under a shorter length threshold of 450 mm, directional selection yields only slightly weaker responses in mean length and weight than under a threshold of 711 mm at high harvest rates; the effect of reducing harvest rate on diminishing response is more pronounced at the lower than at the higher threshold (Figure 11.5A–F). The same patterns are evident for mean growth rate (Figure 11.5G–I). Age is expected to decline by 1 year only at the two highest harvest rates after three to four generations when $\omega \geq 2$, and at the highest harvest rate after five generations when $\omega = 4$ (Figure 11.5J–L). The expected response in spawning date is almost undetectable at all but the highest harvest rate under strong stabilizing selection (Figure 11.5M–O). Again, reducing harvest rate at this threshold yields a sharper reduction in response than at the higher threshold.

2.3.2. Disruptive Selection

Either of the disruptive selection scenarios modeled is predicted to lead to at most minor short-term changes in mean length and weight, even at high harvest rates and with strong stabilizing selection on size. For slot limits of 711–813 mm, there is no appreciable response predicted for any of the traits correlated with length, regardless of the strength of stabilizing natural selection (results not shown). For slot limits of 450–770 mm, at $\omega = 1$ mean length, weight, and growth rate are expected to increase by 10–15% after five generations at the highest harvest rate, with much smaller increases at low to moderate harvest rates (Figure 11.6A–I). These responses are sharply diminished at $\omega = 2$ and undetectable at $\omega = 4$. Responses to selection in mean age and spawn date are trivial under all scenarios (Figure 11.6J–O).

3. Conclusions

That correlations among traits can influence response to selection has long been understood by plant and animal breeders, and in some cases this influence may be considerable. For this population of chinook salmon exposed to size-selective

Figure 11.4. Direct and correlated responses to directional selection on length: I. Predicted trends in mean adult length, adult weight, growth rate, adult age, and spawning date of chinook salmon over 5 generations (25 years) for harvest size thresholds of 711 mm and three different scenarios of stabilizing selection on length at age with an optimum at the current mean (left column: $\omega = 1\sigma$, central column: $\omega = 2\sigma$, right column: $\omega = 4\sigma$). The results are from the discrete-generation model (Equation (1)). Theta (θ) is the probability that fish larger than the size threshold escape the fishery, and represents a fixed proportion of the fish above the threshold length k.

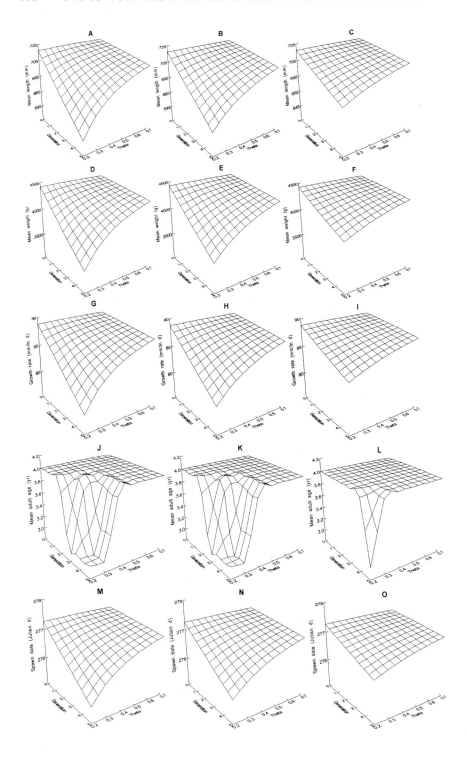

harvest, positive correlations among the monitored traits and weak positive selection on them are generally expected to augment, not retard, the responses to harvest selection on length. For example, predicted values for realized heritabilities of adult length and age, estimated from the ratio of cumulative response to cumulative selection differential, are from 1 to 28 times larger than corresponding REML estimates computed from resemblance among relatives (Table 11.2). Accounting for positive genetic correlations between each of the life history traits and length yields heritability estimates (and predictions of response to selection) that are for this population often substantially larger than initial estimates that do not account for these correlations. Thus, heritability estimates obtained from covariance among relatives before selection would tend to underestimate selection response for most traits (Sheridan 1988; see also Merilä et al. 2001). The positive genetic covariance structure of the traits is predicted to enhance the response to selection for mean weight in particular because the covariances are strongly reinforcing to the direction of selection. Covariances among traits that are antagonistic to the direction of selection (in this case, negative covariances) would be expected to constrain short-term evolutionary change (Etterson and Shaw 2001).

These results lead me to conclude that (1) strong directional selection on size is likely to produce modest short-term reductions in this trait, but the effects depend critically on the harvest rate, harvest size threshold, the strength of stabilizing natural selection on size, and most likely the age structure and heritability of each trait; (2) the capacity of directional selection to reduce size depends on correlations between size and other prominent life history attributes, particularly age and growth rate; (3) disruptive selection can substantially reduce the strength of selection on size if a sufficient proportion of large fish is allowed to escape fishing-related mortality; and (4) although the discrete-generation results do not account for the effects of age structure on selection response, the model predicts that directional selection imposed by size-selective harvest will also reduce age and growth rate (and possibly shift spawning date earlier) detectably within a few generations.

A more comprehensive analysis of the implications of age structure for selection response is important to pursue because age and age-specific growth rates ultimately will determine the actual consequences of harvest. Hamon et al. (2000) demonstrated that a gillnet fishery on neighboring Alaskan sockeye salmon populations imposes disruptive selection on size overall, but that the effects probably differ among populations because of differences in age structure and

Figure 11.5. Direct and correlated responses to directional selection on length: II. Predicted trends in mean adult length, adult weight, growth rate, adult age, and spawning date of chinook salmon over 5 generations (25 years) for harvest size thresholds of 450 mm and three different scenarios of stabilizing selection on length at age with an optimum at the current mean (left column: $\omega = 1\sigma$, central column: $\omega = 2\sigma$, right column: $\omega = 4\sigma$). The results are from the discrete-generation model (Equation (1)). Theta (θ) is the probability that fish larger than the size threshold escape the fishery, and represents a fixed proportion of the fish above the threshold length k.

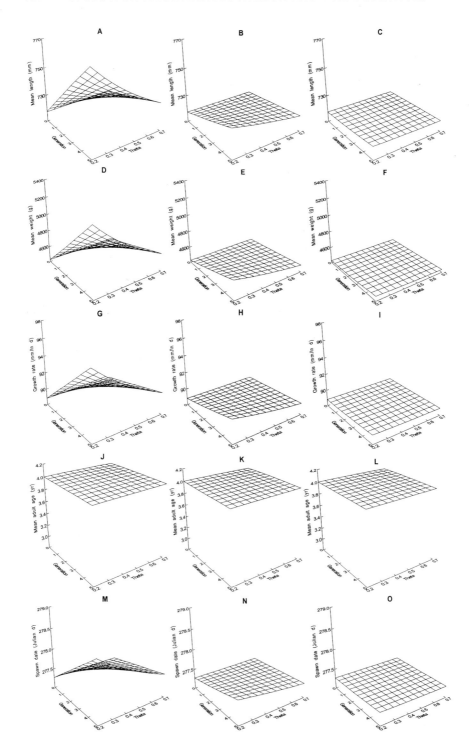

age-specific size. Populations with fewer ages at maturity may experience stronger net directional selection in fisheries that impose disruptive selection than those with more ages at maturity. Hamon et al. (2000) argued that in populations with a greater range of ages at maturity, one consequence of size-selective harvest might be selection on the maturation schedule itself, and the present results are consistent with this hypothesis.

The results indicate that the intensity of stabilizing natural selection on size is critically important in determining response to harvest selection (which is affected largely by the harvest threshold and the harvest rate). Under most of the scenarios considered here, changes in life history resulting from either directional or disruptive selection are at best modest; therefore, it is unlikely that selection imposed by harvest will typically be sufficient to account for declining trends in size as large as those that were observed in British Columbia chinook salmon between the 1950s and 1980s. At least some of the observed declines in size may reflect differential selection among populations that has depleted the abundance of stocks with larger mean sizes, ages, and growth rates. Selection for smaller size, age, and growth within populations may be weaker, owing to the large environmental components of variance for these traits. Perhaps most important is variation in marine or estuarine environmental conditions occurring while these fish are susceptible to harvest (Riddell 1986; Ricker 1995; Hard 2002).

Nevertheless, it should be recognized that the population used in this analysis may not represent accurately the relevant characteristics of many populations, including those imperiled by habitat loss, overfishing, or other factors. On the other hand, substantial declines in size and growth from harvest are possible in species with simple age structure such as pink and coho salmon, because size-selective fisheries act directly on growth rate in these species, and declines have persisted for some populations of these species for decades (Healey 1986; Ricker 1995; Bigler et al. 1996).

A potentially important factor not considered in detail here is the interplay between domestication in hatchery fish and response to harvest selection. Fisheries often target hatchery populations but also capture comigrating wild populations that may differ in their ability to withstand exploitation. The consequences of domestication for harvest selectivity and natural productivity have received almost no research; understanding how hatchery and wild fish might differ in response to fishing and how domestication in hatcheries may degrade

Figure 11.6. Direct and correlated responses to disruptive selection on length. Predicted trends in mean adult length, adult weight, growth rate, adult age, and spawning date of chinook salmon over 5 generations (25 years) for harvest size limits of 450–770 mm and three different scenarios of stabilizing selection on length at age with an optimum at the current mean (left column: $\omega = 1\sigma$, central column: $\omega = 2\sigma$, right column: $\omega = 4\sigma$) (results for size limits of 711–813 mm were nearly undetectable and are not shown). The results are from the discrete-generation model (Equation (1)). Theta (θ) is the probability that fish larger than the size threshold escape the fishery, and represents a fixed proportion of the fish between the lower and upper threshold lengths k_1 and k_2.

Table 11.2. Estimates of the heritability of mean adult traits (with approximate standard errors).

	Covariance h^2 (CAR)	Realized h^2 (RTS)	CAR h^2:RTS h^2
Fork length (mm)	0.336 ± 0.208	0.336	1.00
Adult age (yr^3)	0.349 ± 0.212	0.520	0.67
Adult weight ($\sqrt{g}$)	0.010 ± 0.091	0.284	0.04
Spawn date (ln (Julian d))	0.232 ± 0.172	0.987	0.24
Growth rate (mm $[\ln(d)]^{-1}$)	0.307 ± 0.198	0.348	0.88

Heritabilities estimated by restricted maximum likelihood from covariance among relatives (CAR) were used to parameterize age-structured models of multivariate response to selection. Realized estimates reflect the response to directional selection (RTS) predicted by these models with $\theta = 0.4$ and the legal length threshold at 711 mm. (Note that the CAR and RTS estimates for length are identical because this is the directly selected trait.) The estimates for the other traits depend on the genetic and phenotypic covariance matrix elements in Table 11.1 and are similar for the other values of θ and length threshold.

fitness of wild fish that interbreed with hatchery fish is critical to the development of sustainable hatchery production-harvest systems for salmon. It is tempting to speculate from the results of this chapter that fish spawned in hatcheries may have weaker short-term responses than fish spawning in the wild to directional harvest selection owing to the relaxation of natural selection on adult size in hatchery fish at time of breeding, but this conclusion is a tenuous one and, if such a disparity were to occur, the consequences are likely to be minor unless natural selection in the wild is strong.

The correlated responses in some prominent features of chinook salmon life history to selection on length through fishing indicate that sustainable harvest strategies must consider the selective consequences of fishing to minimize reductions in productivity and yield (Law 2000; Ratner and Lande 2001). For example, reducing harvest selectivity may permit higher sustainable exploitation rates (e.g., Conover and Munch 2002). Because it is not yet known how most salmon populations—which often exhibit remarkable life history variation—respond to harvest selection for any trait, and because populations with different life histories may be exposed to similar fisheries, prudent management would limit selectivity as far as possible in addition to maintaining escapements adequate to sustain production. While opposed to MSY obtainable on an annual basis, such a strategy may be closest to the true concept of sustainable yield.

Both the capacity of natural selection to constrain variation in adult size and the nature of variation in environmental conditions during rapid marine growth are likely to interact with fishing selection in complex ways. Direct empirical estimation of harvest selection differentials and responses to selection over several generations is needed to determine how selective fisheries affect variation in salmon life history and productivity. Until fisheries can be developed that are less selective, that can target populations capable of withstanding high exploitation rates, and that limit mortality in less productive populations and loss of diversity within populations, research to better characterize the genetic and phenotypic architecture of life history for salmon populations susceptible to harvest should be encouraged. Such research would help to identify more conservative management options and minimize human-induced evolutionary change in exploited populations.

Acknowledgments   I am grateful to David Hankin, Russell Lande, Richard Law, Hans Bentsen, Ulf Dieckmann, Gerry Friars, Richard Gomulkiewicz, Mikko Heino, and Jonas Jonasson for helpful discussions; and Jason Miller, Ted Parker, and the staffs of the Conservation Biology Division (Northwest Fisheries Science Center) and the Suquamish Tribal Fisheries Department for assistance with collecting the genetic and phenotypic data used to parameterize the harvest models. Andrew Hendry, Robin Waples, Michael Ford, and four anonymous reviewers provided very helpful comments on drafts of this chapter. This work was supported in part by funds from the Bonneville Power Administration (Project 93-56).

Conservation Units and Preserving Diversity

Michael J. Ford

Spawning pink salmon

12

A major theme in conservation biology is the preservation of diversity. An aspect of this problem that has received considerable attention is how to prioritize conservation of biological diversity below the species level. Many species are subdivided into populations that vary with respect to their genetic or phenotypic characteristics or the habitats they occupy. The taxonomy of some species groups may also be uncertain or out of date, which means that named subspecies and even species may be poor indicators of biological diversity. This diversity and uncertainty poses a challenge for conservation biologists, because a logical first step toward in conservation is describing what is available to be conserved.

1. Theory

To address the problem of categorizing intraspecific diversity, Ryder (1986) first developed the concept of an "Evolutionarily Significant Unit" (ESU), which he defined as a taxonomic unit below the species level that captures a major component of the diversity within the species. Since then, there have been numerous attempts to define ESUs, or other conservation units, more precisely. A representative selection of these concepts is summarized below.

Waples (1991a, 1995)   Waples' (1991a) ESU concept was motivated by the need to develop a policy for listing salmon populations under the United States Endangered Species Act (ESA). The ESA allows listing of, but does not specifically define, "distinct population segments" of vertebrate species. Under Waples' formulation, a population or group of populations is an ESU (and therefore a distinct population segment) if it is (1) substantially reproductively isolated from other populations, and (2) an important component of the "evolutionary legacy" of the species. Waples intended his approach to employ a variety of data sources (genetic, phenotypic, ecological) to address both the isolation and importance criteria (Waples 1995). Waples' ESU concept is "backward looking" in the sense that it describes units that have been created by past evolutionary processes, but the ultimate goal of the approach is "forward looking" in that it is trying to conserve the "genetic variability that is a product of past evolutionary events and which represents the reservoir upon which future evolutionary potential depends" (Waples 1991a). Unlike some later concepts (see below), Waples' concept does not involve attempting to determine which populations are most likely to be important in the future. Rather, it focuses on conserving major components of diversity, some of which may be important for the future evolution of the species. Most ESUs identified under Waples concept contain multiple populations, and McElhany et al. (2000) introduced a framework for Pacific salmon in which intra-ESU diversity can be described and evaluated.

Dizon et al. (1992)   Like Waples' concept, Dizon et al.'s ESU concept was also motivated by a desire to provide a biological meaning for a term that already had a legal meaning under U.S. law. In this case, the term was a "stock," which is used by the Marine Mammal Protection Act to describe management units for marine mammals (Dizon et al. 1992). Dizon et al.'s ESU concept is similar to Waples' concept in that a population must be both reproductively isolated and adaptively diverged from other populations in order to be an ESU. One difference between Waples' and Dizon et al.'s concepts is that the latter explicitly defines a gradation of different population categories based on genetic and geographic distinctiveness: category I populations inhabit geographically disjunct ranges and are characterized by significant genetic divergence; category II populations are genetically distinct but sympatric; category III populations are geographically distinct but genetically similar; and category IV populations are only weakly geographically and genetically differentiated. Under this system, category I populations are most likely to be considered ESUs, and category IV population are least likely to be considered ESUs.

Moritz (1994)   Moritz (1994) defines an ESU as a population or group of populations that forms a monophyletic group based on mtDNA variation, and that has significant frequency differences from other groups at nuclear markers. The goal of Moritz's ESU concept is to identify taxa that have experienced long periods of independent evolution, without attempting to assess the adaptive significance of this evolution. Like Waples' and Dizon et al.'s ESU concepts, Moritz's concept is based on historical information, arguing that populations with distinct evolutionary histories are likely to have different future evolutionary potentials. Moritz also defined a management unit (MU) as a population characterized by significant differences in allele frequency from other populations.

Vogler and Desalle (1994)   Several authors have argued that to be most useful an ESU concept should be well grounded in systematic biology (Rojas 1992; Mayden and Wood 1995). The best example of this is the ESU concept of Vogler and Desalle (1994), who use the framework of the phylogenetic species concept to define a conservation unit as a group of organisms that possess a unique shared character or combination of characters. Their concept is similar to Moritz's concept in that it requires ESUs to be characterized by fixed genetic differences, but under Vogler and Desalle's concept these differences can be at any heritable trait. Like Moritz's concept and unlike Waples' and Dizon et al.'s concepts, Vogler and Desalle's concept focuses strictly on identifying populations with independent evolutionary histories without reference to adaptive divergence.

Bowen (1998)   Bowen (1998) pointed out that although ESU concepts typically have the preservation of future evolutionary potential as a goal, operationally they focus on population history. Bowen proposed a new term, "Germinate Evolutionary Unit" (GEU), to be applied to groups of organisms that are likely progenitors of future biological diversity. Bowen (1998) readily acknowledged that predicting the future is generally more difficult than reconstructing the past, but proposed that traits such as a recent range or niche expansions, recently evolved behavioral barriers, or high levels of morphological or behavioral variation could be used to identify GEUs. GEUs, unlike ESUs, would not necessarily be expected be differentiated at quasi-neutral markers, such as DNA sequences or allozymes.

Crandall et al. (2000) Using a framework similar to that of Dizon et al. (1992), Crandall et al. (2000) proposed abandoning simple ESU concepts altogether in favor of a matrix approach that classifies populations into multiple categories based on their current and historical genetic and ecological "exchangeability." According to Crandall et al.'s criteria, genetic exchangeability would be rejected when estimates of gene flow among populations are low ($Nm < 1$), and ecological exchangeability would be rejected when populations have statistically significant differences at heritable morphological, behavioral, life history, QTL, or allozyme traits under selection, or occupy different habitats. Current ecological and genetic exchangeability is evaluated by examining current patterns of variation, and historical exchangeability is estimated by mapping current patterns of variation on an estimate of the phylogeny of the populations in question.

ESU concepts have sparked a surprising amount of debate in the scientific literature (Pennock and Dimmick 1997; Waples 1998a; Dimmick et al. 1999; Pearman 2001; Young 2001). Some of the debate is about biology. For example, disagreements about how biological diversity should be categorized typically reflect broader debates about how species should be defined and identified (Rojas 1992; Mayden and Wood 1995). Determining to what degree conservation biology is about conserving things (species, ESUs, ecosystems) or about conserving processes (evolution, gene flow, ecosystem functions) is also an important biological conservation issue.

However, in reading over a decade of papers on ESU concepts, one cannot help but be struck by how much of the debate appears to be semantic. For example, Vogler and Desalle (1994) and Goldstein et al. (2000) argue that character data and the phylogenetic species concept should be used to identify species for conservation purposes, but agree that Waples' (1991a) ESU concept is useful for identifying subspecific groups for setting conservation priorities. This is of course exactly the intended purpose of Waples' ESU concept. Likewise, Moritz's (1994) ESU concept requires ESUs to have fixed mtDNA differences compared to other ESUs, but also allows for intra-ESU structure in the form of MUs, many of which may well be analogous to Waples' (1991a) ESUs or Dizon et al.'s (1992) stocks. In a recent review of the matter, Fraser and Bernatchez (2001) concluded that all conservation unit definitions are context dependent, and that it is not possible to develop an ESU concept that is both useful and universal. Waples (1995) also suggested that determining the appropriate level of intraspecific diversity to conserve is in part an arbitrary process that involves social, cultural, and economic, as well as biological criteria.

2. Applications to Salmon

There are several reasons why salmon provide an excellent group of organisms to study the implications of using different ESU concepts. The discussion that follows focuses on Pacific salmon (*Oncorhynchus* sp.), but the concepts apply equally to many other groups of salmonids. First, salmon are highly variable in morphology, behavior, life history, and habitat use (e.g., for Pacific salmon see Groot and Margolis 1991; McElhany et al. 2000; McDowall 2001; Waples et al. 2001; Einum et al. 2003—*this volume*; Hendry et al. 2003b—*this volume*). Conserving this diversity provides motivation for developing methods of categorizing it. Second, salmon have an intrinsically subdivided population structure due to their strong tendency to home to their natal stream (e.g., Quinn 1993; Teel et al. 2000; Hendry et al. 2003a—*this volume*). This structure means that salmon populations are demographically independent from one another on varying time scales, providing a set of naturally replicated and semi-independent experiments that can be used to test hypotheses about the origins of diversity. Third, salmon are the focus of a great deal of conservation, commercial, and cultural interest, and occupy a large variety of habitats affected by many human activities. All aspects of salmon conservation are therefore subject to a great deal of scientific and public scrutiny. This high level of interest illuminates and creates issues that would not be addressed in species that receive less public attention. Finally, in a series of status reviews, over 50 Pacific salmon ESUs have been identified (Weitkamp et al. 1995; Busby et al. 1996; Hard et al. 1996; Gustafson et al. 1997; Johnson et al. 1997; Myers et al. 1998; Johnson et al. 1999), providing an extensive record that can be analyzed, compared, and criticized.

2.1. Intraspecific Patterns of Morphological, Behavioral, and Ecological Diversity

The seven species of Pacific salmon (*Oncorhynchus* sp.) that spawn in North American waters provide an excellent example of the diversity contained within and among salmon populations (Groot and Margolis 1991; Waples et al. 2001). Together, the seven species occupy habitats ranging from the central Pacific ocean to high elevation streams deep in interior mountain ranges. All of the species contain considerable diversity in morphological traits and in behavioral traits such as the timing of spawning migrations. Three of the seven species contain both anadromous and iteroparous populations (McDowall 2001; Stearns and Hendry 2003—*this volume*), and five contain considerable diversity in life history characteristics such as age at smolting and maturity (Groot and Margolis 1991). All seven species are subdivided into hundreds to thousands of semi-independent breeding populations, facilitating the potential for local adaptation and the development of genetic diversity among populations (Taylor 1991b; Wood 1995).

Patterns of juvenile and adult migration timing in chinook salmon provide a good example of intraspecific variation in salmon behavior and life history. In particular, there is considerable diversity both within and among populations with respect to the time of year adults migrate into fresh water to spawn (Figure 12.1). Traditionally, fishery biologists have used these differences to identify distinct populations or "runs" of chinook salmon. For example, the Sacramento River has four chinook salmon runs—"spring," "fall," "late-fall," and "winter" (Healey 1991). Similarly, spring, summer, and fall migrations of chinook salmon are found up and down the Pacific coast (Figure 12.1). There is some evidence that variation in run-timing is adaptive and heritable in salmon. For example, salmon that spawn in the upper parts of watersheds tend to return to fresh water in the spring when high water conditions allow passage to those areas (Miller and Brannon 1981). Moreover, the heritability of run-timing has been found to be high in a study of pink salmon (Smoker et al. 1998) and chinook salmon (Quinn et al. 2000).

Chinook salmon also differ with respect to when juveniles migrate from fresh water to the ocean (Figure 12.2, reviewed by Healey 1983; Healey 1991; Myers et al. 1998). Gilbert (1913) found that chinook salmon life history patterns could be divided into at least two major types: a pattern in which juveniles migrate to the ocean in the summer or fall after they hatch (ocean type) and a pattern in which juveniles spend an entire year in fresh water prior to migrating to the ocean (stream type). This polymorphism in juvenile life history is closely associated with variation in adult run-timing, ocean migration pattern, and other life history traits (Healey 1991; Healey 2001). Populations with fall migrating adults tend to have ocean type juveniles and a relatively near-shore ocean distribution, whereas populations with spring or summer migrating adults tend to have stream-type juveniles and a more offshore ocean distribution. Healey (1991) argued that such highly divergent life histories were evidence for two relatively ancient "races" of chinook salmon: a "stream-type" race spawning in streams north of central British Columbia and Asia, as well as the interior portions of the Fraser and Columbia River systems, and an "ocean-type" race spawning predominantly in streams south of central British Columbia. This insight was later confirmed by molecular genetic studies that found evidence for two highly divergent lineages in chinook salmon (e.g., Teel et al. 2000; Utter et al. 1989). The association between juvenile and adult life history patterns and genetic lineage is not complete, however. For example, southern coastal populations are all of the "ocean-type" lineage, but exhibit a high level of diversity in both juvenile and adult life history patterns (Figure 12.2; Myers et al. 1998). Recently, Beckman (2002) argued that due to extensive variation of juvenile life histories within populations and lineages, the terms "ocean type" and "stream type" are best applied as descriptions of individual fish (consistent with Gilbert's (1913) usage). He then suggested that the terms "northern" and "southern" would be more appropriate labels for the two major lineages within chinook salmon.

The genetic basis of life history variation in Pacific salmon is not well understood. Like many fish species, salmon are developmentally plastic and numerous experiments have demonstrated that behavioral, morphological, and life history traits are all substantially affected by the environment in which a salmon lives.

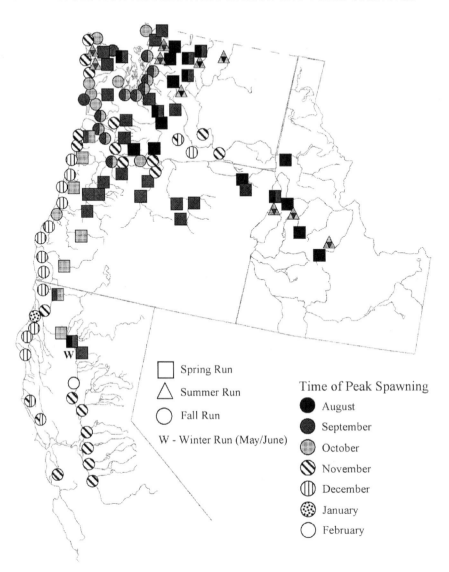

Figure 12.1. Adult migration timing diversity in chinook salmon along the west coast of California and the Pacific Northwest (modified from Myers et al. 1998). Shapes with two shades or patterns indicate that time of peak spawning occurs at the end of the earlier month and the beginning of the later month.

For example, Beckman and Dickhoff (1998) found that for spring chinook salmon, the time of smolting was strongly affected by growth rate, and Berejikian et al. (1997) found that for coho salmon, the rearing environment strongly affected morphology and spawning behavior. However, numerous experiments have also documented a genetic basis for life history variation, including traits such as growth rate (e.g., Tallman 1986), juvenile behavior (e.g., Raleigh 1971;

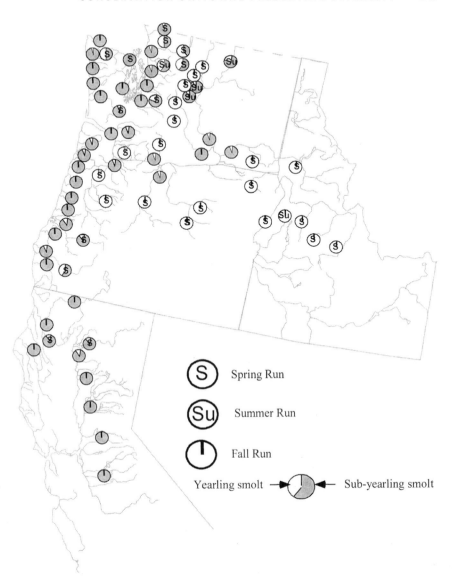

Figure 12.2. Patterns of juvenile life history variation among populations of chinook salmon spawning in California and the Pacific Northwest (modified from Myers et al. 1998).

Berejikian 1995), timing of maturation and migration (e.g., Smoker et al. 1998; Quinn et al. 2000), and disease resistance (e.g., Zinn et al. 1977). Taylor (1991b) and Tave (1993) review the genetic basis for many traits in salmonids and provide many more examples. It appears, therefore, that life history variation in salmon is strongly influenced by both environmental and genetic factors, but that it is reasonable to assume that observed life history variation within and among salmon populations (the vast majority of which have not been subject to detailed genetic analysis) is due in part to genetic variation.

Interpreting patterns of life history diversity is a challenge for conservation biologists interested in defining conservation units for chinook salmon, or any other highly variable species. For example, most salmon populations differ with respect to one or more behavioral or life history traits, and there is considerable evidence that many of these differences are adaptive in contemporary populations (Taylor 1991b). On the other hand, many salmon populations occupy streams that were glaciated as few as 10,000 years ago, so patterns of contemporary diversity must have arisen after that time, either *in situ* or as a result of multiple colonizations. Without some understanding of how rapidly populations can adapt to new environments (Kinnison and Hendry 2003—*this volume*), it is difficult to determine how conservation unit boundaries should be drawn (Allendorf and Waples 1996). As an example, in the case of run-timing diversity in chinook salmon, most populations spawning in coastal streams are fall migrating, whereas most inland populations migrate in the spring or summer (Figure 12.1). On the other hand, there are a few spring-run populations in coastal streams and some fall-run populations spawn hundreds of miles inland—what is the history of these populations? When patterns of variation at multiple traits, such as adult and juvenile migration timing, are not completely correlated, which traits should be considered more important in defining conservation units? Do differences among populations in traits result from unique or rare evolutionary events that are unlikely to be repeated, or have they evolved independently many times? All of these questions have obvious relevance in defining conservation units based on current patterns of phenotypic or life history diversity.

2.2. Patterns of Molecular Genetic Diversity

One way to start making sense of current patterns of phenotypic variation is to put them in the context of evolutionary history, which can often be inferred from patterns of molecular genetic data (Avise 1994). For example, if the history of chinook salmon populations can be approximated by a bifurcating tree in which ancestor populations divide into pairs of daughter populations, then this tree can be estimated from neutrally or quasi-neutrally evolving molecular markers (Felsenstein 1993). Phenotypic traits of interest can then be mapped onto this genealogy of populations, providing insight into the evolutionary history of the traits themselves. A strictly bifurcating tree is certainly not an entirely accurate description of chinook salmon population structure, because gene flow occurs among salmon populations. The level of gene flow among geographically disjunct populations is sufficiently low, however, that a branching tree is a reasonable approximation of their history for the purposes of studying trait evolution.

Chinook salmon have been the subject of numerous molecular population genetic studies, and patterns of variation throughout much of the species range are fairly well documented (e.g., Utter et al. 1989). With some important exceptions, these studies generally find that populations of chinook salmon from the same drainage or geographic region tend to be genetically more similar to each other than they are to populations from other drainages, regardless of their adult run-timing. To illustrate this, we used published (Utter et al. 1989; Bartley et al. 1992b; Utter et al. 1995) and unpublished genetic data (provided by D. Teel and A. Marshal) at 31 polymorphic allozyme loci to estimate the phylogeny of 25 chinook salmon populations spawning in California and the Pacific Northwest (Figure 12.3). Assuming a bifurcating model with minimal gene flow, it is clear from this analysis that run-timing differences must have evolved independently several times over the history of these populations. For example, spring migrating populations in Puget Sound, the Oregon Coast, and the California Coast share a more recent common ancestor with nearby fall migrating populations than they do with other spring migrating populations (Figure 12.3). Similarly, the coastal populations with relatively high proportions of stream-type juveniles share more recent common ancestors with nearby populations containing mostly ocean-type juveniles than they do with geographically distant stream-type populations.

An alternative to the hypothesis that run-timing and juvenile life history differences have arisen independently many times within coastal streams is that these streams were independently colonized by populations that already differed in adult run-timing or juvenile life history type. Under this scenario, the genetic similarity of populations with alternative life histories inhabiting the same stream basins could be explained by high levels of gene flow at quasi-neutral markers (such as allozymes) following the colonization events, combined with continued selection for the alternative life history types. It is difficult to completely refute or confirm this hypothesis without a detailed understanding of the genetic architecture of life history variation in multiple chinook salmon populations. However, several lines of evidence suggest that it is unlikely. First, both adult run-timing and age at smolting are continuously distributed traits and are most likely influenced by many genes of small effect. This means that the evolution of these traits is not expected to depend on unique mutational events, but could occur simply through natural selection on variation that is probably present in many salmon populations. Second, there are empirical examples of natural and artificial selection causing genetic changes in a population's life history distribution, indicating that such changes can occur over short time scales. For example, a single population of chinook salmon introduced to New Zealand early in the twentieth century has subsequently colonized several streams and evolved genetic differences among populations in run-timing and age at maturation, indicating that rapid, independent evolution of life history traits is possible (Kinnison et al. 1998b; Quinn et al. 2000; Quinn et al. 2001b; Kinnison and Hendry 2003—*this volume*;).

One clear example where molecular genetic analyses *do* indicate that a basin was colonized by two divergent lineages involves the evolutionary history of chinook salmon populations in the interior Columbia River Basin. Fall-migrating fish in the Snake River, and fall- and summer-migrating fish in the Middle Columbia River are predominately ocean type, and are evolutionarily more

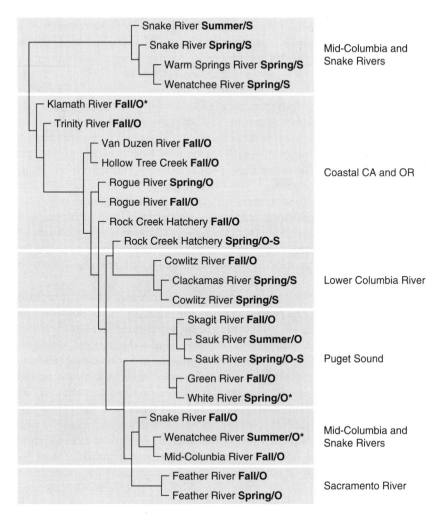

Figure 12.3. Estimated phylogeny of 25 chinook salmon populations sampled from California and the Pacific Northwest. The phylogeny is the maximum likelihood tree under a pure drift model (Felsenstein 1981) estimated from allele frequencies at 31 allozyme loci. The distance scale is in units of genetic variance. Most branches in the tree are significantly greater than zero based on the likelihood surface. Populations are identified by season of peak adult migration timing and proportion of the population smolting at age zero. Codes for smolting at age zero are: O = 90–100%; O* = 75–90%; O-S = 25–75%; S* = 10–25%; S = 0–10%. The genetic data used to construct the tree were compiled by David Teel (NMFS) and originated from Utter et al. (1989), Bartley et al. (1992), Waples et al. (1993); Utter et al. (1995), and unpublished data from D. Teel (NMFS) and A. Marshall (WDFW). Populations are named after the major river basins in which they were sampled and do not always correspond to sample names from the original data sources. The life history data were compiled by Myers et al. (1998) and originated from a variety of sources (see Myers et al. 1998 for details).

closely related to coastal populations than they are to the stream-type spring and summer-run populations spawning in the interior Columbia River Basin (Figure 12.3; Utter et al. 1989; Utter et al. 1995; Myers et al. 1998). Indeed, population samples of the two types have fixed or near-fixed allelic differences at some nuclear genes (Ford 1998). Similar patterns of differentiation between predominantly ocean-type and predominantly stream-type populations have been observed in the Fraser River (Teel et al. 2000). Several investigators have noted these relatively high levels of genetic and life history divergence and speculated that the two types independently colonized these areas as the habitat became ice-free after the most recent Pleistocene glaciation (Healey 1991, Myers et al. 1998; Teel et al. 2000). In the terminology of Crandall et al. (2000), stream- and ocean-type chinook salmon populations in the interior Columbia and Fraser River Basins appear to be ecologically and genetically distinct both recently and historically.

Even with the additional insight gained from analysis of molecular genetic data, defining conservation units for chinook salmon is obviously quite complicated. In the interior river basins, there is evidence for two relatively old lineages that differ in a variety of life history traits. These would almost certainly be classified into separate ESUs under many of the classification schemes currently in use. Despite this evidence of relatively deep divergence, however, some of the most obvious life history differences between these lineages, such as adult and juvenile migration timing, are clearly quite labile in the coastal chinook salmon populations (Figures 12.1 and 12.2). The evolution of these and other traits over short time scales (decades) in chinook salmon transplanted to New Zealand (Kinnison et al. 1998c; Quinn et al. 2001b) or coho salmon raised in artificial settings (Fleming and Gross 1989) provides additional evidence that at least some of the current patterns of phenotypic and life history variation in chinook salmon are likely to be evolutionarily malleable.

This evidence for vagility should not be unduly comforting to conservation biologists, however, because other evidence suggests that replacing extinct salmon populations through natural or artificial recolonization is unlikely to be successful over time scales of interest to society. In a review of efforts to transplant anadromous salmonids, Withler (1982) found only a small number of successful efforts among hundreds of attempts. For example, there have been only three successful transplants of anadromous sockeye salmon. In contrast, there have been numerous successful transplants of kokanee, the non-anadromous life history form of sockeye salmon (Wood 1995). Similar patterns of success and failure have been found for anadromous and non-anadromous steelhead/rainbow trout populations (Withler 1982). In a recent review, Utter (2000) concluded that anadromous salmonids are also less susceptible to conspecific introgression than are resident salmonid populations. The difficulty of transplanting anadromous salmonid populations and their resistance to nonnative introgression suggests that anadromous life history forms might be more locally adapted to their spawning streams than are the more geographically isolated resident forms (Utter 2000). Other data suggest that resident populations may be more locally adapted, however. In particular, resident populations tend to be more genetically differentiated from each other at molecular markers than conspecific anadromous populations (reviewed by Ward et al. 1994), indicating lower levels of gene flow among resident populations (see also Hendry et al. 2003a—*this volume*). Utter (2000) suggested that the apparent contradiction between molecular marker data, which suggest higher gene flow for anadromous than resident populations, and the transplant data, which suggest that resident populations are less likely to be adapted to specific conditions, is the result of a lack of opportunity for gene flow among resident populations.

In summary, patterns of diversity among Pacific salmon populations appear to be evolutionarily plastic over time scales of hundreds to thousands of years, and in a few cases salmonid populations have been observed to evolve new life history strategies over time frames of decades or less. On the other hand, empirical evidence from transplantation efforts suggests that in most cases anadromous salmonid populations that go extinct will not be rapidly recolonized, suggesting that if salmonid diversity is to be protected in the short term, it will be necessary to conserve many diverse populations.

2.3. Chinook Salmon ESUs

In several status reviews published from 1987 to 1998, the National Marine Fisheries Service (NMFS) identified conservation units for chinook salmon spawning in California, Oregon, Idaho and Washington (Matthews and Waples 1991; Waples et al. 1991; Waknitz et al. 1995; Weitkamp et al. 1995; Busby et al. 1996; Hard et al. 1996; Gustafson et al. 1997; Johnson et al. 1997; Myers et al. 1998; Johnson et al. 1999). All but the first review (Sacramento River winter-run—NMFS 1989) explicitly used Waples' (1991a) concept as an operating framework. To identify ESUs, the teams conducting the status reviews first amassed biological information on the populations in question. Specific data available throughout much of the study area included: freshwater distribution, ocean distribution, freshwater habitat characteristics, life history traits (age at smolting and spawning, adult and juvenile migration timing, fecundity, and egg size), and molecular genetic data (primarily allozyme variation). The teams used patterns of genetic diversity, straying, and geographic distribution to address the issue of reproductive isolation (Waples' first criterion), and patterns of life history diversity and environmental attributes of freshwater habitat to assess the degree to which populations contributed to the "evolutionary legacy" of the species (Waples' second criteria). Based on these data, the teams concluded that there were 15 chinook salmon ESUs in the area studied (Figure 12.4; Myers et al. 1998) The number of ESUs was later expanded to 17 (www.nwr.noaa.gov).

In considering these ESUs, it is important to note that the application of Waples' (1991a) ESU concept involves subjective judgment to determine if levels of reproductive isolation and evolutionary divergence among populations are "substantial" enough for the population groups in question to be considered ESUs. It is therefore not surprising that there was some disagreement within the review teams about whether or not certain populations should be considered

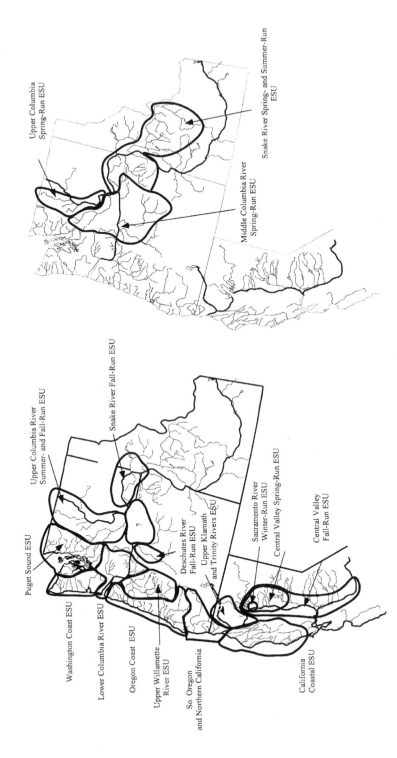

Figure 12.4. Chinook salmon ESUs used by the National Marine Fisheries Service in applying the U.S. Endangered Species Act.

ESUs (Waples et al. 1991, p. 48; Myers et al. 1998, pp. 113–114), and it is possible that a review team with the same mandate but composed of different members would have reached different conclusions about some of the proposed ESUs.

2.3.1. How Would Chinook Salmon ESUs Be Defined under Alternative Conservation Unit Concepts?

Rigorously determining how ESUs would be determined under a variety of ESU concepts is well beyond the scope of this chapter, but is it is possible to make some educated guesses for purposes of illustration. In relation to Moritz's (1994) ESU concept, no fixed mtDNA sequence differences have been observed among chinook salmon populations (e.g., Cronin et al. 1993; Adams et al. 1994), suggesting this species would not be further subdivided into ESUs under Moritz's concept. Similarly, to be considered a conservation unit under Vogler and Desalle's criteria, all members of a population must share at least one heritable character that is entirely absent in all other populations; criteria that would result in few if any subspecific conservation units in chinook salmon. Chinook salmon populations can be found that fall into all four of Dizon et al.'s (1992) categories. If categories I and II (substantially genetically diverged) are considered ESUs, then it seems likely that ESUs identified under Dizon et al.'s concept would be quite similar to the ESUs identified under Waples' concept. In terms of Crandall et al.'s (2000) framework, most of the ESUs identified using Waples' criteria correspond to "Case 5a." These are populations that have recently (within the last 10,000 years) become genetically and ecologically distinct. Some lineages, such as stream-type chinook currently inhabiting the interior Columbia River basin, may be much older, and might be classified as "Case 1" (genetically and ecologically distinct both currently and historically).

3. Evaluating Salmon ESUs: What Has Been Learned?

3.1. Lesson 1: ESUs Are Context Dependent

There are many "reasonable" ways in which biodiversity might be categorized, so any particular manner of defining conservation units will be at least somewhat arbitrary outside of the framework in which it was developed. To be meaningful, conservation units must therefore be defined with respect to a specific goal. In the case of Pacific salmon, for example, Waples (1991a) developed his ESU concept specifically in the context of evaluating subspecific population units for listing under the ESA. The purpose of the ESA is to

> provide a means whereby the ecosystems upon which endangered species and threatened species depend may be conserved, to provide a program for the conservation of such endangered species and threatened

species, and to take such steps as may be appropriate to achieve the purposes of the treaties and conventions set forth in subsection (a) of this section (U.S. code 16:35:1531(b)).

In other words, the overriding purpose of the ESA is to prevent extinctions and allow species to persist in their natural habitats. For vertebrates, the ESA treats "distinct population segments" as though they were true species (U.S. code 16:35:1532:16), but does not provide a biological definition for a distinct population segment. Waples (1991a) used the language of the ESA and other sources of Congressional intent to develop a biological conservation unit concept that met the requirements of the Endangered Species Act. Outside of this context, Waples' ESU definition is simply one of any number of reasonable but arbitrary ways of identifying conservation units.

Would any or all of the other conservation unit concepts summarized above meet the goals of the ESA? Some probably would and some probably would not. For example, it makes little sense to use Vogler and Desalle's concept to identify subspecific groups, because their concept is expressly designed to identify species-level groups. Moritz's concept is used for identifying subspecific taxa, but its focus on mtDNA differentiation seems likely to be considered an overly narrow interpretation of the term "distinct." Alternatively, Moritz's "management units" might be equated with distinct population segments under the ESA. Moritz's (1994) criterion for being considered a management unit is significant differences in allele frequencies from other populations, so most Pacific salmon breeding populations would be considered distinct population segments under this scenario. Although this might reasonably reflect the biology of the species, Waples (1991a, 1995) has argued that this would violate the Congressional intent to list distinct population segments "sparingly." Crandall et al.'s matrix approach might be useful, but would require explicitly identifying which of the many possible combinations of ecological and genetic non-exchangeability correspond to distinct population segments under the ESA. Bowen's GEUs might be reasonably interpreted as "distinct" under the ESA, and indeed some GEU-like populations have been identified as ESUs by the NMFS (see below). Dizon et al.'s ESU concept is similar to Waples' and was formulated in response to a similar legal mandate, and could therefore presumably be used interchangeably with Waples' concept. In general, several alternatives to Waples' (1991a) concept might be reasonably used in the context of the ESA, but this would require explicitly equating one or more of these concepts with the ESA's "distinct population segment" language.

3.2. Lesson 2: Evolutionary Significance Is Not Necessarily Measured by Evolutionary Age

Bowen (1998) and Crandall et al. (2000) have argued that focusing solely on evolutionary divergence in defining conservation units would result in failing to protect recently diverged lineages that might be important to the future persistence of a species. Although Waples' (1991a) ESU concept has at times been criticized for focusing too much on evolutionary history, an examination of the salmon ESUs that have actually been identified using Waples' framework indicates that it has been used to identify both relatively old and young evolutionary lineages. For example, chinook salmon that breed in the Upper Columbia and Snake River Basins have been placed into four ESUs (Figure 12.4): Upper Columbia River spring run (stream type), Upper Columbia River summer–fall run (ocean type), Snake River spring–summer run (stream type), and Snake River fall run (ocean type) (Matthews and Waples 1991; Waples et al. 1991; Waknitz et al. 1995; Myers et al. 1998). The two ocean-type ESUs have diverged from the two stream-type ESUs at a large number of molecular and behavioral traits, and have been estimated to have been on independent evolutionary trajectories for tens of thousands of years (Healey 1991; Myers et al. 1998). However, within each of these lineages, the divergence at molecular markers between the Upper Columbia and Snake River populations is much smaller (Figure 12.4), indicating that these populations are either much more recently diverged (hundreds to thousands of years ago) or that they are connected by relatively high levels of gene flow (Myers et al. 1998). Despite this relatively low degree of molecular genetic divergence, however, the NMFS determined that the Snake and Upper Columbia River populations should be considered as distinct ESUs due to the substantial environmental differences between the two river basins and evidence of behavioral differences between fish from the two systems. In particular, mainstem water temperatures in the Snake River are up to 6°C warmer than Upper Columbia River tributaries, and fall run chinook salmon from the two systems appear to have overlapping but consistently different ocean distributions (Waples et al. 1991). The spawning habitat for the stream-type populations also differs greatly between the two systems, with the Upper Columbia tributaries consisting of relatively cold high-gradient mountain streams and the Snake River spawning areas consisting of somewhat warmer streams meandering through high-elevation meadows (Matthews and Waples 1991). In short, even though the degree of genetic differentiation at molecular markers between these groups is quite low, the review team judged that they were sufficiently ecologically distinct to consider different ESUs.

The flexibility of Waples' ESU concept, illustrated above, is one of its strengths, but this flexibility also illustrates the inherent subjectivity of the approach. Because the degree of reproductive isolation and evolutionary importance necessary to be considered an ESU are not explicitly stated *a priori*, the decision of whether or not to consider a population an ESU rests largely with the judgment of the biologists on the review team making the decision. For example, the level of allozyme diversity among populations of chinook salmon within Puget Sound is comparable to the level of divergence between Snake and Upper Columbia River stream-type chinook, and exceeds the level of divergence between Snake and Upper Columbia River ocean-type chinook (Figure 12.3; Myers et al. 1998). The Puget Sound populations are all considered to be part of the same ESU, however, while the Snake and Columbia River populations were divided into two ESUs within each of the two older lineages (Myers et al. 1998). Presumably, the review teams judged that there was more evidence for evolutionarily significant divergence between the Columbia and Snake River populations than among the Puget Sound populations.

Unfortunately, developing ESU concepts that are useful but neither arbitrary nor subjective is likely to be difficult. For example, Moritz's concept, because its criteria are explicitly measurable, is operationally less subjective than Waples' concept. The price for this greater degree of objectivity, however, is a greater level of arbitrariness. In particular, considering all of the criteria that could reasonably be used, Moritz's focus on mtDNA differentiation is largely arbitrary. At least conceptually, approaches similar to Crandall et al.'s, which use explicitly defined biological criteria to describe where a population falls on a continuum of evolutionary or ecological distinctiveness might prove useful. How such concepts would fit into existing legal framework, such as the ESA, is hard to predict.

3.3. Lesson 3: Arguments Over ESUs Are Not Just Academic

Much of the debate on conservation units in the scientific literature is esoteric, and (as I've argued above) at least in part semantic. In the context of Pacific salmon conservation, however, the debate over how to define conservation units has greatly influenced applied conservation strategies. Two issues that are particularly controversial are the degree to which ESUs should be defined on the basis of evolutionary versus ecological criteria, and under what circumstances artificially propagated populations should be included as part of legally defined conservation units (discussed below). Waples' (1991a) ESU concept is intended to identify evolutionary lineages, and therefore necessarily requires that recovery efforts focus on those lineages. Some have argued, however, that this evolutionary focus unnecessarily constrains recovery strategies. For example, under the NMFS's implementation of Waples' (1991a) ESU concept, introduction of fish from one ESU to increase the abundance of another ESU is generally viewed as a factor that would increase the degree of risk faced by the ESU receiving the introduction (NMFS 1993). Others view such introductions as beneficial if they result in an increase in salmon abundance in a particular ecosystem, and view the risks to the native lineage of such introductions as acceptable (e.g., Mundy et al. 1995). The choice of ESU concept and decisions about ESU boundaries therefore largely influences how one approaches conservation of depressed salmon populations.

3.4. Lesson 4: Pacific Salmon ESUs Have Legal Status, Which Leads to Unintended Consequences

Pacific salmon ESUs have legal as well as biological meaning, and this has led to some unintended consequences and tensions. The clearest example of this comes from debates about when and whether to include artificially propagated salmon in ESUs. Pacific salmon are artificially propagated in hatcheries on a large scale and most ESUs have hatchery-propagated populations that are bred and released within the freshwater range of the ESU. Some of these populations were derived primarily from locally captured adults, whereas others were introduced from distant streams. Under Waples' (1991a) concept of ESUs as evolutionary lineages, it is relatively straightforward to determine whether a given hatchery population should biologically be considered part of an ESU. In particular, the NMFS usually included a hatchery population in an ESU if the hatchery population was primarily derived from fish that originated from that ESU and if the hatchery population had not become obviously diverged from the original ESU due to artificial selection or other factors. However, because most hatchery populations were never intended to be used for conservation purposes and in many cases could not realistically be converted to that role, the NMFS had a policy of only formally listing hatchery populations if they were considered essential for an ESU's recovery (NMFS 1993). In other words, in many cases the natural populations of an ESU were listed under the ESA, but the hatchery populations that were biologically part of the same ESU remained unlisted. This division was useful for both fishery management and conservation purposes, because it allowed greater flexibility in the management of hatchery fish than would have been possible if those fish were listed under the ESA without sacrificing protection when it was needed. However, in 2001, a federal judge ruled that the NMFS's policy was illegal, because the ESA did not authorize the NMFS to list taxa within distinct population segments or ESUs (*Alsea Valley Alliance v. Evans*, 161 F.Supp.2d 1154 (D. Oregon 2001), appeal pending, No. 01-36071 (9th Cir.)). Simply put, the judge ruled that the legal and biological meanings of an ESU had to exactly coincide, implying that either the ESU concept had to be modified to explicitly exclude hatchery populations, or any hatchery populations that were biologically considered part of an ESU must be listed along with the natural populations in the ESU. As of this writing the judge's decision was under appeal, but the lesson is that it can sometimes be difficult to foresee and reconcile all of the potential pitfalls that can arise from the interaction between the legal and biological uses of conservation units.

4. Summary

Since the early 1990s, Pacific salmon have been a sort of test case for developing and applying criteria for defining intraspecific conservation units. The resulting biological and legal scrutiny has served to illuminate critical issues and differences in opinion about how conservation units should be defined and identified. Some of the most important insights gained from the process of identifying Pacific salmon conservation units are: (1) conservation units are most useful if they are defined in response to a particular management goal, and different goals will result in different units; (2) that identifying conservation units can be a flexible process, resulting in both relatively old and relatively young ESUs; (3) that how conservation units are identified and defined can have real effects on not only which populations are protected but also how they are protected, and

(4) that conservation units will often have legal as well as biological meanings, and sometimes the two can conflict.

Acknowledgments   Robin Waples, Jeff Hard, Jerry Johnson, David Teel, Andrew Hendry, and Jim Myers provided valuable comments on earlier drafts of this chapter. David Teel and Anne Marshall kindly provided unpublished genetic data.

Toward Evolutionary Management
Lessons from Salmonids

Kyle A. Young

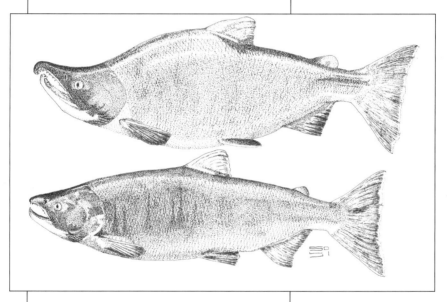

Male and female sockeye salmon

13

The science of fisheries management has traditionally been almost purely ecological (Frank and Leggett 1994; Trippel 1995; Levin and Schiewe 2001). Harvest rates are determined by the demographic indices and vital rates of populations, with little regard for the distribution of phenotypes or how their relative fitnesses are affected by anthropogenic changes to the biotic and abiotic environment. This is not to say that fisheries scientists have not considered evolutionary principles or used managed populations to test evolutionary theory, only that evolutionary principles have not consistently informed management actions. Except for a few notable exceptions (e.g., Chapter 5 in Pitcher and Hart 1982; Stokes et al. 1993; Heino 1998; Mustafa 1999), the striking absence of evolutionary perspectives in most volumes on fisheries management supports this view (Ellis 1977; Everhart and Youngs 1981; Walters 1986; McNeil 1988; Pitcher et al. 1998; Funk et al. 1998). In contrast, conservation (considered here as the collection of actions necessitated by prior mismanagement) is explicitly evolutionary. Conservation efforts to protect harvested species from further mismanagement view molecular genetic and phenotypic data in the context of adaptive landscapes and phylogenetic relationships, and consider the evolutionary consequences of gene flow and effective population size (Waples 1995, 2003—*this volume*; Nielsen 1995; Soulé 1986; Dimmick et al. 1999; Landweber and Dobson 1999; McElhany et al. 2000; Stockwell et al. 2003).

In this chapter, I propose one possible conceptual framework through which evolutionary management might advance. It is based on the simple premise that management actions associated with the exploitation of species can change the force of natural selection by changing the relationship between phenotypes, their environment, and components of fitness (Wade and Kalisz 1990). The framework views changes in this relationship in the context of two evolutionary principles. First, selection can act on organizational levels above the individual, particularly in species with hierarchically structured populations (Sober 1984; Futuyma 1986). Second, there are two fundamentally different types of individual selection, one resulting from interactions between individuals, the other resulting from interactions between individuals and the environment (Wallace 1975).

1. Evolutionary Management: A Contextual Perspective

Although evolutionists have long appreciated that selection can act on organizational levels above the individual, only recently have quantitative approaches for considering the relative importance of different levels and types of selection been clearly articulated (Wade 1985; Heisler and Damuth 1987; Goodnight et al. 1992). "Contextual analysis" recognizes that an individual's fitness can depend on interactions between its own phenotype and the environment, interactions between its own phenotype and the phenotypes of conspecific individuals, and the mean phenotype of the group to which the individual belongs (Goodnight et al. 1992). Originally designed to quantify the relative strength of different levels and types of selection, contextual analysis provides a useful conceptual framework for considering management actions from an evolutionary perspective (Figure 13.1).

1.1. Group Selection, Soft Selection, and Hard Selection

Group selection is traditionally defined as genetic change arising from the differential extinction or proliferation of groups of individuals (Wade 1978). Just as evolution by individual selection requires genetic variation among individuals, evolution by group selection requires genetic variation among groups. Though experiments have demonstrated group selection can operate (Wade 1976; Goodnight 1985), its importance has been questioned because it is unlikely that selection acting on groups will be stronger than selection acting on individuals (Williams 1966a; Wade 1978). I suggest that the concept retains heuristic value, particularly in the present context, and that this most potent criticism is based on an inappropriate criterion. We need not require that group selection be a stronger evolutionary force than individual selection, only that individual fitness can be influenced by group membership (Goodnight et al. 1992). As an example, consider two individuals that successfully migrate through a harvest area at the same time, but whose populations have different mean migration times. If one of the populations is driven to extinction because a fishery is open around its mean migration time, the genes of one individual will be lost, while those of the other will likely persist. Thus, both individuals are favored by individual selection on migration timing, but their fitnesses also depend on the mean migration time of the groups to which they belong (Figure 13.1).

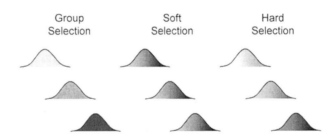

Figure 13.1. Contextual analysis recognizes that three types of selection can affect an individual's relative fitness. For each type of selection the three normal curves represent the distribution of phenotypes within different populations (groups). Shading represents the relative fitness of different phenotypes (darker is higher fitness). Under group selection an individual's relative fitness is influenced by the mean phenotype of the group. Under soft selection, relative fitness depends on the phenotypes of group members, but not on the phenotypes of individuals in other groups. Under hard selection, relative fitness is influenced by an individual's phenotype relative to all other phenotypes in the global population, independently of group membership. (Adapted from Goodnight et al. 1992, *American Naturalist* 140: 743–761, with permission from the University of Chicago Press.)

In addition to considering the level of selection, contextual analysis distinguishes between two kinds of individual selection. The terms "soft" and "hard" selection were first introduced in the context of genetic load (Wallace 1968), and later generalized to describe two types of individual selection (Wallace 1975). Soft selection is frequency- and/or density-dependent selection (Wallace's original definition required that it be both). It is the component of individual fitness that depends on the phenotypes of conspecific competitors within an individual's population (group), not the population or global mean phenotype (Figure 13.1). For this reason, soft selection is also referred to as "local" selection because its contribution to relative fitness arises from interactions within the local population (Wade 1985). For the purposes of this chapter, I will extend the definition of soft selection to include the effects of interspecific interactions on relative fitness. In general, we can view soft selection as the relationship between phenotypes, the biotic environment, and fitness.

Hard selection is frequency- and density-independent selection. The fitness of an individual depends on the interaction between its own phenotype and the environment, not the phenotypes of other group members. Hard selection is "global" in that the relative fitness of an individual depends on the phenotypes of all other conspecific individuals, regardless of group membership (Figure 13.1). Any external selective force acting simultaneously on individuals from one or more groups can produce hard selection. In general, hard selection can be viewed as the relationship between phenotypes, the physical environment, and fitness. Typical examples include variation in relative fitness resulting from temperature extremes or catastrophic disturbance events (e.g., flood, fire, drought).

It is important to appreciate that group and individual selection, and soft and hard selection, are not necessarily opposing or independent. Group selection is often associated with explaining the persistence of traits or altruistic behaviors that reduce individual fitness but increase group fitness (Williams 1966a; Wilson 1975), a phenomenon better explained by the concepts of kin selection and inclusive fitness (Hamilton 1964a,b). There is no reason why a trait (e.g., high fecundity) cannot simultaneously increase individual fitness and promote population persistence. Similarly, the same trait (e.g., body size) may improve competitive ability (soft selection) and increase survivorship during extreme environmental events (hard selection). The operation of one type of individual selection may depend upon that of the other. For example, density-dependent competitive interactions may affect habitat selection, and thus change the form or strength of hard selection by changing the relationship between displaced phenotypes and the physical environment (Wallace 1975).

1.2. Contextual Management

Using a contextual perspective as the conceptual framework for evolutionary management is appealing for several reasons. First, harvested species are often organized hierarchically into local breeding aggregates, populations, and groups of populations. We can view management actions that differentially affect organizational levels above the individual as a type of artificial group selection (even if the selective mechanism itself does not satisfy the traditional definition of group selection). Second, management generally involves two components: environmental manipulation and demographic manipulation. The former may result from management activities not directly related to the target species, whereas the latter may be an indirect result of environmental manipulations or a direct result of harvest. In general, we can view environmental and demographic manipulations as changing hard and soft selection, respectively, but the two types of selection may often be related.

2. Contextual Management: A Salmonid Perspective

2.1. The Salmonid System

Salmonids provide an excellent system with which to illustrate the utility of the contextual approach to evolutionary management. First, salmonids display population genetic structure at spatial scales ranging from stream, to basin, to region (e.g., Campton and Utter 1987; Wehrhahn and Powell 1987; Utter et al. 1989; Nielsen et al. 1997b; National Marine Fisheries Service status reviews). Correlated with this population genetic structure, there is genetically based variation in phenotypic traits (Taylor 1991b; Groot and Margolis 1991) and variation in rates of population decline and extinction (Walters and Cahoon 1985; Nehlsen et al. 1991; National Marine Fisheries Service status reviews). Thus, there are genetically and phenotypically distinct groups whose chances of persistence are differentially affected by various management actions. Second, salmonid management involves two distinct components. We manage salmonids directly through the commercial and sport harvest of individuals, and indirectly by land use activities (e.g., forestry, agriculture, urbanization, hydroelectric development) that change the physical environments used for rearing, migration, and reproduction. A contextual perspective can help identify how different management actions might independently and interactively affect the strength and form of the two types of individual selection.

The goal of this section is not to catalog examples of actual and potential evolutionary responses by salmonids to management actions. Instead I will provide examples that illustrate how the salmonid system and contextual management mutually inform one another. I will focus on wild populations, but use information from hatchery and transplanted populations when appropriate. In some cases, there will be good evidence for an evolutionary response to management actions. In other cases, I will draw upon experimental results to infer the likely evolutionary consequences of various management actions.

2.2. Population Level Responses to Management

None of the seven species of North American Pacific salmon and trout is at risk of global extinction, but all have experienced range contractions and local extinctions, particularly in the contiguous United States (Nehlsen et al. 1991; Slaney et al. 1996; CPMPNAS 1996). From a contextual perspective, management actions have selected against certain populations and groups of populations. The results of some management actions are obvious. Impassable dams and severe environmental degradation can result in immediate local extinction, or at least cause the emigration of an entire population. Cumulative anthropogenic effects are responsible for the dramatic declines and higher extinction rates of Pacific (Nehlsen et al. 1991) and Atlantic salmon (Kellogg 1999) populations in the southern portions of their geographic ranges. A more subtle, but long-recognized mechanism that can differentially affect populations involves mixed stock fisheries (Ricker 1958; Paulik et al. 1967; Hilborn 1985). When large and small populations are harvested in a common fishery, exploitation rates are determined by the larger, more productive populations. As a result, smaller, less productive populations, which already have a higher risk of extinction (Lande 1993; Routledge and Irvine 1999), are exposed to unsustainably high harvest rates.

There is good evidence that mixed stock fisheries can drive smaller, less productive populations to extinction. Walters and Cahoon (1985) found evidence for higher extinction rates in small than in large populations in four species of Pacific salmon harvested in mixed stock fisheries off the coast of British Columbia (sockeye salmon were not included in their analysis). The same pattern was observed in all four species, but was most dramatic for coho and chum. Between 1948 and 1983, the number of stocks contributing to cumulative relative spawner escapement declined. Small, less productive stocks declined and/or went extinct, while large, more productive stocks accounted for an increasing percentage of total escapement. More recently, Walters (unpublished) analyzed historical harvest and spawner escapement data for 984 British Columbia coho populations. His analysis suggests a mean optimum exploitation rate for these populations of approximately 50%, well below the actual harvest rate imposed by mixed stock sport and commercial fisheries between 1950 and 1997 (Figure 13.2). After correcting the data for observation errors (Walters and Ludwig 1981; Ludwig and Walters 1981), he found that the distribution of population growth rates *after* harvest (assuming a constant 70% exploitation rate) was nonnormal and positively skewed (Figure 13.2). There are fewer low productivity populations than expected if populations are as likely to be below as above the mean. This pattern may result from natural extinction–colonization dynamics, or a sampling bias toward more productive stocks (although both effects were corrected for in the analysis), but supports the idea that mixed stock fisheries have driven less productive populations to extinction.

Oregon coast coho populations provide an example where land management practices appear to have altered the expected effect of a mixed stock fishery. The Oregon Department of Fish and Game has collected single "peak count" estimates of spawner abundance ([no. of 3-year-old adults plus no. of 2-year-old jacks]/km) for 30 coastal coho populations since 1950. The populations were harvested in a mixed stock fishery until 1993, when the fishery was closed. I corrected the data for observer bias (as described in Young 1999) and for each population regressed spawner density on time (coded 1 to 51). Between

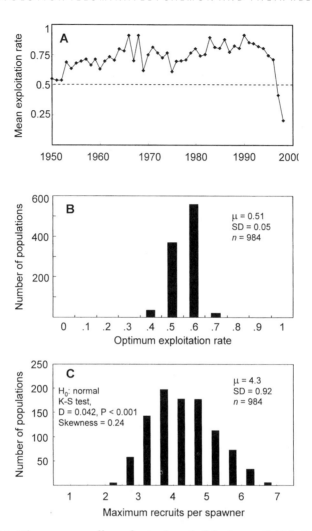

Figure 13.2. The apparent effect of mixed-stock fisheries on British Columbia coho salmon populations. (A) Estimated mean exploitation rate of British Columbia coho salmon between 1950 and 1998. The dotted line is the mean optimum exploitation rate from (B). (B) The distribution of optimum maximum exploitation rates for 984 coho salmon populations with at least five stock–recruitment observations. (C) The distribution of maximum population growth rates for the same 984 populations assuming a historical mean exploitation rate of 70%. There are fewer low productivity stocks than expected if population growth rates were normally distributed.

1950 and 2000, spawner density has declined in all but two of the populations (Figure 13.3). Surprisingly, small populations declined at a slower rate than large populations (Figure 13.4). Because densities early in the time series constrain long-term rates of change (small populations can decline at only so high a rate), I repeated the analysis using standardized regression (Neter et al. 1989, p. 289),

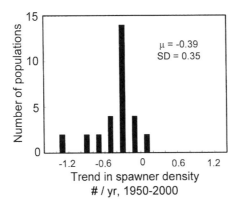

Figure 13.3. The distribution of rates of change in spawner density for 30 coho salmon populations from the Oregon coast between 1950 and 2000.

and by decade using both regression methods. Though the significant negative relationship between population size and rate of decline disappeared, small populations did not decline significantly faster than large populations during any time period between 1950 and 2000.

One explanation for this unexpected pattern is that at low density, small populations have higher maximum population growth rates than large populations. This seems unlikely because when spawner abundance is reduced by harvest, large populations should have more available habitat than small populations. Furthermore, a recent meta-analysis by Myers et al. (1999) found little interpopulation variation in maximum population growth rate at low population

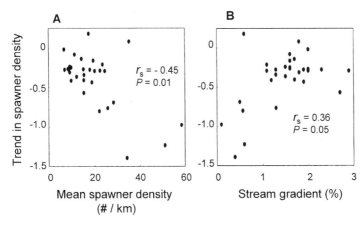

Figure 13.4. (A) The relationship (Spearman rank correlation) between mean spawner abundance and the trend in spawner abundance for 30 coho salmon populations from the Oregon coast between 1950 and 2000. (B) The relationship (Spearman rank correlation) between stream gradient and the trend in spawner abundance for the same populations.

size in salmonids. Another possible explanation is that large populations actually experienced higher exploitation rates, perhaps because they were targeted by time-selective commercial fisheries or sport fisheries.

There is also the possibility that land management activities have affected the freshwater habitats (and thus population growth rates) of small and large populations differently. In these populations, mean spawner density is negatively related to stream elevation ($r_s = -0.49$, $P = 0.02$), and stream gradient ($r_s = -0.43$, $P = 0.02$), which is a proxy for habitat quality (low-gradient streams generally provide better habitat for juvenile coho) (Young 1999; Sharma and Hilborn 2001). Thus, small populations tend to occur in high-gradient, up-basin watersheds, whereas large populations occur in low-gradient, down-basin watersheds. Populations from low-gradient streams declined at a faster rate than populations from high-gradient streams (Figure 13.4). Evidence suggests that the cumulative effects of multiple anthropogenic stressors (e.g., logging, farming, dyking, splash damming, large woody debris removal) are more severe in low-gradient, down-basin watersheds than in high-gradient, up-basin watersheds, where logging is usually the only management activity and occurred more recently (Bisson et al. 1987; Beechie et al. 1994; Ralph et al. 1994). Thus, the expected effect of mixed stock fisheries—more severe declines in small populations—may have been offset by one or more management actions differentially affecting the freshwater habitat of larger populations.

2.3. The Hard and Soft of Age and Size at Maturity

Harvest can impose artificial selection on two related life history traits: age and size at maturity. Both age at maturity (Gjerde 1984a; Silverstein and Hershberger 1992; Hankin et al. 1993; Heath et al. 1994, 2002c) and growth rate (Kincaid et al. 1977; Refstie and Steine 1978; Beacham and Murray 1988; Smoker et al. 1994; Kinnison et al. 1998c) are heritable in salmonids, providing the opportunity for an evolutionary response to artificial selection (see also Hard 2003—*this volume*). Changes in age and size at maturity in response to harvest (or another mortality source) have been modeled using a number of approaches (Law and Grey 1989; Brown and Parman 1993; Kirkpatrick 1993; Heino 1998), and experimentally demonstrated in Atlantic salmon (Gjerde 1984a), guppies (*Poecilia reticulata*) (reviewed in Reznick 1993), and other non-salmonid fishes (Stearns 1992; Conover and Munch 2002). Still, because it is impossible to control environmental conditions throughout the life span of wild salmonids, it requires a bit of Darwinian faith to credit observed changes in age and size at maturity to evolutionary responses to harvest. Indeed, many studies have concluded that such changes are the result of developmental plasticity in response to changing environmental conditions (Peterman 1984; Peterman et al.1986; Stearns and Koella 1986; Ishida et al. 1993; Policansky 1993; McKinnell 1995; Trippel 1995; Cooney and Brodeur 1998; Friedland 1998).

Ricker (1981) compiled data on the age and size of five Pacific salmon species harvested by commercial fisheries in British Columbia, and found that the size of all five species declined (by 8–33%) between 1951 and 1975. The declines in size of pink and coho salmon were most extreme and were not associated with reduced age at maturity, as was the case for chinook. The size of chum salmon declined despite an increase in age at maturity. Early-maturing, fast-growing fish were intercepted by gill net fisheries targeting pink and sockeye salmon (see Hamon et al. (2000) for a recent example documenting artificial selection by gill net fisheries). The decline in size of sockeye was minimal, apparently because the effect of the size-selective fishery was offset by higher growth rates associated with the gradual cooling of the north Pacific Ocean. Similar reductions in the age and size at maturity have been observed in Pacific salmon through 1993 (Bigler et al. 1996), and in harvested populations of Atlantic salmon (Bielak and Power 1986; Sharov and Zubchenko 1993).

Ricker (1981) suggested that changes in age and size at maturity were evolutionary responses to artificial selection. There are a number of ways that harvest could select for decreased age and size at maturity (but see Heino (1998) for the opposite response). Gill net fisheries differentially harvest larger (older) fish, and thus impose artificial selection for small, early-maturing genotypes. Even if fishing gear is not size-selective, the artificial increase in marine mortality rate of late-maturing fish would favor early-maturing genotypes spending less time in the marine environment. Finally, if there is a size threshold for vulnerability to harvest, fast-growing fish may experience increased mortality relative to slower-growing fish because they spend more time in the temporal harvest window.

Because mixed-stock fisheries are abiotic-selective agents acting on phenotypes from multiple populations simultaneously, the evolutionary explanations above invoke hard selection. The effect of harvest on the relative fitness of an individual depends on the phenotypes of all fish in the globally harvested population, not on interactions between phenotypes within a single population. Yet the decline in age and size at maturity observed in Pacific salmon is also consistent with an alternative explanation invoking soft selection operating within populations: density-dependent sexual selection on size during spawning.

Sexual selection in salmonids has been well studied. Female fitness depends on body size via fecundity, the ability to acquire and defend favorable nest sites, and the ability to deposit eggs at depths sufficient to avoid egg loss owing to dig-up by other females or bed load scour during floods (van den Berghe and Gross 1989; Holtby and Healey 1986, 1990; Fleming and Reynolds 2003—*this volume*). Male fitness depends the ability to gain access to spawning females, which is related to body size (Fleming and Reynolds 2003—*this volume*). The opportunity for selection (variance in fitness) on body size in one or both sexes increases with breeding density in coho (van den Berghe and Gross 1989; Chebanov 1990; Fleming and Gross 1994), sockeye (Chebanov 1991), chum (Chebanov 1984), and pink salmon (Chebanov 1986). Competition increases the strength of selection on female (van den Berghe and Gross 1989) and male (Chebanov 1990) body size in coho, and male body size in pink salmon (Chebanov 1986). Though some data are inconclusive or contrary, the balance of evidence suggests that density-dependent sexual selection favors larger body sizes in both male and female salmonids. Thus, harvest may reduce age and size at maturity directly by hard selection, but also indirectly by reducing spawning densities and thus decreasing the strength of soft sexual selection on body size during breeding.

Data from Oregon coast coho populations illustrate the potentially complex relationship between hard selection, soft selection, and changes in age and size at maturity. In these populations, all females and most males mature as 3-year-olds after 18 months in the ocean. Some males mature as 2-year-old precocious "jacks" after only 6 months in the ocean. Jacks use "sneaking" instead of "fighting" tactics to gain access to spawning females (Dominey 1984; Gross 1985; Sandercock 1991). Precocious maturation is heritable (Iwamoto et al. 1984; Silverstein and Hershberger 1992), but, as evidenced from hatchery populations where males mature precociously despite the exclusion of jacks from the breeding population, depends primarily on males reaching a size/growth-rate threshold during freshwater rearing (Hager and Noble 1976; Brannon et al. 1982; Bilton et al. 1982): a threshold-dependent size-based mating polymorphism in the categories of Roff (1996b; see also Hutchings 2003—*this volume*). Numerous authors have suggested that jacks are half of a frequency dependent mixed evolutionary stable strategy, and at frequencies observed in natural populations should have a similar mean fitness (survival × reproductive success) as adult males (Gross 1985, 1996; Bohlin et al. 1990; Charnov 1993; Fleming and Reynolds 2003—*this volume*). One obvious, but untested prediction is that the harvest of adult males should increase the survival advantage of precocious males and thus increase their proportional representation in populations (Myers 1986; Myers et al. 1986; Gross 1991).

Using the same data as above, I calculated the jack proportion (jacks/(jacks + adult males), assuming a 1:1 adult sex ratio (Young 1999)) for 30 populations from 1950 to 1993, the last year of substantial commercial harvest of Oregon coast coho salmon. For each population, I then regressed jack proportion on time (coded 1 to 44) to determine if jack proportions increased in response to artificial selection on adult males. Consistent with the prediction, jack proportion increased in all but three of the populations between 1950 and 1993 (Figure 13.5). Furthermore, there was a significant decrease in jack proportion across all populations following the closure of commercial fisheries in 1993 (ANOVA with 1950–1993 (mean = 0.28, S.D. = 0.23) and 1994–2000 (0.25, 0.24) as two "times"; $F_{1,1308} = 4.0$, $P = 0.04$). If the increases in jack proportion between 1950 and 1993 were simply a result of increased adult mortality due to commercial harvest, the rate of change in a population's jack proportion should be inversely related to the rate of change in adult abundance. To test this possibility, I conducted standardized regression of both demographic indices on time for each population (Neter et al. 1989, p. 289). I found no relationship between the rate of increase in jack proportion and the rate of decline in adult abundance (Figure 13.5).

There are at least two possible explanations for this negative result. First, habitat degradation and over-harvest have reduced stream productivity and the number of nutrient rich carcasses decomposing in streams. Both habitat degradation and low carcass density can reduce juvenile growth rates (Quinn and Peterson 1996; Bilby et al. 1998), which could reduce the percentage of juvenile

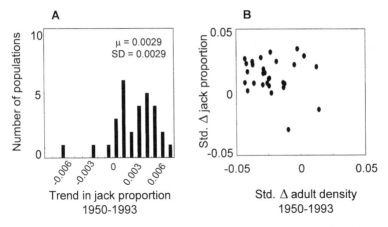

Figure 13.5. (A) The distribution of rates of change in the proportion of male spawners that were 2-year-old jacks (vs. 3-year-old adults) for 30 salmon coho populations from the Oregon coast between 1950 and 1993. (B) The relationship between the standardized rate of change in adult spawner density and the standardized rate of change in jack proportion for the same 30 populations.

males reaching the size threshold necessary for precocious maturation. Second, sneaking tactics are favored at higher breeding densities (Roff 1996b), and as breeding density declines, sexual selection on body size in male coho changes from disruptive to directional (Fleming and Gross 1994). Thus, Oregon coho salmon provide an example where the expected decrease in male age at maturity in response to hard selection (harvest) is observed, but may be mediated by soft selection via the removal of density-dependent disruptive selection on the two size-/age-dependent mating strategies.

2.4. Hard and Soft Selection During Other Life History Stages

Management actions have the potential to change the type, strength, and form of natural selection acting on traits other than age and size at maturity. Here, I summarize evidence suggesting that changes in the abiotic (hard selection) and biotic (soft selection) environment can produce evolutionary changes in juvenile traits. I then review information on three related aspects of juvenile ecology (emergence timing, intraspecific competition, and interspecific competition) that can affect patterns of habitat selection and components of juvenile fitness (growth and survival). I conclude with a contextual perspective on how various management actions might affect hard and soft selection operating during the juvenile stage. Few studies have quantified selection on naturally occurring phenotypic variation among juveniles in natural populations (but see Hendry et al. 2003c). Furthermore, other than studies of hatchery and transplanted populations, I do not know of any studies relating management actions to genetic changes in juvenile traits. Consequently, we will often have to make evolutionary inferences based on studies showing that changes in environmental and demographic conditions are related to changes in the mean and variance of components of juvenile fitness.

There are countless comparative studies that document differences in juvenile traits among natural populations, but relatively few that have demonstrated such differences have a genetic basis (e.g., Riddell and Leggett 1981; Riddell et al. 1981; Rosenau and McPhail 1987; Swain and Holtby 1989; Hendry et al. 1998; Kinnison et al. 1998c). Experiments comparing hatchery and wild populations show that artificial selection associated with changes in physical habitat (hatchery ponds vs. natural streams) and the intensity of intraspecific competition (artificially high vs. natural densities) can produce genetic changes in juvenile traits in only a handful of generations. Though juvenile morphology is highly plastic, coho fry from hatchery and wild populations exhibited heritable morphological divergence after only two to five generations (Swain et al. 1991). After two to seven generations, juveniles from hatchery populations behave more aggressively and outcompete wild conspecifics under experimental conditions (Swain and Riddell 1990; Berejikian et al. 1996, 1999). Using a reciprocal transplant approach, Reisenbichler and McIntyre (1977) documented a fitness tradeoff between a new (three generations) hatchery population of steelhead and its wild source population. When reared in sympatry in hatchery ponds and natural streams, individuals from each population had higher growth and survival rates in their "native" habitat. Fleming and Einum (1997) have documented morphological divergence and a similar fitness tradeoff between wild Atlantic salmon and a seven-generation-old hatchery population.

It is important to appreciate that hatchery "experiments" confound changes in the physical environment (hard selection) with changes in the intensity of intraspecific competition (soft selection). Thus, the contribution of each to evolved changes in morphology, behavior, and competitive ability remains unknown. Nevertheless, these studies provide good evidence that juvenile traits evolve in response to changes in the abiotic and biotic environment, and that such changes can occur rapidly.

Stream-dwelling juvenile salmonids engage in intra- and interspecific interference competition for energetically favorable positions in the water column (Chapman 1966; Fausch 1984). Early emergence can improve competitive ability by at least two mechanisms: prior occupation of favorable habitat, and the size advantage gained by early initiation of exogenous feeding (Chapman 1962; Einum and Fleming 2000b). Early-emerging coho salmon (Chapman 1962) and steelhead (Chandler and Bjornn 1988) displace later-emerging fish from favorable foraging sites. Large juveniles have higher foraging (Abbott et al. 1985; Lahti and Lower 2000) and survival (Quinn and Peterson 1996; Einum and Fleming 2000b) rates than small juveniles. Thus, variation in emergence timing (and thus size) can affect hard selection by changing the relationship between displaced phenotypes and the environment, and the opportunity for soft selection by producing variation in two components of juvenile fitness: growth, and survival.

Density-dependent intraspecific competition can affect the relationship between phenotypes and the physical environment (hard selection) and the evolution of agonistic behavior (soft selection). It can also increase the variance in components of juvenile fitness. Density-dependent habitat selection has been documented in juvenile Atlantic and coho salmon (Bult et al. 1999; Young 2001). The frequency of aggressive encounters increases with the intensity of competition in rainbow trout (Slaney and Northcote 1974). Rosenau and McPhail (1987) controlled for environmental effects and found that juvenile coho from a high-density population outcompeted and initiated more agonistic behaviors than juveniles from a low-density population. Intraspecific competition can also increase the strength of, and/or opportunity for, density-dependent soft selection. Hume and Parkinson (1987) found that the mortality rate of steelhead fry increased with stocking density. Keeley (2001) showed that intraspecific competition in steelhead fry increased the variance in growth rate. Similar to a pattern observed in coho salmon (Fagerlund et al. 1981), Keeley's data also suggest that intraspecific competition results in a negatively skewed distribution of growth rates, meaning that competition reduces the proportion of fish with high growth rates.

The effect of interspecific competition on habitat selection and components of fitness depends on species identity, density, and the size asymmetry between species (Fausch and White 1986; Fausch 1988, Rodríguez 1995; Sabo and Pauley 1997). In two Idaho (U.S.A.) streams, spring chinook salmon and steelhead fry occupied different microhabitats whether in allopatry or sympatry (Everest and Chapman 1972). Consistent with this observation, McMichael and Pearsons (1998) found that chinook salmon juveniles had no effect on steelhead growth rates. Alternatively, Taylor (1991a) found that chinook salmon were outcompeted by coho salmon of similar size in laboratory trials, and displaced by coho salmon from preferred pool habitats in both experimental and natural stream channels. Stein et al. (1972) found that coho salmon reduced the growth rate of similarly sized chinook salmon. Studying populations where chinook salmon emerged earlier and were larger than coho salmon, Lister and Genoe (1970) concluded that larger chinook salmon outcompeted and moved into favorable habitats earlier than coho salmon. I have manipulated the natural size asymmetry between competitively dominant, early-emerging (larger) coho salmon and late-emerging (smaller) steelhead fry to explore the relationship between interspecific size asymmetries and density-dependent habitat selection (Young 2001, In press). In the absence of interspecific competition both species have higher growth rates in pools than in riffles (Young 2001, In press). When coho salmon had their natural size advantage, competition was strongly asymmetric; coho salmon displaced steelhead from pools to riffles, whereas steelhead had no effect on coho salmon habitat selection. When the natural size/competitive asymmetry was removed, the per capita effect of coho salmon on steelhead habitat selection was reduced twofold, and was similar to the per capita effect of steelhead on coho salmon habitat selection.

In total, there is good evidence that intra- and interspecific size asymmetries and density-dependent competition can affect the relationships between phenotypes, the biotic and abiotic environment, and fitness. If selection acts on heritable phenotypic variation, management activities that change emergence times, size distributions, and densities of juveniles will produce evolutionary change. Changes in juvenile size distributions and densities, within or between species, can be viewed as changes in frequency- and density-dependent soft selection, respectively, because the relative fitness of an individual will depend on the relative size and density of con-/heterospecific competitors. When frequency- and/or density-dependent competition results in habitat shifts, soft selection will change the form of hard selection; displaced phenotypes will experience different physical environments than they would in the absence of competition (Wallace 1975).

Harvesting adults reduces the number of spawning females, and thus the densities, mortality rates, and intensity of intraspecific competition in juveniles. The same management action that reduces the strength of density-dependent soft selection on adult body size during breeding, may also remove density-dependent habitat selection, and change the strength and type of selection acting during the juvenile stage. Spawning (and thus emergence) timing in salmonids is heritable (Smoker et al. 1998; Quinn et al. 2000), and harvest regimes that target early or late spawners can change the temporal distribution of spawning and emergence times (Reisenbichler 1997). Because intra- and interspecific size asymmetries affect patterns of habitat selection and the variance in components of juvenile fitness, such harvest regimes may interactively affect hard and soft selection acting on juvenile phenotypes. Land management practices that change water temperature during the egg stage (e.g., riparian logging, dam construction) can change fry emergence time and size (Beacham and Murray 1990). Temperature changes may alter interspecific size asymmetries, because the effect of temperature on alevin emergence time and size varies between species (Beacham and Murray 1990). Finally, land management activities that degrade freshwater habitat can change hard and soft selection by altering the physical environment and changing the absolute and relative abundance of species, respectively (Reeves et al. 1993).

3. Discussion

3.1. Implications for Evolutionary Management

Considering the salmonid system from a contextual perspective suggests that management actions have the potential to produce a variety of evolutionary responses. Harvest and land management activities can affect populations differently depending on their demographic properties and environmental associations. Harvest can impose direct hard selection on adult life history traits. Because it weakens density-dependent effects, harvest also reduces the importance of soft selection during adult and juvenile stages. Changes in intra- and interspecific juvenile size asymmetries can affect hard selection by changing patterns of habitat use, and soft selection by changing the mean and variance of juvenile components of fitness. Land management can directly affect hard selection by changing the relationship between phenotypes and the environment (e.g., changes in adult spawning and juvenile rearing habitat, water temperature, natural disturbance regimes).

Though it would be unrealistic to eliminate the evolutionary effects of management, if one of the goals of management is to minimize our "evolutionary footprint" by maintaining natural evolutionary processes, a contextual perspective suggests three priorities. First, eliminate harvest regimes with the potential to drive small, low-productivity populations to extinction. Second, adjust harvest regimes and land management practices so as to maintain the natural form and strength of hard selection acting on adult and juvenile phenotypes. Third, maintain density- and frequency-dependent soft selection operating during adult breeding and juvenile rearing.

Fisheries management and conservation policies are relatively well attuned to population (group) level effects. In recent years, numerous Pacific salmon fisheries have been closed or reduced in an effort to protect smaller stocks. Conservation plans and recovery strategies are developed for organizational units below the species level (Waples 1991a; McElhany et al. 2000; National Marine Fisheries Service status reviews). Nevertheless, until managers heed repeated warnings concerning the dangers of mixed stock fisheries (Ricker 1958; Hilborn 1985; Walters 1999), smaller, less productive populations will continue to go extinct. In theory, it should be possible to adjust harvest regimes so that hard selection is neutral with regard to age and size at maturity, and spawning time (see also Hard 2003—*this volume*). Land management practices have improved in recent decades, but given the temporal scale of relevant ecological processes, changes in the form and strength of selection associated with habitat alteration will persist for centuries. Finally, a contextual perspective underscores the importance of maintaining density- and frequency-dependent soft selection during all relevant life stages. Doing so will require a fundamental change in the way we view harvest of salmon and other fishes.

3.2. A Requiem for Stock–Recruitment and Sustainable Yield

Harvest rates for salmon are determined using stock–recruitment relationships and the concept of sustainable yield (Ricker 1954, 1958; Beverton and Holt 1957). Stock–recruitment (S-R) curves are empirical relationships derived from time series data relating spawner abundance to subsequent adult recruitment. The curves reveal that maximum recruitment occurs at some intermediate spawner abundance, and that survival/population growth rate (the ratio of recruits to spawners) is maximized at low spawner abundance in the absence of density-dependent competition (Figure 13.6). Harvest rates are set based on idea that recruits beyond those needed for replacement can be harvested, and compensatory effects will result in subsequent recruitment above replacement. This ecological/demographic management paradigm has motivated a remarkable amount of work on: the effects of mixed-stock fisheries (Ricker 1958; Hilborn 1985; Frank and Brickman 2000); the usefulness of different S-R models (Gilbert

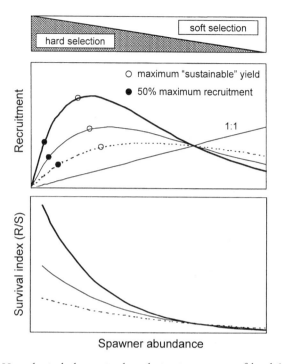

Figure 13.6. Hypothetical change in the relative importance of hard (density-independent) and soft (density-dependent) selection as a function of spawner abundance. Two different harvest rates (the difference between recruitment and the 1:1 replacement line) are shown for populations with high (bold line), medium (solid line), and low (dashed line) productivity. When spawner abundance is reduced by harvest, the survival rate (R/S) increases and the strength of soft selection decreases due to the removal of density dependent competition. The effect is most pronounced for productive populations with high intrinsic growth rates, as estimated by the slope of the S-R curve at the origin. These curves assume a Ricker-type S-R relationship; a similar pattern occurs assuming a Beverton–Holt-type relationship

1997; Barrowman and Myers 2000); predicting maximum reproductive rates at low spawner abundance (Myers et al. 1995, 1999); density dependence during sequential life history stages (Bjorkstedt 2000); accounting for empirical uncertainty (Walters and Ludwig 1981; Ludwig and Walters 1981); correcting for bias in time-series data (Walters 1985); and determining thresholds for recruitment overfishing (Myers et al. 1994). Apparently lost in the methodological and mathematical quibbling, the various evolutionary implications of using demographic criteria to set harvest rates have received almost no attention.

Pacific salmon naturally produce recruits (sometimes far) beyond those needed for replacement. Stock-recruitment curves have been used to justify harvest rates as high as 90% (CPMPNAS 1996). Figure 13.6 shows Ricker type S-R curves, and associated density-dependent survival rates, for three hypothetical populations with different maximum reproductive rates. Two harvest regimes are superimposed onto each S-R curve. Maximum sustainable yield is the point where the difference between the recruitment curve and the 1:1 replacement line is greatest (Ricker 1958). Myers et al. (1994) reviewed various thresholds for recruitment overfishing and found that harvest rates of 50% of the maximum average recruitment provided the most reliable threshold. Because survival rate decreases exponentially with spawner abundance, either harvest regime increases a population's mean survival rate, dramatically so for productive populations. From a contextual perspective, harvest rates based on recruitment and maximum population growth rates remove, or reduce the strength of, density-dependent soft selection during adult breeding and juvenile rearing. In the strictest sense, intraspecific competition is nonexistent in the absence of density dependence (Gill 1974).

Has the removal of density-dependent soft selection resulted in evolutionary changes? There is ample evidence from salmonids for density-dependent changes in the strength of, and opportunity for, selection on adult and juvenile phenotypes. Certain adult (size, mating strategy, competitive ability during breeding) and juvenile (competitive ability) traits appear to be favored by density-dependent soft selection. It seems reasonable that harvest rates based on ecological/demographic criteria have selected for phenotypes that do well at low densities, and against phenotypes that do well at high densities. If so, will there be evolutionary changes that limit the rate at which overharvested populations can return historical abundance? Perhaps in the short term (decades), but the apparent rapidity with which hatchery and transplanted populations (Kinnison et al. 1998c; Quinn et al. 2000; Pascual et al. 2001; Kinnison and Hendry 2003—*this volume*) adapt to new environmental and demographic conditions suggests ecological factors will impart primary control over long-term recovery rates (centuries). Still, a contextual perspective suggests there may be an evolutionary "ghost of management past" that, when combined with ecological depensation, might help explain the apparent inability of some fish populations to recover from low abundance (Shelton and Healey 1999; Walters and Kitchell 2001).

There are also ecological reasons to set escapement goals above the level needed to produce a harvestable surplus of recruits (Bilby et al. 2001). Decomposing salmon carcasses introduce marine-derived nutrients into freshwater, riparian, and terrestrial ecosystems. These nutrients can significantly increase juvenile growth rates (Bilby et al. 1996, 1998; Wipfli et al. 2003), which can in turn increase marine survival rates (Bilton et al. 1982; Ward and Slaney 1988; Holtby et al. 1990). Carcasses provide energy to a wide variety of terrestrial species (Willson and Halupka 1995; Willson et al. 1998), and may improve stream habitat by fertilizing riparian vegetation (Ben-David et al. 1998; Helfield and Naiman 2001). Gresh et al. (2000) estimated that the delivery of marine derived nutrients to watersheds in the northwest United States is less than 10% of historical levels, suggesting that overharvest has contributed to a broad scale, and compounding, productivity deficit in freshwater ecosystems.

In summary, there is reason to doubt that harvest rates and escapement goals based on S-R curves and even the most enlightened ecological/demographic criteria (e.g., Chaput et al. 1998; Bradford et al. 2000) will be evolutionarily or ecologically sustainable (for a possible exception see Hilborn et al. 2003).

3. Toward Evolutionary Management

I have proposed one possible conceptual framework from which evolutionary management might advance. A contextual perspective views management actions as direct and indirect selective forces that can affect different organizational levels, and different types of individual selection. The perspective is well suited to exploited species that inhabit managed landscapes, experience density-dependent intra-/interspecific competition, and display hierarchical population structure. Though the salmonid system illustrates its utility, it seems unlikely that contextual management, or any other evolutionary framework, will ever strongly guide the management of exploited species. Despite the evolutionary implications, the science of management will continue to be a mostly ecological exercise. This may be just as well. The primary impediments to responsible management are not lack of evolutionary or ecological understanding; they are social, political, and economic (Holling and Meffe 1996; Hutchings et al. 1997). The problem is not that we do not know enough, but that we do not allow what we know to constrain our behavior. Nevertheless, incorporating evolutionary thinking into the management of other species may increase the chance of long-term persistence for our own.

Acknowledgments   I thank Carl Walters for providing unpublished data and keen insights on salmon mismanagement. Andrew Hendry, Carl Walters, and an anonymous reviewer provided helpful comments on earlier versions of the manuscript. During the writing of this chapter, I received financial support from a University of British Columbia graduate fellowship and a Natural Sciences and Engineering Research Council (Canada) post-doctoral fellowship.

APPENDIX 1
Straying Rates of
Anadromous Salmonids

Each row in the table following corresponds to a different system (study, species, drainage, lake, or river). Studies that report straying in multiple independent systems (e.g., Quinn et al. 1987; Quinn et al. 1999) are thus afforded one row for each system. The table includes only studies where juveniles were tagged before departure from natal (or release) sites in fresh water. It does not include studies where adults were tagged after spawning or where mature adults were tagged and displaced. We have only included published studies to which we had access. Additional unpublished studies of homing and straying are summarized in Stabell (1984), Altukhov and Salmenkova (1991), and Quinn (1993). Straying rates are the number of tagged adults collected at non-natal sites divided by the total number of tagged adults collected at all sites (i.e., straying from a site). An exception is Schroeder et al. (2001), in which straying is the percentage of adults at a site that were strays from other sites (i.e., straying into a site). For each system, we provide the average straying rate, with the range of straying rates in parentheses. "Geography" indicates whether the source populations were differ-ent tributaries to a single river ("tribs to river"), different rivers flowing into the ocean ("rivers into ocean," followed by "w tribs" if some of the populations were in the same river), or other such categories. "Pops/sites" gives the number of populations (or sites) from which tagged juveniles were released (T), adults were examined for tags (E), and tagged adults were recaptured (R). "Years" gives the number of years that tagging took place (T), adults were examined for tags (E), and tagged adults were recaptured (R). "Number of fish" gives the number of juveniles that were tagged (T), adults that were examined for tags (E), and tagged adults that were recaptured (R). "Spatial scale" gives the shortest water distance between the nearest and farthest pair of sites at which juveniles were tagged (T), adults were examined for tags (E), and tagged adults were recaptured (R). "Tagging method" indicates the type of tag (CWT = coded wire tag). Abbreviations: n.a. = not applicable, + = actual number not known but larger than that shown, ? = unknown or uncertain.

Study	% strays (range)	Geography	Pops/sites	Years	Number of fish	Spatial scale (km)	Tagging method
Coho salmon							
Donaldson and Allen 1957	0.5 (0–1.4)	Rivers into ocean w tribs (Puget Sound, WA)	2 (T) 4 (E) 2 (R)	1 (T) 1 (E) 1 (R)	71,238 (T) 16,451 (E) 194 (R)	~50 (T) ~50–200 (E) ~50 (R)	Fin clips
Labelle 1992	< 2 (0–54.1)	Rivers into ocean (Vancouver Island, BC)	9 (T) 18 (E) 11 (R)	4 (T) 4+ (E) 4+ (R)	? (T) ? (E) 46,979 (R)	~1–180 (T) ~1–300 (E) ~1–300 (R)	CWT
Shapovalov and Taft 1954	21 (15–27)	Rivers into ocean (northern CA)	2 (T) 2 (E) 2 (R)	6 (T) 7 (E) 6 (R)	14,029+ (T) ? (E) 425 (R)	8 (T) 8 (E) 8 (R)	Fin clips
Chinook salmon							
Candy and Beacham 2000	1.2	Rivers into ocean w tribs (BC and WA)	44+ (T) 44+ (E) 44+ (R)	21 (T) 21+ (E) 21+ (R)	47,435,403 (T) millions (E) 61,728 (R)	6–500+ (T) 6–500+ (E) 6–480 (R)	CWT
Hard and Heard 1999	1.2	Rivers into ocean (southeast AK)	1 (T) 37+ (E) 24 (R)	12 (T) 9+(E) 9 (R)	1,862,058 (T) ? (E) 22,193 (R)	n.a. (T) ~5–800+ (E) ~5–780 (R)	CWT
Pascual et al. 1995	41.6 (19.5–63.6)	Tribs to river (lower Columbia, WA and OR)	2 (T) 17+ (E) 17 (R)	4 (T) 4+ (E) 4+ (R)	2,149,509 (T) ? (E) 6379 (R)	~140 (T) ~50–700+ (E) ~50–700+ (R)	CWT
Quinn and Fresh 1984	1.4	Tribs to river (lower Columbia, WA and OR)	1 (T) 11+ (E) 11 (R)	4 (T) 4 (E) 4 (R)	1,200,000 (T) ? (E) 24,139 (R)	n.a. (T) ~50–700+ (E) ~50–700 (R)	CWT
Quinn et al. 1991	17 (9.9–27.5)	Tribs to river (lower Columbia, WA and OR)	5 (T) 23+ (E) 14 (R)	3 (T) 6+ (E) 6 (R)	2,288,400 (T) millions (E) 1929 (R)	~59–199 (T) ~10–250+ (E) ~10–225 (R)	CWT

Reference		Location													Method
Sholes and Hallock 1979	10	Tribs to river (Sacramento, CA)	1 (T)	6+ (E)	2 (R)	3 (T)	6+ (E)	6 (R)	136,545 (T)	? (E)	1703 (R)	n.a. (T)	~10–400 (E)	~200 (R)	CWT & fin clips
Unwin and Quinn 1993	12.1	Rivers into ocean (New Zealand)	1 (T)	12+ (E)	12 (R)	7 (T)	7+ (E)	7+ (R)	1,522,614 (T)	? (E)	17,671 (R)	n.a. (T)	10–580+ (E)	10–580 (R)	CWT
Chum salmon															
Smoker and Thrower 1995	0.2	Rivers into ocean (southeast AK)	1 (T)	9 (E)	2 (R)	3 (T)	3 (E)	3 (R)	~240,000 (T)	18,283 (E)	611 (R)	n.a. (T)	5–30 (E)	10 (R)	CWT
Tallman and Healey 1994	37.9 (17.6–54.0)	Rivers into ocean (Vancouver Island, BC)	3 (T)	3 (E)	3 (R)	1 (T)	2 (E)	2 (R)	356,470 (T)	993 (E)	38 (R)	2 (T)	2 (E)	2 (R)	Fin clips
Pink salmon															
Habicht et al. 1998	34.1	Rivers into ocean (Prince William Sound, AK)	1–9 (T)	33–42 (E)	12–16 (R)	2 (T)	2 (E)	2 (R)	1,421,000 (T)	251,000 (E)	270 (R)	~1–120 (T)	~1–120 (E)	~1–120 (R)	CWT
Mortensen et al. 2002	6.3 (5.7–6.9)	Rivers into ocean (southeast AK)	2 (T)	6 (E)	6 (R)	1 (T)	1 (E)	1 (R)	4,094,166 (T)	? (E)	1758 (R)	14 (T)	2–14 (E)	2–14 (R)	Thermal marks
Thedinga et al. 2000	5.1 (3.7–9.2)	Rivers into ocean (southeast AK)	2 (T)	37 (E)	10 (R)	1 (T)	1 (E)	1 (R)	321,494 (T)	288,492 (E)	3907 (R)	7 (T)	~2–100 (E)	~2–65 (R)	CWT and fin clips
Sockeye salmon															
Quinn et al. 1987	< 1.0	Lakes into river and ocean (Vancouver Island, BC)	3 (T)	3 (E)	3 (R)	n.a. (T)	8 (E)	n.a. (R)	millions (T)	7060+ (E)	n.a. (R)	~15–65 (T)	~15–65 (E)	~15–65 (R)	Natural parasites

(continued)

Appendix 1 (*continued*)

Study	% strays (range)	Geography	Pops/sites	Years	Number of fish	Spatial scale (km)	Tagging method
Quinn et al. 1987	< 1.0	Lakes into ocean (BC)	2 (T) 2 (E) 2 (R)	n.a. (T) 5 (E) n.a. (R)	millions (T) 1203 (E) n.a. (R)	~100 (T) ~100 (E) ~100 (R)	Natural parasites
Quinn et al. 1999	0.1	Beaches vs. creek in lake (Iliamna, AK)	2 (T) 2 (E) 2 (R)	n.a. (T) 1 (E) 1 (R)	millions (T) 100 (E) 100 (R)	~15 (T) ~15 (E) ~15 (R)	Natural otolith marks
Quinn et al. 1999	1.0	Beach vs. creek in lake (Washington, WA)	2 (T) 2 (E) 2 (R)	n.a. (T) 2 (E) 2 (R)	millions (T) 139 (E) 139 (R)	~7 (T) ~7 (E) ~7 (R)	Natural otolith marks
Coastal cutthroat trout							
Michael 1989	35 (30–40)	Rivers into ocean (Puget Sound, WA)	2 (T) 2 (E) 2 (R)	6? (T) 6? (E) 3? (R)	1127? (T) ? (E) 51? (R)	1 (T) 1 (E) 1 (R)	Fin clips and brands
Steelhead trout							
Kenaston et al. 2001	2.4	Tribs to river (Siuslaw, OR)	1 (T) 12 (E) 10 (R)	3 (T) 4 (E) 4 (R)	178,750 (T) 2800 (E) 2084 (R)	n.a. (T) ~1–40 (E) ~1–40 (R)	Fin clips
Schroeder et al. 2001	11 (4–26)	Rivers into ocean (OR)	11 (T) 16 (E) 16 (R)	3 (T) 1–3 (E) 1–3 (R)	4,917,000 (T) 9214 (E) 2535 (R)	~10–500 (T) ~10–500 (E) ~10–500 (R)	Fin clips
Shapovalov and Taft 1954	2.5 (2–3)	Rivers into ocean (northern CA)	2 (T) 2 (E) 2 (R)	9 (T) 9 (E) 9 (R)	? (T) ? (E) 1445 (R)	8 (T) 8 (E) 8 (R)	Fin clips
Atlantic salmon							
Eriksson and Eriksson 1991	4.4	Rivers into ocean (Sweden and Finland)	1 (T) ? (E) ? (R)	3 (T) ? (E) ? (R)	8864 (T) ? (E) 68 (R)	n.a. (T) ? (E) ? (R)	Carlin tags

Reference	Rate	Location					Tag type
Gunnerod et al. 1988	0.0	Rivers into ocean (Norway)	1 (T)	8 (T)	11869 (T)	n.a. (T)	Carlin tags
			? (E)	8? (E)	? (E)	? (E)	
			1 (R)	8? (R)	227 (R)	n.a. (R)	
Heggberget et al. 1991	2.8	Rivers into ocean (Norway)	1 (T)	13 (T)	17,750 (T)	n.a. (T)	External tags
			? (E)	? (E)	? (E)	? (E)	
			? (R)	? (R)	36 (R)	? (R)	
Hvidsten et al. 1994	15.6 (13.7–19.4)	Rivers into ocean (Norway)	3 (T)	13 (T)	78,741 (T)	20–1150 (T)	Carlin tags
			12+ (E)	13+ (E)	? (E)	~20–1500 (E)	
			12 (R)	13+ (R)	260 (R)	~20–1500 (R)	
N. Jonsson et al. 1994a	11.6	Rivers into ocean (Norway)	1 (T)	2 (T)	18,852 (T)	n.a. (T)	Carlin tags
			? (E)	? (E)	? (E)	? (E)	
			? (R)	? (R)	414 (R)	? (R)	
Mills 1994	18.9	Tribs to river (Rossshire, Scotland)	1 (T)	4 (T)	10,482 (T)	n.a. (T)	Disk tags
			4+ (E)	4+ (E)	? (E)	~5–40 (E)	
			4? (R)	4+ (R)	134 (R)	~5–40 (R)	
Potter and Russell 1994	2.2 (0–10)	Rivers into ocean (northeast England)	5 (T)	7 (T)	490,731 (T)	~10–100 (T)	CWT
			5+ (E)	7+ (E)	? (E)	~10–100+ (E)	
			5–8 (R)	7+ (R)	590 (R)	~10–100+ (R)	
Solomon 1973	8	Rivers into ocean	3 (T)	7 (T)	? (T)	? (T)	?
			? (E)	? (E)	? (E)	? (E)	
			3 (R)	? (R)	104 (R)	~75 (R)	

Notes (references to tables and figures refer to each publication):

Candy and Beacham (2000). The number of tagged and recaptured fish includes all releases but the straying rate is only for fish that were not displaced from their rearing site (displaced fish strayed at higher rates). The spatial scale for recaptures is the minimum and maximum straying distance. Most straying was near to the release site (48% within 30 km). A hybrid stock strayed at higher rates than pure stocks.

Donaldson and Allen (1957). Experimental fish were the progeny of fish returning to Soos Creek hatchery. Offspring were reared for a year, clipped, transferred to the Issaquah Creek and University of Washington hatcheries, and then released from those sites. No fish returned to Soos Creek or to another hatchery in Puget Sound (Minter Creek).

(continued)

Appendix 1 (*continued*)

Eriksson and Eriksson (1991). Straying rates are for river-released Bothnian Sea fish only (Table 1). Straying records are based on captures by anglers (the numbers of streams and their locations are not reported).

Gunnerød et al. (1988). Straying rates are only for fish released into rivers (i.e., not released directly into the ocean). Straying records are based on captures by anglers (the numbers of streams and their locations are not reported). No Surna release fish were captured in other rivers and so the spatial scale of tagging and recaptures is not applicable. Fish released from the Surna estuary and in the open sea did stray to other rivers and so the failure to find strays from the within-river releases was not the result of inadequate sampling.

Habicht et al. (1998). Most releases were from hatcheries but some were from natural streams. Tag placement near the olfactory apparatus seemed to increase straying rates.

Hard and Heard (1999). The source fish were from two populations released into a new stream to start a hatchery population. No fish returned to the ancestral stream, which was 250 km distant. Most straying (64%) was within 7 km of the release site. Straying declined with increasing fish age and run size. Males strayed more than females.

Heggberget et al. (1991). Straying rates are for fish released into the River Surna only. Straying records are based on captures by anglers in Norway (the numbers of streams and their locations are not reported).

Hvidsten et al. (1994). Fish were reared in hatcheries and then tagged and released into streams. Recaptures were based on angler surveys.

N. Jonsson et al. (1994a). Straying rates are the number of tagged fish captured in other rivers divided by the total recaptured in the Imsa and other rivers. Data include all release groups (Table 1).

Kenaston et al. (2001). Hatchery reared steelhead were released into Whittaker Creek, where some were then held for 30 days in the stream and others allowed to emigrate immediately. Straying rate is the total number of Whittaker Creek releases that were recaptured in other streams divided by the total number of Whittaker Creek releases recaptured as adults (from Table 2).

Labelle (1992). Straying rates were lower for younger than for older males.

Michael (1989). Straying rate is the average between two populations. One tagged fish was also captured in a different stream. Some returning fish were not adults and therefore may not have been entering the stream to breed (see also Wenburg and Bentzen 2001).

Mills (1994). Smolts were tagged and then transported by vehicle 32 km downstream and past several tributaries before release. Straying was then evaluated as return to tributaries other than the original capture site. Straying rates may be higher than in a natural situation because the fish were not able to sequentially imprint during much of their downstream migration. Recaptures were based on angler surveys and traps in the different tributaries.

Mortensen et al. (2002). Straying rates are within 14 km of Auke Creek only, and are the average of Gastineau Hatchery and Auke Creek for thermal marks only (the tagging method for which samples sizes were largest, Table 3).

Pascual et al. (1995). Straying rates are averages across all release groups and years for each hatchery (from Table 2). Some fish were moved to new sites before release and these had higher straying rates. Release time also had large effects on straying rates.

Potter and Russell (1994). Straying rate is the total number of fish caught in non-natal rivers divided by the total number of recaptures in fresh water. The range is among rivers and age classes with more than 10 recaptures in fresh water. Tag recoveries were based on angler surveys and the total number of rivers sampled was thus probably very high (but not reported). The average straying rate of hatchery fish was 3.0% and of wild fish was 2.2%.

Quinn and Fresh (1984). Straying rates were based on recoveries at hatcheries and in fisheries, and so the number of "surveyed" populations was very large (but not known). A few strays were captured far from the Columbia (e.g., Puget Sound, WA) but 98.3% were caught in the lower Columbia, where the maximum distance among sites where strays were captured was only about 250 km.

Quinn et al. (1987). The tags are natural parasites and differ only in frequency (albeit dramatically) between the lakes. For this reason, individual strays cannot be identified and the straying rate is a maximum. In addition, this makes the number of years of tagging and recapture (as well as the number of tagged fish recaptured) not applicable. The first table entry is for three lakes on Vancouver Island (Great Central, Sproat, Henderson). The second table entry is for two lakes on the coast of British Columbia (Owikeno, Long).

Quinn et al. (1991). Straying rate is the number of strays in the entire sample and the range is for the different release sites. Fish were tagged as juveniles using coded wire tags (CWTs) and released from hatcheries. A large number of additional sites were surveyed without finding any strays. The high straying rate was from the Cowlitz River and was probably the result of ash released during the eruption of Mount St. Helens.

Quinn et al. (1999). In Iliamna Lake, only two populations were sampled although other populations are present in the system. The estimated straying rate is thus an underestimate of the true straying rate. The same is true of Lake Washington, but to a lesser degree because straying to other populations in this system is very low (K. Fresh unpublished). The thermal tags are natural and so are present on all of the juveniles leaving a site. This makes the number of years of tagging not applicable.

Schroeder et al. (2001). Straying rate is the proportion of adults captured at a site that were hatchery fish released at other sites. The authors did not estimate the rate of straying *from* a population because they "could not account for all adult returns of a given release." The number of tagging sites is the number of release sites (from Table 2).

Shapovalov and Taft (1954). This appears to be the only study comparing straying rates for different species from the same streams (Quinn 1993).

Sholes and Hallock (1979). Straying rates are for Feather River hatchery releases only (i.e., excluding fish transported and released near Rio Vista).

Smoker and Thrower (1995). Releases included native fish, fish transplanted from two creeks within 65 km of the release site, and hybrids between these native and transplanted fish. Only one stray was found in the entire study.

Solomon (1973). The author argues that the results support the pheromone hypothesis of Nordeng (1971, 1977).

Tallman and Healey (1994). Straying rate is the average of three populations (WW = 54%, AB = 17.6%, WB = 42.0%). AB and WB are two seasonal runs in the same stream, which enters the ocean 2 km from WW. Spatial scale is the distance between the two creeks where they enter the ocean.

Thedinga et al. (2000). Some of the eggs were incubated away from their natal stream. Most straying (90%) was to three streams and most strays (68%) entered streams within 10 km of their release sites. Straying was higher for the intertidal population (9.2%) than for the upstream population (3.7%).

Unwin and Quinn (1993). New Zeland chinook salmon were introduced 80 years previously, and a homing/straying balance may not have been reached. Fish released at the most appropriate times had lower straying rates.

APPENDIX 2
Genetic Differentiation among Conspecific Salmonid Populations at Nuclear DNA Loci

In the table following, differentiation is reported as the average multilocus F_{ST} (including θ, G_{ST}, Φ_{ST}, and R_{ST}), with the range across loci in parentheses (i.e., not confidence intervals). Multiple values are given for the same study when F_{ST} was reported for more than one separate group of populations. Occasionally F_{ST} was not directly reported and so we estimated it from gene diversity measures (e.g., $1 - H_S/H_T$). When a particular locus was known to be under selection (stated by the authors), it was excluded from the F_{ST} estimate. "Life history" indicates whether the sampled fish were anadromous ("A"), non-anadromous (i.e., freshwater resident, "R"), or both ("A and R"). "Geography" indicates whether the samples were taken from rivers flowing into the ocean ("rivers into ocean") and whether different tributaries were also sampled within rivers ("w tribs"). Other common categories include tributaries flowing into a single river system ("tribs to river") and various lakes and streams in different river systems ("lakes and rivers"). The geographic region is given in parentheses (abbreviations for states and provinces in North America). "Pops/sites" indicates the number of populations or sites sampled. "Spatial scale" indicates the shortest water distance among the nearest and most distant pairs of populations (preceded by "~" when estimated from maps). "GI" indicates that some of the populations in a study were geographically isolated so that gene flow in or out is not possible (e.g., landlocked or non-anadromous populations in different river systems). "Marker type" indicates whether the F_{ST} estimate was based on allozymes ("A"), microsatellites ("M"), or some other kind of marker, with the number of polymorphic loci in parentheses. Studies examining variation among populations that diverged recently (e.g., introductions) are excluded.

Study	F_{ST} (range across loci)	Life history	Geography	Pops/sites	Spatial scale (km)	Marker type (loci)
Coho salmon						
Bartley et al. 1992a	0.158	A	Rivers into ocean w tribs (CA)	27	~5–850	A (23)
Reisenbichler and Phelps 1987	0.022	A	Rivers into ocean w tribs (WA)	12	~10–450	A (21)
Small et al. 1998	0.058 (0.049–0.060)	A	Tribs into large river (Fraser, BC)	16	~1–375	M (3)
Wehrhahn and Powell 1987	0.086 (0.030–0.138)	A	Rivers into ocean w tribs (southern BC)	94	~1–400	A (15)
Chinook salmon						
Banks et al. 2000	0.075	A	Tribs to river (Central Valley, CA)	38	~2–700	M (10)
Banks et al. 2000	0.009	A	Tribs to river (winter run, Central Valley, CA)	8	~10–50	M (10)
Banks et al. 2000	0.017	A	Tribs to river (spring run, Central Valley, CA)	10	~5–75	M (10)
Banks et al. 2000	0.008	A	Tribs to river (fall run, Central Valley, CA)	20	~5–650	M (10)
Bartley and Gall 1990	0.078	A	Rivers into ocean w tribs (Central Valley, CA)	15	~5–600	A (19)
Bartley and Gall 1990	0.099	A	Tribs to river (Kalama-Trinity, CA)	6	~5–100	A (15)
Bartley and Gall 1990	0.147	A	Rivers into ocean (coastal CA)	8	~5–650	A (19)
Bartley and Gall 1990	0.090	A	Tribs to river (Eel, CA)	6	~5–150	A (11)
Bartley and Gall 1990	0.177	A	Rivers into ocean w tribs (CA)	35	~5–1300	A (21)
Bartley et al. 1992b	0.050	A	Rivers into ocean w tribs (OR)	5	~10–250	A (31)
Bartley et al. 1992b	0.022	A	Rivers into ocean w tribs (OR and CA)	6	~30–550	A (31)
Bartley et al. 1992b	0.106	A	Rivers into ocean w tribs (OR and CA)	37	~10–1600	A (31)
Carl and Healey 1984	0.006 (0.000–0.048)	A	Three life history types in one river (Nanaimo, BC)	3	~0–40	A (10)

(continued)

Appendix 2 (continued)

Study	F_{ST} (range across loci)	Life history	Geography	Pops/ sites	Spatial scale (km)	Marker type (loci)
Gharrett et al. 1987	0.059	A	Rivers into ocean w tribs (AK)	37	~10–3000	A (16)
Nelson et al. 2001	0.023 (0.011–0.038)	A	Tribs to large river (Fraser, BC)	20	~25–1100	M (3)
Teel et al. 2000	0.087 (0.051–0.163)	A	Rivers into ocean w tribs (BC)	63	~10–2200	A (21)
Utter et al. 1989	0.123 (0.017–0.310)	A	Rivers into ocean w tribs (BC, WA, OR, and CA)	86	35–2500	A (20)
Chum salmon						
Beacham et al. 1987	0.024 (0.01–0.06)	A	Rivers into ocean (BC)	83	~1–1400	A (7)
Kondzela et al. 1994	0.027	A	Rivers into ocean w tribs (southeast AK and northern BC)	61	~5–1200	A (42)
Phelps et al. 1994	0.028	A	Rivers into ocean w tribs (BC, WA, and OR)	105	~5–800	A (39)
Scribner et al. 1998	0.016 (0–0.360)	A	Tribs to large river (Yukon, AK and YK)	8	15–1500	A (18) and M (5)
Seeb and Crane 1999	0.011	A	Rivers into ocean w tribs (Kotzebue Sound, AK)	4	~1–250	A (40)
Seeb and Crane 1999	0.007	A	Rivers into ocean (Norton Sound, AK)	6	~10–250	A (40)
Seeb and Crane 1999	0.028	A	Tribs to large river (Yukon, AK)	16	~10–1500	A (40)
Seeb and Crane 1999	0.006	A	Tribs to river (Kuskokwim, AK)	6	~75–400	A (40)
Seeb and Crane 1999	0.014	A	Rivers into ocean w tribs (Bristol Bay, AK)	10	~10–600	A (40)
Seeb and Crane 1999	0.023	A	Rivers into ocean (northern Alaska Peninsula, AK)	6	~10–250	A (40)

Reference						
Seeb and Crane 1999	0.015	A	Rivers into ocean (southern Alaska Peninsula, AK)	10	~10–250	A (40)
Seeb and Crane 1999	0.040	A	Rivers into ocean (Kodiak Island, AK)	3	~100–200	A (40)
Tallman and Healey 1994	0.016 (0.000–0.075)	A	Rivers into ocean (Vancouver Island, BC)	3	2	A (24)
Wilmot et al. 1994	0.046	A	Rivers into ocean w tribs (AK, YK, and Russia)	32	~20–7000	A (24)
Winans et al. 1994	0.038 (0.007–0.154)	A	Rivers into ocean (Russia and Japan)	29	~10–3000	A (25)
Pink salmon						
Beacham et al. 1988b	0.019 (0.000–0.059)	A	Rivers into ocean w tribs (even and odd years: BC)	84	~1–1400	A (15)
Gharrett et al. 1988	0.036	A	Rivers into ocean (even years: AK)	19	~50–2500	A (15)
Hawkins et al. 2002	0.051	A	Rivers into ocean (odd years: AK and Asia)	31	~10–5000	A (21)
Hawkins et al. 2002	0.000	A	Rivers into ocean (odd years: AK)	6	~1–10	A (21)
Hawkins et al. 2002	0.006	A	Rivers into ocean (odd years: Asia)	25	~10–3700	A (21)
Hawkins et al. 2002	0.001	A	Rivers into ocean (odd years: Kamchatka and northern Okhotsk Sea)	14	~10–3000	A (21)
Hawkins et al. 2002	0.005	A	Rivers into ocean (odd years: Sakhalin, Kuril, and Hokkaido)	11	~10–700	A (21)
Hawkins et al. 2002	0.013	A	Rivers into ocean (even years: Asia)	13	150–2000	A (21)
Noll et al. 2001	0.028	A	Rivers into ocean (even years: Asia and Alaska)	18	~10–5000	A (17)
Noll et al. 2001	0.015	A	Rivers into ocean (even years: Asia)	13	~10–3700	A (17)
Noll et al. 2001	0.002	A	Rivers into ocean (even years: southern Okhotsk Sea)	6	~10–700	A (17)
Noll et al. 2001	0.002	A	Rivers into ocean (even years: Hokkaido)	2	~30–500	A (17)

(*continued*)

Study	F_{ST} (range across loci)	Life history	Geography	Pops/ sites	Spatial scale (km)	Marker type (loci)
Noll et al. 2001	0.001	A	Rivers into ocean (even years: south Sakhalin)	4	~10–350	A (17)
Noll et al. 2001	0.008	A	Rivers into ocean (even years: western Kamchatka)	5	~10–500	A (17)
Noll et al. 2001	0.005	A	Rivers into ocean (even years: eastern Kamchatka)	2	~5	A (17)
Noll et al. 2001	0.009	A	Rivers into ocean (even years: AK)	5	~1–10	A (17)
Olsen et al. 1998	0.022 (0.007–0.058)	A	Rivers into ocean w tribs (odd years: AK, BC, and WA)	12	~50–4700	M (4)
Olsen et al. 2000	0.020 (0–0.098)	A	Two seasonal runs in one river (odd years: Dungeness, WA)	2	~15	A (24) and M (7)
Seeb et al. 1999	0.007 (0.004–0.012)	A	Rivers into ocean (even years: Prince William Sound, AK)	22	~10–200	A (34)
Shaklee and Varnavskaya 1994	0.008	A	Rivers into ocean (odd years: Russia)	8	~50–3500	A (32)
Varnavskaya and Beacham 1992	<0.006 (0.00–0.006)	A	Rivers into ocean (odd years: Kamchatka, Russia)	8	~50–6000	A (13)
Sockeye salmon						
Allendorf and Seeb 2000	0.070 (0.00–0.188)	A	Rivers into ocean (Cook Inlet, AK)	4	~25–350	A (12), M (8), and RAPD (5)
Beacham and Wood 1999	0.055 (0.022–0.084)	A	Tribs to river (Nass River, BC)	9	25–1200	M (6)
Beacham et al. 2000b	0.094 (0.062–0.158)	A	Tribs to river (Skeena, BC)	17	~50–400	M (6)

Reference		Location	N	Range	Marker (n)	Value (CI)
Beacham et al. 2000c	A	Rivers into ocean w tribs (Barkley Sound, BC)	3	~20–70	M (6)	0.056 (0.013–0.107)
Burger et al. 2000	A	Lakes in separate drainages (AK)	3	~100–1325	M (6)	0.044 (0.015–0.078)
Gustafson and Winans 1999	A	Rivers into ocean w tribs (sea/river type: WA, OR, and BC)	11	10–1806	A (4)	0.004
Seeb et al. 2000	A	Rivers into ocean w tribs (upper Cook Inlet, AK)	44	~5–400	A (29)	0.108
Taylor et al. 2000	R	Beaches and creeks into lake (Okanagan, BC)	6	~5–40	M (8)	0.018 (0.00–0.043)
Utter et al. 1984	A	Rivers and lakes (AK, BC, and WA)	16	35–2500	A (11)	0.038 (0.012–0.073)
Varnavskaya et al. 1994	A	Beaches and tribs in lake (Nachiki, Russia)	14	~1–10	A (2)	0.017 (0.004–0.029)
Varnavskaya et al. 1994	A	Beaches and tribs in lake (Kuril, Russia)	23	~1–15	A (9)	0.017 (0.008–0.028)
Varnavskaya et al. 1994	A	Beaches and tribs in lake (Dvu-Yurta, Russia)	3	~3–5	A (2)	0.007 (0.006–0.008)
Varnavskaya et al. 1994	A	Beaches and tribs in lake (Clark, AK)	3	~15–50	A (5)	0.014 (0.002–0.025)
Varnavskaya et al. 1994	A	Beaches and tribs in lake (Iliamna, AK)	6	~10–90	A (6)	0.012 (0.004–0.026)
Varnavskaya et al. 1994	A	Beaches and tribs in lake (Karluk, AK)	6	~2–20	A (2)	0.011 (0.007–0.014)
Varnavskaya et al. 1994	A	Beaches and tribs in lake (Meziadin, BC)	3	~4–10	A (5)	0.006 (0.001–0.020)
Varnavskaya et al. 1994	A	Beaches and tribs in lake (Babine, BC)	9	~10–140	A (6)	0.004 (0.002–0.008)
Varnavskaya et al. 1994	A	Beaches and tribs in lake (Shuswap, BC)	2	~90	A (6)	0.015 (0.003–0.036)

(continued)

Study	F_{ST} (range across loci)	Life history	Geography	Pops/ sites	Spatial scale (km)	Marker type [loci]
Wilmot and Burger 1985	0.082 (0.036–0.118)	A	Rivers into ocean w tribs (AK)	7	~3–400	A (2)
Winans et al. 1996	0.153 (0.018–0.682)	A & R	Rivers and lakes (BC, WA, MT, ID, and OR)	27	~10–2000	A (16)
Withler 1985	0.227	A & R	Rivers into ocean w tribs (AK, BC, and WA)	66	~1–4000	A (1)
Withler et al. 2000	0.056 (0.033–0.083)	A	Tribs to large river (Fraser, BC)	29	1–1750	M (6)
Withler et al. 2000	0.069	A	Tribs to large river (lower Fraser, BC)	6	~10–150	M (6)
Withler et al. 2000	0.084	A	Tribs to large river (mid Fraser, BC)	3	~20–150	M (6)
Withler et al. 2000	0.022	A	Tribs to large river (upper Fraser, BC)	11	~2–500	M (6)
Withler et al. 2000	0.027	A	Tribs to large river (Thompson region, BC)	9	~1–250	M (6)
Wood et al. 1994	0.172 (0.047–0.305)	A	Rivers into ocean w tribs (BC)	63	~10–3000	A (33)
Woody et al. 2000	0.006	A	Early and late in two creeks of a single lake (Tustemena, AK)	4	20	M (6)
Coastal cutthroat						
Campton and Utter 1987	0.030	A	Rivers into ocean w tribs (northern Puget Sound, WA)	12	~5–75	A (20)
Campton and Utter 1987	0.018	A	Rivers into ocean (Hood Canal, WA)	8	~1–85	A (21)
Campton and Utter 1987	0.024	A	Rivers into ocean w tribs (Puget Sound, WA)	20	~1–285	A (21)
Wenburg and Bentzen 2001	0.030 (0.006–0.039)	A	Rivers into ocean w tribs (Hood Canal, WA)	9	2–100	M (10)

Reference						
Wenburg et al. 1998	0.121 (0.064–0.226)	A	Rivers into ocean (WA)	13	~20–400	M (6)
Williams et al. 1997	0.220	A	Rivers into ocean (CA, OR, WA, BC, and AK)	24	≤ 3,000	A (37)
Interior cutthroat						
Allendorf and Leary 1988	0.037	R	Rivers and lakes (Yellowstone: WY, MT, and ID)	8	GI	A (46)
Allendorf and Leary 1988	0.324	R	Rivers and lakes (Westslope: WA, MT, ID, AB, and BC)	103	GI	A (29)
Loudenslager and Gall 1980	0.445	R	Rivers and lakes (Lahontan: NV and CA)	15	GI	A (14)
Loudenslager and Gall 1980	0.082	R	Rivers and lakes (Yellowstone: WY)	10	GI	A (6)
E. Taylor (unpublished)	0.38 (0.25–0.53)	R	Rivers and lakes (Westslope: upper Kootenay, BC)	18	GI	M (8)
Steelhead						
Beacham et al. 1999	0.076 (0.063–0.143)	A	Rivers into ocean w tribs (BC and WA)	22	~50–1600	M (8)
Beacham et al. 2000a	0.026 (0.008–0.039)	A	Tribs to river (Skeena, BC)	7	~20–300	M (8)
Beacham et al. 2000a	0.024 (0.011–0.033)	A	Tribs to river (Nass, BC)	10	~20–150	M (8)
Heath et al. 2001	0.039	A	Rivers into ocean w tribs (BC)	10	~80–850	M (6)
Heath et al. 2002b	0.041	A	Tribs to river (BC)	3	~120–210	M (7)
Hendry et al. 2002	0.007 (0.000–0.034)	A	Different migratory times in one river (Dean, BC)	2	~20–60	M (10)
Nielsen and Fountain 1999	0.01 (0.00–0.33)	A	Seasonal races in one river (Middle Fork Eel, CA)	2	~0–22	M (16)
Nielsen et al. 1997b	0.07	A & R	Rivers into ocean w tribs (CA)	6	~5–325	M (3)
Reisenbichler and Phelps 1989	0.015	A	Rivers into ocean w tribs (WA)	27	~1–800	A (23)

(continued)

Study	F_{ST} (range across loci)	Life history	Geography	Pops/sites	Spatial scale (km)	Marker type (loci)
Reisenbichler et al. 1992	0.015	A	Rivers into ocean w tribs (northern CA and OR)	19	~20-420	A (8)
Rainbow trout						
Berg and Gall 1988	0.127 (0.002-0.422)	A & R	Rivers into ocean w tribs (CA)	31	~1-1350	A (24)
Nielsen et al. 1997a	0.205	A & R	Rivers into ocean w tribs (CA)	18	~20-1200	M (3)
J. Nielsen et al. 1999	0.173	R	Lakes and rivers (CA)	14	GI	M (11)
Tamkee and E. Taylor (unpublished)	0.44 (0.19-0.77)	A & R	Lakes and rivers into ocean (BC)	32	~1-1000	M (10)
Dolly Varden						
Everett et al. 1997	0.071	A	Tribs to river (Babbage, Beaufort Sea, AK)	3	~30-100	A (6)
Everett et al. 1997	0.019	A	Tribs to river (Hulahula, Beaufort Sea, AK)	2	~10-100	A (12)
Everett et al. 1997	0.029	A	Tribs to river (Sagavanirktok, Beaufort Sea, AK)	3	~30-140	A (11)
Everett et al. 1997	0.090	A	Rivers into ocean w tribs (Beaufort Sea, AK)	16	~10-800	A (21)
Leary and Allendorf 1997	0.520	A	Rivers into ocean (WA)	2	~100-400	A (5)
Bull trout						
Kanda and Allendorf 2001	0.195	R	Tribs to river (Flathead, MT)	14	~20-270	M (3)
Leary and Allendorf 1997	0.210	R	Rivers into ocean (WA)	6	~100-400	A (2)
Leary et al. 1993	0.261 (0.000-1.000)	R	Lakes and rivers (BC, ID, MN, OR, and WA)	16	GI	A (10)

Reference	F_{ST} (95% CI)	A/R	Location	No.	Range	Marker (no.)
Latham and Taylor 2001	0.11 (0.04–0.18)	R	Rivers into lake (Kootenay region, BC)	14	~1–260	M (4)
Neraas and Spruell 2001	0.137	R	Tribs to river and lake (Clark Fork, ID and MT)	19	~5–225	M (8)
Spruell et al. 1999	0.063	R	Tribs to river (Lightening, Pend Oreille, ID)	5	~5–30	M (6)
Taylor et al. 2001	0.33 (0.07–0.42)	R	Different watersheds w tribs (AB and BC)	8	GI	M (5)
Atlantic salmon						
Beacham and Dempson 1998	0.011 (0.004–0.033)	A	Tribs to river (Conne, NF)	3	~10	M (4)
Bourke et al. 1997	0.084 (0.009–0.163)	A	Rivers into ocean (Europe)	14	~150–4000	A (12)
Bourke et al. 1997	0.087 (0.018–0.161)	A	Rivers into ocean (eastern Europe)	5	~500–2000	A (12)
Bourke et al. 1997	0.221 (0.004–1.000)	A	Rivers into ocean (Atlantic)	15	~150–6000	A (14)
Danielsdottir et al. 1997	0.062	A	Rivers into ocean w tribs (Iceland)	32	~1–1000	A (7)
Fontaine et al. 1997	0.072 (0.019–0.207)	A	Rivers into ocean (PQ)	6	~35–400	M (5)
Galvin et al. 1996	0.034 (0.023–0.049)	A	Tribs to river (Ireland)	10	~25–130	Minisats (3)
Garant et al. 2000	0.034	A	Tribs to river (Sainte-Marguerite, PQ)	7	~3–112	M (5)
Jordan et al. 1992	0.016	A	Rivers into ocean (Scotland)	17	~5–500	A (14)
Jordan et al. 1992	0.007	A	Tribs to river (Tweed, Scotland)	7	~20–150	A (14)
Jordan et al. 1997	0.024 (0.012–0.053)	A	Rivers into ocean (Atlantic Scandinavia)	5	~50–2700	A (2)

(continued)

Study	F_{ST} (range across loci)	Life history	Geography	Pops/sites	Spatial scale (km)	Marker type (loci)
Jordan et al. 1997	0.124 (0.067–0.147)	A	Rivers into ocean (Baltic Sea)	3	~100–1000	A (2)
Jordan et al. 1997	0.042 (0.022–0.053)	A	Rivers into ocean (Britain and Ireland)	28	~10–1600	A (3)
Jordan et al. 1997	<0.001	A	Rivers into ocean (Spain and France)	3	~100–800	A (2)
Jordan et al. 1997	0.068 (0.001–0.075)	A	Rivers into ocean (Atlantic Canada)	4	~200–1200	A (3)
King et al. 2001	0.114	A	Rivers into ocean (Europe)	13	~10–4000	M (12)
King et al. 2001	0.061	A & R	Rivers into ocean (North America)	16	GI	M (12)
Koljonen et al. 1999	0.161	A	Rivers into ocean (Baltic Sea)	14	~10–1350	A (7)
McConnell et al. 1997	0.054 (0.037–0.124)	A	Rivers into ocean (NS and NF)	15	~5–1300	M (8)
McElligott and Cross 1991	0.043	A	Rivers into ocean (Ireland)	9	~5–1000	A (7)
McElligott and Cross 1991	0.034	A	Tribs to river (Blackwater, Ireland)	9	~5–100	A (7)
Nielsen et al. 2001	0.056 (0.027–0.093)	A	Rivers into ocean (Jutland, Denmark)	6	~25–200	M (6)
E. Nielsen et al. 1999	0.061 (0.028–0.146)	A	Rivers into ocean (Jutland, Denmark)	5	30–300	M (6)
O'Connell et al. 1995	0.016	A	Tribs to river (Wye, Wales)	6	~5–40	A (5)
O'Connell et al. 1995	0.025	A	Rivers into ocean w tribs (Wales)	7	~5–500	A (5)
Sánchez et al. 1996	0.092	A	Rivers into ocean (Ireland and Spain)	7	~120–1200	A (6) and M (3)
Spidle et al. 2001	0.082	A	Tribs to river (Penobscot, ME)	3	~15–30	M (12)
Ståhl 1987	0.013 (0.010–0.026)	A	Tribs to river (Alta, Norway)	5	≤ 65	A (2)
Ståhl 1987	0.021 (0.009–0.034)	A	Tribs to river (Kalix, Norway)	5	≤ 250	A (2)

Study	Value	Type	Description	n	Range	Marker
Ståhl and Hindar 1988	0.047 (0.015–0.048)	A	Tribs to river (Tana, Norway)	10	≤ 1000	A (3)
Tessier and Bernatchez 2000	0.006	A	Rivers into ocean (Saguenay, PQ)	4	~10–80	M (7)
Tessier et al. 1997	0.143 (0.031–0.370)	R	Rivers into lake (St.-Jean, PQ)	4	~1–100	M (7)
Brook charr						
Angers and Bernatchez 1998	0.370 (0.213–0.687)	R	Sites in river w tribs (St. Maurice, PQ)	26	~1–50	M (5)
Castric et al. 2001	0.182 (0.099–0.369)	R	Sites in river w tribs (St. John, ME)	11	20–575	M (6)
Castric et al. 2001	0.216 (0.145–0.396)	R	Sites in river w tribs (Penobscot, ME)	10	50–385	M (6)
Castric et al. 2001	0.222 (0.106–0.348)	R	Sites in river w tribs (Kennebec, ME)	4	125–400	M (6)
Hébert et al. 2000	0.033	A & R	Rivers into ocean (Kouchibouguac, NB)	6	~4–15	M (6)
Hébert et al. 2000	0.110	R	Rivers into ocean w tribs (Forillon, PQ)	8	GI	M (6)
Hébert et al. 2000	0.145	R	Rivers into ocean w tribs (Fundy, NB)	10	GI	M (6)
Jones et al. 1996	0.232	A & R	Lakes and rivers (NS and NB)	34	GI	A (11)
Jones et al. 1997	0.029	A & R	Rivers into ocean w tribs (NB)	8	~5–160	A (7)
Perkins et al. 1993	0.375	R	Lakes and rivers (NY and PA)	21	GI	A (32)
Brown trout						
Antunes et al. 1999	0.027	A & R	Tribs to river (Lima, Portugal)	3	~10–30	A (4)
Apostolidis et al. 1996	0.562 (0.117–1.000)	R	Rivers w tribs (Greece and southern Europe)	14	GI	A (16)
Bernatchez and Osinov 1995	0.150	A & R	Lakes and rivers (White, Barent and Baltic seas)	4	GI	A (23)
Bernatchez and Osinov 1995	0.392	A & R	Lakes and rivers (Caspian, Black, and Aral seas)	7	GI	A (23)

(continued)

Appendix 2 (*continued*)

Study	F_{ST} (range across loci)	Life history	Geography	Pops/ sites	Spatial scale (km)	Marker type (loci)
Bouza et al. 2001	0.455 (0.016–0.599)	R	Tribs to large river (Duero, Spain)	39	~5–900	A (15)
Carlsson et al. 1999	0.026	R	Sites in river w one trib (Nordre Finnvikelv, Norway)	4	0.25–1.5	M (6)
Carlsson and Nilsson 2000	0.025 (0.011–0.055)	R	Tribs to river (Färsån, Sweden)	7	1–10	M (5)
Carlsson and Nilsson 2001	0.285 (0.141–0.409)	R	Tribs to river (Ammerån, Sweden)	7	GI	M (5)
Chakraborty et al. 1982	0.056 (0.009–0.106)	R	Sites in river w tribs (Indalsälven and Lången, Norway)	9	~2–30	A (5)
Crozier and Ferguson 1986	0.307	R	Rivers into lake w tribs (Neagh, Ireland)	34	~2–180	A (12)
Estoup et al. 1998	0.229 (0.005–0.375)	R	Tribs to river (Moselle, France)	9	2.5–328	A (8) and M (7)
García-Marín and Pla 1996	0.640 (0.030–1)	A & R	Rivers into ocean w tribs (Spain)	24	GI	A (25)
García-Marín et al. 1999	0.482 (0.208–0.807)	A & R	Rivers and lakes (Europe)	232	GI	A (11)
Hansen and Mensberg 1996	0.008	A & R	Sites in river w tribs (Odder, Jutland, Denmark)	7	~1–10	A (11)
Hansen et al. 1993	0.047	R	Rivers into lake w tribs (Hald, Jutland, Denmark)	7	~1–10	A (12)
Hansen et al. 1993	0.026	A & R	Rivers into ocean (Jutland, Denmark)	3	~20–100	A (12)
Heggenes et al. 2002	0.022 (0.002–0.054)	R	Rivers into lake (Mosvatn, Norway)	4	~1–12	A (8)
Hindar et al. 1991a	0.172 (0–0.674)	A & R	Tribs to river (Voss, Norway)	5	~5–40	A (6)

Reference					
Morán et al. 1995	0.215 (0.035–0.348)	A & R	Rivers into ocean (Spain)	7	~10–70	A (8)
Morán et al. 1995	0.077 (0.032–0.126)	A & R	Tribs to river (Esva, Spain)	12	~3–45	A (9)
Ruzzante et al. 2001	0.049 (0.037–0.065)	A & R	Rivers into ocean w tribs (Jutland, Denmark)	32	1–153	M (7)
Sanz et al. 2000	0.398 (0.063–0.634)	R	Tribs to river (Tajo, Spain)	6	~5–350	A (6)
Sanz et al. 2000	0.062 (0.005–0.166)	R	Tribs to river (Duero, Spain)	8	~5–500	A (15)
Sanz et al. 2000	0.058 (0.012–0.134)	R	Tribs to river (Sil, Spain)	4	~5–75	A (11)
Sanz et al. 2000	0.604 (0.011–0.960)	A & R	Rivers into ocean w tribs (Cantabrian, Spain)	5	~3–200	A (10)
Sanz et al. 2000	0.645 (0–0.92)	A & R	Rivers into ocean w tribs (Spain)	25	5–2000	A (21)
White spotted charr						
Koizumi et al. (unpublished)	0.0561 (0.028–0.124)	R	Tribs to river (Hokkaido, Japan)	6	0.5–34.2	M (4)
Yamamoto et al. (unpublished)	0.101 (0.027–0.156)	A	Rivers into ocean (Hokkaido, Japan)	3	2–148	M (5)
Yamamoto et al. (unpublished)	0.102 (0.023–0.219)	A & R	Sites in rivers, separated by recent dams (Hokkaido, Japan)	9	GI	M (5)

(continued)

Study	F_{ST} (range across loci)	Life history	Geography	Pops/sites	Spatial scale (km)	Marker type (loci)
Arctic charr						
Bernatchez et al. 1998	0.039	A	Rivers into ocean (Labrador, Canada)	7	~10–600	M (6)
Brunner et al. 1998	0.370 (0.087–0.657)	R	Lakes and rivers (Central Alpine region, Europe)	15	GI	M (6)
Primmer et al. 1999	0.360 (0.022–0.627)	R	Lakes and rivers (Norway and Finland)	4	GI	M (8)
Lake whitefish						
Lu et al. 2001	0.161	R	Lakes (ME and PQ)	35	~0–1500	M (6)
Douglas et al. 1999	0.088	R	Lakes and rivers (Central Alpine region, Europe)	20	GI	M (6)
European grayling						
Koskinen et al. 2001	0.261	R	Tribs and beaches into lake (Saimaa, Finland)	3	~55–350	M (17)
Koskinen et al. 2001	0.365	R	Rivers and lakes (Finland and Germany)	4	GI	M (17)
Koskinen et al. 2002a	0.24 (0.025–0.47)	R	Sites in lake (Saimaa, Finland)	9	~10–200	M (10)
Koskinen et al. 2002c	0.40	R	Lakes and rivers (Europe)	17	GI	M (17)
Arctic grayling						
Stamford and Taylor (unpublished)	0.232 (0–0.687)	R	Rivers into lakes or larger rivers (BC)	8	25–625	M (5)
Cisco						
Turgeon and Bernatchez 2001b	0.159	R	Lakes (Canada)	22	GI	M (7)

Notes for some of the above references (references to tables and figures refer to each publication):

Allendorf and Leary (1988). F_{ST} values are from Table 5 (1 – RGD within samples) for the Yellowstone and Westslope subspecies. The other subspecies are not reported because they are from other papers in this appendix.

Allendorf and Seeb (2000). F_{ST} is the average of the single locus values from Table 4, after excluding sAH, which showed markedly higher divergence ($F_{ST} = 0.713$) than all the other loci ($F_{ST} \leq 0.188$).

Angers and Bernatchez (1998). The area was colonized after the last glaciation by at least two different lineages.

Antunes et al. (1999). F_{ST} is for the three sites within the Lima River only. Some anadromous fish are apparently present within this system.

Banks et al. (2000). F_{ST} values are from Table 4 after adjustment for kinship and run admixture. Multiple samples were taken from each site and these are included as separate subpopulations in the F_{ST} calculation. The late fall run category was excluded because it includes only one natural population, with the two others from hatcheries.

Bartley et al. (1992a). Although 23 loci were polymorphic, most showed little variation.

Bartley et al. (1992b). We report F_{ST} only for regions that were not reported in Bartley and Gall (1990). The total number of polymorphic loci is 31 but the number that are polymorphic within each region may be lower. This paper was initially published with the following author order: Gall, G. A., D. Bartley, B. Bentley, J. Brodziak, R. Gomulkiewicz, and M. Mangel.

Beacham and Dempson (1998). One of the tributaries (Bernard Brook) was colonized in 1963.

Beacham et al. (1988b). Odd and even year samples were counted as separate populations even if they came from the same river. Most of the genetic variation is due to differences between odd and even year samples.

Beacham et al. (2000b). Spatial scale is from T. Beacham (pers. comm.).

Beacham and Wood (1999). Spatial scale is from T. Beacham (pers. comm.).

Berg and Gall (1988). Most of the populations were formerly anadromous but most (22/31) were landlocked by twentieth century water control projects.

Bernatchez et al. (1998). F_{ST} excludes the Newfoundland population, which is landlocked and far from the others.

Bourke et al. (1997). F_{ST} values are the average of single locus values from group ii in Table 4 after excluding MEP-2^*, which appears to be under directional selection (Jordan et al. 1997).

Bouza et al. (2001). F_{ST} values are from Table 4 and include only the 39 sites assayed at all polymorphic loci.

Brunner et al. (1998). Samples are from multiple lakes within three river systems (Rhine, Rhône, Danube) that drain into different seas. Includes different morphs within lakes. F_{ST} is based on Φ_{ST}.

Burger et al. (2000). F_{ST} is among donor populations only. Derived populations in Frazer Lake show much lower differentiation.

Campton and Utter (1987). F_{ST} excludes the one landlocked population.

Carl and Healey (1984). Only 1 of 10 loci showed substantial differentiation (6-PGD, $F_{ST} = 0.048$), with $F_{ST} < 0.005$ for all the others.

Carlsson and Nilsson (2000). F_{ST} was nearly identical in each of 2 years (Table 4) and differences between years within sites were very small. Both resident and migratory trout are present in the system but the migratory trout are not anadromous. Instead they move between rivers. All samples were collected within a single river.

Carlsson and Nilsson (2001). Some of the populations are separated by large waterfalls. The distance among non-isolated populations is ~4–50 km.

(continued)

Appendix 2 (continued)

Carlsson et al. (1999). Significant differentiation was mainly attributable to one population sampled in a tributary.

Crozier and Ferguson (1986). F_{ST} was estimated from Table 3.

Danielsdottir et al. (1997). F_{ST} is for wild populations only.

Douglas et al. (1999). Sampled the same region and many of the same sites as Brunner et al. (1998).

Estoup et al. (1998). F_{ST} is average of microsatellite and allozyme averages for the Moselle drainage only (from Table 2).

Everett et al. (1997). Numbers of polymorphic loci were determined using Table 3.

Fontaine et al. (1997). F_{ST} is based on Φ_{ST} and excludes Koskoa, which is far away from the other populations.

Galvin et al. (1996). F_{ST} does not include variation among age classes, which is about 0.2% of the total genetic diversity. Extensive stocking has occurred in the system and a dam was built in 1929.

Garant et al. (2000). Interannual variation at sites was about 2/3 as great as variation among sites.

Garcia-Marin and Pla (1996). The collections straddle the boundary south of which anadromous fish are no longer present.

Garcia-Marin et al. (1999). This study represents a compilation of data sets from across Europe.

Gustafson and Winans (1999). F_{ST} was provided by R. Gustafson and G. Winans (pers. comm.) and includes 11 river/sea-type populations (i.e., excluding lake populations and Alsek river/sea type, which was an outlier). Spatial scale was provided by R. Gustafson (pers. comm.).

Hansen and Mensberg (1996). No mention of whether anadromous fish were included in the samples, but anadromy does occur in the system. Some of the populations recently underwent extinction/recolonization events.

Hansen et al. (1993). Stocking has taken place in this system but may not have been successful. Recent dams have increasingly isolated some of the populations.

Hawkins et al. (2002). Spatial scale for the Asian even-year broodline was provided by A.J. Gharrett (pers. com.)

Heath et al. (2002b). F_{ST} is from Table 3 and is averaged over populations separated by 5 or fewer years.

Hébert et al. (2000). The authors state that anadromy is not observed in Forillon and Fundy but is present in the Kouchibouguac.

Heggenes et al. (2002). The stocking population (Tunhovd) was removed to compute the mean F_{ST} from Table 2. Stocking has occurred but does not seem to have had a large impact.

Hendry et al. (2002). F_{ST} is based on a comparison of early and late migrating fish captured in the lower part of the river. These two groups likely represent populations that migrate at different times. Spatial scale is the minimum and maximum possible distance between potential spawning locations.

Jones et al. (1996). No mention of whether anadromous fish were included in the samples but anadromy does occur within the region and some samples were collected in coastal areas.

Jones et al. (1997). Includes resident, anadromous, and hatchery samples.

Jordan et al. (1997). This study used published and unpublished data to estimate F_{ST} over various spatial scales with and without $_m$MEP-2*, which appears to be under selection. We have reported the values without $_m$MEP-2*.

Kanda and Allendorf (2001). F_{ST} is the sum of the relative variance due to differences among drainages and among samples within drainages.

King et al. (2001). F_{ST} is from Table 5. F_{ST} for North America includes two landlocked populations, which were also the most divergent in that sample.

Koljonen (1999). F_{ST} is for wild populations only.

Kondzela et al. (1994). F_{ST} is the sum of within-region and among-region relative gene diversity. Of the 42 variable loci, 21 had a common allele frequency < 0.95.

Koskinen et al. (2001). The larger F_{ST} value includes the three Lake Saimaa populations and an outgroup from Germany.

Latham and Taylor (2001). Dams have recently isolated some populations and hatchery influence may be reducing variation.

Leary et al. (1993). F_{ST} is the average of single locus F_{ST} values from Table 3.

Loudenslager and Gall (1980). F_{ST} values are from Table 6 and are for the Lahontan subspecies (*S. c. henshawi*) and the Yellowstone subspecies (*S. c. bouvieri*).

Lu et al. (2001). Three glacial races are included in the calculation. Most samples are from the contact zone between the Acadian and Atlantic lineages.

McConnell et al. (1997). One of the populations (Newfoundland) is much farther from the others (Nova Scotia).

Neraas and Spruell (2001). Dams built in the last 90 years have increasingly separated some populations but the genetic differences are so large that at least some of them must have predated the dams.

E. Nielsen et al. (1999). F_{ST} is based on historical samples collected in 1913.

Nielsen et al. (2001). This system has been subjected to considerable stocking.

Nielsen and Fountain (1999). F_{ST} is between two seasonal races within a single river.

Nielsen et al. (1997a). Sites were selected for minimal hatchery influence. F_{ST} is based on R_{ST}.

Nielsen et al. (1997b). F_{ST} is between six "habitats," each of which includes several populations. Non-migrating fish were collected but access to ocean was possible for some sites.

J. Nielsen et al. (1999). F_{ST} equals one minus the proportion of molecular variance within natural populations.

Noll et al. (2001). Even-year populations only.

Olsen et al. (1998). Odd-year populations only.

Olsen et al. (2000). Odd-year populations only.

Primmer et al. (1999). F_{ST} is for wild populations only. Many additional populations with varying degrees of hatchery influence were sampled and $F_{ST} = 0.317$ for them.

Reisenbichler and Phelps (1987). F_{ST} is from Table 6 after excluding "between broods" diversity. F_{ST} excludes *Pnp-1*, which was not screened in all samples. $F_{ST} = 0.048$ when *Pnp-1* is included (7 sites). Many more sites were sampled but excluded from the analysis owing to small sample sizes (see caption for Fig. 6). Although 21 loci were polymorphic only 3 had common allele frequencies < 0.95 in most samples.

Reisenbichler and Phelps (1989). F_{ST} is from Table 6 after excluding "between broods" diversity.

Reisenbichler et al. (1992). F_{ST} is from Table 5 (excluding variation between broods) and is for wild populations from Trask (OR) to Trinity (CA) only.

Ruzzante et al. (2001). Spatial scale provided by D. Ruzzante (pers. comm.).

Sánchez et al. (1996). F_{ST} is from Table 5. Two groups of samples: Cantabrian and Ireland. Mean value is 0.079 for microsatellites and 0.104 for allozymes.

Sanz et al. (2000). F_{ST} is from Table 4 and was "corrected" by the authors for hatchery influence. Anadromy occurs within this region but samples collected from far upstream are considered resident fish.

Scribner et al. (1998). Mean F_{ST} was 0.015 for microsatellites and 0.010 for allozymes.

Seeb et al. (1999). Some of the 22 streams had separate collections from estuarine and upstream sites.

Seeb et al. (2000). F_{ST} is the sum of relative gene diversity among sites within lakes, among nursery lakes, and among regions.

(continued)

Appendix 2 (continued)

Shaklee and Varnavskaya (1994). Of 32 polymorphic loci, 14 had common allele frequencies < 0.95.

Small et al. (1998). Upper and lower Fraser River populations are probably derived from two distinct glacial races.

Spidle et al. (2001). F_{ST} is for the three Penobscot River populations only, including the mainstem (Veazie Dam), Kenduskeag Stream, and Cove Brook. Some stocking has occurred in this system.

Ståhl and Hindar (1988). F_{ST} was reported in Jordan et al. (1997) and the spatial scale was provided by K. Hindar (pers. comm.).

Ståhl (1987). F_{ST} was reported in Jordan et al. (1997) and is computed without $_m MEP-2^*$, which appears to be under selection.

Tallman and Healey (1994). One of the two creeks includes two seasonal races, which were kept separate in the genetic analysis. The number of loci is the number of loci/tissue combinations from Table 7.

Taylor et al. (2000). Minimum F_{ST} is from E. Taylor (pers. comm.). Taylor et al. (1997) studied the same populations and found that $F_{ST} = 0.01$ between beach and stream ecotypes (allozymes and minisatellites).

Taylor et al. (2001). Some hybridization occurs between Dolly Varden and bull trout, but only allopatric samples were included in the F_{ST} estimate.

Tessier et al. (1997). The group as a whole is landlocked but all populations are in tributaries to the same lake. F_{ST} is Φ_{ST} and is the average of the single-locus means for pairwise comparison of wild fish from Table 4.

Tessier and Bernatchez (2000). Only anadromous populations from the Saguenay Fjord were included. F_{ST} is the mean of pairwise values from Table 2.

Turgeon and Bernatchez (2001). Contemporary differentiation patterns result from post-glacial contact of two glacial races.

Utter et al. (1984). Average F_{ST} was recalculated after excluding PGM1 which has a much higher F_{ST} (0.247) than all the other loci. Spatial scale is from F. Utter (pers. comm.). Mean F_{ST} including PGM1 was 0.096.

Utter et al. (1989). Spatial scale is from F. Utter (pers. comm.).

Varnavskaya and Beacham (1992). Average F_{ST} not reported. One population is very far from the others (near Vancouver, BC). Odd-year populations only.

Varnavskaya et al. (1994). F_{ST} values are estimated from Table 3: first averaged across years within loci and then across loci. Ranges are min and max F_{ST} by locus (averaged across years).

Wehrhahn and Powell (1987). F_{ST} is from Table 3 and is the mean of the single-locus values.

Williams et al. (1997). The total number of loci is 37 but some were monomorphic.

Wilmot and Burger (1985). Most of the variation was attributable to differences between the two lake systems. Includes collections of early and late spawners.

Wilmot et al. (1994). F_{ST} was calculated from Table 3.

Winans (1996). F_{ST} includes 12 sockeye salmon collections and 15 kokanee collections.

Winans et al. (1994). F_{ST} is from Table 4 and includes only the 25 loci for which the common allele frequency was < 0.95 (a total of 61 loci were variable). More locations were sampled but only 29 were included in the analysis.

Withler et al. (2000). F_{ST} values are mean pairwise values within groups of populations. Includes populations descended from two distinct glacial races.

Wood et al. (1994). F_{ST} is from Table 6 and may include a few kokanee populations (although not landlocked). The minimum single locus F_{ST} is only for the eight most polymorphic loci (i.e., those shown in Table 6). Of the 33 polymorphic loci, 14 had common allele frequencies < 0.95.

Woody et al. (2000). F_{ST} is the average of all pairwise values (early and late samples from each of two creeks, from Table 4). One of the populations is only about 2000 years old.

APPENDIX 3
Review of Differences between Anadromous and Non-Anadromous Salmonids

In the table following, the attributes of anadromous and non-anadromous salmon are reviewed. For each type of trait ("Life history" or "Other"), comparisons are intended to be made between the two forms within a row. Sometimes the two forms may be distinct gene pools that spawn sympatrically (subscript s), sometimes they may be isolated by a barrier to anadromous migration (subscript i), sometimes they may be found in nearby systems (subscript n), and sometimes they may be polymorphic forms within a single population (subscript p). The data represent values for individuals captured from the wild (subscript w) and or raised under common-garden conditions (subscript g). Genetic differences are always for wild-caught individuals. The data may be for females only (subscript f), males only (subscript m), or both sexes combined (subscript b). Following each data point is a letter corresponding to the notes for that species: notes follow the table.

Genetic difference	Life history traits		Other traits	
	Anadromous	Non-anadromous	Anadromous	Non-anadromous
Oncorhynchus nerka				
Nei's D_s = 0.115 (c)	$AM_{s,w,b}$ = 4 y (a)	$AM_{s,w,b}$ = 4 y (a)	$GRN_{s,w,b}$ = 36.2 (a)	$GRN_{s,w,b}$ = 39.5 (a)
Nei's D_s = 0.100 (c)	$L_{s,w,b}$ > 380 mm (a)	$L_{s,w,b}$ < 220 mm (a)	$GRN_{s,w,b}$ = 36.5 (d)	$GRN_{s,w,b}$ = 39.7 (d)
Nei's D_s = 0.010 (c)	$L_{s,w,b}$ = 563 mm (d)	$L_{s,w,b}$ = 185 mm (d)	$GRN_{s,g,b}$ = 35.0 (d)	$GRN_{s,g,b}$ = 37.3 (d)
RGD_i = 0.012 (c)	$L_{s,w,m}$ = 582 mm (e)	$L_{s,w,m}$ = 194 mm (e)	$GRL_{s,g,b}$ = 5.94 mm (d)	$GRL_{s,g,b}$ = 5.00 mm (d)
RGD_s = 0.175 (a)	$L_{s,w,f}$ = 541 mm (e)	$L_{s,w,f}$ = 191 mm (e)	$CSV_{s,g,b}$ = 8.3 bl·s⁻¹ (b)	$CSV_{s,g,b}$ = 7.3 bl·s⁻¹ (b)
Nm_s = 1.4 (a)	$GSI_{s,w,f}$ = 15.3–15.6% (a)	$GSI_{s,w,f}$ = 13.0–14.0% (a)	$SkCol_{s,w,m}$ = 21 (e)	$SkCol_{s,w,m}$ = 19 (e)
	$EN_{s,w}$ = 3067–3438 (a)	$EN_{s,w}$ = 139–148 (a)	$SkCol_{s,w,f}$ = 18 (e)	$SkCol_{s,w,f}$ = 16 (e)
	$ES_{s,w}$ = 83.2–93.7 mg (a)	$ES_{s,w}$ = 71.8–82.0 mg (a)	$SkCol_{s,g,b}$ = -0.01 (e)	$SkCol_{s,g,b}$ = 7.2 (e)
	$ES_{s,w}$ = 5.21 mm (f)	$ES_{s,w}$ = 4.84 mm (f)	$SkCar_{s,g,b}$ = 11.2 mg·kg⁻¹ (e)	$SkCar_{s,g,b}$ = 33.4 mg·kg⁻¹ (e)
Oncorhynchus mykiss				
F_{STp} = 0.000 (b)	$L_{p,w,b}$ = 622 mm (b)	$L_{p,w,b}$ = 344 mm (b)		
F_{STp} = 0.00–0.33 (a)				
Nei's D_i = 0.021 (a)				

Salmo trutta

$F_{STp} = 0.000$ (b) $\quad AM_{p,w,m} = 4.8\text{-}5.6$ y (f) $\quad AM_{p,w,m} = 3.4\text{-}3.8$ y (f)

$F_{STi} = 0.022\text{-}0.029$ (b) $\quad L_{p,w,m} = 392\text{-}392$ mm (c) $\quad L_{p,w,m} = 228\text{-}229$ mm (c) $\quad SoEnSp_{p,w,b} = 5.0$ kJ $\cdot$ g^{-1} (a) $\quad SoEnSp_{p,w,b} = 4.5$ kJ $\cdot$ g^{-1} (a)

Nei's $D_p = 0.001$ (d) $\quad L_{p,w,f} = 413\text{-}448$ mm (c) $\quad L_{p,w,f} = 260$ mm (c) $\quad SoFtSp_{p,w,b} = 4.9\%$ (a) $\quad SoFtSp_{p,w,b} = 3.5\%$ (a)

Nei's $D_i = 0.045$ (d) $\quad L_{i,w,m} = 392\text{-}392$ mm (c) $\quad L_{i,w,m} = 229\text{-}244$ mm (c) $\quad SoPrSp_{p,w,b} = 18.5\%$ (a) $\quad SoPrSp_{p,w,b} = 18.8\%$ (a)

Nei's $D_p = 0.001$ (c) $\quad L_{i,w,f} = 413\text{-}448$ mm (c) $\quad L_{i,w,f} = 251\text{-}259$ mm (c) $\quad SoEn_{p,w,m} = 600\text{-}4500$ kJ (a) $\quad SoEn_{p,w,m} = 70\text{-}3500$ kJ (a)

Nei's $D_i = 0.022$ (c) $\quad L_{p,w,f} = 412$ mm (e) $\quad L_{p,w,f} = 259$ mm (e) $\quad SoEn_{p,w,f} = 700\text{-}10{,}000$ kJ (a) $\quad SoEn_{p,w,f} = 210\text{-}2200$ kJ (a)

$RGD_p = 0.002$ (c) $\quad L_{n,w,f} = 412\text{-}528$ mm (e) $\quad L_{n,w,f} = 162\text{-}266$ mm (e) $\quad GoEn_{p,w,m} = 15\text{-}170$ kJ (a) $\quad GoEn_{p,w,m} = 1\text{-}70$ kJ (a)

$RGD_i = 0.321$ (c) $\quad L_{p,w,m} = 356\text{-}440$ mm (f) $\quad L_{p,w,m} = 206\text{-}239$ mm (f) $\quad GoEn_{p,w,f} = 200\text{-}2500$ kJ (a) $\quad GoEn_{p,w,f} = 50\text{-}700$ kJ (a)

$M_{n,w,b} = 427\text{-}622$ g (g) $\quad M_{n,w,b} = 110\text{-}115$ g (g) $\quad GoEn_{n,w,f} = 457\text{-}2331$ kJ (g) $\quad GoEn_{n,w,f} = 78\text{-}143$ kJ (g)

$GSI_{p,w,m} = 3\%$ (a) $\quad GSI_{p,w,m} = 3\%$ (a) $\quad ToEn_{n,w,f} = 1254\text{-}7665$ kJ (g) $\quad ToEn_{n,w,f} = 463\text{-}886$ kJ (g)

$GSI_{p,w,f} = 34\%$ (a) $\quad GSI_{p,w,f} = 29\%$ (a)

$GSI_{n,w,f} = 25.0\text{-}26.0\%$ (g) $\quad GSI_{n,w,f} = 10.0\%$ (g)

$EN_{p,w} = 1023$ (e) $\quad EN_{p,w} = 210$ (e)

$EN_{n,w} = 1023\text{-}2707$ (e) $\quad EN_{n,w} = 102\text{-}510$ (e)

$EN_{n,w} = 512\text{-}1698$ (g) $\quad EN_{n,w} = 155\text{-}239$ (g)

$ES_{p,w} = 79.3$ mg (e) $\quad ES_{p,w} = 84.5$ mg (e)

$ES_{n,w} = 79.3\text{-}86.3$ mg (e) $\quad ES_{n,w} = 41.5\text{-}84.5$ mg (e)

$ES_{n,w} = 100\text{-}140$ mg (g) $\quad ES_{n,w} = 45\text{-}65$ mg (g)

Salmo salar

Nei's $D_s = 0.057$ (b) $\quad L_{i,w,b} = 798$ mm (f) $\quad L_{i,w,b} = 157\text{-}176$ mm (f)

$RGD_i = 0.183$ (a) $\quad L_{s,w,b} = 458\text{-}640$ mm (g) $\quad L_{s,w,b} = 298\text{-}396$ mm (g) $\quad ChCells_{s,g,b} = 27$ (g) $\quad ChCells_{s,g,b} = 1.3$ (g)

$M_{n,w,f} = 1000$ g (c) $\quad M_{n,w,f} = 21$ g (c) $\quad ChCell_{s,g,b} = 16.0$ μm (g) $\quad ChCell_{s,g,b} = 13.5$ μm (g)

$M_{p,w,m} = 16{,}250$ g (d) $\quad M_{p,w,m} = 24$ g (d) $\quad GillATPase_{s,g,b} = 34.5$ (g) $\quad GillATPase_{s,g,b} = 18.0$ (g)

$GSI_{p,w,m} = 2.3\%$ (d) $\quad GSI_{p,w,m} = 4.7\%$ (d)

$GSI_{p,w,m} = 4.4\%$ (e) $\quad GSI_{p,w,m} = 9.1\%$ (e)

$EN_{n,w,f} = 2000$ (c) $\quad EN_{n,w,f} = 55$ (c)

$ES_{n,w,f} = 98$ mm^3 (c) $\quad ES_{n,w,f} = 51$ mm^3 (c)

(continued)

Appendix 3 (*continued*)

Genetic difference	Life history traits		Other traits	
	Anadromous	Non-anadromous	Anadromous	Non-anadromous
Salvelinus alpinus/malma				
	$AM_{p,w,m} = 6.9$ y (d)	$AM_{p,w,m} = 3.7$ y (d)	$GRN_{p,w,b} = 12.6$ (b)	$GRN_{p,w,b} = 12.9$ (b)
	$AM_{n,w,m} = 6.9$ y (d)	$AM_{n,w,m} = 4.3$ y (d)	$GRN_{i,w,b} = 12.7$ (b)	$GRN_{i,w,b} = 12.1$ (b)
	$AM_{n,w,f} = 6.4$ y (d)	$AM_{n,w,f} = 4.6$ y (d)	$PyC_{p,w,b} = 31.5$ (b)	$PyC_{p,w,b} = 29.7$ (b)
	$AM_{p,w,m} = 9.7$ y (c)	$AM_{p,w,m} = 6.6{-}10.4$ y (c)	$PyC_{i,w,b} = 28.7$ (b)	$PyC_{i,w,b} = 23.1$ (b)
	$AM_{p,w,f} = 10.3$ y (c)	$AM_{p,w,f} = 7.1$ y (c)		
	$L_{p,w,m} = 415$ mm (d)	$L_{p,w,m} = 205$ mm (d)		
	$L_{n,w,m} = 415$ mm (d)	$L_{n,w,m} = 171$ mm (d)		
	$L_{n,w,f} = 421$ mm (d)	$L_{n,w,f} = 187$ mm (d)		
	$L_{n,w,f} = 383{-}385$ mm (a)	$L_{n,w,f} = 114$ mm (a)		
	$L_{p,w,m} = 417$ mm (c)	$L_{p,w,m} = 123{-}238$ mm (c)		
	$L_{p,w,f} = 465$ mm (c)	$L_{p,w,f} = 122$ mm (c)		
	$M_{n,w,f} = 635{-}635$ g (a)	$M_{n,w,f} = 17.3$ g (a)		
	$GM_{n,w,f} = 96{-}104$ g (a)	$GM_{n,w,f} = 1.8$ g (a)		
	$EN_{n,w} = 1846{-}1931$ (a)	$EN_{n,w} = 66$ (a)		
	$ES_{n,w} = 4.4{-}4.7$ mm (a)	$ES_{n,w} = 3.6$ mm (a)		
Salvelinus fontinalis				
$F_{STs} = 0.019$ (c)	$Amax_{i,w,b} = 7$ y (a)	$Amax_{i,w,b} = 4$ y (a)	$GRN_{p,w,m} = 17.4$ mm (b)	$GRN_{p,w,b} = 17.3$ mm (b)
$F_{STs} = 0.165$ (c)	$L_{i,w,b} = 140{-}559$ mm (a)	$L_{i,w,b} = 140{-}339$ mm (a)	$SnL_{p,w,m} = 13.1$ mm (b)	$SnL_{p,w,m} = 14.0$ mm (b)
	$L_{p,w,m} = 234$ mm (b)	$L_{p,w,m} = 192$ mm (b)	$SnL_{p,w,f} = 12.3$ mm (b)	$SnL_{p,w,f} = 12.8$ mm (b)
	$L_{p,w,f} = 231$ mm (b)	$L_{p,w,f} = 197$ mm (b)	$BD_{p,w,m} = 43.7$ mm (b)	$BD_{p,w,m} = 45.5$ mm (b)
	$M_{p,w,b} = 246.8$ g (b)	$M_{p,w,b} = 124.1$ g (b)	$BD_{p,w,m} = 43.1$ mm (b)	$BD_{p,w,m} = 43.5$ mm (b)
			GillATPase $= 1.65$ (c)	GillATPase $= 0.8$ (c)

Salvelinus leucomaenis

$L_{p,w,f} = 308$ mm (a) $L_{p,w,f} = 200$ mm (a)
$EN_{p,w} = 544$ (a) $EN_{p,w} = 238$ (a)
$ES_{p,w} = 5.9$ mm (a) $ES_{p,w} = 5.7$ mm (a)

Notes (references to tables and figures refer to each publication):
Oncorhynchus nerka:

(a) Wood and Foote (1996). *Location:* various creeks in Takla Lake, British Columbia, Canada. Relative gene diversity (RGD) is between sockeye salmon and kokanee, relative to the total gene diversity within the Takla Lake system (Table 3A). The effective number of migrants (*Nm*) is between sympatric morphs in five creeks, calculated from G_{ST} using the island model (Method 1 in Table 3B). Body length (*L*) is the maximum or minimum fork length for breeding fish in three creeks (reported in the text). Age at maturity (AM) is the modal age for breeding adults in three creeks. Egg size (ES) is mean water-hardened mass, reported as the range across three creeks (Table 4). Egg number (EN) and gonadosomatic index (GSI) are ranges across three creeks (Table 4). The number of gill-rakers on the first left arch (GRN) is averaged across three creeks (reported in the text).

(b) Taylor and Foote (1991). *Location:* Pierre Creek, Babine Lake, British Columbia, Canada. Critical swimming velocities in body lengths per second (CSV) are for similar sized juveniles raised in a common-garden hatchery environment.

(c) Foote et al. (1989). *Location:* lakes in British Columbia, Canada. Nei's genetic distance (Nei's *D*) is estimated from Figure 2 (polymorphic loci only), and represents differences between sockeye salmon and kokanee within each of three lake systems: Takla, Shuswap, and Babine (from top to bottom). Relative gene diversity (RGD) is from Table 4 and represents the diversity between sockeye salmon and kokanee relative to the total diversity, which includes multiple years, multiple drainages (Skeena, Columbia, Fraser, and others), multiple lakes within drainages, and multiple creeks within lakes.

(d) Foote et al. (1999). *Location:* Narrows Creek, Takla Lake, British Columbia, Canada. Length (*L*) is the fork length of breeding adults in the wild. The number of gill-rakers on the first left arch (GRN) is for breeding adults captured from the wild (subscript w) and for 2-year-old juveniles under common-garden conditions (subscript g). Gill-raker length (GRL) is for 2-year-old juveniles under common-garden conditions, and is allometrically adjusted to a common fork length of 264.4 mm. All values are from Table 1. Hybrids between sockeye salmon and kokanee had intermediate gill-raker numbers and lengths.

(e) Craig and Foote (2001). *Location:* Narrows Creek, Takla Lake, British Columbia, Canada. Length (*L*) is the fork length of breeding adults captured from the wild. The intensity of red coloration in the skin (SkCol) is given for breeding adults captured from the wild (subscript w; Figure 1A) and for fish raised to maturity in a common-garden hatchery environment (subscript g; Figure 3). Units are an international color standard, a*, for which increasing values are associated with more intense red color. Carotenoid content in the skin (SkCar) is for fish raised to maturity in a common-garden environment (Figure 3). Hybrids between sockeye salmon and kokanee raised under the same conditions had intermediate skin color and carotenoid content.

(f) Wood and Foote (1990). *Location:* Lower Shuswap River, British Columbia, Canada. Egg diameter (ES) is for breeding females in the wild.

(*continued*)

Appendix 3 (*continued*)

Oncorhynchus mykiss:

(a) Currens et al. (1990). *Location:* Deschutes River, Oregon. Nei's genetic distance is for White River fish, which are above a waterfall, versus the rest of the Deschutes River (estimated from Fig. 2). F_{ST} values are the range among loci for comparisons between native fish in the White River and native fish elsewhere in the Deschutes River. The largest F_{ST} was for *Ldh-B2*, whereas $F_{ST} = 0.00$–0.09 for the other allozyme loci. Samples from the Deschutes River downstream of the barrier may include anadromous and non-anadromous fish.

(b) Pascual et al. (2001). *Location:* Santa Cruz River, Patagonia, Argentina. Non-anadromous rainbow trout were introduced into the system starting in the early 1900s and anadromous fish appear to have arisen subsequently. Fish were collected from two sites on the river, with anadromous fish coming exclusively from the downstream site and resident fish primarily from the upstream site. Average fork length (*L*) in the two forms was calculated using Appendix 1. Even though the two forms came primarily from different locations in the watershed, we consider them polymorphic because they show no genetic differentiation.

(c) Neave (1944). *Location:* Cowichan River, British Columbia, Canada. Traits are the number of scales in a row above the lateral line (Scales) for fish caught in the wild (subscript w) and raised in a common-garden hatchery environment (subscript g). The two forms appear to be reproductively isolated because the scale number differences are hereditary and the progeny of anadromous fish are much more likely to migrate to the ocean than are the progeny of non-anadromous fish.

Salmo trutta:

(a) Jonsson and Jonsson (1997). *Location:* three small streams (Helldalsbekken, Skagestadbekken, Vesbekken) in southern Norway. Traits are mean gonadosomatic index (GSI, Fig. 2), mean mass-specific somatic energy (SoEnSp, Table 1), mean mass-specific somatic fat content (SoFt, Table 1), mean mass-specific somatic protein content (SoPrSp, Table 1), within-population range in total somatic energy (SoEn, Figure 3), and within-population range in total gonad energy (GoEn, Figure 3). All values are for resident versus anadromous trout pooled across the three streams.

(b) Pettersson et al. (2001). *Location:* River Jörlanda, western Sweden. Comparisons are for resident versus anadromous fish sampled within the lower parts of the stream (subscript p) and for upstream landlocked fish versus downstream fish of both forms (subscript i).

(c) Hindar et al. (1991a). *Location:* Voss River system, western Norway. Nei's D (estimated from Figure 3), relative gene diversities (RGD, from Table 3) and natural tip lengths (*L*, from Table 1) are for anadromous versus non-anadromous forms in areas with access to the ocean (subscript p), and for fish in areas with access to the ocean versus fish landlocked above barriers (subscript i). Lengths are ranges of means across different sampling locations.

(d) Cross et al. (1992). *Location:* Burrishoole system, western Ireland. Comparisons are for two landlocked populations above barriers versus all collections from areas with access to the ocean (subscript i) and for anadromous versus non-anadromous fish from Lough Feeagh (subscript p—because no genetic differences were evident). Nei's genetic distances were estimated from Figure 2.

(e) Olofsson and Mosegaard (1999). *Location:* six streams in Sweden. Traits include natural total length (*L*), egg number (EN), and egg wet weight (ES), all from Table 1. Comparisons are between anadromous and non-anadromous fish in the same stream (Jörlandaån, subscript p) and ranges for anadromous forms in two streams and non-anadromous forms in five streams (subscript n). We assume anadromous and non-anadromous fish in the Jörlandaån are part of a single polymorphic population but no genetic data are available for confirmation.

(f) Jonsson (1985). *Location*: Vangsvatnet Lake and its tributaries, Norway. Data represent ranges among tributaries and years and include average age at maturity (AM), calculated for anadromous fish as average smolt age plus average sea age (Table 2); and tip lengths (L). Resident and anadromous fish were seen spawning together and the population is assumed to be polymorphic.

(g) Elliott (1988). *Location*: Two creeks located 3 km from each other in northwestern England (Wilfin Beck, non-anadromous trout; and Black Brows Beck, anadromous trout). Trait values are given in three different forms: (1) within-population ranges of body mass (M, 4-year-old spawners) and wet egg mass (ES, all spawning females, from Figure 2); (2) population averages for gonadosomatic index (GSI); and (3) predicted ranges of egg number (EN), total amount of energy (ToEn), and total amount of gonad energy (GoEn) for spawning females of a typical range of sizes in each stream (25–45 cm for anadromous females in Black Brows Beck and 20–25 cm for non-anadromous females in Wilfin Beck, from Table 2).

Salmo salar:

(a) Vuorinen and Berg (1989). *Location*: River Namsen, central Norway. The two forms are separated by an impassible waterfall. Non-anadromous samples were taken from several locations, themselves separated by waterfalls. Relative gene diversity (RGD) is the proportion of the total genetic diversity attributable to differences between the two forms.

(b) Verspoor and Cole (1989). *Location*: Little Gull Lake, Gander River, Newfoundland, Canada. Fish were collected from the lake and partitioned by the authors into resident and anadromous gene pools based on life history characteristics and genetic profiles. As a result, Nei's genetic distance (Nei's D) may upwardly biased.

(c) Sutterlin and MacLean (1984). *Location*: anadromous fish from the Exploits River (Newfoundland) and non-anadromous fish from Five Mile Pond East (Newfoundland), with the later "selected as an extreme representative of salmon maturing at a small size." Data were estimated by eye from Figure 4 and include body mass (M), egg number (EN), and individual egg volume (ES).

(d) Gage et al. (1995). *Location*: North Tyne system, United Kingdom. Anadromous males and mature male parr were captured from the wild and allowed to mature in a hatchery. Data include body mass (M) and gonadosomatic index (GSI, Figure 3).

(e) Fleming (1996). *Location*: River Imsa, southern Norway. Gonadosomatic index (GSI) is for anadromous males and mature male parr captured from the wild (Table 1).

(f) N. Jonsson et al. (1991a) and Berg and Gausen (1988). *Location*: River Namsen, central Norway. The non-anadromous form is isolated above a barrier to migration. Data for anadromous fish is natural tip length (L), calculated using Table 2 in N. Jonsson et al. (1991a). Data for non-anadromous fish is total length (L) from Berg and Gausen (1988), shown as mean values for males and females.

(g) Birt et al. (1991b). *Location*: Gambo River, Newfoundland, Canada. Fork length (L) is for wild mature anadromous fish from Triton Brook and wild mature non-anadromous fish from Gambo Pond. Chloride cells per gill lamella (ChCellS), chloride cell length (ChCellL), and gill Na$^+$-K$^+$ ATPase (GillATPase, units = μmol P_i per mg protein per hour) are in May for fish reared in captivity. The two forms are considered sympatric because they are in the same system and no physical barriers prevent their mixing. They are also genetically distinct from each other (Birt et al. 1991a).

(*continued*)

Appendix 3 (continued)

Salvelinus alpinus/malma:

(a) Blackett (1973). *Location:* anadromous populations in two creeks (Hood Bay and Bear) and a non-anadromous population above a waterfall in another creek (Falls), all within 10 km of each other in southeastern Alaska. Traits include fork length (*L*), body mass (*M*), egg number (EN), egg diameter (ES), and total gonad mass (GM).

(b) Reist (1989). *Location:* tributaries to the Beaufort Sea, Arctic Canada. Traits include the number of lower gill-rakers (GRN, Table 3) and the number of pyloric caeca (PyC, Table 3). The first comparison is for anadromous and non-anadromous individuals in a polymorphic population (no genetic differences between morphs) in Joe Creek (subscript p). The second comparison is between anadromous individuals in Canoe River and landlocked individuals further upstream in the Babbage River (subscript i).

(c) Gulseth and Nilssen (2001). *Location:* Dieset River system, Svalbard, Norway. All data are for first-time spawners only (Table 2) and include age at maturity (AM) and total length (*L*). Values for non-anadromous males are given as those for "small residents" – "large residents." We assume the population is polymorphic (subscript p) because the authors state "There was no evidence of reproductive isolation between these High Arctic charr morphs." However, genetic studies have not been performed.

(d) Maekawa et al. (1993). *Location:* Steep Creek and Tiekel River, Alaska. Traits are age at maturity (AM) and total length (*L*). Comparisons are between anadromous and stream resident males in the same population (Steep Creek, subscript p) and between anadromous fish in Steep Creek and landlocked fish in Tiekel River (subscript n).

Salvelinus fontinalis:

(a) Castonguay et al. (1982). *Location:* St. Jean River, Gaspé, Québec, Canada. The anadromous form was from the main St. Jean River. The non-anadromous form was from a small lake (McLaren) above a waterfall in the same system. These McLaren fish were descended from non-native anadromous fish stocked in 1932. Maximum age (Amax) is from Figure 3. Fork length (*L*) is the range for maturing fish from Figure 4.

(b) Wilder (1952). *Location:* Moser River, Nova Scotia, Canada. Traits are standard length (*L*, Tables 14 and 15), body mass (*M*, Table 10), number of gill-rakers on the first arch on one side (GRN, Table 8), snout length standardized to a common body length of 200 mm (SnL, Table 3), and body depth standardized to a common body length of 200 mm (BD, Table 3). Comparisons are for fresh-run sea trout caught at a weir and "migratory-sized" freshwater trout caught in the river.

(c) Boula et al. (2002). *Location:* Laval River, Québec, Canada. F_{ST} values are for anadromous brook charr versus non-anadromous brook charr in the Adam River ($F_{ST} = 0.1645$) and in Alexis Pond ($F_{ST} = 0.019$). Gill Na^+-K^+ ATPase (GillATPase, units = µmol P_i per mg protein per hour) is for fish caught in the spring (from Figure 3).

Salvelinus leucomaenis:

(a) Morita and Takashima (1998). *Location:* Kame River, southern Hokkaido, Japan. Fish were identified as resident or sea-run forms using scale analysis. The two forms are probably part of the same polymorphic population but this has not been confirmed using genetic analyses. Traits are fork length (*L*), egg number (EN), and egg diameter (ES). EN and ES were calculated using mean fork length and the regression equations in the caption of Figure 1.

REFERENCES

Numbers in square brackets indicate the chapter numbers in which a reference is cited (also: I = Introduction, A1 = Appendix 1, A2 = Appendix 2, A3 = Appendix 3).

Abbott, J. C., R. L. Dunbrack, and C. D. Orr. 1985. The interaction of size and experience in dominance relationships of juvenile steelhead trout (*Salmo gairdneri*). Behaviour 92:241–253. [13]

Adams, B. K. 1999. Variation in genetic structure and life histories among populations of brook char, *Salvelinus fontinalis*, from insular Newfoundland, Canada. M.Sc. thesis, Dalhousie Univ., Halifax, NS, Canada. [5]

Adams, N. S., W. J. Spearman, C. V. Burger, K. P. Currens, C. B. Schreck, and H. W. Li. 1994. Variation in mitochondrial DNA and allozymes discriminates early and late forms of chinook salmon (*Oncorhynchus tshawytscha*) in the Kenai and Kasilof Rivers, Alaska. Can. J. Fish. Aquat. Sci. 51(Suppl. 1):172–181. [12]

Adkison, M. D. 1995. Population differentiation in Pacific salmon: local adaptation, genetic drift, or the environment? Can. J. Fish. Aquat. Sci. 52:2762–2777. [2, 3, 7]

Adriaensen, F., and A. A. Dhondt. 1990. Population dynamics and partial migration of the European robin (*Erithacus rubecula*) in different habitats. J. Anim. Ecol. 59:1077–1090. [3]

Ahnesjö, I., A. Vincent, R. Alatalo, T. Halliday, and W. J. Sutherland. 1993. The role of females in influencing mating patterns. Behav. Ecol. 4:187–189. [9]

Allendorf, F. W., and R. F. Leary. 1988. Conservation and distribution of genetic variation in a polytypic species, the cutthroat trout. Cons. Biol. 2:170–184. [A2]

Allendorf, F. W., and S. R. Phelps. 1981. Use of allelic frequencies to describe population structure. Can. J. Fish. Aquat. Sci. 38:1507–1514. [2]

Allendorf, F. W., and L. W. Seeb. 2000. Concordance of genetic divergence among sockeye salmon populations at allozyme, nuclear DNA, and mitochondrial DNA markers. Evolution 54:640–651. [A2]

Allendorf, F. W., and G. H. Thorgaard. 1984. Tetraploidy and the evolution of salmonid fishes. Pp. 1–53 *in* B. J. Turner, ed. Evolutionary genetics of fishes. Plenum Press, New York. [7, 8]

Allendorf, F. W., and R. S. Waples. 1996. Conservation and genetics of salmonid fishes. Pp. 238–280 *in* J. C. Avise and J. L. Hamrick, eds. Conservation genetics: case histories from nature. Chapman and Hall, New York. [12]

Allendorf, F. W., R. F. Leary, P. Spruell, and J. K. Wenburg. 2001. The problems with hybrids: setting conservation guidelines. Trends Ecol. Evol. 16:613–622. [I, 8]

Alm, G. 1959. Connection between maturity, size and age in fishes. Rep. Inst. Freshwater Res., Drottingholm 40:1–145. [1, 5]

Altukhov, Y. P. 1994. Genetic consequences of selective fishing. Genetika 30:5–21 (in Russian with English abstract, tables, and figures). [11]

Altukhov, Y. P., and E. A. Salmenkova. 1991. The genetic structure of salmon populations. Aquaculture 98:11–40. [3, A1]

Altukhov, Y. P., E. A. Salmenkova, and V. T. Omelchenko. 2000. Salmonid fishes: population biology, genetics and management. Blackwell Science, Oxford, U.K. [3]

Alverson, D. L, and G. T. Ruggerone. 1997. Escaped farm salmon: environmental and ecological concerns. Discussion paper. Volume 3, Part B. Salmon Aquaculture Review Rept. B.C. Environmental Assessment Office. Victoria, BC, Canada. Available from www.eao.gov.bc.ca/PROJECT/AQUACULT/SALMON/Report/final/vol1/toc.htm. [8]

Amundsen, T. 2000. Why are female birds ornamented? Trends Ecol. Evol. 15:149–155. [9]

Anderson, E. C. 2001. Monte Carlo methods for inference in population genetic models. Ph.D. dissertation, Univ. of Washington, Seattle, WA. [10]

Andersson, L. 2001. Genetic dissection of phenotypic diversity in farm animals. Nature Rev. Genet. 2:130–138. [6]

Andersson, M. 1986. Evolution of condition-dependence sex ornaments and mating preferences: sexual selection based on viability differences. Evolution 40:804–816. [9]

Andersson, M. B. 1994. Sexual selection. Princeton Univ. Press, Princeton, NJ. [I, 9]

Angers, B., and L. Bernatchez. 1998. Combined use of SMM and non-SMM methods to infer fine structure and evolutionary history of closely related brook charr (*Salvelinus fontinalis*, Salmonidae) populations from microsatellites. Mol. Biol. Evol. 15:143–159. [A2]

Antunes, A., P. Alexandrino, and N. Ferrand. 1999. Genetic characterization of Portuguese brown trout (*Salmo trutta* L.) and comparisons with other European populations. Ecol. Freshwater Fish 8:194–200. [A2]

Apostolidis, A., Y. Karakousis, and C. Triantaphyllidis. 1996. Genetic divergence and phylogenetic relationships among *Salmo trutta* L. (brown trout) populations from Greece and other European countries. Heredity 76:551–560. [A2]

Arendt, J. D. 1997. Adaptive intrinsic growth rates: an integration across taxa. Quart. Rev. Biol. 72:149–177. [3]

Arendt, J. D, and D. S. Wilson. 2000. Population differences in the onset of cranial ossification in pumpkinseed (*Lepomis gibbosus*), a potential cost of growth. Can. J. Fish. Aquat. Sci. 57:351–356. [3]

Arendt, J., D. S. Wilson, and E. Stark. 2001. Scale strength as a cost of rapid growth in sunfish. Oikos 93:95–100. [3]

Arkush, K. D., A. R. Giese, H. L. Mendonca, A. M. McBride, G. D. Marty, and P. W. Hedrick. 2002. Resistance to three pathogens in the endangered winter-run chinook salmon (*Oncorhynchus tshawytscha*): effects of inbreeding and major histocompatability complex genotypes. Can. J. Fish. Aquat. Sci. 59:966–975. [2]

Armstrong, J. D., F. A. Huntingford, and N. A. Herbert. 1999. Individual space use strategies of wild juvenile Atlantic salmon. J. Fish Biol. 55:1201–1212. [2]

Armstrong, R. H., and J. E. Morrow. 1980. The Dolly Varden charr, *Salvelinus malma*. Pp. 99–140 *in* E. K. Balon, ed. Charrs: salmonid fishes of the genus *Salvelinus*. Dr. W. Junk Publishers, Dordrecht, The Netherlands. [9]

Arnold M. L. 1997. Natural hybridization and evolution. Oxford Univ. Press, Oxford, U.K. [1, 8]

Arnold, M. L., and S. A. Hodges. 1995. Are natural hybrids fit or unfit relative to their parents? Trends Ecol. Evol. 10:67–71. [8]

Arnold, S. J., and M. J. Wade. 1984a. On the measurement of natural and sexual selection: theory. Evolution 38:709–719. [10, 11]

Arnold, S. J., and M. J. Wade. 1984b. On the measurement of natural and sexual selection: applications. Evolution 38:720–734. [11]

Arnold, S. J., M. E. Pfrender, and A. G. Jones. 2001. The adaptive landscape as a conceptual bridge between micro- and macroevolution. Genetica 112/113:9–32. [2, 6, 11]

Ashley, M. V., M. F. Willson, O. R. W. Pergams, D. J. O'Dowd, S. M. Gende, and J. S. Brown. 2003. Evolutionarily enlightened management. Biol. Cons. 111:115–123. [1]

Asmussen, M. A. 1983. Evolution of dispersal in density regulated populations: a haploid model. Theor. Pop. Biol. 23:281–299. [2]

Avise, J. C. 1994. Molecular markers, natural history and evolution. Chapman and Hall, New York. [8, 12]

Bagenal, T. B. 1969. Relationship between egg size and fry survival in brown trout *Salmo trutta* L. J. Fish Biol. 1:349–353. [4]

Bagenal, T. B. 1971. The interrelation of the size of fish eggs, the date of spawning and the production cycle. J. Fish Biol. 3:207–219. [4]

Balkau, B. J., and M. W. Feldman. 1973. Selection for migration modification. Genetics 74:171–174. [2]

Balloux, F. 2001. EASYPOP (Version 1.7): A computer program for population genetics simulations. J. Heredity 92:301–302. [2]

Balloux, F., and J. Goudet. 2002. Statistical properties of population differentiation estimators under stepwise mutation in a finite island model. Mol. Ecol. 11:771–783. [2]

Balon, E. K., ed. 1980. Charr, salmonid fishes of the genus *Salvelinus*. Dr. W. Junk Publishers, Dordrecht, The Netherlands. [2]

Bams, R. A. 1976. Survival and propensity for homing as affected by presence or absence of locally adapted paternal genes in two transplanted populations of pink salmon (*Oncorhynchus gorbuscha*). J. Fish. Res. Board Can. 33:2716–2725. [2, 8]

Banks, M. A., V. K. Rashbrook, M. J. Calavetta, C. A. Dean, and D. Hedgecock. 2000. Analysis of microsatellite DNA resolves genetic structure and diversity of chinook salmon (*Oncorhynchus tshawytscha*) in California's Central Valley. Can. J. Fish. Aquat. Sci. 57:915–927. [A2]

Barker, J. S. F., and R. H. Thomas. 1987. A quantitative genetic perspective on adaptive evolution. Pp. 3–23 *in* V. Loeschke, ed. Genetic constraints on adaptive evolution. Springer-Verlag, Berlin, Germany. [11]

Barlaup, B. T., H. Lura, H. Sægrov, and R. C. Sundt. 1994. Inter- and intra-specific variability in female salmonid spawning behaviour. Can. J. Zool. 72:636–642. [1]

Barraclough, T. G., and S. Nee. 2001. Phylogenetics and speciation. Trends Ecol. Evol. 16:391–399. [6]

Barrett, S. C. H., and D. Charlesworth. 1991. Effects of a change in the level of inbreeding on the genetic load. Nature 352:522–524. [2]

Barrett-Hamilton, G. E. H. 1900. A suggestion as to the possible mode of origin of some of the secondary sexual characters in animals as afforded by observation on certain salmonids. Cambridge Phil. Soc. Proc. 10:279–285. [9]

Barrowman, N. J., and R. A. Myers. 2000. Still more spawner-recruitment curves: the hockey stick and its generalizations. Can. J. Fish. Aquat. Sci. 57:665–676. [13]

Bartley, D. M., and G. A. E. Gall. 1990. Genetic structure and gene flow in chinook salmon populations of California. Trans. Am. Fish. Soc. 119:55–71. [A2]

Bartley, D., M. Bagley, G. Gall, and B. Bentley. 1992c. Use of linkage disequilibrium data to estimate effective size of hatchery and natural fish populations. Cons. Biol. 6:365–375. [10]

Bartley, D. M., B. Bentley, P. G. Olin, and G. A. E. Gall. 1992a. Population genetic structure of coho salmon (*Oncorhynchus kisutch*) in California. Calif. Fish Game 78:88–104. [A2]

Bartley, D., B. Bentley, J. Brodziak, R. Gomulkiewicz, M. Mangel, and G. A. E. Gall. 1992b. Geographic variation in population genetic structure of chinook salmon from California and Oregon. Fishery Bull. 90:77–100. (Originally published as: Gall, Bartley, Bentley, Brodziak, Gomulkiewicz and Mangel.) [12, A2]

Bartley, D. M., G. A. E. Gall, and B. Bentley. 1990. Biochemical genetic detection of natural and artificial hybridization of chinook and coho salmon in northern California. Trans. Am. Fish. Soc. 119:431–437. [8]

Barton, N. H. 1998. The geometry of adaptation. Nature 395:751–752. [6]

Barton, N. H. 2001. The role of hybridization in evolution. Mol. Ecol. 10:551–568. [8]

Barton, N. H., and G. M. Hewitt. 1985. Analysis of hybrid zones. Annu. Rev. Ecol. Syst. 16:113–148. [8]

Barton, N. H., and M. Turelli. 1989. Evolutionary quantitative genetics: how little do we know? Annu. Rev. Genet. 23:337–370. [11]

Bateson, P. 1998. The active role of behaviour in evolution. Pp. 191–207 *in* M.-W. Ho and S. Fox, eds. Process and metaphors in evolution. Wiley, Chichester, U.K. [I]

Baxter, J. S. 1997. Aspects of the reproductive ecology of bull trout (*Salvelinus confluentus*) in the Chowade River, British Columbia. M.Sc. thesis, Univ. of British Columbia, Vancouver, Canada. [8]

Baxter, J. S., E. B. Taylor, R. H. Devlin, J. Hagen, and J. D. McPhail. 1997. Evidence for natural hybridization between Dolly Varden (*Salvelinus malma*) and bull trout (*Salvelinus confluentus*) in a northcentral British Columbia watershed. Can. J. Fish. Aquat. Sci. 54:421–429. [I, 8]

Beacham, T. D., and J. B. Dempson. 1998. Population structure of Atlantic salmon from the Conne River, Newfoundland as determined from microsatellite DNA. J. Fish Biol. 52:665–676. [A2]

Beacham, T. D., and C. B. Murray. 1985. Effect of female size, egg size, and water temperature on developmental biology of chum salmon (*Oncorhynchus keta*) from the Nitinat River, British Columbia. Can. J. Fish. Aquat. Sci. 42:1755–1765. [5]

Beacham, T. D., and C. B. Murray. 1988. A genetic analysis of body size in pink salmon (*Oncorhynchus gorbuscha*). Genome 30:31–35. [13]

Beacham, T. D., and C. B. Murray. 1989. Variation in developmental biology of sockeye salmon (*Oncorhynchus nerka*) and chinook salmon (*O. tshawytscha*) in British-Columbia. Can. J. Zool. 67:2081–2089. [7]

Beacham, T. D., and C. B. Murray. 1990. Temperature, egg size, and development of embryos and alevins of five species of Pacific salmon: a comparative analysis. Trans. Am. Fish. Soc. 119:927–945. **[13]**

Beacham, T. D., and C. B. Murray. 1993. Fecundity and egg size variation in North American Pacific salmon (*Oncorhynchus*). J. Fish Biol. 42:485–508. **[3, 4]**

Beacham, T. D., and C. C. Wood. 1999. Application of microsatellite DNA variation to estimation of stock composition and escapement of Nass River sockeye salmon (*Oncorhynchus nerka*). Can. J. Fish. Aquat. Sci. 56:297–310. **[A2]**

Beacham, T. D., A. P. Gould, R. E. Withler, C. B. Murray, and L.W. Barner. 1987. Biochemical genetic survey and stock identification of chum salmon (*Oncorhynchus keta*) in British Columbia. Can. J. Fish. Aquat. Sci. 44:1702–1713. **[A2]**

Beacham, T. D., K. D. Le, M. R. Raap, K. Hyatt, W. Luedke, and R. E. Withler. 2000c. Microsatellite DNA variation and estimation of stock composition of sockeye salmon, *Oncorhynchus nerka*, in Barkley Sound, British Columbia. Fishery Bull. 98:14–24. **[A2]**

Beacham, T. D., C. B. Murray, and R. E. Withler. 1988a. Age, morphology, developmental biology, and biochemical genetic variation of Yukon River fall chum salmon, *Oncorhynchus keta*, and comparisons with British Columbia populations. Fishery Bull. 86:663–674. **[3]**

Beacham, T. D., S. Pollard, and K. D. Le. 1999. Population structure and stock identification of steelhead in southern British Columbia, Washington, and the Columbia River based on microsatellite DNA variation. Trans. Am. Fish. Soc. 128:1068–1084. **[A2]**

Beacham, T. D., S. Pollard, and K. D. Le. 2000a. Microsatellite DNA population structure and stock identification of steelhead trout (*Oncorhynchus mykiss*) in the Nass and Skeena Rivers in northern British Columbia. Mar. Biotechnol. 2:587–600. **[A2]**

Beacham, T. D., R. E. Withler, C. B. Murray, and L. W. Barner. 1988b. Variation in body size, morphology, egg size, and biochemical genetics of pink salmon in British Columbia. Trans. Am. Fish. Soc. 117:109–126. **[8, A2]**

Beacham, T. D., C. C. Wood, R. E. Withler, K. D. Le, and K. M. Miller. 2000b. Application of microsatellite DNA variation to estimation of stock composition and escapement of Skeena River sockeye salmon (*Oncorhynchus nerka*). North Pacific Anadromous Fish Comm. Bull. 2:263–276. **[A2]**

Beamesderfer, R. C. P., H. A. Schaller, M. P. Zimmerman, C. E. Petrosky, O. P. Langness, and L. LaVoy. 1998. Spawner-recruit data for spring and summer chinook salmon populations in Idaho, Oregon, and Washington. Section 2, Chapter 1 *in* D. R. Marmorek and C. N. Peters, eds. Plan for Analyzing and Testing Hypotheses (PATH): retrospective and prospective analyses of spring/summer chinook reviewed in FY 1997. ESSA Technologies, Vancouver, Canada. **[10]**

Beckman, B. 2002. Growth and the plasticity of smolting in chinook salmon. Ph.D. dissertation, Univ. of Washington, Seattle, WA. **[12]**

Beckman, B. R., and W. W. Dickhoff. 1998. Plasticity of smolting in spring chinook salmon: relation to growth and insulin-like growth factor-I. J. Fish Biol. 53:808–826. **[12]**

Beechie, T. J., and T. H. Sibley. 1997. Relationships between channel characteristics, woody debris, and fish habitat in northwestern Washington streams. Trans. Am. Fish. Soc. 126:217–229. **[2]**

Beechie, T., E. Beamer, and L. Wasserman. 1994. Estimating coho salmon rearing habitat and smolt production losses in a large river basin, and implications for habitat restoration. N.A. J. Fish. Manage. 14:797–811. [13]

Beerli, P., and J. Felsenstein. 1999. Maximum-likelihood estimation of migration rates and effective population numbers in two populations using a coalescent approach. Genetics 152:763–773. [2]

Beerli, P., and J. Felsenstein. 2001. Maximum likelihood estimation of a migration matrix and effective population sizes in *n* subpopulations by using a coalescent approach. Proc. Natl. Acad. Sci. USA 98:4563–4568. [2]

Behnke, R. J. 1972. The systematics of salmonid fishes of recently glaciated lakes. J. Fish. Res. Board Can. 29:639–671. [8]

Behnke, R. J. 1980. A systematic review of the genus *Salvelinus*. Pp. 441–481 *in* E. K. Balon, ed. Charrs: salmonid fishes of the genus *Salvelinus*. Dr. W. Junk Publishers, Dordrecht, The Netherlands. [8]

Behnke, R. J. 1992. Native trout of western North America. American Fisheries Society Monograph, No. 6. American Fisheries Society, Bethesda, MD. [I, 8]

Behnke, R. J., T. P. Koh, and P. R. Needham. 1962. Status of the landlocked salmonid fishes of Formosa with a review of *Oncorhynchus masou* (Brevoort). Copeia 1962:400–407. [7]

Bekkevold, D. 2002. Reproductive behaviour in fishes: patterns of parentage and effects on population genetic structure. Ph.D. dissertation, Univ. of Aarhus, Denmark. [2]

Belding, D. I. 1934. The cause of high mortality in Atlantic salmon after spawning. Trans. Am. Fish. Soc. 64:219–224. [1]

Bélichon, S., J. Clobert, and M. Massot. 1996. Are there differences in fitness components between philopatric and dispersing individuals? Acta (Ecologica) 17:503–517. [2]

Bell, G. 1976. On breeding more than once. Am. Nat. 110:5777. [I]

Bell, M. A., and S. A. Foster, eds. 1994. The evolutionary biology of the threespine stickleback. Oxford Univ. Press, Oxford, U.K. [I]

Bellman, R. 1957. Dynamic programming. Princeton Univ. Press. Princeton, NJ. [1]

Ben-David, M., T. A. Hanley, and D. M. Schell. 1998. Fertilization of terrestrial vegetation by spawning Pacific salmon: the role of flooding and predator activity. Oikos 83:47–55. [13]

Bengtsson, B. O. 1978. Avoiding inbreeding: at what cost? J. theor. Biol. 73:439–444. [2]

Bentzen, P., J. B. Olsen, J. E. McLean, T. R. Seamons, and T. P. Quinn. 2001. Kinship analysis of Pacific salmon: insights into mating, homing, and timing of reproduction. J. Heredity 92:127–136. [I, 8, 10]

Berejikian, B. A. 1995. The effects of hatchery and wild ancestry and experience on the relative ability of steelhead trout fry (*Oncorhynchus mykiss*) to avoid a benthic predator. Can. J. Fish. Aquat. Sci. 52:2476–2482. [12]

Berejikian, B. A., S. B. Mathews, and T. P. Quinn. 1996. Effects of hatchery and wild ancestry and rearing environments on the development of agonistic behavior in steelhead trout (*Oncorhynchus mykiss*) fry. Can. J. Fish. Aquat. Sci. 53:2004–2014. [13]

Berejikian, B. A., E. P. Tezak, and A. L. LaRae. 2000. Female mate choice and spawning behaviour of chinook salmon under experimental conditions. J. Fish Biol. 57:647–661. [I, 3, 9]

Berejikian, B. A., E. P. Tezak, L. Park, E. LaHood, S. L. Schroder, and E. Beall. 2001. Male competition and breeding success in captively reared and wild coho salmon *Oncorhynchus kisutch*). Can. J. Fish. Aquat. Sci. 58:804–810. [10]

Berejikian, B. A., E. P. Tezak, S. L. Schroder, T. A. Flagg, and C. M. Knudsen. 1999. Competitive differences between newly emerged offspring of captive-reared and wild coho salmon. Trans. Am. Fish. Soc. 128:832–839. [13]

Berejikian, B. A., E. P. Tezak, S. L. Schroder, C. M. Knudsen, and J. J. Hard. 1997. Reproductive behavioral interactions between wild and captively reared coho salmon (*Oncorhynchus kisutch*). ICES J. Mar. Sci. 54:1040–1050. [12]

Berg, O. K. 1985. The formation of non-anadromous populations of Atlantic salmon, *Salmo salar* L., in Europe. J. Fish Biol. 27:805–815. [I, 3, 7]

Berg, O. K., and D. Gausen. 1988. Life history of a riverine, resident Atlantic salmon *Salmo salar* L. Fauna Norv. Ser. A 9:63–68. [A3]

Berg, W. J., and G. A. E. Gall. 1988. Gene flow and genetic differentiation among California coastal rainbow trout populations. Can. J. Fish. Aquat. Sci. 45:122–131. [A2]

Berglund, I. 1992. Growth and early sexual maturation in Baltic salmon (*Salmo salar*) parr. Can. J. Zool. 70:205–211. [3]

Bergstrand, E. 1982. The diet of four sympatric whitefish species in Lake Parkijaure. Rep. Inst. Freshwater Res., Drottningholm 60:5–14. [6]

Berlocher, S. H. 1998. Origins: a brief history of research on speciation. Pp. 3–15 *in* D. J. Howard and S. H. Berlocher, eds. Endless forms. Species and speciation. Oxford Univ. Press, Oxford, U.K. [8]

Bernardo, J. 1996a. Maternal effects in animal ecology. Am. Zool. 36:83–105. [4]

Bernardo, J. 1996b. The particular maternal effect of propagule size, especially egg size: patterns, models, quality of evidence and interpretations. Am. Zool. 36:216–236. [4]

Bernatchez, L. 1995. Réseau de suivi environnemental du complexe La Grande. Caractérisation génétique des formes naines et normales de grand corégone du réservoir Caniapiscau et du lac Sérigny à l'aide de marqueurs microsatellites. Rapport présenté par l'université Laval à la vice-présidence Environnement et Collectivités d'Hydro-Québec, Québec, Canada. [6]

Bernatchez, L. 2001. The evolutionary history of brown trout (*Salmo trutta* L.) inferred from phylogeographic, nested clade, and mismatch analyses of mitochondrial DNA variation. Evolution 55:351–379. [2, 6]

Bernatchez, L. 2003. Ecological theory of adaptive radiation: an empirical assessment from Coregonine fishes (Salmoniformes). Pp. 175–207 *in* A. P. Hendry and S. C. Stearns, eds. Evolution illuminated: salmon and their relatives. Oxford Univ. Press, New York, NY—*this volume* [I, 3, 7, 8]

Bernatchez, L., and J. J. Dodson. 1987. Relationship between bioenergetics and behaviour in anadromous fish migrations. Can. J. Fish. Aquat. Sci. 44:399–407. [3]

Bernatchez, L., and J. J. Dodson. 1990. Allopatric origin of sympatric populations of Lake whitefish (*Coregonus clupeaformis*) as revealed by mitochondrial-DNA restriction analysis. Evolution 44:1263–1271. [6]

Bernatchez, L., and J. J. Dodson. 1991. Phylogeographic structure in mitochondrial DNA of the Lake whitefish (*Coregonus clupeaformis*) and its relation to Pleistocene glaciations. Evolution 45:1016–1035. [6]

Bernatchez, L., and C. Landry. 2003. MHC studies in nonmodel vertebrates: what have we learned about natural selection in 15 years? J. Evol. Biol. 16:363–377. [9]

Bernatchez, L., and A. Osinov. 1995. Genetic diversity of trout (genus *Salmo*) from its most eastern native range based on mitochondrial DNA and nuclear gene variation. Mol. Ecol. 4:285–297. [A2]

Bernatchez, L., and C. C. Wilson. 1998. Comparative phylogeography of Nearctic and Palearctic fishes. Mol. Ecol. 7:431–452. [6, 7]

Bernatchez, L., A. Chouinard, and G. Lu. 1999. Integrating molecular genetics and ecology in studies of adaptive radiation: whitefish, *Coregonus* sp., as a case study. Biol. J. Linn. Soc. 68:173–194. [I, 6]

Bernatchez, L., F. Colombani, and J. J. Dodson. 1991. Phylogenetic relationships among the subfamily Coregoninae as revealed by mitochondrial DNA restriction analysis. J. Fish Biol. 39(Suppl. A):283–290. [6]

Bernatchez, L., J. B. Dempson, and S. Martin. 1998. Microsatellite gene diversity analysis in anadromous Arctic char, *Salvelinus alpinus*, from Labrador, Canada. Can. J. Fish. Aquat. Sci. 55:1264–1272. [A2]

Bernatchez, L., H. Glémet, C. C. Wilson, and R. G. Danzmann. 1995. Introgression and fixation of Arctic char (*Salvelinus alpinus*) mitochondrial genome in an allopatric population of brook trout (*Salvelinus fontinalis*). Can. J. Fish. Aquat. Sci. 52:179–185. [I, 8]

Bernatchez, L., J. A. Vuorinen, R. A. Bodaly, and J. J. Dodson. 1996. Genetic evidence for reproductive isolation and multiple origins of sympatric trophic ecotypes of whitefish (*Coregonus*). Evolution 50:624–635. [6]

Bernoulli, D. 1738. Specimen theoriae novae de mensura sortis. Papers Imp. Acad. Sci. St. Petersburg 5:175–192. (*English translation* in Bernoulli, D. 1954. Exposition of a new theory on the measurement of risk. Econometrica 22:23–36.) [I]

Bert, T. M., and W. S. Arnold. 1995. An empirical test of predictions of two competing models for the maintenance and fate of hybrid zones: both models are supported in a hard-clam hybrid zone. Evolution 49:276–289. [8]

Beverton, R. J. H., and S. J. Holt. 1957. On the dynamics of exploited fish populations, Vol. Fisheries Investigations, Series 2, Volume 19. United Kingdom Ministry of Agriculture and Fisheries, London, U.K. [13]

Bielak, A. T., and G. Power. 1986. Changes in mean weight, sea-age composition, and catch-per-unit-effort of Atlantic salmon (*Salmo salar*) angled in the Godbout River, Québec, 1859–1983. Can. J. Fish. Aquat. Sci. 43:281–287. [5, 7, 11, 13]

Bigler, B. S., D. W. Welch, and J. H. Helle. 1996. A review of size trends among North Pacific salmon (*Oncorhynchus* spp.). Can. J. Fish. Aquat. Sci. 53:455–465. [I, 7, 11, 13]

Bilby, R. E., B. R. Fransen, and P. A. Bisson. 1996. Incorporation of nitrogen and carbon from spawning coho salmon into the trophic system of small streams: evidence from stable isotopes. Can. J. Fish. Aquat. Sci. 53:164–173. [2, 13]

Bilby, R. E., B. R. Fransen, P. A. Bisson, and J. K. Walter. 1998. Response of juvenile coho salmon (*Oncorhynchus kisutch*) and steelhead (*Oncorhynchus mykiss*) to the addition of salmon carcasses to two streams in southwestern Washington, U.S.A. Can. J. Fish. Aquat. Sci. 55:1909–1918. [2, 13]

Bilby, R. E., B. R. Fransen, J. K. Walter, C. J. Cederholm, and W. J. Scarlett. 2001. Preliminary evaluation of the use of nitrogen stable isotope ratios to establish escapement levels for Pacific salmon. Fisheries (Bethesda) 26(1):6–14. [13]

Billerbeck, J. M., T. E. Lankford, Jr., and D. O. Conover. 2001. Evolution of intrinsic growth and energy acquisition rates. I. Trade-offs with swimming performance in *Menidia menidia*. Evolution 55:1863–1872. [3]

Bilton, H. T., D. F. Alderdice, and J. T. Schnute. 1982. Influence of time and size at release of juvenile coho salmon (*Oncorhynchus kisutch*) on returns at maturity. Can. J. Fish. Aquat. Sci. 39:426–447. [13]

Birkhead, T. R., and A. P. Møller. 1998. Sperm competition and sexual selection. Academic Press, New York. [9]

Birt, T. P., J. M. Green, and W. S. Davidson. 1991a. Mitochondrial DNA variation reveals genetically distinct sympatric populations of anadromous and nonanadromous Atlantic salmon, *Salmo salar*. Can. J. Fish. Aquat. Sci. 48:577–582. [3, A3]

Birt, T. P., J. M. Green, and W. S. Davidson. 1991b. Contrasts in development and smolting of genetically distinct sympatric anadromous and nonanadromous Atlantic salmon, *Salmo salar*. Can. J. Zool. 69:2075–2084. [3, A3]

Bisson, P. A., R. E. Bilby, M. D. Bryant, C. A. Dolloff, G. B. Grette, R. A. House, M. L. Murphy, K. V. Koski, and J. R. Sedell. 1987. Large woody debris in forested streams in the Pacific Northwest: past, present, and future. Pp. 141–190 *in* E. O. Salo and T. W. Cundy, eds. Steamside management: forestry and fishery interactions. Univ. of Washington, Seattle, WA. [13]

Bjorkstedt, E. P. 2000. Stock-recruitment relationships for life cycles that exhibit concurrent density dependence. Can. J. Fish. Aquat. Sci. 57:459–467. [13]

Blackett, R. F. 1973. Fecundity of resident and anadromous Dolly Varden (*Salvelinus malma*) in southeastern Alaska. J. Fish. Res. Board Can. 30:543–548. [A3]

Blair, G. R., D. E. Rogers, and T. P. Quinn. 1993. Variation in life history characteristics and morphology of sockeye salmon in the Kvichak River system, Bristol Bay, Alaska. Trans. Am. Fish. Soc. 122:550–559. [3, 7, 9]

Blanchfield, P. J., and M. S. Ridgway. 1997. Reproductive timing and use of redd sites by lake-spawning brook trout (*Salvelinus fontinalis*). Can. J. Fish. Aquat. Sci. 54:747–756. [9]

Blanchfield, P. J., and M. S. Ridgway. 1999. The cost of peripheral males in a brook trout mating system. Anim. Behav. 57:537–544. [I, 9]

Blanckenhorn, W. U. 2000. The evolution of body size: what keeps organisms small? Quart. Rev. Biol. 75:385–407. [3]

Blaxter, J. H. S. 1969. Development: eggs and larvae. Pp. 177–252 *in* W. S. Hoar, and K. J. Randall, eds. Fish physiology. Vol. 3. Academic Press, New York. [4]

Blumer, L. S. 1982. A bibliography and categorization of bony fishes exhibiting parental care. Zool. J. Linn. Soc. Lond. 75:1–22. [9]

Bodaly, R. A. 1979. Morphological and ecological divergence within the lake whitefish (*Coregonus clupeaformis*) species complex in Yukon territory. J. Fish. Res. Board Can. 36:1214–1222. [6]

Bodaly, R. A., J. W. Clayton, and C. C. Lindsey. 1988. Status of the Squanga Whitefish, *Coregonus* sp., in the Yukon Territory, Canada. Can. Field-Nat. 102:114–125. [6]

Bodaly, R. A., J. Vuorinen, R. D. Ward, M. Luczynski, and J. D. Reist. 1991. Genetic comparisons of New and Old World coregonid fishes. J. Fish Biol. 38:37–51. [6]

Bohlin, T., C. Dellefors, and U. Faremo. 1986. Early sexual maturation of male sea trout and salmon—an evolutionary model and some practical implications. Rep. Inst. Freshwater Res., Drottningholm 63:17–25. [9]

Bohlin, T., C. Dellefors, and U. Faremo. 1990. Large or small at maturity—theories on the choice of alternative male strategies in anadromous salmonids. Ann. Zool. Fennici 27:139–147. [5, 9, 13]

Bohlin, T., C. Dellefors, and U. Faremo. 1993. Optimal time and size for smolt migration in wild sea trout (*Salmo trutta*). Can. J. Fish. Aquat. Sci. 50:224–232. [3]

Bohlin, T., C. Dellefors, and U. Faremo. 1994. Probability of first sexual maturation of male parr in wild sea-run brown trout (*Salmo trutta*) depends on condition factor one year in advance. Can. J. Fish. Aquat. Sci. 51:1920–1926. [3, 5]

Bohlin, T., J. Pettersson, and E. Degerman. 2001. Population density of migratory and resident brown trout (*Salmo trutta*) in relation to altitude: evidence for a migration cost. J. Anim. Ecol. 70:112–121. [I, 3]

Bohlin, T., L. F. Sundström, J. I. Johnsson, J. Höjesjö, and J. Pettersson. 2002. Density-dependent growth in brown trout: effects of introducing wild and hatchery fish. J. Anim. Ecol. 71:683–692. [2]

Bone, E., and A. Farres. 2001. Trends and rates of microevolution in plants. Genetica 112/113:165–182. [I, 7]

Bonner, J. T. 1965. Size and cycle: an essay on the structure of biology. Princeton Univ. Press. Princeton, NJ. [1]

Boula, D., V. Castric, L. Bernatchez, and C. Audet. 2002. Physiological, endocrine, and genetic basis of anadromy in the brook charr, *Salvelinus fontinalis*, of the Laval River (Québec, Canada). Environ. Biol. Fishes 64:229–242. [3, A3]

Bourke, E. A., J. Coughlan, H. Jansson, P. Galvin, and T. F. Cross. 1997. Allozyme variation in populations of Atlantic salmon located throughout Europe: diversity that could be compromised by introductions of reared fish. ICES J. Mar. Sci. 54:974–985. [A2]

Bouza, C., J. Castro, L. Sánchez, and P. Martínez. 2001. Allozymic evidence of parapatric differentiation of brown trout (*Salmo trutta* L.) within an Atlantic river basin of the Iberian Peninsula. Mol. Ecol. 10:1455–1469. [A2]

Bowen, B. W. 1998. What is wrong with ESUs?: the gap between evolutionary theory and conservation principles. J. Shellfish Res. 17:1355–1358. [12]

Bradford, M. J. 1995. Comparative review of Pacific salmon survival rates. Can. J. Fish. Aquat. Sci. 52:1327–1338. [I, 4, 11]

Bradford, M. J., R. A. Myers, and J. R. Irvine. 2000. Reference points for coho salmon (*Oncorhynchus kisutch*) harvest rates and escapement goals based on freshwater production. Can. J. Fish. Aquat. Sci. 57:677–686. [13]

Bradshaw, A. D. 1965. Evolutionary significance of phenotypic plasticity in plants. Adv. Genet. 13:115–155. [5]

Bradshaw, H. D., K. G. Otto, B. E. Frewen, J. K. McKay, and D. W. Schemske. 1998. Quantitative trait loci affecting differences in floral morphology between two species of monkeyflower (*Mimulus*), Genetics 149:367–382. [6]

Bradshaw, H. D., S. M. Wilbert, K. G. Otto, and D. W. Schemske. 1995. Genetic mapping of floral traits associated with reproductive isolation in monkeyflowers (*Mimulus*). Nature 376:762–765. [6]

Brännäs, E. 1995. First access to territorial space and exposure to strong predation pressure: a conflict in early emerging Atlantic salmon (*Salmo salar* L.) fry. Evol. Ecol. 9:411–420. [9]

Brannon, E. L. 1987. Mechanisms stabilizing salmonid fry emergence timing. Can. Spec. Publ. Fish. Aquat. Sci. 96:120–124. [7, 9]

Brannon, E. L., and T. P. Quinn. 1990. Field test of the pheromone hypothesis for homing by Pacific salmon. J. Chem. Ecol. 16:603–609. [2]

Brannon, E., C. Feldman, and L. Donaldson. 1982. University of Washington zero-age coho salmon smolt production. Aquaculture 28:195–200. [13]

Brett, J. R. 1995. Energetics. Pp. 3–68 *in* C. Groot, L. Margolis, and W. C. Clarke. Physiological ecology of Pacific salmon. UBC Press, Vancouver, BC, Canada. [3, 9]

Brockelman, W. Y. 1975. Competition, the fitness of offspring, and optimal clutch size. Am. Nat. 109:677–699. **[4]**

Brockmann, H. J., C. Nguyen, and W. Potts. 2000. Paternity in horseshoe crabs when spawning in multiple-male groups. Anim. Behav. 60:837–849. **[9]**

Brosseau, D. J., and J. A. Baglivo. 1988. Life tables for two field populations of softshell clam, *Mya arenaria*, (Mollusca: Pelecypoda) from Long Island Sound. Fishery Bull. 86:567–579. **[4]**

Brown, C. J. D. 1966. Natural hybrids between *Salmo trutta* X *Salvelinus fontinalis*. Copeia 1966:600–601. **[8]**

Brown, G. E., and J. A. Brown. 1993. Social dynamics in salmonid fishes: do kin make better neighbours? Anim. Behav. 45:863–871. **[2]**

Brown, G. E., J. A. Brown, and A. M. Crosbie. 1993. Phenotype matching in juvenile rainbow trout. Anim. Behav. 46:1223–1225. **[2]**

Brown, G. E., J. A. Brown, and W. R. Wilson. 1996. The effects of kinship on the growth of juvenile Arctic charr. J. Fish Biol. 48:313–320. **[2]**

Brown, J. L., and E. R. Brown. 1984. Parental facilitation: parent-offspring relations in communally breeding birds. Behav. Ecol. Sociobiol. 14:203–209. **[2]**

Brown, J. S., and A. O. Parman. 1993. Consequences of size-selective harvesting as an evolutionary game. Pp. 248–261 *in* T. M. Stokes, J. M. McGlade, and R. Law, eds. The exploitation of evolving resources, Springer-Verlag, Berlin, Germany. **[13]**

Brunner, P. C., M. R. Douglas, and L. Bernatchez. 1998. Microsatellite and mitochondrial DNA assessment of population structure and stocking effects in Arctic charr *Salvelinus alpinus* (Teleostei: Salmonidae) from central Alpine lakes. Mol. Ecol. 7:209–223. **[A2]**

Brunner, P. C., M. R. Douglas, A. Osinov, C. C. Wilson, and L. Bernatchez. 2001. Holarctic phylogeography of Arctic charr (*Salvelinus alpinus* L.) inferred from mitochondrial DNA sequences. Evolution 55:573–586. **[8]**

Bryant, M. D., B. J. Frenette, and S. J. McCurdy. 1999. Colonization of a watershed by anadromous salmonids following the installation of a fish ladder in Margaret Creek, southeast Alaska. N.A. J. Fish. Manage. 19:1129–1136. **[2]**

Bull, J. J., C. Thompson, D. Ng, and R. Moore. 1987. A model for natural selection of genetic migration. Am. Nat. 129:143–157. **[2]**

Bulmer, M. G. 1983. Models for the evolution of protandry in insects. Theor. Pop. Biol. 23:314–322. **[9]**

Bult, T. P., S. C. Riley, R. L. Haedrich, R. J. Gibson, and J. Heggenes. 1999. Density-dependent habitat selection by juvenile Atlantic salmon (*Salmo salar*) in experimental riverine habitats. Can. J. Fish. Aquat. Sci. 56:1298–1306. **[13]**

Burger, C. V., K. T. Scribner, W. J. Spearman, C. O. Swanton, and D. E. Campton. 2000. Genetic contribution of three introduced life history forms of sockeye salmon to colonization of Frazer Lake, Alaska. Can. J. Fish. Aquat. Sci. 57:2096–2111. **[A2]**

Burger, C. V., W. J. Spearman, and M. A. Cronin. 1997. Genetic differentiation of sockeye salmon subpopulations from a geologically young Alaskan lake system. Trans. Am. Fish. Soc. 126:926–938. **[2]**

Bürger, R., and M. Lynch. 1995. Evolution and extinction in a changing environment: a quantitative-genetic analysis. Evolution 49:151–163. **[I]**

Burgner, R. L. 1987. Factors influencing age and growth of juvenile sockeye salmon (*Oncorhynchus nerka*) in lakes. Can. Spec. Publ. Fish. Aquat. Sci. 96:129–142. **[3]**

Burgner, R. L. 1991. Life history of sockeye salmon (*Oncorhynchus nerka*). Pp. 3–117 *in* C. Groot and L. Margolis, eds. Pacific salmon life histories. UBC Press, Vancouver, BC, Canada. **[2, 3, 8]**

Burley, N. 1983. The meaning of assortative mating. Ethol. Sociobiol. 4:191–203. **[9]**

Busby, P. J., T. C. Wainwright, G. J. Bryant, L. J. Lierheimer, R. S. Waples, F. W. Waknitz, and I. V. Lagomarino. 1996. Status review of West Coast steelhead from Washington, Idaho, Oregon, and California. NOAA Tech. Memo. NMFS-NWFSC-27. **[9, 12]**

Bush, G. L. 1994. Sympatric speciation in animals: new wine in old bottles. Trends Ecol. Evol. 9:285–288. **[I]**

Butlin, R. 1989. Reinforcement of premating isolation. Pp. 158–179 *in* D. Otte and J. Endler, eds. Speciation and its consequences. Sinauer Assoc., Sunderland, MA. **[8]**

Calder, W. A., III. 1984. Size, function, and life history. Harvard Univ. Press. Cambridge, MA. **[1]**

Caldwell, J. P., J. H. Thorp, and T. O. Jervey. 1980. Predator-prey relationships among larval dragonflies, salamanders, and frogs. Oecologia 46:285–289. **[4]**

Campton, D. E. 1987. Natural hybridization and introgression in fishes: methods of detection and genetic interpretations. Pp. 161–192 *in* N. Ryman and F. Utter, eds. Population genetics and fishery management. Univ. of Washington Press, Seattle, WA. **[8]**

Campton, D. E., and F. M. Utter. 1985. Natural hybridization between steelhead trout (*Salmo gairdneri*) and coastal cutthroat trout (*S. clarki*) in two Puget Sound streams. Can. J. Fish. Aquat. Sci. 42:110–119. **[I, 8]**

Campton, D. E., and F. M. Utter. 1987. Genetic structure of anadromous cutthroat trout (*Salmo clarki clarki*) populations in the Puget Sound area: evidence for restricted gene flow. Can. J. Fish. Aquat. Sci. 44:573–582. **[13, A2]**

Candy, J. R., and T. D. Beacham. 2000. Patterns of homing and straying in southern British Columbia coded-wire tagged chinook salmon (*Oncorhynchus tshawytscha*) populations. Fish. Res. 47:41–56. **[2, A1]**

Carl, L. M., and M. C. Healey. 1984. Differences in enzyme frequency and body morphology among three juvenile life history types of chinook salmon (*Oncorhynchus tshawytscha*) in the Nanaimo River, British Columbia. Can. J. Fish. Aquat. Sci. 41:1070–1077. **[A2]**

Carlsson, J., and J. Nilsson. 2000. Population genetic structure of brown trout (*Salmo trutta* L.) within a northern boreal forest stream. Hereditas 132:173–181. **[A2]**

Carlsson, J., and J. Nilsson. 2001. Effects of geomorphological structures on genetic differentiation among brown trout populations in a northern boreal river drainage. Trans. Am. Fish. Soc. 130:36–45. **[A2]**

Carlsson, J., K. H. Olsén, J. Nilsson, Ø. Øverli, and O. B. Stabell. 1999. Microsatellites reveal fine-scale genetic structure in stream-living brown trout. J. Fish Biol. 55:1290–1303. **[2, A2]**

Cass, A., and B. Riddell. 1999. A life history model for assessing alternative management policies for depressed chinook salmon. ICES J. Mar. Sci. 56:414–421. **[10]**

Castonguay, M., G. J. FitzGerald, and Y. Côté. 1982. Life history and movements of anadromous brook charr, *Salvelinus fontinalis*, in the St-Jean River, Gaspé, Québec. Can. J. Zool. 60:3084–3091. **[A3]**

Castric, V., F. B. Bonney, and L. Bernatchez. 2001. Landscape structure and hierarchical genetic diversity in the brook charr, *Salvelinus fontinalis*. Evolution 55:1016–1028. **[A2]**

Caswell, H. 1979. A general formula for the sensitivity of population growth rate to changes in life history parameters. Theor. Pop. Biol. 14:215–230. [1]

Caswell, H., R. J. Naiman, and R. Morin. 1984. Evaluating the consequences of reproduction in complex salmonid life cycles. Aquaculture 43:123–134. [3]

Caswell, J. 1980. On the equivalence of maximizing reproductive value and maximizing fitness. Ecology 61:19–24. [1]

Chakraborty, R., M. Haag, N. Ryman, and G. Ståhl. 1982. Heirarchical gene diversity analysis and its application to brown trout data. Hereditas 97:17–21. [A2]

Chambers, R. C. 1997. Environmental influences on egg and propagule sizes in marine fishes. Pp. 62–102 in R. C. Chambers, and E. A. Trippel, eds. Early life history and recruitment in fish populations. Chapman and Hall, London, U.K. [4]

Chandler, G. L., and T. C. Bjornn. 1988. Abundance, growth, and interactions of juvenile steelhead relative to time of emergence. Trans. Am. Fish. Soc. 117:432–443. [9, 13]

Chapman, D. W. 1962. Aggressive behavior in juvenile coho salmon as a cause of emigration. J. Fish. Res. Board Can. 19:1047–1080. [13]

Chapman, D. W. 1966. Food and space as regulators of salmonid populations in streams. Am. Nat. 100:345–357. [2, 13]

Chapman, D. W. 1988. Critical review of variables used to define effects of fines in redds of large salmonids. Trans. Am. Fish. Soc. 117:1–21. [I, 2, 4, 9]

Chapman, T., G. Arnqvist, J. Bangham, and L. Rowe. 2003. Sexual conflict. Trends Ecol. Evol. 18:41–47. [1]

Chaput, G., J. Allard, F. Caron, J. B. Dempson, C. C. Mullins, and M. F. O'Connell. 1998. River-specific target spawning requirements for Atlantic salmon (*Salmo salar*) based on a generalized smolt production model. Can. J. Fish. Aquat. Sci. 55:246–261. [13]

Charlesworth, B. 1990. Optimization models, quantitative genetics, and mutation. Evolution 44:520–538. [5]

Charlesworth, B. 1994. Evolution in age-structured populations, 2nd ed. Cambridge Univ. Press, Cambridge, U.K. [11]

Charlesworth, D., and B. Charlesworth. 1987. Inbreeding depression and its evolutionary consequences. Annu. Rev. Ecol. Syst. 18:237–268. [2]

Charlesworth, B., and J. A. Williamson. 1975. The probability of survival of a mutant gene in an age-structured population and implications for the evolution of life histories. Genetical Res. (Cambridge) 26:1–10. [1]

Charlesworth, B., R. Lande, and M. Slatkin. 1982. A neo-Darwinian commentary on macroevolution. Evolution 36:474–498. [7]

Charnov, E. L. 1976. Optimal foraging, the marginal value theorem. Theor. Pop. Biol. 9:129–136. [1, 4]

Charnov, E. L. 1982. The theory of sex allocation. Princeton Univ. Press, Princeton, NJ. [I]

Charnov, E. L. 1993. Life history invariants: some explorations of symmetry in evolutionary ecology. Oxford Univ. Press, Oxford, U.K. [I, 13]

Charnov, E. L., and W. M. Schaffer. 1973. Life history consequences of natural selection: Cole's result revisited. Am. Nat. 106:791–793. [I, 1]

Chebanov, N. A. 1984. Effect of spawner length and age on the viability of progeny during early ontogeny in some species of the genus *Oncorhynchus* (Salmonidae). J. Ichthyol. 24(3):82–93. [13]

Chebanov, N. A. 1986. Factors controlling spawning success in pink salmon, *Oncorhynchus gorbuscha*. J. Ichthyol. 26(3):69–78. [13]

Chebanov, N. A. 1990. Spawning behavior, assortative mating, and spawning success of coho salmon, *Oncorhynchus kisutch*, under natural and experimental conditions. J. Ichthyol. 30(6):1–12. [13]

Chebanov, N. A. 1991. The effect of spawner density on spawning success, egg survival, and size structure of the progeny of the sockeye salmon, *Oncorhynchus nerka*. J. Ichthyol. 31(2):103–109. [13]

Chebanov, N. A., N. V. Varnavskaya, and V. S. Varnavskiy, 1984. Effectiveness of spawning of male sockeye salmon, *Oncorhynchus nerka* (Salmonidae), of differing hierarchical rank by means of genetic-biochemical markers. J. Ichthyol. 23(5):51–55. [3, 9]

Chernavin, V. V. 1918. [Nuptial changes in the skeleton of salmon.] Izv. otd. ryboved. uchencogo s-kh. komiteta 1. [9]

Chernavin, V. V. 1921. [Origin of the nuptial color in salmon.] Zh. Petrogradsk. Agronom. In-ta. No. 3–4. [9]

Chevassus, B. 1979. Hybridization in salmonids: results and perspectives. Aquaculture 17:113–128. [8]

Cheverud, J. M. 1984. Quantitative genetics and developmental constraints on evolution by selection. J. theor. Biol. 110:155–172. [6]

Cheverud, J. M. 1988. A comparison of genetic and phenotypic correlations. Evolution 42:958–968. [6]

Chiariello, N., and J. Roughgarden. 1983. Storage allocation in seasonal races of an annual plant: optimal versus actual allocation. Ecology 65:1290–1301. [1]

Chouinard, A., and L. Bernatchez. 1998. A study of trophic niche partitioning between larval populations of reproductively isolated whitefish (*Coregonus* sp.) ecotypes. J. Fish Biol. 53:1231–1242. [6]

Chouinard, A., D. Pigeon, and L. Bernatchez. 1996. Lack of specialization in trophic morphology between genetically differentiated dwarf and normal forms of lake whitefish (*Coregonus clupeaformis* Mitchill) in Lac de l'Est, Québec. Can. J. Zool. 74:1989–1998. [6]

Churikov, D., and A. J. Gharrett. 2002. Comparative phylogeography of the two pink salmon broodlines: an analysis based on a mitochondrial DNA genealogy. Mol. Ecol. 11:1077–1101. [2]

Clarke, W. C., R. E. Withler, and J. E. Shelbourn. 1992. Genetic control of juvenile life-history pattern in chinook salmon (*Oncorhynchus tshawytscha*). Can. J. Fish. Aquat. Sci. 49:2300–2306. [7]

Clutton-Brock, T. H. 1988a. Reproductive success. Pp. 472–485 *in* T. H. Clutton-Brock, ed. Reproductive success: Studies of individual variation in contrasting breeding systems. Univ. of Chicago Press, Chicago, IL. [9, 10]

Clutton-Brock, T. H., ed. 1988b. Reproductive success: Studies of individual variation in contrasting breeding systems. Univ. of Chicago Press, Chicago, IL. [10]

Clutton-Brock, T. H. 1991. The evolution of parental care. Princeton Univ. Press, Princeton, NJ. [9]

Clutton-Brock, T. H., and G. A. Parker. 1992. Potential reproductive rates and the operation of sexual selection. Quart. Rev. Biol. 67:437–456. [9]

Cohen, D., and S. A. Levin. 1991. Dispersal in patchy environments—the effects of temporal and spatial structure. Theor. Pop. Biol. 26:165–191. [2]

Cole, L. C. 1954. The population consequences of life history phenomena. Quart. Rev. Biol. 29:103–137. [I, 1]

Collis, K., D. D. Roby, D. P. Craig, B. A. Ryan, and R. D. Ledgerwood. 2001. Colonial waterbird predation on juvenile salmonids tagged with passive integrated transponders in the Columbia River estuary: vulnerability of different salmonid species, stocks, and rearing types. Trans. Am. Fish. Soc. 130:385–396. [3]

Comins, H. N. 1982. Evolutionarily stable strategies for localized dispersal in two dimensions. J. theor. Biol. 94:579–606. [2]

Comins, H. N., W. D. Hamilton, and R. M. May. 1980. Evolutionarily stable dispersal strategies. J. theor. Biol. 82:205–230. [2]

Congdon, J. D., J. W. Gibbons, and J. L. Greene. 1983. Parental investment in the chicken turtle (*Deirochelys reticularia*). Ecology 64:419–425. [4]

Connell, J. H. 1970. A predator-prey system in the marine intertidal region. I. *Balanus glandula* and several predatory species of *Thais*. Ecol. Monogr. 40:49–78. [4]

Conover, D. O. 2000. Darwinian fishery science. Mar. Ecol. Prog. Ser. 208:303–307. [5, 7]

Conover, D. O., and S. B. Munch. 2002. Sustaining fisheries yields over evolutionary time scales. Science 297:94–96. [5, 7, 11, 13]

Conover, D. O., and E. T. Schultz. 1995. Phenotypic similarity and the evolutionary significance of countergradient variation. Trends. Ecol. Evol. 10:248–252. [3, 5]

Cooney, R. T., and R. D. Brodeur. 1998. Carrying capacity and North Pacific salmon production: stock enhancement implications. Bull. Mar. Sci. 62:443–464. [13]

Cooper, A. B., and M. Mangel. 1999. The dangers of ignoring metapopulation structure for the conservation of salmonids. Fishery Bull. 97:213–226. [1]

Cooper, J. C., A. T. Scholz, R. M. Horrall, A. D. Hasler, and D. M. Madison. 1976. Experimental confirmation of the olfactory hypothesis with homing, artificially imprinted coho salmon (*Oncorhynchus kisutch*). J. Fish. Res. Board Can. 33:703–710. [2]

Côté, I. M., and W. Hunte. 1989. Male and female choice in the redlip blenny: why bigger is better. Anim. Behav. 38:78–88. [9]

Courchamp, F., T. Clutton-Brock, and B. Grenfell. 1999. Inverse density dependence and the Allee effect. Trends. Ecol. Evol. 14:405–410. [2]

Courtenay, S. C., T. P. Quinn, H. M. C. Dupuis, C. Groot, and P. A. Larkin. 1997. Factors affecting the recognition of population-specific odours by juvenile coho salmon. J. Fish Biol. 50:1042–1060. [2]

Courtenay, S. C., T. P. Quinn, H. M. C. Dupuis, C. Groot, and P. A. Larkin. 2001. Discrimination of family-specific odours by juvenile coho salmon: roles of learning and odour concentration. J. Fish Biol. 58:107–125. [2]

Couturier, C. Y., L. Clarke, and A. M. Sutterlin. 1986. Identification of spawning areas of two forms of Atlantic salmon (*Salmo salar* L.) inhabiting the same watershed. Fish. Res. 4:131–144. [3]

Cox, C. R., and B. J. Le Beouf. 1977. Female incitation of male competition: a mechanism in sexual selection. Am. Nat. 111:317–335. [9]

Cox, S. P., and S. G. Hinch. 1997. Changes in size at maturity of Fraser River sockeye salmon (*Oncorhynchus nerka*) (1952–1993) and associations with temperature. Can. J. Fish. Aquat. Sci. 54:1159–1165. [1, 7]

Coyne, J. A., and H. A. Orr. 1989. Two rules of speciation. Pp. 180–297 *in* D. Otte and J. A. Endler, eds. Speciation and its consequences. Sinauer Assoc., Sunderland, MA. [8]

Coyne, J. A., and H. A. Orr. 1997. "Patterns of speciation in *Drosophila*" revisited. Evolution 51:295–303. [6, 8]

CPMPNAS (Committee on Protection and Management of Pacific Northwest Anadromous Salmonids). 1996. Upstream. Natl. Acad. Press, Washington, D.C. [1, 2, 13]

Craig, J. K., and C. J. Foote. 2001. Countergradient variation and secondary sexual color: phenotypic convergence promotes genetic divergence in carotenoid use between sympatric anadromous and nonanadromous morphs of sockeye salmon (*Oncorhynchus nerka*). Evolution 55:380–391. [I, 3, 7, 8, 9, A3]

Crandall, K. A., O. R. P. Bininda-Emonds, G. M. Mace, and R. K. Wayne. 2000. Considering evolutionary processes in conservation biology. Trends Ecol. Evol. 15:290–295. [11, 12]

Crandell, P. A., and G. A. E. Gall. 1993. The genetics of body weight and its effect on early maturity based on individually tagged rainbow trout (*Oncorhynchus mykiss*). Aquaculture 117:77–93. [I]

Crespi, B. J., and P. D. Taylor. 1990. Dispersal rates under variable patch density. Am. Nat. 135:48–62. [2]

Crespi, B. J., and R. Teo. 2002. Comparative phylogenetic analysis of the evolution of semelparity and life-history in salmonid fishes. Evolution 56:1008–1020. [I, 1, 4, 7, 9]

Crisp, D. T., and P. A. Carling. 1989. Observations on siting, dimensions and structure of salmonid redds. J. Fish Biol. 34:119–134. [9]

Crittenden, L. B. 1961. An interpretation of familial aggregation based on multiple genetic and environmental factors. Ann. NY Acad. Sci. 91:769–780. [11]

Cronin, M. A., W. J. Spearman, R. L. Wilmot, J. C. Patton, and J. W. Bickham. 1993. Mitochondrial DNA variation in chinook (*Oncorhynchus tshawytscha*) and chum salmon (*O. keta*) detected by restriction enzyme analysis of polymerase chain reaction (PCR) products. Can. J. Fish. Aquat. Sci. 50:708–715. [12]

Cross, T. F., C. P. R. Mills, and M. de Courcy Williams. 1992. An intensive study of allozyme variation in freshwater resident and anadromous trout, *Salmo trutta* L., in western Ireland. J. Fish Biol. 40:25–32. [3, A3]

Crossin, G. T., S. G. Hinch, A. P. Farrell, D. Higgs, A. Lotto, J. Oakes, and M. C. Healey. In review. Energetics and morphology of sockeye salmon (*Oncorhynchus nerka*): effects of upriver migratory distance and elevation. J. Anim. Ecol. [3, 7]

Crow, J. F. 1954. Breeding structure of populations. II. Effective population number. Pp. 543–556 *in* O. Kempthorne, T. Bancroft, J. Gowen, and J. Lush, eds. Statistics and mathematics in biology. Iowa State Univ. Press, Ames, IA. [10]

Crow, J. F. 1958. Some possibilities for measuring selection intensities in man. Hum. Biol. 30:1–13. [10, 11]

Crow, J. F. 1986. Basic concepts in population, quantitative, and evolutionary genetics. Freeman, New York. [11]

Crow, J. F., and C. Denniston. 1988. Inbreeding and variance effective population numbers. Evolution 42:482–495. [10]

Crow, J. F., and M. Kimura. 1970. An introduction to population genetics theory. Harper and Row, New York. [1, 10]

Crow, J. F., and N. E. Morton. 1955. Measurement of gene frequency drift in small populations. Evolution 9:202–214. [10]

Crozier, W. W., and A. Ferguson. 1986. Electrophoretic examination of the population structure of brown trout, *Salmo trutta* L., from the Lough Neagh catchment, Northern Ireland. J. Fish Biol. 28:459–477. [A2]

Cunningham, J. T. 1900. Sexual dimorphism in the animal kingdom. Adam and Charles Black, London, U.K. [9]

Currens, K. P., C. B. Schreck, and H. W. Li. 1990. Allozyme and morphological divergence of rainbow trout (*Oncorhynchus mykiss*) above and below waterfalls in the Deschutes River, Oregon. Copeia 1990:730–746. [A3]

Cutts, C. J., B. Brembs, N. B. Metcalfe, and A. C. Taylor. 1999a. Prior residence, territory quality and life-history strategies in juvenile Atlantic salmon (*Salmo salar* L.). J. Fish Biol. 55:784–794. [2]

Cutts, C. J., N. B. Metcalfe, and A. C. Taylor. 1999b. Competitive asymmetries in territorial juvenile Atlantic salmon, *Salmo salar*. Oikos 86:479–486. [4]

Danielsdottir, A. K., G. Marteinsdottir, F. Arnason, and S. Gudjonsson. 1997. Genetic structure of wild and reared Atlantic salmon (*Salmo salar* L.) populations in Iceland. ICES J. Mar. Sci. 54:986–997. [A2]

Danzmann, R. G., and K. Gharbi. 2001. Gene mapping in fishes: a means to an end. Genetica 111:3–23. [6, 8]

Darwin, C. 1871. The descent of man, and selection in relation to sex. J. Murray, London, U.K. [9]

Davidson, F. A. 1935. The development of the secondary sexual characters in the pink salmon (*Oncorhynchus gorbuscha*). J. Morphol. 57:169–183. [9]

Daye, P. G., and B. D. Glebe. 1984. Fertilization success and sperm motility of Atlantic salmon (*Salmo salar* L.) in acidified water. Aquaculture 43:307–312. [9]

de Gaudemar, B., and E. P. Beall. 1998. Effects of overripening on spawning behaviour and reproductive success of Atlantic salmon females spawning in a controlled flow channel. J. Fish Biol. 53:434–446. [9]

de Gaudemar, B., and E. P. Beall. 1999. Reproductive behavioural sequences of single pairs of Atlantic salmon in an experimental stream. Anim. Behav. 57:1207–1217. [9]

de Gaudemar, B., J. M. Bonzom, and E. P. Beall. 2000b. Effects of courtship and relative mate size on sexual motivation in Atlantic salmon. J Fish Biol. 57:502–515. [I, 3, 9]

de Gaudemar, B., S. L. Schroder, and E. P. Beall. 2000a. Nest placement and egg distribution in Atlantic salmon redds. Environ. Biol. Fishes 57:37–47. [9]

Debat, V., and P. David. 2001. Mapping phenotypes: canalization, plasticity and developmental stability. Trends Ecol. Evol. 16:555–561. [5]

Deevey, E. S., Jr. 1947. Life tables for natural populations of animals. Quart. Rev. Biol. 22:283–314. [1]

Dellefors, C. and U. Faremo. 1988. Early sexual maturation in males of wild sea trout, *Salmo trutta* L., inhibits smoltification. J. Fish Biol. 33:741–749. [3]

DeMarais, B. D., T. E. Dowling, M. E. Douglas, W. L. Minckley, and P. C. Marsh. 1992. Origin of *Gila seminuda* (Teleostei: Cyprinidae) through introgressive hybridization: implications for evolution and conservation. Proc. Natl. Acad. Sci. USA 89:2747–2751. [8]

Devlin, R. H., B. K. McNeil, T. D. D. Groves, and E. M. Donaldson. 1991. Isolation of a Y-chromosomal DNA probe capable of determining genetic sex in chinook salmon (*Oncorhynchus tshawytscha*). Can. J. Fish. Aquat. Sci. 48:1606–1612. [8]

Devlin, R. H., T. Y. Yesaki, C. A. Biagi, E. M. Donaldson, P. Swanson, and W.-H. Chan. 1994. Extraordinary salmon growth. Nature 371:209–210. [8]

Dickerson, B. R., T. P. Quinn, and M. F. Willson. 2002. Body size, arrival date, and reproductive success of pink salmon, *Oncorhynchus gorbuscha*. Ethol. Ecol. Evol. 14:29–44. [9]

Dickerson, B. R., M. F. Willson, P. Bentzen, and T. P. Quinn. 2003. Size-assortative mating in salmonids: negative evidence for pink salmon in natural conditions. Anim. Behav.: in press [9]

Dickerson, G. E. 1955. Genetic slippage in response to selection for multiple objectives. Cold Spring Harbor Symp. Quant. Biol. 20:213–224. [11]

Dieckmann, U., and M. Doebeli. 1999. On the origin of species by sympatric speciation. Nature 400:354–357. [I]

Dimmick, W. W., M. J. Ghedotti, M. J. Grose, A. M. Maglia, D. J. Meinhardt, and D. S. Pennock. 1999. The importance of systematic biology in defining units of conservation. Cons. Biol. 13:653–660. [12, 13]

Dingle, H. 1984. Behavior, genes, and life histories: complex adaptations in uncertain environments. Pp. 170–194 in P. W. Price, C. N. Slobodchikoff, and W. S. Gaud, eds. A new ecology: novel approaches to interactive systems. Wiley, New York. [11]

Dingle, H. 1996. Migration: the biology of life on the move. Oxford Univ. Press, Oxford, U.K. [3]

Dittman, A. H., and T. P. Quinn. 1996. Homing in Pacific salmon: mechanisms and ecological basis. J. Exp. Biol. 199:83–91. [2]

Dittman, A. H., T. P. Quinn, and E. C. Volk. 1998. Is the distribution, growth and survival of juvenile salmonids sex biased? Negative results for coho salmon in an experimental stream channel. J. Fish Biol. 53:1360–1364. [2]

Dizon, A. E., C. Lockyer, W. F. Perrin, D. P. Demaster, and J. Sisson. 1992. Rethinking the stock concept: a phylogeographic approach. Cons. Biol. 6:24–36. [12]

Dobzhansky, T. 1933. On the sterility of the interracial hybrids in *Drosophila pseudoobscura*. Proc. Natl. Acad. Sci. USA 19:397–403. [8]

Dobzhansky, T. 1937. Genetics and the origin of species. Columbia Univ. Press, New York. [2, 8]

Dobzhansky, T. 1940. Speciation as a stage in evolutionary divergence. Am. Nat. 74:312–321. [8]

Dobzhansky, T. 1951. Genetics and the origins of species. 3rd ed. Columbia Univ. Press, New York. [6]

Docker, M. F., and D. D. Heath. 2003. Genetic comparison between sympatric anadromous steelhead and freshwater resident rainbow trout in British Columbia, Canada. Cons. Genet. 4:227–231. [3]

Dominey, W. J. 1984. Alternative mating tactics and evolutionarily stable strategies. Am. Zool. 24:385–396. [13]

Donaldson, L. R., and G. H. Allen. 1957. Return of silver salmon, *Oncorhynchus kisutch* (Walbaum) to point of release. Trans. Am. Fish. Soc. 87:13–22. [A1]

Doucett, R. R., W. Hooper, and G. Power. 1999. Identification of anadromous and nonanadromous adult brook trout and their progeny in the Tabusintac River, New Brunswick, by means of multiple-stable-isotope analysis. Trans. Am. Fish. Soc. 128:278–288. [3]

Douglas, M. R., and P. C. Brunner. 2002. Biodiversity of central alpine *Coregonus* (Salmoniformes): impact of one-hundred years of management. Ecol. Appl. 12:154172. [6]

Douglas M. R., P. C. Brunner, and L. Bernatchez. 1999. Do assemblages of *Coregonus* (Teleostei: Salmoniformes) in the Central Alpine region of Europe represent species flocks? Mol. Ecol. 8:589–604. [6, A2]

Dowling, T. E., and M. R. Childs. 1992. Impact of hybridization on a threatened trout of the southeastern United States. Cons. Biol. 6:355–364. [8]

Doyon, J.-F., L. Bernatchez, M. Gendron, R. Verdon, and R. Fortin. 1998. Comparison of normal and dwarf populations of lake whitefish (*Coregonus clupeaformis*) with reference to hydroelectric reservoirs in Northern Québec. Arch. Hydro. Spec. Iss. Limn. 50:97–108. [6]

Dressler, R. L. 1990. The orchids: natural history and classification. Harvard Univ. Press, Cambridge, MA. [I]

Ducharme, L. J. A. 1969. Atlantic salmon returning for their fifth and sixth consecutive spawning trips. J. Fish. Res. Board Can. 26:1661–1664. [1, 9]

Duchesne, P., and L. Bernatchez. 2002. An analytical investigation of the dynamics of inbreeding in multi-generation supportive breeding. Cons. Genet. 3:47–60. [I, 10]

Duellman, W. E. and L. Trueb. 1986. Biology of amphibians. McGraw-Hill, New York. [I]

Dupont, P.-P. 2000. Étude comparative du partage de niches trophiques entre quatre écotypes de cisco (*Coregonus* spp) du lac Nipigon. Initiation à la recherche, Université Laval, Québec City, Québec, Canada. 22pp. [6]

Dutil, J.-D. 1986. Energetic constraints and spawning interval in the anadromous Arctic charr (Salvelinus alpinus). Copeia 1986:945–955. [9]

Duvernell, D. D., and N. Aspinwall. 1995. Introgression of *Luxilus cornutus* mtDNA into allopatric populations of *Luxilus chrysocephalus* (Teleostei: Cyprinidae) in Missouri and Arkansas. Mol. Ecol. 4:173–181. [8]

Eady, P. E. 2001. Postcopulatory, prezygotic reproductive isolation. J. Zool. 253:47–52. [8]

Ebert, D., C. Haag, M. Kirkpatrick, M. Riek, J. W. Hottinger, and V. I. Pajunen. 2002. A selective advantage to immigrant genes in a *Daphnia* metapopulation. Science 295:485–488. [2]

Edge, T. A., D. E. McAllister, and S. U. Qadri. 1991. Meristic and morphometric variation between the endangered Acadian whitefish, *Coregonus huntsmani*, and the lake whitefish, *Coregonus clupeaformis*, in the Canadian Maritime Provinces and the State of Maine, USA. Can. J. Fish. Aquat. Sci. 48:2140–2151. [6]

Edmundson, J. A., and A. Mazumder. 2001. Linking growth of juvenile sockeye salmon to habitat temperature in Alaskan lakes. Trans. Am. Fish. Soc. 130:644–662. [2]

Einum, S., and I. A. Fleming. 1999. Maternal effects of egg size in brown trout (*Salmo trutta*): norms of reaction to environmental quality. Proc. R. Soc. Lond. B 266:2095–2100. [I, 4, 5]

Einum, S., and I. A. Fleming. 2000a. Highly fecund mothers sacrifice offspring survival to maximize fitness. Nature 405:565–567. [I, 4]

Einum, S., and I. A. Fleming. 2000b. Selection against late emergence and small offspring in Atlantic salmon (*Salmo salar*). Evolution 54:628–639. [3, 4, 9, 13]

Einum, S., and I. A. Fleming. 2002. Does within-population variation in fish egg size reflect maternal influences on optimal values? Am. Nat. 160:756–765. [4]

Einum, S., I. A. Fleming, I. M. Côté, and J. D. Reynolds. In press. Population stability in salmon species: effects of population size and female reproductive allocation. J. Anim. Ecol. In press [9]

Einum, S., A. P. Hendry, and I. A. Fleming. 2002. Egg-size evolution in aquatic environments: does oxygen availability constrain size? Proc. R. Soc. Lond. B. 269:2325–2330. [3, 4]

Einum, S., M. T. Kinnison, and A. P. Hendry. 2003. Evolution of egg size and number. Pp. 127–153 *in* A. P. Hendry and S. C. Stearns, eds. Evolution illuminated: salmon and their relatives. Oxford Univ. Press, New York, NY—*this volume* [I, 1, 3, 9, 10, 12]

Eisenberg, J. F. 1981. The mammalian radiations. Univ. of Chicago Press, Chicago, IL. [I]

Elgar, M. A. 1990. Evolutionary compromise between a few large and many small eggs: comparative evidence in teleost fish. Oikos 59:283–287. **[4]**

Elliott, J. M. 1988. Growth, size, biomass and production in contrasting populations of trout *Salmo trutta* in two lake district streams. J. Anim. Ecol. 57:49–60. **[3, A3]**

Elliott, J. M. 1994. Quantitative ecology and the brown trout. Oxford Univ. Press, Oxford, U.K. **[I, 2, 9]**

Elliott, S. R., T. A. Coe, J. M. Helfield, and R. J. Naiman. 1998. Spatial variation in environmental characteristics of Atlantic salmon (*Salmo salar*) rivers. Can. J. Fish. Aquat. Sci. 55(Suppl. 1):267–280. **[2]**

Ellis, D. V., ed. 1977. Pacific salmon: management for people. Univ. of Victoria, Victoria, BC, Canada. **[13]**

Ellstrand, N. C. 1983. Why are juveniles smaller than their parents? Evolution 37:1091–1094. **[4]**

Emlen, J. M. 1970. Age specificity and ecological theory. Ecology 51:588–601. **[1]**

Emlen, S. T., and L. W. Oring. 1977. Ecology, sexual selection, and the evolution of animal mating systems. Science 197:215–223. **[9]**

Endler, J. A. 1973. Gene flow and population differentiation. Science 179:243–250. **[2]**

Endler, J. A. 1977. Geographical variation, speciation, and clines. Princeton Univ. Press, Princeton, NJ. **[2, 3]**

Endler, J. A. 1986. Natural selection in the wild. Princeton Univ. Press, Princeton, NJ. **[2, 11]**

Eriksson, T., and L.-O. Eriksson. 1991. Spawning migratory behaviour of coastal-released Baltic salmon (*Salmo salar*). Effects on straying frequency and time of river ascent. Aquaculture 98:79–87. **[A1]**

Essington, T. E. 2001. The precautionary approach in fisheries management: the devil is in the details. Trends Ecol. Evol. 16:121–122. **[11]**

Essington, T. E., T. P. Quinn, and V. E. Ewert. 2000. Intra- and inter-specific competition and the reproductive success of sympatric Pacific salmon. Can. J. Fish. Aquat. Sci. 57:205–213. **[2]**

Essington, T. E., P. W. Sorensen, and D. G. Paron. 1998. High rate of redd superimposition by brook trout (*Salvelinus fontinalis*) and brown trout (*Salmo trutta*) in a Minnesota stream cannot be explained by habitat availability alone. Can. J. Fish. Aquat. Sci. 55:2310–2316. **[9]**

Estoup, A., F. Rousset, Y. Michalakis, J.-M. Cornuet, M. Adriamanga, and R. Guyomard. 1998. Comparative analysis of microsatellite and allozyme markers: a case study investigating microgeographic differentiation in brown trout (*Salmo trutta*). Mol. Ecol. 7:339–353. **[A2]**

Etterson, J. R., and R. G. Shaw. 2001. Constraint to adaptive evolution in response to global warming. Science 294:151–154. **[11]**

Evans, D. M. 1994. Observations on the spawning behaviour of male and female sea trout, *Salmo trutta* L., using radio-telemetry. Fish. Manag. Ecol. 1:91–105. **[9]**

Everest, F. H., and D. W. Chapman. 1972. Habitat selection and spatial interaction by juvenile chinook salmon and steelhead trout in two Idaho streams. J. Fish. Res. Board Can. 29:91–100. **[13]**

Everett, R. J., R. L. Wilmot, and C. C. Krueger. 1997. Population genetic structure of Dolly Varden from Beaufort Sea drainages of northern Alaska and Canada. Am. Fish. Soc. Symp. 19:240–249. **[A2]**

Everhart, W. H., and W. D. Youngs. 1981. Principles of fishery science. Cornell Univ. Press, London, U.K. **[13]**

Ewens, W. J. 1979. Mathematical population genetics. Springer-Verlag, Berlin, Germany. [10, 11]

Fagen, R. M. 1972. An optimal life history in which reproductive effort decreases with age. Am. Nat. 106:258–261. [1]

Fagerlund, U. H. M., J. R. McBride, and E. T. Stone. 1981. Stress-related effects of hatchery rearing density on coho salmon. Trans. Am. Fish. Soc. 110:644–649. [13]

Falconer, D. S. 1965. The inheritance of liability to certain diseases, estimated from the incidence among relatives. Ann. Hum. Genet. 29:51–76. [11]

Falconer, D. S., and T. F. C. Mackay. 1996. Introduction to quantitative genetics, 4th ed. Addison Wesley Longman, Harlow, Essex. [11]

Fausch, K. D. 1984. Profitable stream positions for salmonids: relating specific growth rate to net energy gain. Can. J. Zool. 62:441–451. [13]

Fausch, K. D. 1988. Tests of competition between native and introduced salmonids in streams: what have we learned? Can. J. Fish. Aquat. Sci. 45:2238–2246. [13]

Fausch, K. D., and R. J. White. 1986. Competition among juveniles of coho salmon, brook trout, and brown trout in a laboratory stream, and implications for Great Lakes tributaries. Trans. Am. Fish. Soc. 115:363–381. [13]

Felsenstein, J. 1971. Inbreeding and variance effective numbers in populations with overlapping generations. Genetics 68:581–597. [10]

Felsenstein, J. 1981. Skepticism towards santa Rosalia, or why are there so few kinds of animals? Evolution 35:124–138. [6]

Felsenstein, J. 1993. PHYLIP (Phylogeny inference package) version 3.5c. Department of Genetics, SK-50, Univ. of Washington, Seattle, WA. [6, 12]

Fenderson, O. 1964. Evidence of subpopulations of lake whitefish, *Coregonus clupeaformis*, involving a dwarfed form. Trans. Am. Fish. Soc. 93:77–94. [6]

Ferguson, A., and J. B. Taggart. 1991. Genetic differentiation among the sympatric brown trout (*Salmo trutta*) populations of Lough Melvin, Ireland. Biol. J. Linn. Soc. 43:221–237. [I]

Ferguson, M. M., R. G. Danzmann, and J. A. Hutchings 1991. Incongruent estimates of population differentiation among brook charr, *Salvelinus fontinalis*, from Cape Race, Newfoundland, Canada, based upon allozyme and mitochondrial DNA variation. J. Fish Biol. 39(Suppl. A):79–85. [5]

Ferris, S. D., S. L. Portnoy, and G. S. Whitt. 1979. The roles of speciation and divergence time in the loss of duplicate gene expression. Theor. Pop. Biol. 15:114–139. [8]

Ferris, S. D. 1984. Tetraploidy and the evolution of the catostomid fishes. Pp. 55–88 *in* B. Turner, ed. Evolutionary genetics of fishes. Plenum Press, New York. [7]

Field-Dodgson, M. J. 1988. Size characteristics and diet of emergent chinook salmon in a small, stable, New Zealand stream. J. Fish Biol. 32:27–40. [5]

Filchak, K. E., J. B. Roethele, and J. L. Feder. 2000. Natural selection and sympatric divergence in the apple maggot *Rhagoletis pomonella*. Nature 407:739–742. [I]

Finney, B. P., I. Gregory-Eaves, J. Sweetman, M. S. V. Douglas, and J. P. Smol. 2000. Impacts of climatic change and fishing on Pacific salmon abundance over the past 300 years. Science 290:795–798. [2]

Finstad, B., and T. G. Heggberget. 1995. Seawater tolerance, migration, growth and recapture rates of wild and hatchery-reared Arctic charr (*Salvelinus alpinus* (L.)). Nordic J. Freshwater Res. 71:229–236. [3]

Fisher, R. A. 1930. The genetical theory of natural selection. Clarendon Press, Oxford, U.K. [1, 6, 11]

Fitch, W. M., and F. J. Ayala. 1995. Tempo and mode in evolution: genetics and paleontology 50 years after Simpson. National Academy Press, Washington, D.C. [7]

Fleming, I. A. 1996. Reproductive strategies of Atlantic salmon: ecology and evolution. Rev. Fish Biol. Fish. 6:379–416. [5, 9, A3]

Fleming, I. A. 1998. Pattern and variability in the breeding system of Atlantic salmon (*Salmo salar*), with comparisons to other salmonids. Can. J. Fish. Aquat. Sci. 55(Suppl. 1):59–76. [I, 4, 7, 9]

Fleming, I. A., and S. Einum. 1997. Experimental tests of genetic divergence of farmed from wild Atlantic salmon due to domestication. ICES J. Mar. Sci. 54:1051–1063. [13]

Fleming, I. A., and M. R. Gross. 1989. Evolution of adult female life history and morphology in a Pacific salmon (coho: *Oncorhynchus kisutch*). Evolution 43:141–157. [1, 3, 4, 7, 9, 12]

Fleming, I. A., and M. R. Gross. 1990. Latitudinal clines: a trade-off between egg number and size in Pacific salmon. Ecology 71:1–11. [4, 5]

Fleming, I. A., and M. R. Gross. 1993. Breeding success of hatchery and wild coho salmon (*Oncorhynchus kisutch*) in competition. Ecol. Appl. 3:230–245. [9]

Fleming, I. A., and M. R. Gross. 1994. Breeding competition in a Pacific salmon (coho: *Oncorhynchus kisutch*): measures of natural and sexual selection. Evolution 48:637–657. [I, 1, 3, 4, 9, 13]

Fleming, I. A., and E. Petersson. 2001. The ability of released, hatchery salmonids to breed and contribute to the natural productivity of wild populations. Nordic J. Freshwater Res. 75:71–98. [9]

Fleming, I. A., and J. D. Reynolds. 2003. Salmonid breeding systems. Pp. 264–294 *in* A. P. Hendry and S. C. Stearns, eds. Evolution illuminated: salmon and their relatives. Oxford Univ. Press, Oxford, U.K.—*this volume* [I, 1, 3, 5, 7, 8, 13]

Fleming, I. A., S. Einum, B. Jonsson and N. Jonsson. 2003. Comment on "rapid evolution of egg size in captive salmon." Science. In press. [4].

Fleming, I. A., K. Hindar, I. B. Mjølnerød, B. Jonsson, T. Balstad, and A. Lamberg. 2000. Lifetime success and interactions of farm salmon invading a native population. Proc. R. Soc. Lond. B 267:1517–1523. [I, 2, 9]

Fleming, I. A., B. Jonsson, and M. R. Gross. 1994. Phenotypic divergence of sea-ranched, farmed, and wild salmon. Can. J. Fish. Aquat. Sci. 51:2808–2824. [6, 9]

Fleming, I. A., A. Lamberg, and B. Jonsson. 1997. Effects of early experience on the reproductive performance of Atlantic salmon. Behav. Ecol. 8:470–480. [9]

Fleming, I. A., B. Jonsson, M. R. Gross, and A. Lamberg. 1996. An experimental study of the reproductive behaviour and success of farmed and wild Atlantic salmon (*Salmo salar*). J. Appl. Ecol. 33:893–905. [3, 9]

Flick, W. A., and Webster, D. A. 1964. Comparative first year survival and production in wild and domestic strains of brook trout (*Salvelinus fontinalis*). Trans. Am. Fish. Soc. 93:58–69. [7]

Foerster, R. E. 1947. Experiment to develop sea-run from landlocked sockeye salmon (*Oncorhynchus nerka kennerlyi*). J. Fish. Res. Board Can. 7:88–93. [3]

Foerster, R. E. 1968. The sockeye salmon, *Oncorhynchus nerka*. Fish. Res. Board Can. Bull. 162. [1, 9]

Folmar, L. C., and W. W. Dickhoff. 1980. The parr-smolt transformation (smoltification) and seawater adaptation in salmonids: A review of selected literature. Aquaculture 21:1–37. [7]

Fontaine, P.-M., and J. J. Dodson. 1999. An analysis of the distribution of juvenile Atlantic salmon (*Salmo salar*) in nature as a function of relatedness using microsatellites. Mol. Ecol. 8:189–198. **[2]**

Fontaine, P.-M., J. J. Dodson, L. Bernatchez, and A. Slettan. 1997. A genetic test of metapopulation structure in Atlantic salmon (*Salmo salar*) using microsatellites. Can. J. Fish. Aquat. Sci. 54:2434–2442. **[A2]**

Foote, C. J. 1988. Male mate choice dependent on male size in salmon. Behaviour 106: 63–80. **[I, 3, 8, 9]**

Foote, C. J. 1989. Female mate preference in Pacific salmon. Anim. Behav. 38: 721–723. **[9]**

Foote, C. J. 1990. An experimental comparison of male and female spawning territoriality in a Pacific salmon. Behaviour 115:283–314. **[I, 2, 3, 4, 9]**

Foote, C. J., and P. A. Larkin. 1988. The role of male choice in the assortative mating of anadromous and non-anadromous sockeye salmon (*Oncorhynchus nerka*). Behaviour 106:43–62. **[I, 3]**

Foote, C. J., G. S. Brown, and C. C. Wood. 1997. Spawning success of males using alternative mating tactics in sockeye salmon, *Oncorhynchus nerka*. Can. J. Fish. Aquat. Sci. 54:1785–1795. **[I, 3, 9]**

Foote, C. J., I. Mayer, C. C. Wood, W. C. Clarke, and J. Blackburn. 1994. On the developmental pathway to nonanadromy in sockeye salmon, *Oncorhynchus nerka*. Can. J. Zool. 72:397–405. **[3]**

Foote, C. J., K. Moore, K. Stenberg, K. J. Craig, J. K. Wenburg, and C. C. Wood. 1999. Genetic differentiation in gill raker number and length in sympatric anadromous and nonanadromous morphs of sockeye salmon, *Oncorhynchus nerka*. Environ. Biol. Fishes 54:263–274. **[A3]**

Foote, C. J., C. C. Wood, and R. E. Withler. 1989. Biochemical genetic comparison of sockeye salmon and kokanee, the anadromous and nonanadromous forms of *Oncorhynchus nerka*. Can. J. Fish. Aquat. Sci. 46:149–158. **[3, A3]**

Foote, C. J., C. C. Wood, W. C. Clarke, and J. Blackburn. 1992. Circannual cycle of seawater adaptability in *Oncorhynchus nerka*: genetic differences between sympatric sockeye salmon and kokanee. Can. J. Fish. Aquat. Sci. 49:99–109. **[3, 7]**

Forbes, L. S. 1999. Within-clutch variation in propagule size: the double-fault model. Oikos 85:146–150. **[4]**

Forbes, S. H., and F. W. Allendorf. 1991. Associations between mitochondrial and nuclear genotypes in cutthroat trout hybrid swarms. Evolution 45:1332–1349. **[8]**

Ford, M. J. 1998. Testing models of migration and isolation among populations of chinook salmon (*Oncorhynchus tshawytscha*). Evolution 52:539–557. **[12]**

Ford, M. J. 2003. Conserving units and preserving diversity. Pp. 338–357 *in* A. P. Hendry and S. C. Stearns, eds. Evolution illuminated: salmon and their relatives. Oxford Univ. Press, New York, NY—*this volume* **[I, 7, 10]**

Forsgren, E., J. D. Reynolds, and A. Berglund. 2002. Behavioural ecology of reproduction in fish. Pp. 225–247 *in* P. J. B. Hart and J. D. Reynolds, eds. Handbook of fish biology and fisheries: Volume 1, Fish biology. Blackwell Publications, Oxford, U.K. **[9]**

Fortier, L., and W. C. Leggett. 1985. A drift study of larval fish survival. Mar. Ecol. Prog. Ser. 25:245–257. **[4]**

Frank, K. T., and D. Brickman. 2000. Allee effects and compensatory population dynamics within a stock complex. Can. J. Fish. Aquat. Sci. 57:513–517. **[13]**

Frank, K. T., and W. C. Leggett. 1994. Fisheries ecology in the context of ecological and evolutionary theory. Annu. Rev. Ecol. Syst. 25:401–422. **[13]**

Frank, S. A. 1986. Dispersal polymorphisms in subdivided populations. J. theor. Biol. 122:303–309. [2]

Frankel, R. 1983. Heterosis: reappraisal of theory and practice. Springer-Verlag, Berlin, Germany. [11]

Frankham, R. 1995. Effective population size/adult population size ratios in wildlife: a review. Genetical Res. (Cambridge) 66:95–107. [10]

Frary, A., and 10 other authors. 2000. *fw2.2*: A quantitative trait locus key to the evolution of tomato fruit size. Science 289:85–88. [6]

Fraser, D. J., and L. Bernatchez. 2001. Adaptive evolutionary conservation: towards a unified concept for defining conservation units. Mol. Ecol. 10:2741–2752. [12]

Fretwell, S. D., and H. L. Lucas, Jr. 1970. On territorial behavior and other factors influencing habitat distribution in birds. I. Theoretical development. Acta Biotheor. 19:16–36. [9]

Friedland, K. D. 1998. Ocean climate influences on critical Atlantic salmon (*Salmo salar*) life history events. Can. J. Fish. Aquat. Sci. 55(Suppl. 1):119–130. [13]

Fukushima, M., and W. W. Smoker. 1998. Spawning habitat segregation of sympatric sockeye and pink salmon. Trans. Am. Fish. Soc. 127:253–260. [I, 8]

Fulton, R. J., and J. T. Andrews, eds. 1987. La calotte glaciaire laurentidienne. *Géographie physique et Quaternaire*. Les Presses de l'Université de Montréal, Montréal, Québec, Canada. [6]

Funk, F., T. P. Quinn II, J. Heifetz, J. N. Ianelli, J. E. Powers, J. F. Schweigert, P. J. Sullivan, and C.-I. Zhang, eds. 1998. Fishery stock assessment models, Alaska Sea Grant College Program Report No. AK-SG-98–01. Univ. of Alaska, Fairbanks, AK. [13]

Futuyma, D. J. 1986. Evolutionary Biology. Sinauer Assoc., Sunderland, MA. [13]

Gadgil, M. 1971. Dispersal: population consequences and evolution. Ecology 52:253–261. [2]

Gadgil, M. D., and W. Bossert. 1970. Life historical consequences of natural selection. Am. Nat. 104:1–24. [1]

Gage, M. J. G., P. Stockley, and G. A. Parker. 1995. Effects of alternative male mating strategies on characteristics of sperm production in the Atlantic salmon (*Salmo salar*): theoretical and empirical investigations. Phil. Trans. R. Soc. Lond. B 350:391–399. [9, A3]

Galloway, S. M., and 12 other authors. 2000. Mutations in an oocyte-derived growth factor gene (*BMP15*) cause increased ovulation rate and infertility in a dosage-sensitive manner. Nature Genet. 25:279–283. [6]

Galvin, P., J. Taggart, A. Ferguson, M. O'Farrell, and T. Cross. 1996. Population genetics of Atlantic salmon (*Salmo salar*) in the River Shannon system in Ireland: an appraisal using single locus minisatellite (VNTR) probes. Can. J. Fish. Aquat. Sci. 53:1933–1942. [A2]

Gandon, S. 1999. Kin competition, the cost of inbreeding and the evolution of dispersal. J. theor. Biol. 200:345–364. [2]

Gandon, S., and Y. Michalakis. 1999. Evolutionarily stable dispersal rate in a metapopulation with extinctions and kin competition. J. theor. Biol. 199:275–290. [2]

Gandon, S., and Y. Michalakis. 2001. Multiple causes of the evolution of dispersal. Pp. 155–167 *in* J. Clobert, E. Danchin, A. A. Dhondt, and J. D. Nichols, eds. Dispersal. Oxford Univ. Press, Oxford, U.K. [2]

Gandon, S., and F. Rousset. 1999. Evolution of stepping-stone dispersal rates. Proc. R. Soc. Lond. B 266:2507–2513. [2]

Gantmacher, F. R. 1959. The theory of matrices. Chelsea Publ. Co., New York.
[1]

Garant, D., J. J. Dodson, and L. Bernatchez. 2000. Ecological determinants and temporal stability of the within-river population structure in Atlantic salmon (*Salmo salar* L.). Mol. Ecol. 9:615–628. [2, A2]

Garant, D., J. J. Dodson, and L. Bernatchez. 2001. A genetic evaluation of mating system and determinants of individual reproductive success in Atlantic salmon (*Salmo salar* L.). J. Hered. 92:137–145. [I, 3, 9, 10]

Garant, D., J. J. Dodson, and L. Bernatchez. 2003a. Differential reproductive success and heritability of alternative reproductive tactics in wild Atlantic salmon (*Salmo salar* L.) Evolution 57:1133–1141. [I, 5]

Garant, D., I. A. Fleming, S. Einum, and L. Bernatchez. 2003b. Alternative male life-history tactics as potential vehicles for speeding introgression of farm salmon traits into wild populations. Ecol. Letters 6:541–549. [9]

Garant, D., P.-M. Fotaine, S. P. Good, J. J. Dodson, and L. Bernatchez. 2002. The influence of male parental identity on growth and survival of offspring in Atlantic salmon (*Salmo salar*). Evol. Ecol. Res. 4:537–549. [5, 9]

García-Marín, J. L., and C. Pla. 1996. Origins and relationships of native populations of *Salmo trutta* (brown trout) in Spain. Heredity 77:313–323. [A2]

García-Marín, J. L., F. M. Utter, and C. Pla. 1999. Postglacial colonization of brown trout in Europe based on distribution of allozymes variants. Heredity 82:46–56. [A2]

Garcia-Vazquez, E., P. Moran, J. L. Martinez, J. Perez, B. de Gaudemar, and E. Beall. 2001. Alternative mating strategies in Atlantic salmon and brown trout. J. Heredity 92:146–149. [9, 10]

Gausen, D., and V. Moen. 1991. Large-scale escapes of Atlantic salmon (*Salmo salar*) into Norwegian rivers threaten natural populations. Can. J. Fish. Aquat. Sci. 48:426–428. [I]

Geiger, H. J., W. W. Smoker, L. A. Zhivotovsky, and A. J. Gharrett. 1997. Variability of family size and marine survival in pink salmon (*Oncorhynchus gorbuscha*) has implications for conservation biology and human use. Can. J. Fish. Aquat. Sci. 54:2684–2690. [10]

Geritz, S. A. H. 1995. Evolutionarily stable seed polymorphism and small-scale spatial variation in seedling density. Am. Nat. 146:685–707. [4]

Gharrett, A. J., and W. W. Smoker. 1991. Two generations of hybrids between even- and odd-year pink salmon (*Oncorhynchus gorbuscha*): a test for outbreeding depression? Can. J. Fish. Aquat. Sci. 48:1744–1749. [8]

Gharrett, A. J., and W. W. Smoker. 1993. Genetic components in life history traits con-tribute to population structure. Pp. 197–202 *in* J. G. Cloud and G. H. Thorgaard, eds. Genetic conservation of salmonid fishes. Plenum Press, New York. [9]

Gharrett, A. J., S. Lane, A. J. McGregor, and S. G. Taylor. 2001. Use of a genetic marker to examine genetic interaction among subpopulations of pink salmon (*Oncorhynchus gorbuscha*). Genetica 111:259–267. [2]

Gharrett, A. J., S. M. Shirley, and G. R. Tromble. 1987. Genetic relationships among populations of Alaskan chinook salmon (*Oncorhynchus tshawytscha*). Can. J. Fish. Aquat. Sci. 44:765–774. [A2]

Gharrett, A. J., W. W. Smoker, R. R. Reisenbichler, and S. G. Taylor. 1999. Outbreeding depression in hybrids between odd- and even-broodyear pink salmon. Aquaculture 173:117–129. [8]

Gharrett, A. J., C. Smoot, A. J. McGregor, and P. B. Holmes. 1988. Genetic relationships of even-year northwestern Alaskan pink salmon. Trans. Am. Fish. Soc. 117:536–545. [A2]

Giberson, D. J., and D. Caissie. 1998. Stream habitat hydraulics: interannual variability in three reaches of Catamaran Brook, New Brunswick. Can. J. Fish. Aquat. Sci. 55:485–494. [2]

Gibson, R. J. 1993. The Atlantic salmon in fresh water: spawning, rearing and production. Rev. Fish Biol. Fish. 3:39–73. [9]

Gilbert, C. H. 1913. Age at maturity of the Pacific coast salmon of the genus *Oncorhynchus*. Bull. Bur. Fisheries (U.S.) 32:1–22. [12]

Gilbert, D. J. 1997. Towards a new recruitment paradigm for fish stocks. Can. J. Fish. Aquat. Sci. 54:969–977. [13]

Gilchrist, G. W., R. B. Huey, and L. Serra. 2001. Rapid evolution of wing size clines in *Drosophila subobscura*. Genetica 112/113:273–286. [7]

Gilhousen, P. 1980. Energy sources and expenditures in Fraser River sockeye salmon during their spawning migration. Int. Pacific Salmon Fish. Comm. Bull. 22. [3, 9]

Gilhousen, P. 1990. Prespawning mortalities of sockeye salmon in the Fraser River system and possible causal factors. Int. Pacific Salmon Fish. Comm. Bull. 26. [3]

Gill, D. E. 1974. Intrinsic rate of increase, saturation density, and competitive ability. II. the evolution of competitive ability. Am. Nat. 108:103–116. [13]

Gingerich, P. D. 1993. Quantification and comparison of evolutionary rates. Am. J. Sci. 293-A:453–478. [7]

Gislason, D, M. M. Ferguson, S. Skúlason, and S. S. Snorrason. 1999. Rapid and coupled phenotypic and genetic divergence in Icelandic Arctic char (*Salvelinus alpinus*) Can. J. Fish. Aquat. Sci. 56:2229–2234. [I, 6]

Giuffra, E., R. Guyomard, and G. Forneris. 1996. Phylogenetic relationships and introgression patterns between incipient parapatric species of Italian brown trout (*Salmo trutta* L. complex). Mol. Ecol. 5:207–220. [8]

Gjedrem, T. 1983. Genetic variation in quantitative traits and selective breeding in fish and shellfish. Aquaculture 33:51–72. [5]

Gjerde, B. 1984a. Response to individual selection for age at sexual maturity in Atlantic Salmon. Aquaculture 38:229–240. [5, 13]

Gjerde, B. 1984b. Variation in semen production of farmed Atlantic salmon and rainbow trout. Aquaculture 40:109–114. [9]

Gjerde, B. 1986. Growth and reproduction in fish and shellfish. Aquaculture 57: 37–55. [5]

Gjerde, B., and L. R. Schaeffer. 1989. Body traits in rainbow trout. II. Estimates of heritabilities and of phenotypic and genetic correlations. Aquaculture 80:25–44. [I]

Glebe, B. D., and R. L. Saunders. 1986. Genetic factors in sexual maturity of cultured Atlantic salmon (*Salmo salar*) parr and adults reared in sea cages. Can. Spec. Publ. Fish. Aquat. Sci. 89:24–29. [5, 9]

Glémet, H., P. Blier, and L. Bernatchez. 1998. Geographical extent of Arctic char (*Salvelinus alpinus*) mtDNA introgression in brook char populations (*S. fontinalis*) from eastern Québec, Canada. Mol. Ecol. 7:1655–1662. [I, 8]

Goldstein, P. Z., R. DeSalle, G. Amato, and A. P. Vogler. 2000. Conservation genetics at the species boundary. Cons. Biol. 14:120–131. [12]

Good, S. P., J. J. Dodson, M. G. Meekan, and D. A. J. Ryan. 2001. Annual variation in size-selective mortality of Atlantic salmon (*Salmo salar*) fry. Can. J. Fish. Aquat. Sci. 58:1187–1195. [2]

Goodnight, C. J. 1985. The influence of environmental variation on group and individual selection in a cress. Evolution 39:545–558. [13]

Goodnight, C. J., J. M. Schwartz, and L. Stevens. 1992. Contextual analysis of models of group selection, soft selection, hard selection, and the evolution of altruism. Am. Nat. 140:743–761. [13]

Gould, S. J. 1989. Wonderful life. Norton and Company Inc., New York. [7]

Gould, S. J., and N. Eldredge. 1977. Punctuated equilibria: the tempo and mode of evolution reconsidered. Paleobiology 3:115–151. [7]

Gould, S. J., and R. C. Lewontin. 1979. The spandrels of San Marco and the Panglossian paradigm—a critique of the adaptationist programme. Proc. R. Soc. Lond. B 205:581–598. [1]

Grant, J. W. A., and D. L. Kramer. 1990. Territory size as a predictor of the upper limit to population density of juvenile salmonids in streams. Can. J. Fish. Aquat. Sci. 47:1724–1737. [2]

Grant, P. R., and B. R. Grant. 1992. Hybridization of bird species. Science 256:193–197. [I]

Grant, P. R., and B. R. Grant. 1995. Predicting microevolutionary responses to directional selection on heritable variation. Evolution 49:241–251. [7, 11]

Green, C. W. 1952. Results from stocking brook trout of wild and hatchery strains at Stillwater Pond. Trans. Am. Fish. Soc. 81:43–52. [7]

Greenwood, P. J. 1980. Mating systems, philopatry and dispersal in birds and mammals. Anim. Behav. 28:1140–1162. [2]

Greenwood, P. J. 1983. Mating systems and the evolutionary consequences of dispersal. Pp. 116–131 in I. R. Swingland and P. J. Greenwood, eds. The ecology of animal movement. Clarendon Press, Oxford, U.K. [2]

Gresh, T., J. Lichatowich, and P. Schoonmaker. 2000. An estimation of historic and current levels of salmon production in the Northeast Pacific ecosystem: evidence of a nutrient deficit in the freshwater systems of the Pacific Northwest. Fisheries (Bethesda) 25(1):15–21. [13]

Griffith, J. N., A. P. Hendry, and T. P. Quinn. 1999. Straying of adult sockeye salmon, Oncorhynchus nerka, entering a non-natal hatchery. Fishery Bull. 97:713–716. [2]

Griffiths, S. W., and J. D. Armstrong 2000. Differential responses of kin and nonkin salmon to patterns of water flow: does recirculation influence aggression? Anim. Behav. 59:1019–1023. [2]

Griffiths, S. W., and J. D. Armstrong. 2001. The benefits of genetic diversity outweigh those of kin association in a territorial animal. Proc. R. Soc. Lond. B 268:1293–1296. [2]

Griffiths, S. W., and J. D. Armstrong. 2002. Kin-biased territory overlap and food sharing among Atlantic salmon juveniles. J. Anim. Ecol. 71:480–486. [2]

Groot, E. P., and D. F. Alderdice. 1985. Fine structure of the external egg membrane of five species of Pacific salmon and steelhead trout. Can. J. Zool. 63:552–566. [8]

Groot, C., and L. Margolis, eds. 1991. Pacific salmon life histories. UBC Press, Vancouver, BC, Canada. [I, 1, 2, 4, 5, 8, 9, 10, 12, 13]

Groot, C., L. Margolis, and W. C. Clarke, eds. 1995. Physiological ecology of Pacific salmon. UBC Press, Vancouver, BC, Canada. [I]

Groot, C., T. P. Quinn, and T. J. Hara. 1986. Responses of migrating adult sockeye salmon (Oncorhynchus nerka) to population-specific odours. Can. J. Zool. 64:926–932. [2]

Gross, M. R. 1984. Sunfish, salmon, and the evolution of alternative reproductive strategies and tactics in fishes. Pp. 55–75 *in* G. Potts and R. J. Wootton, eds. Fish reproduction: strategies and tactics. Academic Press, New York. [9]

Gross, M. R. 1985. Disruptive selection for alternative life histories in salmon. Nature 313:47–48. [I, 5, 8, 9, 13]

Gross, M. R. 1987. Evolution of diadromy in fishes. Am. Fish. Soc. Symp. 1:14–25. [3,7]

Gross, M. R. 1991. Salmon breeding behavior and life history evolution in changing environments. Ecology 72:1180–1186. [3, 9, 13]

Gross, M. R. 1996. Alternative reproductive strategies and tactics: diversity within sexes. Trends Ecol. Evol. 11:92–98. [I, 1, 3, 5, 9, 13]

Gross, M. R. 1998. One species with two biologies: Atlantic salmon (*Salmo salar*) in the wild and in aquaculture. Can. J. Fish. Aquat. Sci. 55(Suppl. 1):131–144. [9]

Gross, M. R., and J. Repka. 1998. Stability with inheritance in the conditional strategy. J. theor. Biol. 192:445–453. [I, 3, 5, 9]

Gross, M. R., and R. C. Sargent. 1985. The evolution of male and female parental care in fishes. Am. Zool. 25:807–822. [9]

Gross, M. R., R. M. Coleman, and R. M. McDowall. 1988. Aquatic productivity and the evolution of diadromous fish migration. Science 239:1291–1293. [3, 9]

Gulseth, O. A., and K. J. Nilssen. 2000. The brief period of spring migration, short marine residence, and high return rate of a northern Svalbard population of Arctic char. Trans. Am. Fish. Soc. 129:782–796. [3]

Gulseth, O. A., and K. J. Nilssen. 2001. Life-history traits of charr, *Salvelinus alpinus*, from a high Arctic watercourse on Svalbard. Arctic 54:1–11. [A3]

Gunnerød, T. B., N. A. Hvidsten, and T. G. Heggberget. 1988. Open sea releases of Atlantic salmon smolts, *Salmo salar*, in central Norway, 1973–83. Can. J. Fish. Aquat. Sci. 45:1340–1345. [A1]

Gustafson, R. G., and G. A. Winans. 1999. Distribution and population genetic structure of river- and sea-type sockeye salmon in western North America. Ecol. Freshwater Fish 8:181–193. [2, A2]

Gustafson, R. G., T. C. Wainwright, G. A. Winans, F. W. Waknitz, L. T. Parker, and R. S. Waples. 1997. Status review of sockeye salmon from Washington and Oregon. NOAA Tech. Memo. NMFS-NWFSC-33. [12]

Gustafsson, L. 1986. Lifetime reproductive success and heritability: empirical support for Fisher's fundamental theorem. Am. Nat. 128:761–764. [11]

Habicht, C., S. Sharr, D. Evans, and J. E. Seeb. 1998. Coded wire tag placement affects homing ability of pink salmon. Trans. Am. Fish. Soc. 127:652–657. [A1]

Hagen, J., and E. B. Taylor. 2001. Resource partitioning as a factor limiting gene flow in hybridizing populations of Dolly Varden char (*Salvelinus malma*) and bull trout (*S. confluentus*). Can. J. Fish. Aquat. Sci. 58:2037–2047. [8]

Hager, R. C., and R. E. Noble. 1976. Relation of size at release of hatchery-reared coho salmon to age, size, and sex composition of returning adults. Prog. Fish-Cult. 38:144–147. [13]

Haig, D. 2000. The kinship theory of genomic imprinting. Annu. Rev. Ecol. Syst. 31: 9–32. [I]

Hairston, N. G., Jr., and C. E. Caceres. 1996. Distribution of crustacean diapause: micro- and macroevolutionary pattern and process. Hydrobiologia 320:27–44. [10]

Haldane, J. B. S. 1949. Suggestions as to quantitative measurement of rates of evolution. Evolution 3:51–56. [7]

Haldane, J. B. S. 1954. The measurement of natural selection. Caryologia (Suppl. Vol. 6) 1:480–487. [11]

Hamilton, W. D. 1964a. The genetical evolution of social behavior. I. J. theor. Biol. 7:1–16. [2, 13]

Hamilton, W. D. 1964b. The genetical evolution of social behavior. II. J. theor. Biol. 7:17–52. [13]

Hamilton, W. D. 1966. The moulding of senescence by natural selection. J. theor. Biol. 12:12–45. [1]

Hamilton, W. D., and R. M. May. 1977. Dispersal in stable habitats. Nature 269:578–581. [2]

Hammar, J., J. B. Dempson, and E. Skuld. 1989. Natural hybridization between Arctic char (*Salvelinus alpinus*) and lake char (*S. namaycush*): evidence from northern Labrador. Nordic J. Freshwater Res. 65:54–70. [8]

Hammond, D. S., and V. K. Brown. 1995. Seed size of woody-plants in relation to disturbance, dispersal, and soil type in wet neotropical forests. Ecology 76:2544–2561. [4]

Hamon, T. R., C. J. Foote, and G. S. Brown. 1999. Use of female nest characteristics in the sexual behaviour of male sockeye salmon. J. Fish Biol. 55:459–471. [I, 9]

Hamon, T. R., C. J. Foote, R. Hilborn, and D. E. Rogers. 2000. Selection on morphology of spawning wild sockeye salmon by a gill-net fishery. Trans. Am. Fish. Soc. 129:1300–1315. [11, 13]

Handford, P., G. Bell, and T. Reimchen. 1977. A gillnet fishery considered as an experiment in artificial selection. J. Fish. Res. Board Can. 34:954–961. [5]

Hankin, D. G., and M. C. Healey. 1986. Dependence of exploitation rates for maximum yield and stock collapse on age and sex structure of chinook salmon (*Oncorhynchus tshawytscha*) stocks. Can. J. Fish. Aquat. Sci. 43:1746–1759. [11]

Hankin, D. G., J. W. Nicholas, and T. W. Downey. 1993. Evidence for inheritance of age of maturity in chinook salmon (*Oncorhynchus tshawytscha*). Can. J. Fish. Aquat. Sci. 50:347–358. [11, 13]

Hansen, M. M., and K.-L. D. Mensberg. 1996. Founder effects and genetic population structure of brown trout (*Salmo trutta*) in a Danish river system. Can. J. Fish. Aquat. Sci. 53:2229–2237. [I, 2, A2]

Hansen, M. M., E. Kenchington, and E. E. Nielsen. 2001. Assigning individual fish to populations using microsatellite DNA markers. Fish Fish. 2:93–112. [2]

Hansen, M. M., V. Loeschcke, G. Rasmussen, and V. Simonsen. 1993. Genetic differentiation among Danish brown trout (*Salmo trutta*) populations. Hereditas 118:177–185. [A2]

Hansen, M. M., E. E. Nielsen, and K.-L. D. Mensberg. 1997. The problem of sampling families rather than populations: relatedness among individuals in samples of juvenile brown trout *Salmo trutta* L. Mol. Ecol. 6:469–474. [2]

Hansen, M. M., E. E. Nielsen, D. E. Ruzzante, C. Bouza, and K.-L. D. Mensberg. 2000a. Genetic monitoring of supportive breeding in brown trout (*Salmo trutta* L.) using microsatellite DNA markers. Can. J. Fish. Aquat. Sci. 57:2130–2139. [2, 10]

Hansen, M. M., D. E. Ruzzante, E. E. Nielsen, and K.-L. D. Mensberg. 2000b. Microsatellite and mitochondrial DNA polymorphism reveals life-history dependent interbreeding between hatchery and wild brown trout (*Salmo trutta* L.). Mol. Ecol. 9:583–594. [2, 8]

Hansen, M. M., D. E. Ruzzante, E. E. Nielsen, D. Bekkevold, and K.-L. D. Mensberg. 2002. Long-term effective population sizes, temporal stability of genetic composition and potential for local adaptation in anadromous brown trout (*Salmo trutta*) populations. Mol. Ecol. 11:2523–2535. [2]

Hanski, I. 1999. Metapopulation ecology. Oxford Univ. Press, Oxford, U.K. [I]

Hanski, I., and D. Simberloff. 1997. The metapopulation approach, its history, conceptual domain, and application to conservation. Pp. 5–26 *in* I. A. Hanski and M. E. Gilpin, eds. Metapopulation biology: ecology, genetics, and evolution. Academic Press, New York. [2]

Hanson, A. J., and H. D. Smith. 1967. Mate selection in a population of sockeye salmon (*Oncorhynchus nerka*) of mixed age-groups. J. Fish. Res. Board Can. 24:1955–1977. [9]

Harache, Y. 1992. Pacific salmon in Atlantic waters. ICES Mar. Sci. Symp. 194:31–55. [2]

Hard, J. J. 1995. A quantitative genetic perspective on the conservation of intraspecific diversity. Am. Fish. Soc. Symp. 17:304–326. [11]

Hard, J. J. In press. Case study of Pacific salmon. *In* U. Dieckmann, O. R. Godø, M. Heino, and J. Mork, eds. Fisheries-induced adaptive change. Cambridge studies in adaptive dynamics, Cambridge Univ. Press, Cambridge, U.K. In press. [11]

Hard, J. J. 2003. Evolution of chinook salmon life history under size-selective harvest. Pp. 314–337 *in* A. P. Hendry and S. C. Stearns, eds. Evolution illuminated: salmon and their relatives. Oxford Univ. Press, New York, NY—*this volume* [I, 5, 9, 10, 13]

Hard, J. J., and W. R. Heard. 1999. Analysis of straying variation in Alaskan hatchery chinook salmon (*Oncorhynchus tshawytscha*) following transplantation. Can. J. Fish. Aquat. Sci. 56:578–589. [2, A1]

Hard, J. J., W. E. Bradshaw, and C. M. Holzapfel. 1993. Genetic coordination of demography and phenology in the pitcher-plant mosquito, *Wyeomyia smithii*. J. Evol. Biol. 6:707–723. [11]

Hard, J. J., L. Connell, W. K. Hershberger, and L. W. Harrell. 2000. Genetic variation in mortality of chinook salmon during a bloom of the marine alga *Heterosigma akashiwo*. J. Fish Biol. 56:1387–1397. [10]

Hard, J. J., R. G. Kope, W. S. Grant, F. W. Waknitz, L. T. Parker, and R. S. Waples. 1996. Status review of pink salmon from Washington, Oregon, and California. NOAA Tech. Memo. NMFS-NWFSC-25. [12]

Hard, J. J., G. A. Winans, and J. C. Richardson. 1999. Phenotypic and genetic architecture of juvenile morphometry in chinook salmon. J. Hered. 90:597–606. [11]

Harden Jones, F. R. 1968. Fish migration. Edward Arnold (Publ.), London, U.K. [2]

Harper, J. L. 1977. Population biology of plants. Academic Press, New York. [1]

Harrison, R. G. 1990. Hybrid zones: windows on the evolutionary process. Oxford Surv. Evol. Biol. 7:69–128. [8]

Harrison, R. G., and S. M. Bogdanowicz. 1997. Patterns of variation and linkage disequilibrium in a field cricket hybrid zone. Evolution 51:493–505. [8]

Hartman, G. F. 1965. The role of behavior in the ecology and interaction of underyearling coho salmon (*Oncorhnynchus kisutch*) and steelhead trout (*Salmo gairdneri*). J. Fish. Res. Board Can. 22:1035–1081. [8]

Hartman, G. F., and C. A. Gill. 1968. Distributions of juvenile steelhead and cutthroat trout (*Salmo gairdneri* and *S. clarki clarki*) within streams in southwestern British Columbia. J. Fish. Res. Board Can. 25:33–48. [8]

Hartman, W. L., and R. F. Raleigh. 1964. Tributary homing of sockeye salmon at Brooks and Karluk Lakes, Alaska. J. Fish. Res. Board Can. 21:485–504. [2]

Harvey, P. H., and M. D. Pagel. 1991. The comparative method in evolutionary biology. Oxford Univ. Press, Oxford, U.K. [7]

Hasler, A. D., and A. T. Scholz. 1983. Olfactory imprinting and homing in salmon: investigations into the mechanism of the imprinting process. Zoophysiology, Volume 14. Springer-Verlag, New York. [2]

Hastings, A. 1983. Can spatial variation alone lead to selection for dispersal? Theor. Pop. Biol. 24:244–251. [2]

Haugen, T. O. 2000a. Early survival and growth in populations of grayling with recent common ancestors—field experiments. J. Fish Biol. 56:1173–1191. [5]

Haugen, T. O. 2000b. Growth and survival effects on maturation pattern in populations of grayling with recent common ancestors. Oikos 90:107–118. [5]

Haugen, T. O. 2000c. Life-history evolution in grayling: evidence for adaptive phenotypic divergence during 8–28 generations. Ph.D. dissertation, Univ. of Oslo, Oslo, Norway. [5]

Haugen, T. O., and L. A. Vøllestad. 2000. Population differences in early life-history traits in grayling. J. Evol. Biol. 13:897–905. [5]

Haugen, T. O., and L. A. Vøllestad, 2001. A century of life-history evolution in grayling. Genetica 112/113:475–491. [I, 2, 5, 7]

Hauser, L., G. J. Adcock, P. J. Smith, J. H. B. Ramírez, and G. R. Carvalho. 2002. Loss of microsatellite diversity and low effective population size in an overexploited population of New Zealand snapper (Pagrus auratus). Proc. Natl. Acad. Sci. USA 99:11742–11747. [10]

Hawkins, D. K., and C. J. Foote. 1998. Early survival and development of coastal cutthroat trout (Oncorhynchus clarki clarki), steelhead (Oncorhynchus mykiss) and reciprocal hybrids. Can. J. Fish. Aquat. Sci. 55:2097–2104. [I]

Hawkins, D. K., and T. P. Quinn. 1996. Critical swimming velocity and associated morphology of juvenile coastal cutthroat trout (Oncorhynchus clarki clarki), steelhead trout (Oncorhynchus mykiss) and their hybrids. Can. J. Fish. Aquat. Sci. 53:1487–1496. [8]

Hawkins, S. L., N. V. Varnavskaya, E. A. Matzak, V. V. Efremov, C. M. Guthrie III, R. L. Wilmot, H. Mayama, F. Yamazaki, and A. J. Gharrett. 2002. Population structure of odd-broodline Asian pink salmon and its contrast to the even-broodline structure. J. Fish Biol. 60:370–388. [A2]

Hawthorne, D. J., and S. Via. 2001. Genetic linkage of ecological specialization and reproductive isolation in pea aphids. Nature 412:904–907. [I, 6]

Hazel, W. N., R. Smock, and M. D. Johnson. 1990. A polygenic model for the evolution and maintenance of conditional strategies. Proc. R. Soc. Lond. B 242:181–187. [5, 9]

Healey, M. C. 1983. Coastwide distribution and ocean migration patterns of stream- and ocean-type chinook salmon, Oncorhynchus tshawytscha. Can. Field-Nat. 97:427–433. [7, 12]

Healey, M. C. 1986. Optimum size and age at maturity in Pacific salmon and effects of size-selective fisheries. Can. Spec. Publ. Fish. Aquat. Sci. 89:39–52. [11]

Healey, M. C. 1987. The adaptive significance of age and size at maturity in female sockeye salmon (Oncorhynchus nerka). Can. Spec. Publ. Fish. Aquat. Sci. 96:110–117. [I]

Healey, M. C. 1991. Life history of chinook salmon (Oncorhynchus tshawytscha). Pp. 313–393 in C. Groot and L. Margolis, eds. Pacific salmon life histories. UBC Press, Vancouver, BC, Canada. [2, 8, 12]

Healey, M. C. 2001. Patterns of gametic investment by female stream- and ocean-type chinook salmon. J. Fish Biol. 58:1545–1556. **[3, 4, 12]**

Healey, M. C., and W. R. Heard. 1984. Inter- and intra-population variation in the fecundity of chinook salmon (*Oncorhynchus tshawytscha*) and its relevance to life history theory. Can. J. Fish. Aquat. Sci. 41:476–483. **[I]**

Heard, W. R. 1991. Life history of pink salmon (*Oncorhynchus gorbuscha*). Pp. 121–230 *in* C. Groot and L. Margolis, eds. Pacific salmon life histories, UBC Press, Vancouver, BC, Canada. **[2, 8]**

Heath, D. D., C. A. Bryden, J. M. Shrimpton, G. K. Iwama, J. Kelly, and J. W. Heath. 2002a. Relationships between heterozygosity, allelic distance (d^2), and reproductive traits in chinook salmon, *Oncorhynchus tshawytscha*. Can. J. Fish. Aquat. Sci. 59:77–84. **[2]**

Heath, D. D., C. Busach, J. Kelly, and D. Y. Atagi. 2002b. Temporal change in genetic structure and effective population size in steelhead trout (*Oncorhynchus mykiss*). Mol. Ecol. 11:197–214. **[2, 10, A2]**

Heath, D. D., L. Rankin, C. A. Bryden, J. W. Heath, and J. M. Shrimpton. 2002c. Heritability and Y-chromosome influence in the jack male life history of chinook salmon (*Oncorhynchus tshawytscha*). Heredity 89:311–317. **[5, 9, 13]**

Heath, D. D., R. H. Devlin, J. W. Heath, and G. K. Iwama. 1994. Genetic, environmental and interaction effects on the incidence of jacking in *Oncorhynchus tshawytscha* (chinook salmon). Heredity 72:146–154. **[5, 9, 13]**

Heath, D. D., R. H. Devlin, J. W. Heath, R. M. Sweeting, B. A. McKeown, and G. K. Iwama. 1996. Growth and hormonal changes associated with precocious sexual maturation in male chinook salmon (*Oncorhynchus tshawytscha* (Walbaum)). J. Exp. Mar. Biol. Ecol. 208:239–250. **[9]**

Heath, D. D., C. W. Fox, and J. W. Heath. 1999. Maternal effects on offspring size: variation through early development of chinook salmon. Evolution 53:1605–1611. **[4]**

Heath, D. D., J. W. Heath, C. A. Bryden, R. M. Johnson, and C. W. Fox. 2003. Rapid evolution of egg size in captive salmon. Science 299:1738–1740. **[1, 4, 7]**

Heath, D. D., S. Pollard, and C. Herbinger. 2001. Genetic structure and relationships among steelhead trout (*Oncorhynchus mykiss*) populations in British Columbia. Heredity 86:618–627. **[A2]**

Hébert, C., R. G. Danzmann, M. W. Jones, and L. Bernatchez. 2000. Hydrography and population genetic structure in brook charr (*Salvelinus fontinalis*, Mitchill) from eastern Canada. Mol. Ecol. 9:971–982. **[A2]**

Hebert, K. P., P. L. Goddard, W. W. Smoker, and A. J. Gharrett. 1998. Quantitative genetic variation and genotype by environment interaction of embryo development rate in pink salmon (*Oncorhynchus gorbuscha*). Can. J. Fish. Aquat. Sci. 55:2048–2057. **[I]**

Hedenström, A., and T. Alerstam. 1998. Optimum fuel loads in migratory birds: distinguishing between time and energy minimization. J. theor. Biol. 189:227–234. **[3]**

Hedgecock, D. 1994. Does variance in reproductive success limit effective population sizes of marine organisms? Pp. 122–134 *in* A. R. Beaumont, ed. Genetics and evolution of aquatic organisms. Chapman and Hall, London, U.K. **[10]**

Hedgecock, D., V. Chow, and R. S. Waples. 1992. Effective population numbers of shellfish broodstocks estimated from temporal variance in allele frequencies. Aquaculture 108:215–232. **[10]**

Hedrick, P. W. 1999. Highly variable loci and their interpretation in evolution and conservation. Evolution 53:313–318. [2]

Hedrick, P. W. 2000. Genetics of populations, 2nd ed. Jones and Bartlett Publishers, Sudbury, MA. [10]

Hedrick, P. W., and P. S. Miller. 1992. Conservation genetics: techniques and fundamentals. Ecol. Appl. 2:30–46. [11]

Hedrick, P. W., D. Hedgecock, and S. Hamelberg. 1995. Effective population size in winter-run chinook salmon. Cons. Biol. 9:615–624. [10]

Hedrick, P. W., D. Hedgecock, S. Hamelberg, and S. J. Croci. 2000a. The impact of supplementation in winter-run chinook salmon on effective population size. J. Heredity 91:112–116. [10]

Hedrick, P. W., V. K. Rashbrook, and D. Hedgecock. 2000b. Effective population size of winter-run chinook salmon based on microsatellite analysis of returning spawners. Can. J. Fish. Aquat. Sci 57:2368–2373. [I, 2, 10]

Heggberget, T. G. 1988. Timing of spawning in Norwegian Atlantic salmon (*Salmo salar*). Can. J. Fish Aquat. Sci. 45:845–849. [9]

Heggberget, T. G., L. P. Hansen, and T. F. Næsje. 1988. Within river spawning migration of Atlantic salmon (*Salmo salar*). Can. J. Fish. Aquat. Sci. 45:1691–1698. [9]

Heggberget, T. G., N. A. Hvidsten, T. B. Gunnerød, and P. I. Møkkelgjerd. 1991. Distribution of adult recaptures from hatchery-reared Atlantic salmon (*Salmo salar*) smolts released in and off-shore of the River Surna, western Norway. Aquaculture 98:89–96. [A1]

Heggenes, J., K. H. Røed, B. Høyheim, and L. Rosef. 2002. Microsatellite diversity assessment of brown trout (*Salmo trutta*) population structure indicate limited genetic impact of stocking in a Norwegian alpine lake. Ecol. Freshwater Fish 11:93–100. [A2]

Heino, M. 1998. Management of evolving fish stocks. Can. J. Fish. Aquat. Sci. 55:1971–1982. [11, 13]

Heisler, I. L., and J. Damuth. 1987. A method for analyzing selection in hierarchically structured populations. Am. Nat. 130:582–602. [13]

Helfield, J. M., and R. J. Naiman. 2001. Effects of salmon-derived nitrogen on riparian forest growth and implications for stream productivity. Ecology 82:2403–2409. [13]

Heming, T. A. 1982. Effects of temperature on utilization of yolk by chinook salmon (*Oncorhynchus tshawytscha*) eggs and alevins. Can. J. Fish. Aquat. Sci. 39:184–190. [4]

Henderson, R., J. L. Kershner, and C. A. Toline. 2000. Timing and location of spawning by nonnative wild rainbow trout and native cutthroat trout in the South Fork Snake River, Idaho, with implications for hybridization. N.A. J. Fish. Manage. 20:584–596. [8]

Hendry, A. P. 2001. Adaptive divergence and the evolution of reproductive isolation in the wild: an empirical demonstration using introduced sockeye salmon. Genetica 112/113:515–534. [I, 2, 7, 8]

Hendry, A. P., and O. K. Berg. 1999. Secondary sexual characters, energy use, senescence, and the cost of reproduction in sockeye salmon. Can. J. Zool. 77:1663–1675. [3, 7, 9]

Hendry, A. P., and T. Day. 2003. Revisiting the positive correlation between female size and egg size. Evol. Ecol. Res. 5:421–429 [1, 4]

Hendry, A. P., and M. T. Kinnison. 1999. The pace of modern life: measuring rates of contemporary microevolution. Evolution 53:1637–1653. [I, 4, 5, 7]

Hendry, A. P., O. K. Berg, and T. P. Quinn. 1999. Condition dependence and adaptation-by-time: breeding date, life history, and energy allocation within a population of salmon. Oikos 85:499–514. [9]

Hendry, A. P., O. K. Berg, and T. P. Quinn. 2001c. Breeding location choice in salmon: causes (habitat, competition, body size, energy stores) and consequences (life span, energy stores). Oikos 93:407–418. [9]

Hendry, A. P., T. Bohlin, B. Jonsson, and O. K. Berg. 2003b. To sea or not to sea? Anadromy vs. non-anadromy in salmonids. Pp. 92–126 *in* A. P. Hendry and S. C. Stearns, eds. Evolution illuminated: salmon and their relatives. Oxford Univ. Press, New York, NY—*this volume* [I, 1, 2, 4, 7, 8, 9, 12]

Hendry, A. P., V. Castric, M. T. Kinnison, and T. P. Quinn. 2003a. The evolution of philopatry and dispersal: homing vs. straying in salmonids. Pp. 52–91 *in* A. P. Hendry and S. C. Stearns, eds. Evolution illuminated: salmon and their relatives. Oxford Univ. Press, New York, NY—*this volume* [I, 1, 3, 4, 12]

Hendry, A. P., T. Day, and A. B. Cooper. 2001b. Optimal size and number of propagules: allowance for discrete stages and effects of maternal size on reproductive output and offspring fitness. Am. Nat. 157:387–407. [I, 1, 3, 4]

Hendry, A. P., T. Day, and E. B. Taylor. 2001a. Population mixing and the adaptive divergence of quantitative traits in discrete populations: a theoretical framework for empirical tests. Evolution 55:459–466. [2, 6]

Hendry, A. P., A. H. Dittman, and R. W. Hardy. 2000a. Proximate composition, reproductive development, and a test for trade-offs in captive sockeye salmon. Trans. Am. Fish. Soc. 129:1082–1095. [9]

Hendry, A. P., J. E. Hensleigh, and R. R. Reisenbichler. 1998. Incubation temperature, developmental biology, and the divergence of sockeye salmon (*Oncorhynchus nerka*) within Lake Washington. Can. J. Fish. Aquat. Sci. 55:1387–1394. [4, 5, 7, 13]

Hendry, A. P., F. E. Leonetti, and T. P. Quinn. 1995. Spatial and temporal isolating mechanisms: the formation of discrete breeding aggregations of sockeye salmon (*Oncorhynchus nerka*). Can. J. Zool. 73:339–352. [2, 3, 9]

Hendry, A. P., B. H. Letcher, and G. Gries. 2003. Estimating natural selection acting on stream-dwelling Atlantic salmon: implications for the restoration of extirpated populations. Cons. Biol. 17:795–805. [13]

Hendry, A. P., J. K. Wenburg, P. Bentzen, E. C. Volk, and T. P. Quinn. 2000b. Rapid evolution of reproductive isolation in the wild: evidence from introduced salmon. Science 290:516–518. [I, 2, 7, 9]

Hendry, M. A., J. K. Wenburg, K. W. Myers, and A. P. Hendry. 2002. Genetic and phenotypic variation through the migratory season provides evidence for multiple populations of wild steelhead in the Dean River, British Columbia. Trans. Am. Fish. Soc. 131:418–434. [A2]

Herbinger, C. M. 1987. A study of Atlantic salmon (*Salmo salar*) maturation using individually identified fish. Ph.D. dissertation, Dalhousie Univ., Halifax, NS, Canada. [5]

Hewitt, G. M. 1993. After the ice: *parallelus* meets *erythropus* in the Pyrenees. Pp. 140–164 *in* R. G. Harrison, ed. Hybrid zones and the evolutionary process. Oxford Univ. Press, Oxford, U.K. [8]

Hewitt, G. M. 2000. The genetic legacy of the Quaternary ice ages. Nature. 405:907–913. [8]

Hilborn, R. 1985. Apparent stock recruitment relationships in mixed stock fisheries. Can. J. Fish. Aquat. Sci. 42:718–723. [11, 13]

Hilborn, R., T. P. Quinn, D. E. Schindler, and D. E. Rogers. 2003. Biocomplexity and fisheries sustainability. Proc. Natl. Acad. Sci. USA. In press. [13]

Hill, W. G. 1979. A note on effective population size with overlapping generations. Genetics 92:317–322. [10]

Hill, W. G. 1981. Estimation of effective population size from data on linkage disequilibrium. Genetical Res. (Cambridge) 38:209–216. [10]

Hinch, S. G., and J. Bratty. 2000. Effects of swim speed and activity pattern on success of adult sockeye salmon migration through an area of difficult passage. Trans. Am. Fish. Soc. 129:598–606. [3]

Hinch, S. G., and P. S. Rand. 1998. Swim speeds and energy use of upriver-migrating sockeye salmon (*Oncorhynchus nerka*): role of local environment and fish characteristics. Can. J. Fish. Aquat. Sci. 55:1821–1831. [3]

Hinch, S. G., and P. S. Rand. 2000. Optimal swimming speeds and forward-assisted propulsion: energy-conserving behaviours of upriver-migrating adult salmon. Can. J. Fish. Aquat. Sci. 57:2470–2478. [3]

Hindar, K., B. Jonsson, N. Ryman, and G. Ståhl. 1991a. Genetic relationships among landlocked, resident, and anadromous brown trout, *Salmo trutta* L. Heredity 66:83–91. [3, A2, A3]

Hindar, K., N. Ryman, and F. Utter. 1991b. Genetic effects of cultured fish on natural fish populations. Can. J. Fish. Aquat. Sci. 48:945–957. [10]

Hipfner, M. J., and A. J. Gaston. 1999. The relationship between egg size and posthatching development in the thick-billed murre. Ecology 80:1289–1297. [4]

Hocutt, C. H., and E. O. Wiley, eds. 1986. The zoogeography of North American freshwater fishes. John Wiley and Sons, New York. [6, 7]

Hoekstra H. E., J. M. Hoekstra, D. Berrigan, S. N. Vignieri, A. Hoang, C. E. Hill, P. Beerli, and J. G. Kingsolver. 2001. Strength and tempo of directional selection in the wild. Proc. Natl. Acad. Sci. USA 98:9157–9160. [11]

Hoelzer, G. A., and D. J. Melnick. 1994. Patterns of speciation and limits to phylogenetic resolution. Trends Ecol. Evol. 9:104–107. [7]

Höglund, J., and R. V. Alatalo. 1995. Leks. Princeton Univ. Press, Princeton, NJ. [I]

Holčík, J., K. Hensel, J. Nieslanik, and L. Skácel. 1988. The Eurasian Huchen *Hucho hucho*: largest salmon of the world. Dr. W. Junk Publishers, Dordrecht, The Netherlands. [I]

Holling, C. S., and G. K. Meffe. 1996. Command and control and the pathology of natural resource management. Cons. Biol. 10:328–337. [13]

Holt, R. D. 1985. Population dynamics in two-patch environments: some anomalous consequences of an optimal habitat distribution. Theor. Pop. Biol. 28:181–208. [2]

Holt, R. D., and M. A. McPeek. 1996. Chaotic population dynamics favors the evolution of dispersal. Am. Nat. 148:709–718. [2]

Holtby, L. B., and M. C. Healey. 1986. Selection for adult size in female coho salmon (*Oncorhynchus kisutch*). Can. J. Fish. Aquat. Sci. 43:1946–1959. [3, 4, 13]

Holtby, L. B., and M. C. Healey. 1990. Sex-specific life history tactics and risk-taking in coho salmon. Ecology 71:678–690. [9, 13]

Holtby, L. B., B. C. Anderson, and R. K. Kadowaki. 1990. Importance of smolt size and early ocean growth to interannual variability in marine survival of coho salmon (*Oncorhynchus kisutch*). Can. J. Fish. Aquat. Sci. 47:2181–2194. [13]

Houde, A. E. 1997. Sex, color and mate choice in guppies. Princeton Univ. Press, Princeton, NJ. [9]

Houle, D. 1992. Comparing evolvability and variability of quantitative traits. Genetics 130:195–204. **[I, 4]**

Howard, D. J. 1986. A zone of overlap and hybridization between two ground cricket species. Evolution 40:34–43. **[8]**

Howard, D. J., and S. H. Berlocher, eds. 1998. Endless forms: species and speciation. Oxford Univ. Press, Oxford, U.K. **[8]**

Howard, W. E. 1960. Innate and environmental dispersal of individual vertebrates. Am. Midland Nat. 63:152–161. **[2]**

Hoysak, D. J. 2001. Factors influencing reproductive success of male sockeye salmon. Ph.D. dissertation, Univ. of British Columbia, Vancouver. **[9]**

Hoysak, D. J., and N. R. Liley. 2001. Fertilization dynamics in sockeye salmon and a comparison of sperm from alternative male phenotypes. J. Fish Biol. 58:1286–1300. **[9]**

Hubbs, C. L. 1955. Hybridization between fish species in nature. Syst. Zool. 4:1–20. **[8]**

Hughes, T. C., D. C. Josephson, C. C. Krueger, and P. J. Sullivan. 2000. Comparison of large and small visible implant tags: retention and readability in hatchery brook trout. N.A. J. Fish. Manage. 62:273–278. **[2]**

Hughes, T. R., and D. D. Shoemaker. 2001. DNA microarrays for expression profiling. Curr. Opin. Chem. Biol. 5:21–25. **[6]**

Hume, J. M. B., and E. A. Parkinson. 1987. Effect of stocking density on the survival, growth, and dispersal of steelhead trout fry (*Salmo gairdneri*). Can. J. Fish. Aquat. Sci. 44:271–281. **[13]**

Hunstman, A. G. 1939. Salmon for angling in the Margaree river. Fish. Res. Board Can. Bull. 57:1–75. **[1]**

Hurst, L. D., and J. R. Peck. 1996. Recent advances in understanding of the evolution and maintenance of sex. Trends Ecol. Evol. 11:46–52. **[I]**

Hurst, L. D., A. Atlan, and B. O. Bengtsson. 1996. Genetic conflicts. Quart. Rev. Biol. 71:317–364. **[I]**

Husband, B. C., and S. C. H. Barrett. 1992. Effective population size and genetic drift in tristylous *Eichhornia paniculata* (Pontederiaceae). Evolution 46:1875–1890. **[10]**

Hutchings, J. A. 1990. The evolutionary significance of life history divergence among brook trout, *Salvelinus fontinalis*, populations. Ph.D. dissertation, Memorial Univ. of Newfoundland, St. John's, NF, Canada. **[5]**

Hutchings, J. A. 1991. Fitness consequences of variation in egg size and food abundance in brook trout, *Salvelinus fontinalis*. Evolution 45:1162–1168. **[I, 4, 5]**

Hutchings, J. A. 1993a. Adaptive life histories effected by age-specific survival and growth rate. Ecology 74:673–684. **[I, 5, 9]**

Hutchings, J. A. 1993b. Reaction norms for reproductive traits in brook trout and their influence on life history evolution effected by size-selective harvesting. Pp. 107–125 *in* T. K. Stokes, J. M. McGlade, and R. Law, eds. The exploitation of evolving resources. Springer-Verlag, Berlin, Germany. **[5]**

Hutchings, J. A. 1994. Age- and size-specific costs of reproduction within populations of brook trout, *Salvelinus fontinalis*. Oikos 70:12–20. **[5]**

Hutchings, J. A. 1996. Adaptive phenotypic plasticity in brook trout, *Salvelinus fontinalis*, life histories. EcoScience 3:25–32. **[I, 5]**

Hutchings, J. A. 1997. Life history responses to environmental variability in early life. Pp. 139–168 *in* R. C. Chambers, and E. A. Trippel, eds. Early life history and recruitment in fish populations. Chapman and Hall, London, U.K. **[4, 5]**

Hutchings, J. A. 2000. Numerical assessment in the front seat, ecology and evolution in the back seat: time to change drivers in fisheries and aquatic sciences? Mar. Ecol. Prog. Ser. 208:299–303. [5]

Hutchings, J. A. 2002. Life histories of fish. Pp. 149–174 *in* P. J. B. Hart and J. D. Reynolds, eds. Handbook of fish and fisheries: Volume 1. Fish biology. Blackwell, Oxford, U.K. [5]

Hutchings, J. A. 2003. Norms of reaction and phenotypic plasticity in salmonid life histories. Pp. 154–174 *in* A. P. Hendry and S. C. Stearns, eds. Evolution illuminated: salmon and their relatives. Oxford Univ. Press, New York, NY—*this volume* [I, 1, 2, 3, 9, 13]

Hutchings, J. A., and L. Gerber. 2002. Sex-biased dispersal in a salmonid fish. Proc. Roy. Soc. Lond. B. 269:2487–2493.

Hutchings, J. A., and M. E. B. Jones. 1998. Life history variation and growth rate thresholds for maturity in Atlantic salmon, *Salmo salar*. Can. J. Fish. Aquat. Sci. 55(Suppl. 1):22–47. [I, 1, 3, 5]

Hutchings, J. A., and D. W. Morris. 1985. The influence of phylogeny, size and behaviour on patterns of covariation in salmonid life histories. Oikos 45:118–124. [4, 9]

Hutchings, J. A., and R. A. Myers. 1988. Mating success of alternative maturation phenotypes in male Atlantic salmon, *Salmo salar*. Oecologia 75:169–174. [I, 3, 5, 9]

Hutchings, J. A., and R. A. Myers. 1994. The evolution of alternative mating strategies in variable environments. Evol. Ecol. 8:256–268. [I, 3, 5, 9]

Hutchings, J. A., C. Walters, and R. L. Haedrich. 1997. Is scientific inquiry incompatible with government information control? Can. J. Fish. Aquat. Sci. 54:1198–1210. [13]

Huxley, J. 1942. Evolution, the modern synthesis. Allen and Unwin, London, U.K. [6]

Hvidsten, N. A., T. G. Heggberget, and L. P. Hansen. 1994. Homing and straying of hatchery-reared Atlantic salmon, *Salmo salar* L., released in three rivers in Norway. Aquacult. Fish. Manage. 25(Suppl. 2):9–16. [A1]

Iizuka, M. 2001. The effective size of fluctuating populations. Theor. Pop. Biol. 59:281–286. [10]

Ims, R. A., and D. Ø. Hjermann. 2001. Condition-dependent dispersal. Pp. 203–216 *in* J. Clobert, E. Danchin, A. A. Dhondt, and J. D. Nichols, eds. Dispersal. Oxford Univ. Press, Oxford, U.K. [2]

Ingvarsson, P. K., and M. C. Whitlock. 2000. Heterosis increases the effective migration rate. Proc. R. Soc. Lond. B 267:1321–1326. [2]

Intrilligator, M. D. 1971. Mathematical optimization and economic theory. Prentice-Hall, New York. [1]

Irwin, A. J., and P. D. Taylor. 2000. Evolution of dispersal in a stepping-stone population with overlapping generations. Theor. Pop. Biol. 58:321–328. [2]

Ishida, Y., S. Ito, M. Kaeriyama, S. McKinnell, and K. Nagasawa. 1993. Recent changes in age and size of chum salmon (*Oncorhynchus keta*) in the North Pacific Ocean and possible causes. Can. J. Fish. Aquat. Sci. 50:290–295. [13]

Itô, Y. 1980. Comparative ecology. Cambridge Univ. Press, Cambridge, U.K. [4]

Itô, Y., and Y. Iwasa. 1981. Evolution of litter size. I. Conceptual reexamination. Res. Pop. Ecol. 23:344–356. [4]

Iverson, J. B. 1991. Patterns of survivorship in turtles (Order Testudines). Can. J. Zool. 69:385–391. [4]

Iwamoto, R. N., B. A. Alexander, and W. K. Hershberger. 1984. Genotypic and environmental effects on the incidence of sexual precocity in coho salmon (*Oncorhynchus kisutch*). Aquaculture 43:105–121. [9, 13]

Jablonski, D. 2000. Micro- and macroevolution: scale and hierarchy in evolutionary biology and paleobiology. Paleobiology 26(Suppl.):15–52. [7]

Jamieson, I. 1995. Do female fish prefer to spawn in nests with eggs for reasons of mate choice copying or egg survival? Am. Nat. 145:824–832. [9]

Jánosi, I. M., and I. Scheuring. 1997. On the evolution of density dependent dispersal in a spatially structured population model. J. theor. Biol. 187:397–408. [2]

Järvi, T. 1990. The effects of male dominance, secondary sexual characteristics and female mate choice on the mating success of male Atlantic salmon *Salmo salar*. Ethology 84:123–132. [9]

Jenkins, T. M., Jr., S. Diehl, K. W. Kratz, and S. D. Cooper. 1999. Effects of population density on individual growth of brown trout in streams. Ecology 80:941–956. [2, 3]

Jennions, M. D., A. P. Møller, and M. Petrie. 2001. Sexually selected traits and adult survival: a meta-analysis. Quart. Rev. Biol. 76:3–36. [9]

Jensen, A. J., T. Forseth, and B. O. Johnsen. 2000. Latitudinal variation in growth of young brown trout *Salmo trutta*. J. Anim. Ecol. 69:1010–1020. [4]

Jensen, A. J., and 10 other authors. 1999. Cessation of the Norwegian drift net fishery: changes observed in Norwegian and Russian populations of Atlantic salmon. ICES J. Mar. Sci. 56:84–95. [11]

Jiggins, C. D., and J. Mallet. 2000. Bimodal hybrid zones and speciation. Trends Ecol. Evol. 15:250–255. [8]

Jin, L., and R. Chakraborty. 1995. Population structure, stepwise mutations, heterozygote deficiency and their implications in DNA forensics. Heredity 74:274–285. [2]

Johnson, M. L., and M. S. Gaines. 1990. Evolution of dispersal: theoretical models and empirical tests using birds and mammals. Annu. Rev. Ecol. Syst. 21:449–480. [2]

Johnson, O. W., W. S. Grant, R. G. Kope, K. Neely, F. W. Waknitz, and R. S. Waples. 1997. Status review of chum salmon from Washington, Oregon, and California. NOAA Tech. Memo. NMFS-NWFSC-32. [12]

Johnson, O. W., M. H. Ruckelshaus, W. S. Grant, F. W. Waknitz, A. M. Garrett, G. J. Bryant, K. Neely, and J. J. Hard. 1999. Status review of coastal cutthroat trout from Washington, Oregon and California. NOAA Tech. Memo. NMFS-NWFSC-37. [12]

Johnsson, J. I., F. Nöbbelin, and T. Bohlin. 1999. Territorial competition among wild brown trout fry: effects of ownership and body size. J. Fish Biol. 54:469–472. [4]

Johnstone, R. A., J. D. Reynolds, and J. C. Deutsch. 1996. Mutual mate choice and sex differences in choosiness. Evolution 50:1382–1391. [9]

Jones, J. W. 1959. The salmon. Collins, London, U.K. [1, 5, 9]

Jones, M. W., and J. A. Hutchings. 2001. The influence of male parr body size and mate competition on fertilization success and effective population size in Atlantic salmon. Heredity 86:675–684. [I, 5, 9, 10]

Jones, M. W., and J. A. Hutchings. 2002. Individual variation in Atlantic salmon fertilization success: implications for effective population size. Ecol. Appl. 12:184–193. [I, 5, 9, 10]

Jones, M. W., D. Clay, and R. G. Danzmann. 1996. Conservation genetics of brook trout (*Salvelinus fontinalis*): population structuring in Fundy National Park, New Brunswick, and eastern Canada. Can. J. Fish. Aquat. Sci. 53:2776–2791. [A2]

Jones, M. W., R. G. Danzmann, and D. Clay. 1997. Genetic relationships among populations of wild resident, and wild and hatchery anadromous brook charr. J. Fish Biol. 51:29–40. [3, A2]

Jonsson, B. 1982. Diadromous and resident trout *Salmo trutta*: is their difference due to genetics? Oikos 38:297–300. [3]

Jonsson, B. 1985. Life history patterns of freshwater resident and sea-run migrant brown trout in Norway. Trans. Am. Fish. Soc. 114:182–194. [I, A3]

Jonsson, B., and K. Hindar. 1982. Reproductive strategy of dwarf and normal Arctic charr (*Salvelinus alpinus*) from Vangsvatnet Lake, Western Norway. Can. J. Fish. Aquat. Sci. 39:1404–1413. [9]

Jonsson, B., and N. Jonsson. 1993. Partial migration: niche shift versus sexual maturation in fishes. Rev. Fish Biol. Fish. 3: 348–365. [I, 3, 9]

Jonsson, B., and N. Jonsson. 2001. Polymorphism and speciation in Arctic charr. J. Fish Biol. 58:605–638. [I, 6]

Jonsson, B., and J. H. L'Abée-Lund. 1993. Latitudinal clines in life-history variables of anadromous brown trout in Europe. J. Fish Biol. 43(Suppl. A): 1–16. [9]

Jonsson, B., T. Forseth, A. J. Jensen, and T. F. Næsje. 2001b. Thermal performance of juvenile Atlantic salmon, *Salmo salar* L. Func. Ecol. 15:701–711. [3, 5, 7]

Jonsson, B., N. Jonsson, E. Brodtkorb, and P.-J. Ingebrigtsen. 2001a. Life-history traits of brown trout vary with the size of small streams. Func. Ecol. 15:310–317. [3, 7]

Jonsson, B., N. Jonsson, and L. P. Hansen. 1990. Does juvenile experience affect migration and spawning of adult Atlantic salmon? Behav. Ecol. Sociobiol. 26:225–230. [9]

Jonsson, N., and B. Jonsson. 1997. Energy allocation in polymorphic brown trout. Func. Ecol. 11: 310–317. [3, 9, A3]

Jonsson, N., and B. Jonsson. 1999. Trade-off between egg mass and egg number in brown trout. J. Fish Biol. 55:767–783. [4]

Jonsson, N., and B. Jonsson. 2003. Energy allocation among developmental stages, age groups, and types of Atlantic salmon (*salmo salar*) spawners. Can. J. Fish. Aquat. Sci. 60:506–516. [9]

Jonsson, N., L. P. Hansen, and B. Jonsson. 1991a. Variation in age, size and repeat spawning of adult Atlantic salmon in relation to river discharge. J. Anim. Ecol. 60:937–947. [1, 3, 9, A3]

Jonsson, N., L. P. Hansen, and B. Jonsson. 1994a. Juvenile experience influences timing of adult river ascent in Atlantic salmon. Anim. Behav. 48:740–742. [A1]

Jonsson, N., B. Jonsson, and I. A. Fleming. 1996. Does early growth cause a phenotypically plastic response in egg production of Atlantic salmon? Func. Ecol. 10:89–96. [4, 5]

Jonsson, N., B. Jonsson, and L. P. Hansen. 1991b. Energetic cost of spawning in male and female Atlantic salmon (*Salmo salar* L.). J. Fish Biol. 39:739–744. [9]

Jonsson, N., B. Jonsson, and L. P. Hansen. 1997. Changes in proximate composition and estimates of energetic costs during upstream migration and spawning in Atlantic salmon, *Salmo salar*. J. Anim. Ecol. 66:425–436. [9]

Jonsson, N., B. Jonsson, J. Skurdal, and L. P. Hansen. 1994b. Differential response to water current in offspring of inlet- and outlet-spawning brown trout *Salmo trutta*. J. Fish Biol. 45:356–359. [3]

Jordan, W. C., and A. F. Youngson. 1992. The use of genetic marking to assess the reproductive success of mature male Atlantic salmon parr (*Salmo salar*, L.) under natural spawning conditions. J. Fish Biol. 41:613–618. [3, 5, 9]

Jordan, W. C., E. Verspoor, and A. F. Youngson. 1997. The effect of natural selection on estimates of genetic divergence among populations of the Atlantic salmon. J. Fish Biol. 51:546–560. [2, A2]

Jordan, W. C., A. F. Youngson, D. W. Hay, and A. Ferguson. 1992. Genetic protein variation in natural populations of Atlantic salmon (*Salmo salar*) in Scotland: temporal and spatial variation. Can. J. Fish. Aquat. Sci. 49:1863–1872. [A2]

Jorde, P. E., and N. Ryman. 1995. Temporal allele frequency change and estimation of effective size in populations with overlapping generations. Genetics 139:1077–1090. [10]

Jorde, P. E., and N. Ryman. 1996. Demographic genetics of brown trout (*Salmo trutta*) and estimation of effective population size from temporal change of allele frequency. Genetics 143:1369–1381. [10]

Kaev, A. M., and V. E. Kaeva. 1987. Variability in fecundity and egg size in chum salmon, *Oncorhynchus keta*, and pink salmon, *Oncorhynchus gorbuscha*, in relation to the size-age structure of spawning populations. J. Ichthyol. 27(1):76–86. [I]

Kaitala, A., V. Kaitala, and P. Lundberg. 1993. A theory of partial migration. Am. Nat. 142: 59–81. [3]

Kaitala, V. 1990. Evolutionary stable migration in salmon: a simulation study of homing and straying. Ann. Zool. Fennici 27:131–138. [2]

Kaitala, V., and W. M. Getz. 1995. Population dynamics and harvesting of semelparous species with phenotypic and genotypic variability in reproductive age. J. Math. Biol. 33:521–556. [11]

Kalinowski, S. T., and R. S. Waples. 2002. Relationship of effective to census size in fluctuating populations. Cons. Biol. 16:129–136. [I, 10]

Kanda, N., and F. W. Allendorf. 2001. Genetic population structure of bull trout from the Flathead River basin as shown by microsatellites and mitochondrial DNA markers. Trans. Am. Fish. Soc. 130:92–106. [A2]

Karlson, R. H., and H. M. Taylor. 1992. Mixed dispersal strategies and clonal spreading of risk: predictions from a branching process model. Theor. Pop. Biol. 42:218–233. [2]

Karlson, R. H., and H. M. Taylor. 1995. Alternative predictions for optimal dispersal in response to local catastrophic mortality. Theor. Pop. Biol. 47:321–330. [2]

Kazakov, R. V. 1981. Peculiarities of sperm production by anadromous and parr Atlantic salmon (*Salmo salar* L.) and fish cultural characteristics of such sperm. J. Fish Biol. 18:1–8. [9]

Keeley, E. R. 2001. Demographic responses to food and space competition by juvenile steelhead trout. Ecology 82:1247–1259. [3, 13]

Keenleyside, M. H. A., and H. M. C. Dupuis. 1988. Courtship and spawning competition in pink salmon (*Oncorhynchus gorbuscha*). Can. J. Zool. 66:262–265. [9]

Keightley, P. D., and A. Eyre-Walker. 2000. Deleterious mutations and the evolution of sex. Science 290:331–333. [I]

Keller, L. F., and D. M. Waller. 2002. Inbreeding effects in wild populations. Trends Ecol. Evol. 17:230–241. [2]

Kellogg, K. A. 1999. Salmon on the edge. Trends Ecol. Evol. 14:45–46. [13]

Kelso, B. W., T. G. Northcote, and C. F. Wehrhahn. 1981. Genetic and environmental aspects of the response to water current by rainbow trout (*Salmo gairdneri*) originating from inlet and outlet streams of two lakes. Can. J. Zool. 59:2177–2185. [3]

Kenaston, K. R., R. B. Lindsay, and R. K. Schroeder. 2001. Effect of acclimation on the homing and survival of hatchery winter steelhead. N.A. J. Fish. Manage. 21:765–773. [A1]

Kendeigh, S. C., T. C. Kramer, and F. Hamerstrom. 1956. Variations in egg characteristics of the House Wren. Auk 73:42–65. [4]

Kennedy B. P., J. D. Blum, C. L. Folt, and K. H. Nislow. 2000. Using natural strontium isotopic signatures as fish markers: methodology and application. Can. J. Fish. Aquat. Sci. 57:2280–2292. [2]

Kerfoot, W. C. 1974. Egg size cycle of a cladoceran. Ecology 55:1259–1270. [4]

Ketterson, E. D., and V. Nolan, Jr. 1976. Geographic variation and its climatic correlates in the sex ratio of eastern-wintering dark-eyed juncos (*Junco hyemalis hyemalis*). Ecology 57:679–693. [9]

Keyfitz, N. 1968. Introduction to the mathematics of population. Addison-Wesley. Reading, Mass. [1]

Kim, S.-C., and L. H. Rieseberg. 1999. Genetic architecture of species differences in annual sunflowers: Implications for adaptive trait introgression. Genetics 153:965–977. [6]

Kim, S.-C., and L. H. Rieseberg. 2001. The contribution of epistasis to species differences in annual sunflowers. Mol. Ecol. 10:683–690. [6]

Kimura, M., and T. Ohta. 1971. Theoretical aspects of population genetics. Princeton Univ. Press, Princeton, NJ. [10]

Kimura, M., and G. H. Weiss. 1964. The stepping stone model of population structure and the decrease of genetic correlation with distance. Genetics 49:561–576. [2]

Kincaid, H. L., W. R. Bridges, and B. von Limbach. 1977. Three generation of selection for growth rate in fall-spawning rainbow trout. Trans. Am. Fish. Soc. 106:621–628. [13]

King, T. L., S. T. Kalinowski, W. B. Schill, A. P. Spidle, and B. A. Lubinski. 2001. Population structure of Atlantic salmon (*Salmo salar* L.): a range-wide perspective from microsatellite DNA variation. Mol. Ecol. 10:807–821. [A2]

Kinghorn, B. P. 1983. A review of quantitative genetics in fish breeding. Aquaculture 31:283–304. [5]

Kingsland, S. 1985. Modeling nature. Univ. Chicago Press. Chicago, IL. [1]

Kingsolver, J. G., H. E. Hoekstra, J. M. Hoekstra, D. Berrigan, S. N. Vignieri, C. E. Hill, A. Hoang, P. Gibert, and P. Beerli. 2001. The strength of phenotypic selection in natural populations. Am. Nat. 157:245–261. [I, 11]

Kinnison, M. T., and A. P. Hendry. 2001. The pace of modern life II: from rates of contemporary microevolution to pattern and process. Genetica 112/113:145–164. [I, 4, 5, 6, 7, 8]

Kinnison, M. T., and A. P. Hendry. 2003. Tempo and mode of evolution in salmonids. Pp. 208–231 *in* A. P. Hendry and S. C. Stearns, eds. Evolution illuminated: salmon and their relatives. Oxford Univ. Press, New York, NY—*this volume* [I, 1, 3, 5, 8, 12]

Kinnsion, M. T., P. Bentzen, M. J. Unwin, and T. P. Quinn. 2002. Reconstructing recent divergence: evaluating nonequilibrium population structure in New Zealand chinook salmon. Mol. Ecol. 11:739–754. [2]

Kinnison, M., M. Unwin, N. Boustead, and T. Quinn. 1998b. Population-specific variation in body dimensions of adult chinook salmon (*Oncorhynchus tshawytscha*) from New Zealand and their source population, 90 years after introduction. Can. J. Fish. Aquat. Sci. 55:554–563. [12]

Kinnison, M. T., M. J. Unwin, A. P. Hendry, and T. P. Quinn. 2001. Migratory costs and the evolution of egg size and number in introduced and indigenous salmon populations. Evolution 55:1656–1667. [I, 2, 3, 4, 5, 7]

Kinnison, M. T., M. J. Unwin, W. K. Hershberger, and T. P. Quinn. 1998a. Egg size, fecundity, and development rate of two New Zealand chinook salmon (*Oncorhychus tshawytscha*) populations. Can. J. Fish. Aquat. Sci. 55:1946–1953. [4]

Kinnison, M. T., M. J. Unwin, and T. P. Quinn. 1998c. Growth and salinity tolerance of juvenile chinook salmon (*Oncorhynchus tshawytscha*) from two introduced New Zealand populations. Can. J. Zool. 76:2219–2226. [13]

Kinnison, M. T., M. J. Unwin, and T. P. Quinn. In review. Travel and sex on a tight budget: migratory costs and contemporary evolution of reproductive allocation in male salmon. J. Evol. Biol. [3, 7]

Kirkpatrick, M. 1993. The evolution of size and growth in harvested natural populations. Pp. 145–154 in T. K. Stokes, J. M. McGlade, and R. Law, eds. The exploitation of evolving resources, Springer-Verlag, Berlin, Germany. [13]

Kirkpatrick, M. 2000. Reinforcement and divergence under assortative mating. Proc. R. Soc. Lond. B 267:1649–1655. [6]

Kirkpatrick, M. 2001. Reinforcement during ecological speciation. Proc. R. Soc. Lond. Ser. B 268: 1259–1263. [6]

Kitano, S. 1996. Size-related factors causing individual variation in seasonal reproductive success of fluvial male Dolly Varden (*Salvelinus malma*). Ecol. Freshwater Fish 5:59–67. [9]

Kitano, S., K. Maekawa, S. Nakano, and K. D. Fausch. 1994. Spawning behaviour of bull trout in the Upper Flathead Drainage, Montana, with special reference to hybridization with brook trout. Trans. Am. Fish. Soc. 123:988–992. [8]

Kleckner, C. A., W. A. Hawley, W. E. Bradshaw, C. M. Holzapfel, and I. J. Fisher. 1995. Protandry in *Aedes sierrensis*: the significance of temporal variation in female fecundity. Ecology 76:1242–1250. [9]

Kline, T. C., Jr., J. J. Goering, O. A. Mathisen, P. H. Poe, P. L. Parker, and R. S. Scalan. 1993. Recycling of elements transported upstream by runs of Pacific salmon: II. δ^{15}N and δ^{13}C evidence in the Kvichak River watershed, Bristol Bay, southwestern Alaska. Can. J. Fish. Aquat. Sci. 50:1–16. [2]

Klomp, H. 1970. The determination of clutch-size in birds: a review. Ardea 58:1–124. [4]

Knutsen, H., J. A. Knutsen, and P. E. Jorde. 2001. Genetic evidence for mixed origin of recolonized sea trout populations. Heredity 87:207–214. [2]

Koelz, W. 1927. Coregonid fishes of the Great Lakes. Bull. U.S. Bur. Fish. 43:297–643. [6]

Koenig, W. D., D. Van Vuren, and P. N. Hooge. 1996. Detectability, philopatry, and the distribution of dispersal distances in vertebrates. Trends Ecol. Evol. 11:514–517. [2]

Kokko, H., and M. Jennions. 2003. It takes two to tango. Trends Ecol. Evol. 18:103–104 [9]

Kokko, H., and P. Lundberg. 2001. Dispersal, migration, and offspring retention in saturated habitats. Am. Nat. 157:188–202. [2]

Kölding, S., and T. M. Fenchel. 1981. Patterns of reproduction in different populations of five species of the amphipod genus *Gammarus*. Oikos 37:167–172. [4]

Koljonen, M.-L., H. Jansson, T. Paaver, O. Vasin, and J. Koskiniemi. 1999. Phylogeographic lineages and differentiation pattern of Altantic salmon (*Salmo salar*) in the Baltic Sea with management implications. Can. J. Fish. Aquat. Sci. 56:1766–1780. [A2]

Kondolf, G. M., M. J. Sale, and M. G. Wolman. 1993. Modification of fluvial gravel size by spawning salmonids. Water Resources Res. 29:2265–2274. [2]

Kondrashov, A. S., and F. A. Kondrashov. 1999. Interactions among quantitative traits in the course of sympatric speciation. Nature 400:351–354. [1]

Kondrashov, A. S., and M. V. Mina. 1986. Sympatric speciation, when is it possible? Biol. J. Linn. Soc. 27:201–223. [6]

Kondzela, C. M., C. M. Guthrie, S. L. Hawkins, C. D. Russell, J. H. Helle, and A. J. Gharrett. 1994. Genetic relationships among chum salmon populations in southeast Alaska and northern British Columbia. Can. J. Fish. Aquat. Sci. 51(Suppl. 1):50–64. [A2]

Koops, M. A., J. A. Hutchings, and B. K. Adams. 2003. Environmental predictability and the cost of imperfect information: influences on offspring size variability. Evol. Ecol. Res. 5:29–42. [1, 4]

Kornfield, I., and P. F. Smith. 2000. African cichlid fishes: model systems for evolutionary biology. Annu. Rev. Ecol. Syst. 31:163–196. [7]

Koseki, Y., I. Koizumi, H. Kobayashi, and K. Maekawa. 2002. Does the refuge availability influence the spawning behavior of mature male parr in salmonids? A test in the Miyabe charr. Environ. Biol. Fishes 64:87–93. [9]

Koskinen, M. T., T. O. Haugen, and C. R. Primmer. 2002b. Contemporary Fisherian life-history evolution in small salmonid populations. Nature 419:826–830. [1, 2]

Koskinen, M. T., J. Nilsson, A. Veselov, A. G. Potutkin, E. Ranta, and C. R. Primmer. 2002c. Microsatellite data resolve phylogeographic patterns in European grayling, *Thymallus thymallus*, Salmonidae. Heredity 88:391–401. [A2]

Koskinen, M. T., J. Piironen, and C. R. Primmer. 2001. Interpopulation genetic divergence in European grayling (*Thymallus thymallus*, Salmonidae) at a microgeographic scale: implications for conservation. Cons. Genet. 2:133–143. [A2]

Koskinen, M. T., P. Sundell, J. Piironen, and C. R. Primmer. 2002a. Genetic assessment of spatiotemporal evolutionary relationships and stocking effects in grayling (*Thymallus thymallus*, Salmonidae). Ecol. Lett. 5:193–205. [A2]

Kosswig, C. 1963. Ways of speciation in fishes. Copeia 1963:238–244. [8]

Kottelat, M. 1997. European freshwater fishes. An heuristic checklist of the freshwater fishes of Europe (exclusive of former USSR), with an introduction for non-systematists and comments on nomemclature and conservation. Biol. Sec. Zool. 52(Suppl. 5):1–271. [6]

Kristjánsson, L. T., and L. A. Vøllestad. 1996. Individual variation in progeny size and quality in rainbow trout, *Oncorhynchus mykiss* (Walbaum). Aquacult. Res. 27:335–343. [4]

Kristoferrsen, K. 1994. The influence of physical watercourse parameters on the degree of anadromy in different lake populations of Arctic charr (*Salvelinus alpinus* (L.)) in northern Norway. Ecol. Freshwater Fish 3:80–91. [3]

Kristoffersen, K., M. Halvorsen, and L. Jørgensen. 1994. Influence of parr growth, lake morphology, and freshwater parasites on the degree of anadromy in different populations of Arctic char (*Salvelinus alpinus*) in northern Norway. Can. J. Fish. Aquat. Sci. 51:1229–1246. [3]

Krogh, A. 1959. The comparative physiology of respiratory mechanisms. Univ. of Pennsylvania Press, Philadelphia, PA. [4]

Krogius, F. V. 1982. The role of resident fish in the reproduction of anadromous sockeye salmon, *Oncorhynchus nerka*. J. Ichthyol. 21(6):14–21. [3]

Kruuk, L. E. B., T. H. Clutton-Brock, J. Slate, J. M. Pemberton, S. Brotherstone, and F. E. Guinness. 2000. Heritability of fitness in a wild mammal population. Proc. Natl. Acad. Sci. USA 97:698–703. [11]

Kvarnemo, C., and I. Ahnesjö. 1996. The dynamics of operational sex ratios and competition for mates. Trends Ecol. Evol. 11:404–408. [9]

Kwain, W., and A. H. Lawrie. 1981. Pink salmon in the Great Lakes. Fisheries (Bethesda) 6(2):2–6. [2, 7]

L'Abée-Lund, J. H. 1991. Variation within and between rivers in adult size and sea age at maturity of anadromous brown trout, *Salmo trutta*. Can. J. Fish. Aquat. Sci. 48:1015–1021. [1, 3]

L'Abée-Lund, J. H., and L. A. Vøllestad. 1985. Homing precision of roach *Rutilus rutilus* in Lake Arungen, Norway. Environ. Biol. Fishes 13:235–239. [2]

L'Abée-Lund, J. H., A. J. Jensen, and B. O. Johnsen. 1990. Interpopulation variation in male parr maturation of anadromous brown trout (*Salmo trutta*) in Norway. Can. J. Zool. 68:1983–1987. [3, 9]

Labelle, M. 1992. Straying patterns of coho salmon (*Oncorhynchus kisutch*) stocks from southeast Vancouver Island, British Columbia. Can. J. Fish. Aquat. Sci. 49:1843–1855. [2, A1]

Lacasse, S., and P. Magnan. 1994. Distribution post-glaciaire des poissons dans le bassin hydrographique du fleuve Saint-Laurent: impact des interventions humaines, Université du Québec à Trois-Rivières, pour le ministère du Loisir, de la Chasse et de la Pêche du Québec. Rapp. Tech. [6]

Lack, D. 1947a. The significance of clutch size. I. Intraspecific variations. Ibis 89:302–352. [4]

Lack, D. 1947b. Darwin's finches. Cambridge Univ. Press, Cambridge, U.K. [6]

Lack, D. L. 1966. Population studies of birds. Oxford Clarendon Press, Oxford, U.K. [1]

Lahti, K., and N. Lower. 2000. Effects of size asymmetry on aggression and food acquisition in Arctic charr. J. Fish Biol. 56:915–922. [13]

Laikre, L., and N. Ryman. 1996. Effects on intraspecific biodiversity from harvesting and enhancing natural populations. Ambio 25:504–509. [11]

Laikre, L., T. Järvi, L. Johansson, S. Palm, J-F. Rubin, C. E. Glimsäter, P. Landergren, and N. Ryman. 2002. Spatial and temporal population structure of sea trout at the Island of Gotland, Sweden, delineated from mitochondrial DNA. J. Fish Biol. 60:49–71. [2, 10]

Laikre, L., P. E. Jorde, and N. Ryman. 1998. Temporal change of mitochondrial DNA haplotype frequencies and female effective size in a brown trout (*Salmo trutta*) population. Evolution 52:910–915. [10]

Lamoureux, J., and J. Sylvain. 1986. L'exploitation du Grand Corégone au lac Témiscouata: Bilan et recommandations. Ministère du Loisir, de la Chasse et de la Pêche, Québec, Canada. [6]

Lande, R. 1976. Natural selection and random genetic drift in phenotypic evolution. Evolution 30:314–334. [11]

Lande, R. 1979. Quantitative genetic analysis of multivariate evolution, applied to brain:body size allometry. Evolution 33:402–416. [6, 11]

Lande, R. 1980. The genetic covariance between characters maintained by pleiotropic mutations. Genetics 94:203–215. [6]

Lande, R. 1981. The minimum number of genes contributing to quantitative variation between and within populations. Genetics 99:541–553. [11]

Lande, R. 1987. Genetic correlations between the sexes in the evolution of sexual dimorphism and mating preferences. Pp. 83–94 *in* J. W. Bradbury and M. B. Andersson, eds. Sexual selection: testing the alternatives. John Wiley and Sons, New York. [9]

Lande, R. 1992. Neutral theory of quantitative genetic variance in an island model with local extinction and recolonization. Evolution 46:381–389. [6]

Lande, R. 1993. Risks of population extinction from demographic and environmental stochasticity and random catastrophes. Am. Nat. 142:911–927. [13]

Lande, R. 1994. Risk of population extinction from fixation of new deleterious mutations. Evolution 48:1460–1469. [I, 2]

Lande, R., and S. J. Arnold. 1983. The measurement of selection on correlated characters. Evolution 37:1210–1226. [11]

Landry, C., D. Garant, P. Duchesne, and L. Bernatchez. 2001. 'Good genes as heterozygosity': the major histocompatibility complex and mate choice in Atlantic salmon (*Salmo salar*). Proc. R. Soc. Lond. B 268:1279–1285. [I, 2, 9]

Landweber, L. F., and A. P. Dobson, eds. 1999. Genetics and the extinction of species. Princeton Univ. Press, Princeton, NJ. [13]

Lankford, T. E., Jr., J. M. Billerbeck, and D. O. Conover. 2001. Evolution of intrinsic growth and energy acquisition rates. II. Trade-offs with vulnerability to predation in *Menidia menidia*. Evolution 55:1873–1881. [3]

Lapointe, M., B. Eaton, S. Driscoll, and C. Latulippe. 2000. Modeling the probability of salmonid egg pocket scour due to floods. Can. J. Fish. Aquat. Sci. 57:1120–1130. [2]

Largiadèr, C. R., and A. Scholl. 1996. Genetic introgression between native and introduced brown trout *Salmo trutta* L. populations in the Rhône River basin. Mol. Ecol. 5:417–426. [8]

Larkin, P. A. 1978. Fisheries management—an essay for ecologists. Annu. Rev. Ecol. Syst. 9:57–73. [11]

Larkin, P. A. 1981. A perspective on population genetics and salmon management. Can. J. Fish. Aquat. Sci. 38:1469–1475. [11]

Latham, S. J., and E. B. Taylor. 2001. Hatchery-mediated reduction of genetic differentiation among bull trout populations: assessment of mechanisms and consequences using neutral molecular markers. Bull Trout II Conference Proceedings: 89–97. [A2]

Law, R. 1991. On the quantitative genetics of correlated characters under directional selection in age-structured populations. Phil. Trans. R. Soc. Lond. B 331:213–223. [11]

Law, R. 2000. Fishing, selection, and phenotypic evolution. ICES J. Mar. Sci. 57:659–668. [11]

Law, R., and D. R. Grey. 1989. Evolution of yields from populations with age-specific cropping. Evol. Ecol. 3:343–359. [11, 13]

Lawrence, M. J. 1984. The genetical analysis of ecological traits. Pp. 27–63 *in* B. Shorrocks, ed. Evolutionary ecology. Blackwell, Oxford, U.K. [11]

Le Cren, E. D. 1973. The population dynamics of young trout (*Salmo trutta*) in relation to density and territorial behaviour. Rapports et procès-verbaux des réunions (Conseil international pour l'exploration de la mer) 164:241–246. [3]

Leary, R. F., and F. W. Allendorf. 1997. Genetic confirmation of sympatric bull trout and Dolly Varden in western Washington. Trans. Am. Fish. Soc. 126:715–720. [A2]

Leary, R. F., F. W. Allendorf, and S. H. Forbes. 1993. Conservation genetics of bull trout in the Columbia and Kalamath River drainages. Cons. Biol. 7:856–865. [8, A2]

Leary, R. F., F. W. Allendorf, and K. L. Knudsen. 1984. Introgression between westslope cutthroat trout and rainbow trout in Clark Fork River drainage, Montana. Proc. Montana Acad. Sci. 43:1–18. [8]

Leary, R. F., F. W. Allendorf, S. R. Phelps, and K. L. Knudsen. 1987. Genetic divergence and identification of seven cutthroat trout subspecies and rainbow trout. Trans. Am. Fish. Soc. 116:580–587. [8]

Leary, R. F., F. W. Allendorf, and G. K. Sage. 1995. Hybridization and introgression between introduced and native fish. Am. Fish. Soc. Symp. 15:91–101. [8]

Leck, M. A., V. T. Parker, and R. L. Simpson, eds. 1989. Ecology of soil and seed banks. Academic Press, New York. [10]

Legendre, P., and V. Legendre. 1984. Postglacial dispersal of freshwater fishes in the Québec peninsula. Can. J. Fish. Aquat. Sci. 41:1781–1802. [6]

Leider, S. A. 1989. Increased straying by adult steelhead trout, *Salmo gairdneri*, following the 1980 eruption of Mount St. Helens. Environ. Biol. Fishes 24:219–229. [2]

Leider, S. A., M. W. Chilcote, and J. J. Loch. 1984. Spawning characteristics of sympatric populations of steelhead trout (*Salmo gairdneri*): evidence for partial reproductive isolation. Can. J. Fish. Aquat. Sci. 41:1454–1462. [I]

Leider, S. A., P. L. Hulett, J. J. Loch, and M. W. Chilcote. 1990. Electrophoretic comparison of the reproductive success of naturally spawning transplanted and wild steelhead trout through the returning adult stage. Aquaculture 88:239–252. [2]

Leitmann, G. 1966. An introduction to optimal control. McGraw-Hill, New York. [1]

Lemel, J.-Y., S. Belichon, J. Clobert, and M. E. Hochberg. 1997. The evolution of dispersal in a two-patch system: some consequences of differences between migrants and residents. Evol. Ecol. 11:613–629. [2]

Lenormand, T. 2002. Gene flow and the limits to natural selection. Trends Ecol. Evol. 17:183–189. [2]

Leon, J. A. 1976. Life histories as adaptive strategies. J. theor. Biol. 60:301–335. [1]

Leonardsson, K., and P. Lundberg. 1986. The choice of reproductive tactics as mixed evolutionarily stable strategy: the case of male Atlantic salmon (*Salmo salar* L.) Rep. Inst. Freshwater Res., Drottningholm 63:69–76. [5]

Leslie, P. H. 1945. On the use of matrices in certain population mathematics. Biometrika 33:183–212. [1]

Leslie, P. H. 1948. Some further notes on the use of matrices in certain population mathematics. Biometrika 35:213–245. [1]

Letcher, B. H., and T. L. King. 2001. Parentage and grandparentage assignment with known and unknown matings: application to Connecticut River Atlantic salmon restoration. Can. J. Fish. Aquat. Sci. 58:1812–1821. [2, 10]

Lever, C. 1996. Naturalized fishes of the world. Academic Press, New York. [I, 2, 7]

Levin, P. S., and M. H. Schiewe. 2001. Preserving salmon biodiversity. Am. Sci. 89:220–227. [11, 13]

Levin, S. A., D. Cohen, and A. Hastings. 1984. Dispersal strategies in patchy environments. Theor. Pop. Biol. 26:165–191. [2]

Levins, R. 1966. The strategy of model building in population biology. Am. Sci. 54:421–431. [1]

Levins, R. 1968. Evolution in changing environments. Princeton Univ. Press, Princeton, NJ. [1, 5]

Lewontin, R. C. 1991. Twenty-five years ago in *Genetics*. Electrophoresis in the development of evolutionary genetics: milestone or millstone? Genetics 128:657–662. [11]

Lindsey, C. C. 1981. Stocks are chameleons: plasticity in gill rakers of coregonid fishes. Can. J. Fish. Aquat. Sci. 38:1497–1506. [6]

Lindsey, C. C. 1988. The relevance of systematics and nomenclature to coregonid management. Finnish Fish. Res. 9:1–10. [6]

Lindsey, C. C., J. W. Clayton, and W. G. Franzin. 1970. Zoogeographic problems and protein variation in the *Coregonus clupeaformis* whitefish species complex. Pp. 127–146 *in* C. C. Lindsey and C. S. Woods, eds. Biology of coregonid fishes. Univ. of Manitoba Press, Winnipeg, Manitoba, Canada. [6]

Lindsey, C. C., T. G. Northcote, and G. F. Hartman. 1959. Homing of rainbow trout to inlet and outlet spawning streams at Loon Lake, British Columbia. J. Fish. Res. Board Can. 16:695–719. [2]

Liou, L. W., and T. D. Price. 1994. Speciation by reinforcement of premating isolation. Evolution 48:1451–1459. [8]

Lister, D. B., and H. S. Genoe. 1970. Stream habitat utilization by cohabitating underyearlings of chinook (*Oncorhynchus tshawytscha*) and coho (*O. kisutch*) salmon in the Big Qualicum River, British Columbia. J. Fish. Res. Board Can. 27:1215–1224. [13]

Lloyd, D. G. 1987. Selection of offspring size at independence and other size-versus-number strategies. Am. Nat. 129:800–817. [4]

Lobón-Cerviá, J., C. G. Utrilla, P. A. Rincón, and F. Amezcua. 1997. Environmentally induced spatio-temporal variations in the fecundity of brown trout *Salmo trutta* L.: trade-offs between egg size and number. Freshwater Biol. 38:277–288. [4]

Losos, J. B., T. R. Jackman, A. Larson, K. de Queiroz, and L. Rodriguez-Schettino. 1998. Contingency and determinism in replicated adaptive radiations of island lizards. Science 279:2115–2118. [7]

Loudenslager, E. J., and G. A. E. Gall. 1980. Geographic patterns of protein variation and subspeciation in the cutthroat trout, *Salmo clarki*. Syst. Zool. 29:27–42. [A2]

Lu, G., and L. Bernatchez. 1998. Experimental evidence for reduced hybrid viability between dwarf and normal ecotypes of lake whitefish (*Coregonus clupeaformis* Mitchill). Proc. R. Soc. Lond. Ser. B 265:1025–1030. [6, 8]

Lu, G., and L. Bernatchez. 1999. Correlated trophic specialization and genetic divergence in sympatric lake whitefish ecotypes (*Coregonus clupeaformis*): support for the ecological speciation hypothesis. Evolution 53:1491–1505. [I, 6, 8]

Lu, G., D. J. Basley, and L. Bernatchez. 2001. Contrasting patterns of mitochondrial DNA and microsatellite introgressive hybridization between lineages of lake whitefish (*Coregonus clupeaformis*); relevance for speciation. Mol. Ecol. 10:965–985. [6, 8, A2]

Ludwig, D., and C. J. Walters. 1981. Measurement errors and uncertainty in parameter estimates for stock and recruitment. Can. J. Fish. Aquat. Sci. 38:711–720. [13]

Ludwig, D., R. Hilborn, and C. Walters. 1993. Uncertainty, resource exploitation, and conservation: lessons from history. Science 260:17–36. [11]

Lundberg, P. 1988. The evolution of partial migration in birds. Trends Ecol. Evol. 3:172–175. [3]

Lynch, M. 1984. The selective value of alleles underlying polygenic traits. Genetics 108:1021–1033. [11]

Lynch, M. 1990. The rate of morphological evolution in mammals from the standpoint of the neutral expectation. Am. Nat. 136:727–741. [6, 7]

Lynch, M. 1996. A quantitative-genetic perspective on conservation issues. Pp. 471–501 *in* J. C. Avise and J. L. Hamrick, eds. Conservation genetics: case histories from nature. Chapman and Hall, New York. [11]

Lynch, M., and B. Walsh. 1998. Genetics and analysis of quantitative traits. Sinauer Assoc., Sunderland, MA. [6, 11]

Lynch, M., J. Conery, and R. Bürger. 1995. Mutation accumulation and the extinction of small populations. Am. Nat. 146:489–518. [I, 2]

Lynch, M., M. Pfrender, K. Spitze, N. Lehman, J. Hicks, D. Allen, L. Latta, M. Ottene, F. Bogue, and J. Colbourne. 1999. The quantitative and molecular genetic architecture of a subdivided species. Evolution 53:100–110. [6]

MacArthur, R. H. 1958. Population ecology of some warblers of northeastern coniferous forests. Ecology 39:599–619. [1]

MacArthur, R. H. 1962. Some generalized theorems of natural selection. Proc. Natl. Acad. Sci. USA 48:1893–1897. [1]

MacArthur, R. H. 1969. Species packing and what competition minimizes. Proc. Natl. Acad. Sci. USA 64:1369–1371. [1]

MacArthur, R. H. 1970. Species packing and competitive equilibrium for many species. Theor. Pop. Biol. 1:1–11. [1]

MacArthur, R. H. 1972. Geographical ecology. Harper and Row, New York. [1]

MacArthur, R. H., and R. Levins. 1967. The limiting similarity, convergence, and divergence of coexisting species. Am. Nat. 101:377–385. [1]

MacArthur, R. H., and E. O. Wilson. 1967. Theory of island biogeography. Princeton Univ. Press. Princeton, NJ. [1]

MacCrimmon, H. R., and B. L. Gots. 1979. World distribution of Atlantic salmon, *Salmo salar*. J. Fish Res. Board Can. 36:422–457. [I]

Maekawa, K., and T. Hino. 1990. Spawning tactics of female Miyabe charr (*Salvelinus malma miyabei*) against egg cannibalism. Can. J. Zool. 68:889–894. [9]

Maekawa, K., and H. Onozato. 1986. Reproductive tactics and fertilization success of mature male Miyabe charr, *Salvelinus malma miyabei*. Environ. Biol. Fishes 15:119–129. [5, 9]

Maekawa, K., T. Hino, S. Nakano, and W. W. Smoker. 1993. Mate preference in anadromous and landlocked Dolly Varden (*Salvelinus malma*) females in two Alaskan streams. Can. J. Fish. Aquat. Sci. 50:2375–2379. [3, 8, A3]

Maekawa, K., N. Shigeru, and S. Yamamoto. 1994. Spawning behaviour and size-assortative mating of Japanese charr in an artificial lake-inlet stream system. Environ. Biol. Fishes 39:109–117. [8, 9]

Magee, W. T. 1965. Estimating response to selection. J. Anim. Sci. 24:242–247. [11]

Mahnken, C. V. W., G. T. Ruggerone, F. W. Waknitz, and T. A. Flagg. 1998. A historical perspective on salmonid production from Pacific Rim hatcheries. N. Pacific Anadromous Fish. Comm. Bull. 1:38–53. [10]

Mangel, M. 1996. Life history invariants, age at maturity and the ferox trout. Evol. Ecol. 10:249–263. [I]

Manly, B. F. J. 1985. The statistics of natural selection: population and community biology. Chapman and Hall, London, U.K. [11]

Marschall, E. A., T. P. Quinn, D. A. Roff, J. A. Hutchings, N. B. Metcalfe, T. A. Bakke, R. L. Saunders, and N. L. Poff. 1998. A framework for understanding Atlantic salmon (*Salmo salar*) life history. Can. J. Fish. Aquat. Sci. 55(Suppl. 1):48–58. [5]

Martin, K. 1995. Patterns and mechanisms for age-dependent reproduction and survival in birds. Am. Zool. 35:340–348. [2]

Martinez, J. L., P. Moran, J. Perez, B. de Gaudemar, E. Beall, and E. Garcia-Vazquez. 2000. Multiple paternity increases effective size of southern Atlantic salmon populations. Mol. Ecol. 9:293–298. [3, 9]

Martínez-Garmendia, J. 1998. Simulation analysis of evolutionary response of fish populations to size-selective harvesting with the use of an individual-based model. Ecol. Model. 111:37–60. [11]

Martinson, D. G., N. G. Pisias, J. D. Hays, J. Imbrie, T. C. Moore, Jr., and N. J. Shackleton. 1987. Age, dating and orbital theory of the Ice Ages: development of a high resolution 0–300,000 year chronostratigraphy. Quat. Res. 27:1–29. [6]

Mathias, A., E. Kisdi, and I. Olivieri. 2001. Divergent evolution of dispersal in a heterogeneous landscape. Evolution 55:246–259. [2]

Matthews, G. M., and R. S. Waples. 1991. Status review for Snake River spring and summer chinook salmon. NOAA Tech. Memo. NMFS-F/NWC-200. [12]

Mauricio, R. 2001. Mapping quantitative trait loci in plants: uses and caveats for evolutionary biology. Nature Rev. Genet. 2:370–381. [6]

May, B., F. M. Utter, and F. W. Allendorf. 1975. Biochemical genetic-variation in pink and chum salmon—Inheritance of intraspecies variation and apparent absence of interspecies introgression following massive hybridization of hatchery stocks. J. Heredity 66:227–232. [8]

May, R. M. 1976. Simple mathematical models with very complicated dynamics. Nature 261:459–467. [1]

May, R. M., and R. H. MacArthur. 1972. Niche overlap as a function of environmental variability. Proc. Natl. Acad. Sci. USA 69:1109–1113. [1]

Mayden, R. L., and R. M. Wood. 1995. Systematics, species concepts, and the evolutionarily significant unit in biodiversity and conservation biology. Am. Fish. Soc. Symp. 17:58–113. [12]

Maynard Smith, J. 1982. Evolution and the theory of games. Cambridge Univ. Press, Cambridge, U.K. [4]

Mayo, O. 1980. The theory of plant breeding. Clarendon Press, Oxford, U.K. [11]

Mayr, E. 1942. Systematics and the origins of species. Columbia Univ. Press, New York. [6]

Mayr, E. 1963. Animal species and evolution. Belknap Press, Cambridge, MA. [I, 2, 8]

Mayr, E. 1982. The growth of biological thought. Belknap Press, Cambridge, MA. [8]

Mazer, S. J., and J. Damuth. 2001. Evolutionary significance of variation. Pp. 16–28 in C.W. Fox, D. A. Roff, and D. J. Fairbairn, eds. Evolutionary ecology: concepts and case studies. Oxford Univ. Press, Oxford, U.K. [5]

McAllister, M. K., R. M. Peterman, and D. M. Gillis. 1992. Statistical evaluation of a large-scale fishing experiment designed to test for a genetic effect of size-selective fishing on British Columbia pink salmon (Oncorhynchus gorbuscha). Can. J. Fish. Aquat. Sci. 49:1294–1304. [11]

McConnell, S. K. J., D. E. Ruzzante, P. T. O'Reilly, L. Hamilton, and J. M. Wright. 1997. Microsatellite loci reveal highly significant genetic differentiation among Atlantic salmon (Salmo salar L.) stocks from the east coast of Canada. Mol. Ecol. 6:1075–1089. [A2]

McCormick, S. D., and R. L. Saunders. 1987. Preparatory physiological adaptations for marine life of salmonids: osmoregulation, growth, and metabolism. Am. Fish. Soc. Symp. 1:211–229. [4]

McCusker, M. R., E. Parkinson, and E. B. Taylor. 2000. Mitochondrial DNA variation in rainbow trout (Oncorhynchus mykiss) across its native range: testing biogeographical hypotheses and their relevance to conservation. Mol. Ecol. 9:2089–2108. [2]

McDowall, R. M. 1988. Diadromy in fishes. Croom Helm, London, U.K. [3, 7]

McDowall, R. M. 1994. The origins of New Zealand's chinook salmon, *Oncorhynchus tshawytscha*. Mar. Fish. Rev. 56:1–7. [4, 7]

McDowall, R. M. 2001. Anadromy and homing: two life-history traits with adaptive synergies in salmonid fishes? Fish Fish. 2:78–85. [3, 12]

McDowall, R. M. 2002. The origin of the salmonid fishes: marine, freshwater ... or neither? Rev. Fish Biol. Fish. 11:171–179. [3, 7]

McElhany, P., M. H. Ruckleshaus, M. J. Ford, T. C. Wainwright, and E. P. Bjorkstedt. 2000. Viable salmonid populations and the recovery of evolutionary significant units. U.S. Department of Commerce, NOAA Tech. Memo. NMFS-NWFSC-42. [12, 13]

McElligott, E. A., and T. F. Cross. 1991. Protein variation in wild Atlantic salmon, with particular reference to southern Ireland. J. Fish Biol. 39(Suppl. A):35–42. [A2]

McElroy, D. M., and I. Kornfield. 1993. Novel jaw morphology in hybrids between *Pseudotropheus zebra* and *Labeotropheus fuelleborni* (Teleostei: Cichlidae) from Lake Malawi, Africa. Copeia 4:933–945. [8]

McGinley, M. A. 1989. The influence of a positive correlation between clutch size and offspring fitness on the optimal offspring size. Evol. Ecol. 3:150–156. [1, 4]

McGinley, M. A., D. H. Temme, and M. A. Geber. 1987. Parental investment in offspring in variable environments: theoretical and empirical considerations. Am. Nat. 130:370–398. [4]

McGowan, C., and W. S. Davidson. 1992a. Unidirectional natural hybridization between brown trout (*Salmo trutta*) and Atlantic salmon (*S. salar*) in Newfoundland. Can. J. Fish. Aquat. Sci. 49:1953–1958. [8]

McGowan, C., and W. S. Davidson. 1992b. Artificial hybridization of Newfoundland brown trout and Atlantic salmon: hatchability survival and growth to first feeding. Aquaculture 106:117–125. [8]

McGuigan, K. L. 2001. Evolutionary genetics of rainbowfish: phylogeny, adaptation and constraint. PhD. dissertation, Univ. of Queensland, St. Lucia. [6]

McGurk, M. D. 2000. Comparison of fecundity-length-latitude relationships between nonanadromous (kokanee) and anadromous sockeye salmon (*Oncorhynchus nerka*). Can. J. Zool. 78:1791–1805. [3]

McIsaac, D. O., and T. P. Quinn. 1988. Evidence for a hereditary component in homing behavior of chinook salmon (*Oncorhynchus tshawytscha*). Can. J. Fish. Aquat. Sci. 45:2201–2205. [2, 7]

McKay, S. J., R. H. Devlin, and M. J. Smith. 1996. Phylogeny of Pacific salmon and trout based on growth hormone type-2 and mitochondrial NADH dehydrogenase subunit 3 DNA sequences. Can. J. Aquat. Sci. 53:1165–1176. [1]

McKinnell, S. 1995. Age-specific effects of sockeye abundance on adult body size of selected British Columbia sockeye stocks. Can. J. Fish. Aquat. Sci. 52:1050–1063. [13]

McKinnell, S., A. J. Thomson, E. A. Black, B. L. Wing, C. M. Guthrie, III, J. F. Koerner, and J. H. Helle. 1997. Atlantic salmon in the North Pacific. Aquacult. Res. 28:145–157. [I]

McLeod, C. L., and J. P. O'Neil. 1983. Major range extensions of anadromous salmonids and first record of chinook salmon in the Mackenzie River drainage. Can. J. Zool. 61:2183–2184. [2]

McMichael, G. A., and T. N. Pearsons. 1998. Effects of wild juvenile spring chinook salmon on growth and abundance of wild rainbow trout. Trans. Am. Fish. Soc. 127:261–274. [13]

McNeil, W. J. 1969. Survival of pink salmon eggs and alevins. Pp. 101–119 *in* T. G. Northcote, ed. Symposium on Salmon and Trout in Streams. H. R. MacMillan Lectures in Fisheries, Univ. of British Columbia, Vancouver, BC, Canada. [9]

McNeil, W. J., ed. 1988. Salmon production, management, and allocation. Oregon State Univ. Press, Corvallis, OR. [13]

McPeek, M. A., and R. D. Holt. 1992. The evolution of dispersal in spatially and temporally varying environments. Am. Nat. 140:1010–1027. [2]

McPhail, J. D. 1961. A systematic study of the *Salvelinus alpinus* complex in N.A. J. Fish. Res. Board Can. 18:793–816. [8]

McPhail, J. D., and J. S. Baxter. 1996. A review of bull trout (*Salvelinus confluentus*) life-history and habitat use in relation to compensation and improvement opportunities. BC Ministry of Environment, Lands and Parks, Fisheries Branch, Fish. Manage. Rep. 104:1–31. [8]

McPhail, J. D., and C. C. Lindsey. 1970. Freshwater Fishes of Northwestern Canada and Alaska. Bull. Fish. Res. Board Can. 173. [6]

McPhail, J. D., and C. C. Lindsey. 1986. Zoogeography of the freshwater fishes of Cascadia (the Columbia River and rivers north to the Stikine). Pp. 615–637 *in* C. H. Hocutt and E. O. Wiley, eds. The zoogeography of North American freshwater fishes. John Wiley and Sons, New York. [8]

McVeigh, H. P., and W. S. Davidson. 1991. A salmonid phylogeny inferred from mitochondrial cytochrome b gene sequences. J. Fish Biol. 39(Suppl. A):277–282. [7]

Merilä, J. 1997. Quantitative trait and allozyme divergence in the greenfinch (*Carduelis chloris*, Aves: Fringillidae). Biol. J. Lin. Soc. 61:243–266. [6]

Merilä, J., B. C. Sheldon, and L. E. B. Kruuk. 2001. Explaining stasis: microevolutionary studies in natural populations. Genetica 112/113:199–222. [I, 11]

Mertz, G., and R. A. Myers. 1996. Influence of fecundity on recruitment variability of marine fish. Can. J. Fish. Aquat. Sci. 53:1618–1625. [I]

Metcalfe, N. B. 1998. The interaction between behaviour and physiology in determining life history patterns in Atlantic salmon (*Salmo salar*). Can. J. Fish. Aquat. Sci. 55(Suppl. 1):93–103. [5, 9]

Metcalfe, N. B., and J. E. Thorpe. 1990. Determinants of geographical variation in the age of seaward-migrating salmon, *Salmo salar*. J. Anim. Ecol. 59:135–145. [5]

Michael, J. H., Jr. 1989. Life history of anadromous coastal cutthroat trout in Snow and Salmon creeks, Jefferson County, Washington, with implications for management. Calif. Fish Game 75:188–203. [A1]

Miles, D. B., B. Sinervo, and W. A. Frankino. 2000. Reproductive burden, locomotor performance, and the cost of reproduction in free ranging lizards. Evolution 54:1386–1395. [4]

Milkman, R. 1982. Toward a unified selection theory. Pp. 105–118 *in* R. Milkman, ed. Perspectives on evolution. Sinauer Assoc., Sunderland, MA. [11]

Miller, R. J., and E. L. Brannon. 1981. The origin and development of life history patterns in Pacific salmonids. Proceedings of the Salmon and Trout Migratory Behavior Symposium, School of Fisheries, Univ. of Washington, Seattle, WA: 296–309. [12]

Miller, R. M., J. D. Williams, and J. E. Williams. 1989. Extinction of North American fishes during the past century. Fisheries (Bethesda) 14(6):22–38. [I]

Mills, D. 1994. Evidence of straying from wild Atlantic salmon, *Salmo salar* L., smolt transportation experiments in northern Scotland. Aquacult. Fish. Manage. 25(Suppl. 2):3–8. [A1]

Mills, L. S., and F. W. Allendorf. 1996. The one-migrant-per-generation rule in conservation and management. Cons. Biol. 10:1509–1518. [2]

Milner, A. M., and R. G. Bailey. 1989. Salmonid colonization of new streams in Glacier Bay National Park, Alaska. Aquacult. Fish. Manage. 20:179–192. [2, 3]

Milner, A. M., E. E. Knudsen, C. Soiseth, A. L. Robertson, D. Schell, I. T. Phillips, and K. Magnusson. 2000. Colonization and development of stream communities across a 200-year gradient in Glacier Bay National Park, Alaska, U.S.A. Can. J. Fish. Aquat. Sci. 57:2319–2335. [2, 3]

Mina, M. V. 1992. An interpretation of diversity in the salmonid genus *Brachymystax* (Pisces: Salmonidae) as the possible result of multiple hybrid speciation. J. Ichthyol. 32 (6):117–122. [8]

Mitton, J. B., and M. C. Grant. 1984. Associations among protein heterozygosity, growth rate, and developmental homeostasis. Annu. Rev. Ecol. Syst. 15:479–499. [11]

Mjølnerød, I. B., I. A. Fleming, U. H. Refseth, and K. Hindar. 1998. Mate and sperm competition during multiple-male spawnings of Atlantic salmon. Can. J. Zool. 76:70–75. [9]

Mjølnerød, I. B., U. H. Refseth, and K. Hindar. 1999. Spatial association of genetically similar Atlantic salmon juveniles and sex bias in spatial patterns in a river. J. Fish Biol. 55:1–8. [2]

Mock, D. W., and G. A. Parker. 1997. The evolution of sibling rivalry. Oxford Univ. Press, Oxford, U.K. [I]

Moegenburg, S. M. 1996. Sabal palmetto seed size: causes of variation, choices of predators, and consequences for seedlings. Oecologia 106:539–543. [4]

Montgomery, D. R. 2000. Coevolution of the Pacific salmon and the Pacific Rim topography. Geology 28:1107–1110. [8]

Montgomery, D. R., J. M. Buffington, N. P. Peterson, D. Schuett-Hames, and T. P. Quinn. 1996. Stream-bed scour, egg burial depths, and the influence of salmonid spawning on bed surface mobility and embryo survival. Can. J. Fish. Aquat. Sci. 53:1061–1070. [2]

Moore, K. 1996. The adaptive significance of body size and shape in sexually mature sockeye salmon (*Oncorhynchus nerka*). M.Sc. thesis, Univ. of Washington, Seattle, WA. [3]

Moore, W. S. 1977. Evaluation of narrow hybrid zones in vertebrates. Quart. Rev. Biol. 52:263–277. [8]

Morán, P., A. M. Pendás, E. Beall, and E. García-Vázquez. 1996. Genetic assessment of the reproductive success of Atlantic salmon precocious parr by means of VNTR loci. Heredity 77:655–660. [3, 9]

Morán, P., A. M. Pendás, E. García-Vázquez, J. I. Izquierdo, and J. Lobón-Cerviá. 1995. Estimates of gene flow among neighbouring populations of brown trout. J. Fish Biol. 46:593–602. [A2]

Morbey, Y. 2000. Protandry in Pacific salmon. Can. J. Fish. Aquat. Sci. 57:1252–1257. [I, 9]

Morbey, Y. E. 2002. The mate-guarding behaviour of male kokanee *Oncorhynchus nerka*. Behaviour 139:507–528. [I, 2, 9]

Morbey, Y. E. 2003. Pair formation, pre-spawning waiting, and protandry in kokanee, *Oncorhynchus nerka*. Behav. Ecol. Sociobiol. 54:127–135 [9]

Morbey, Y., and R. C. Ydenberg. 2001. Protandrous arrival timing to breeding areas: a review. Ecol. Lett. 4:663–673. [9]

Morbey, Y. E., and R. C. Ydenberg. 2003. Timing games in the reproductive phenology of female Pacific salmon (*Oncorhynchus* spp.). Am. Nat. 161:284–298. [9]

Morinville, G. R., and J. B. Rasmussen. 2003. Early juvenile bioenergetic differences between anadromous and resident brook trout (*Salvelinus fontinalis*). Can. J. Fish. Aquat. Sci. 60:401–410. [3].

Morita, K., and Y. Takashima. 1998. Effect of female size on fecundity and egg size in white-spotted charr: comparison between sea-run and resident forms. J. Fish Biol. 53:1140–1142. [A3]

Morita, K., S. Yamamoto, and N. Hoshino. 2000. Extreme life history change of white-spotted char (*Salvelinus leucomaenis*) after damming. Can. J. Fish. Aquat. Sci. 57:1300–1306. [3]

Morita, K., S. Yamamoto, Y. Takashima, T. Matsuishi, Y. Kanno, and K. Nishimura. 1999. Effect of maternal growth history on egg number and size in wild white-spotted char (*Salvelinus leucomaenis*). Can. J. Fish. Aquat. Sci. 56:1585–1589. [4, 5]

Moritz, C. 1994. Defining 'evolutionarily significant units' for conservation. Trends Ecol. Evol. 9:373–375. [12]

Morris, D. W. 1987. Optimal allocation of parental investment. Oikos 49:332–339. [4]

Morris, D. W. 1991. On the evolutionary stability of dispersal to sink habitats. Am. Nat. 137:907–911. [2]

Mortensen, D. G., A. C. Wertheimer, J. M. Maselko, and S. G. Taylor. 2002. Survival and straying of Auke Creek, Alaska, pink salmon marked with coded wire tags and thermally induced otolith marks. Trans. Am. Fish. Soc. 131:14–26. [2, A1]

Motro, U. 1982a. Optimal rates of dispersal. I. Haploid populations. Theor. Pop. Biol. 21:394–411. [2]

Motro, U. 1982b. Optimal rates of dispersal. II. Diploid populations. Theor. Pop. Biol. 21:412–429. [2]

Motro, U. 1983. Optimal rates of dispersal. III. Parent-offspring conflict. Theor. Pop. Biol. 23:159–168. [2]

Motro, U. 1991. Avoiding inbreeding and sibling competition: the evolution of sexual dimorphism for dispersal. Am. Nat. 137:108–115. [2]

Mousseau, T. A., and C. W. Fox. 1998a. Maternal effects as adaptations. Oxford Univ. Press, Oxford, U.K. [4]

Mousseau, T. A., and C. W. Fox. 1998b. The adaptive significance of maternal effects. Trends Ecol. Evol. 13:403–407. [4]

Mousseau, T. A., and D. A. Roff. 1987. Natural selection and the heritability of fitness components. Heredity 59:181–197. [I, 4, 11]

Mundy, P. R., T. W. H. Backman, and J. M. Berkson. 1995. Selection of conservation units for Pacific salmon: Lessons from the Columbia River. Am. Fish. Soc. Symp. 17:28–38. [12]

Murata, S., N. Takasaki, M. Saitoh, and N. Okada. 1993. Determination of the phylogenetic relationships among Pacific salmonids by using short interspersed elements (SINEs) as temporal landmarks of evolution. Proc. Natl. Acad. Sci. USA 90:6995–6999. [7]

Mustafa, S., ed. 1999. Genetics in sustainable fisheries management. Blackwell Science, Oxford, U.K. [13]

Myers, J. M., P. O. Heggelund, G. Hudson, and R. N. Iwamoto. 2001. Genetics and broodstock management of coho salmon. Aquaculture 197:43–62. [2]

Myers, J. M., and 10 others. 1998. Status review of chinook salmon from Washington, Idaho, Oregon and California. NOAA Tech. Memo. NMFS-NWFSC-35. [7, 12]

Myers, R. A. 1984. Demographic consequences of precocious maturation of Atlantic salmon (Salmo salar). Can. J. Fish. Aquat. Sci. 41:1349–1353. [5]

Myers, R. A. 1986. Game theory and the evolution of Atlantic salmon (*Salmo salar*) age at maturity. Can. Spec. Publ. Fish. Aquat. Sci. 89:53–61. **[3, 5, 9, 13]**

Myers, R. A., and J. A. Hutchings. 1986. Selection against parr maturation in Atlantic salmon. Aquaculture 53:313–320. **[5]**

Myers, R. A., and J. A. Hutchings. 1987a. A spurious correlation in an interpopulation comparison of Atlantic salmon life histories. Ecology. 68:1839–1843. **[I, 1]**

Myers, R. A., and J. A. Hutchings. 1987b. Mating of anadromous Atlantic salmon, *Salmo salar* L., with mature male parr. J. Fish Biol. 31:143–146. **[5]**

Myers, R. A., N. J. Barrowman, J. A. Hutchings, and A. A. Rosenberg. 1995. Population dynamics of exploited fish stocks at low population levels. Science 269:1106–1108. **[13]**

Myers, R. A., K. G. Bowen, and N. J. Barrowman. 1999. Maximum reproductive rate of fish at low population sizes. Can. J. Fish. Aquat. Sci. 56:2404–2419. **[13]**

Myers, R. A., J. A. Hutchings, and R. J. Gibson. 1986. Variation in male parr maturation within and among populations of Atlantic salmon, *Salmo salar*. Can. J. Fish. Aquat. Sci. 43:1242–1248. **[3, 5, 9, 13]**

Myers, R. A., A. A. Rosenberg, P. M. Mace, N. Barrowman, and V. R. Restrepo. 1994. In search of thresholds for recruitment overfishing. ICES J. Mar. Sci. 51:191–205. **[13]**

Nagata, M., and J. R. Irvine. 1997. Differential dispersal patterns of male and female masu salmon fry. J. Fish Biol. 51:601–606. **[2]**

National Marine Fisheries Service. All status reviews can be accessed on line at: http://research.nwfsc.noaa.gov/pubs/nwfscpubs.html. **[13]**

Neave, F. 1944. Racial characteristics and migratory habits in *Salmo gairdneri*. J. Fish. Res. Board Can. 6:245–251. **[3, A3]**

Neave, F. 1958. The origin and speciation of *Oncorhynchus*. Trans. R. Soc. Can. 52:25–39. **[1]**

Nehlsen, W., J. E. Williams, and J. A. Lichatowich. 1991. Pacific salmon at the crossroads: stocks at risk from California, Oregon, Idaho, and Washington. Fisheries (Bethesda) 16(2):4–21. **[2, 13]**

Nei, M. 1978. Estimation of average heterozygosity and genetic distance from a small number of individuals. Genetics 89:583–590. **[6]**

Nei, M. 1987a. Genetic distance and molecular phylogeny. Pp. 193–223 *in* N. Ryman and F. Utter, eds. Population genetics & fishery management. Univ. of Washington Press, Seattle, WA. **[8]**

Nei, M. 1987b. Molecular evolutionary genetics. Columbia Univ. Press, New York. **[6, 11]**

Nei, M., and A. Chakravarti. 1977. Drift variances of F_{ST} and G_{ST} statistics obtained from a finite number of isolated populations. Theor. Pop. Biol. 11:307–325. **[2]**

Nei, M., and F. Tajima. 1981. Genetic drift and estimation of effective population size. Genetics 98:625–640. **[10]**

Neigel, J. E. 2002. Is F_{ST} obsolete? Cons. Genet. 3:167–173. **[2]**

Nelson, J. S. 1968. Distribution and nomenclature of North American kokanee, *Oncorhynchus nerka*. J. Fish. Res. Board Can. 25:409–414. **[3]**

Nelson, K., and M. Soulé. 1987. Genetical conservation of exploited fishes. Pp. 345–368 *in* N. Ryman and F. Utter, eds. Population genetics & fishery management. Univ. of Washington Press, Seattle, WA. **[11]**

Nelson, R. J., M. P. Small, T. D. Beacham, and K. J. Supernault. 2001. Population structure of Fraser River chinook salmon (*Oncorhynchus tshawytscha*): an analysis using microsatellite DNA markers. Fishery Bull. 99:94–107. **[A2]**

Neraas, L. P., and P. Spruell. 2001. Fragmentation of riverine systems: the genetic effects of dams on bull trout (*Salvelinus confluentus*) in the Clark Fork River system. Mol. Ecol. 10:1153–1164. [A2]

Neter, J., W. Wasserman, and M. H. Kutner. 1989. Applied linear regression models. 2nd ed. Irwin, Boston, MA. [13]

Newman, D., and D. A. Tallmon. 2001. Experimental evidence for beneficial fitness effects of gene flow in recently isolated populations. Cons. Biol. 15:1054–1063. [2]

Nicieza, A. G., Reiriz, L., and F. Braña. 1994. Variation in digestive performance between geographically disjunct populations of Atlantic salmon: countergradient in passage time and digestion rate. Oecologia 99:243–251. [5]

Nielsen, C., G. Holdensgaard, H. C. Petersen, B. Th. Björnsson, and S. S. Madsen. 2001. Genetic differences in physiology, growth hormone levels and migratory behaviour of Atlantic salmon smolts. J. Fish Biol. 59:28–44. [3]

Nielsen, E. E., M. M. Hansen, and L. A. Bach. 2001. Looking for a needle in a haystack: discovery of indigenous Atlantic salmon (*Salmo salar* L.) in stocked populations. Cons. Genet. 2:219–232. [A2]

Nielsen, E. E., M. M. Hansen, and V. Loeschcke. 1999. Genetic variation in time and space: microsatellite analysis of extinct and extant populations of Atlantic salmon. Evolution 53:261–268. [2, A2]

Nielsen, J. L., ed. 1995. Evolution and the aquatic ecosystem: defining unique units in population conservation. American Fisheries Society Symposium 17. American Fisheries Society, Bethesda, MD. [13]

Nielsen, J. L., and M. C. Fountain. 1999. Microsatellite diversity in sympatric reproductive ecotypes of Pacific steelhead (*Oncorhynchus mykiss*) from the Middle Fork Eel River, California. Ecol. Freshwater Fish 8:159–168. [A2]

Nielsen, J. L., C. Carpanzano, M. C. Fountain, and C. A. Gan. 1997b. Mitochondrial DNA and nuclear microsatellite diversity in hatchery and wild *Oncorhynchus mykiss* from freshwater habitats in southern California. Trans. Am. Fish. Soc. 126:397–417. [13, A2]

Nielsen, J. L., K. D. Crow, and M. C. Fountain. 1999. Microsatellite diversity and conservation of a relic trout population: McCloud River redband trout. Mol. Ecol. 8:S129–S142. [A2]

Nielsen, J. L., M. C. Fountain, and J. M. Wright. 1997a. Biogeographic analysis of Pacific trout (*Oncorhynchus mykiss*) in California and Mexico based on mitochondrial DNA and nuclear microsatellites. Pp. 53–73 *in* T. D. Kocher and C. A. Stepien, eds. Molecular systematics of fishes. Academic Press, New York. [A2]

Nielsen, L. A. 1992. Methods of marking fish and shellfish. American Fisheries Society, Bethesda, MD. [2]

Nieminen, M., M. C. Singer, W. Fortelius, K. Schöps, and I. Hanski. 2001. Experimental confirmation that inbreeding depression increases extinction risk in butterfly populations. Am. Nat. 157:237–244. [2]

Nilsson, J. 1994. Genetics of growth of juvenile Arctic char. Trans. Am. Fish. Soc. 123:430–434. [5]

NMFS (National Marine Fisheries Service). 1989. Endangered and threatened species; critical habitat; winter-run chinook salmon. Federal Register 54:32085–32088. [12]

NMFS (National Marine Fisheries Service). 1993. Interim policy on artificial propagation of Pacific salmon under the Endangered Species Act. Federal Register 58:17573–17576. [12]

Noll, C., and 12 other authors. 2001. Analysis of contemporary genetic structure of even-broodyear populations of Asian and western Alaskan pink salmon, *Onchorhynchus gorbuscha*. Fishery Bull. 99:123–138. [A2]

Noor, M. 1999. Reinforcement and other consequences of sympatry. Heredity 83:503–508. [6]

Noor, M. A. 1995. Speciation driven by natural selection in *Drosophila*. Nature 375:674–675. [8]

Nordeng, H. 1971. Is the local orientation of anadromous fishes determined by pheromones? Nature 233:411–413. [2]

Nordeng, H. 1977. A pheromone hypothesis for homeward migration in anadromous salmonids. Oikos 28:155–159. [2]

Nordeng, H. 1983. Solution to the "Char Problem" based on Arctic char (*Salvelinus alpinus*) in Norway. Can. J. Fish. Aquat. Sci. 40:1372–1387. [I, 3]

Northcote, T. G. 1978. Migratory strategies and production in freshwater fishes. Pp. 326–359 *in* S. D. Gerking, ed. Ecology of freshwater fish production. Blackwell Scientific Publications, Oxford, U.K. [9]

Northcote, T. G. 1984. Mechanisms of fish migration in rivers. Pp. 317–355 *in* J. D. McLeave, G. P. Arnold, J. J. Dodson, and W. H. Neill, eds. Mechanisms of migration in fishes. Plenum Press, New York. [5]

Nunney, L. 1993. The influence of mating system and overlapping generations on effective population size. Evolution 47:1329–1341. [10]

Nunney, L. 1996. The influence of variation in female fecundity on effective population size. Biol. J. Linn. Soc. 59:411–425. [10]

Nunney, L. 2002. The effective size of annual plant populations: The interaction of a seed bank with fluctuating population size in maintaining genetic variation. Am. Nat. 160:195–204. [10]

Oakley, T. H., and R. B. Phillips. 1999. Phylogeny of salmonine fishes based on growth hormone introns: Atlantic (*Salmo*) and Pacific (*Oncorhynchus*) salmon are not sister taxa. Mol. Phyl. Evol. 11:381–393. [1, 7]

O'Connell, M., D. O. F. Skibinski, and J. A. Beardmore. 1995. Mitochondrial DNA and allozyme variation in Atlantic salmon (*Salmo salar*) populations in Wales. Can. J. Fish. Aquat. Sci. 52:171–178. [A2]

O'Donald, P. 1970. Change of fitness by selection for a quantitative character. Theor. Pop. Biol. 1:219–232. [11]

Økland, F., B. Jonsson, A. J. Jensen, and L. P. Hansen. 1993. Is there a threshold size regulating seaward migration of brown trout and Atlantic salmon? J. Fish Biol. 42:541–550. [3]

Olivieri, I., and P.-H. Gouyon. 1997. Evolution of migration rate and other traits: the metapopulation effect. Pp. 293–323 *in* I. Hanski and M. E. Gilpin, eds. Metapopulation biology: ecology, genetics, and evolution. Academic Press, New York. [2]

Olivieri, I., Y. Michalakis, and P.-H. Gouyon. 1995. Metapopulation genetics and the evolution of dispersal. Am. Nat. 146:202–228. [2]

Olofsson, H., and H. Mosegaard. 1999. Larger eggs in resident brown trout living in sympatry with anadromous brown trout. Ecol. Freshwater Fish 8:59–64. [A3]

Olsen, J. B., P. Bentzen, M. A. Banks, J. B. Shaklee, and S. Young. 2000. Microsatellites reveal population identity of individual pink salmon to allow supportive breeding of a population at risk of extinction. Trans. Am. Fish. Soc. 129:232–242. [A2]

Olsen, J. B., L. W. Seeb, P. Bentzen, and J. E. Seeb. 1998. Genetic interpretation of broad-scale microsatellite polymorphism in odd-year pink salmon. Trans. Am. Fish. Soc. 127:535–550. [A2]

Olsén, K. H., and T. Järvi. 1997. Effects of kinship on aggression and RNA content in juvenile Arctic charr. J. Fish Biol. 51:422–435. [2]

Olsén, K. H., M. Grahn, J. Lohm, and A. Langefors. 1998. MHC and kin discrimination in juvenile Arctic charr, *Salvelinus alpinus* (L.). Anim. Behav. 56:319–327. [2]

Olsson, G. 1960. Some relations between number of seeds per pod, seed size and oil content and the effects of selection for these characters in *Brassica* and *Sinapis*. Hereditas 46:29–70. [4]

Oohara, I., K. Sawano, and T. Okazaki. 1997. Mitochondrial DNA sequence analysis of the masu salmon—phylogeny in the genus *Oncorhynchus*. Mol. Phyl. Evol. 7:71–78. [7]

Orell, M., K. Lahti, K. Koivula, S. Rytkönen, and P. Welling. 1999. Immigration and gene flow in a northern willow tit (*Parus montanus*) population. J. Evol. Biol. 12:283–295. [2]

Orr, H. A. 1998. The population genetics of adaptation: the distribution of factors fixed during adaptive evolution. Evolution 52:935–949. [6]

Orr, H. A. 2001. The genetics of species differences. Trends Ecol. Evol. 16:343–350. [6]

Osinov, A. G., and V. S. Lebedev. 2000. Genetic divergence and phylogeny of the Salmonidae based on allozyme data. J. Fish Biol. 57:354–381. [1, 7]

Otto, S. P., and J. Whitton. 2000. Polyploid incidence and evolution. Annu. Rev. Genet. 34:401–437. [7]

Paine, M. D. 1984. Ecological and evolutionary consequences of early ontogenies of darters (Etheostomatini). Environ. Biol. Fishes 11:97–106. [4]

Pakkasmaa, S., N. Peuhkuri, A. Laurila, H. Hirvonen, and E. Ranta. 2001. Female and male contribution to egg size in salmonids. Evol. Ecol. 15:143–153. [9]

Palmer, A. R. 1990. Predator size, prey size, and the scaling of vulnerability: hatching gastropods vs barnacles. Ecology 71:759–775. [4]

Palumbi, S. R. 1998. Species formation and the evolution of gamete recognition loci. Pp. 271–278 *in* D. J. Howard and S. H. Berlocher, eds. Endless forms: species and speciation. Oxford Univ. Press, Oxford, U.K. [8]

Palumbi, S. R. 1999. All males are not created equal: fertility differences depend on gamete recognition polymorphisms in sea urchins. Proc. Natl. Acad. Sci. USA 96:12632–12637. [9]

Pante, M. J. R., B. Gjerde, and I. McMillan. 2001. Effect of inbreeding on body weight at harvest in rainbow trout, *Oncorhynchus mykiss*. Aquaculture 192:201–211. [2]

Paredes, R., and C. B. Zavalaga. 2001. Nesting sites and nest types as important factors for the conservation of Humboldt penguins (*Sphensicus humboldti*). Biol. Cons. 100:199–205. [2]

Parker, G. A. 1979. Sexual selection and sexual conflict. Pp. 123–166 *in* M. S. Blum and N. B. Blum, eds. Sexual selection and reproductive competition in insects. Academic Press, New York. [I]

Parker, G. A. 1984. Sperm competition and the evolution of animal mating strategies. Pp. 1–60 *in* R. L. Smith, ed. Sperm competition and the evolution of animal mating systems. Academic Press, New York. [9]

Parker, G. A., and M. Begon. 1986. Optimal egg size and clutch size: effects of environment and maternal phenotype. Am. Nat. 128:573–592. [1, 4]

Parker, G. A., and M. R. MacNair. 1978. Models of parent-offspring conflict. I. Monogamy. Anim. Behav. 26:97–100. [4]

Parker, G. A., R. R. Baker, and V. G. F. Smith. 1972. The origin and evolution of gamete dimorphism and male-female phenomenon. J. theor. Biol. 36:529–553. [9]

Parrish, D. L., R. J. Behnke, S. R. Gephard, S. D. McCormick, and G. H. Reeves. 1998. Why aren't there more Atlantic salmon. Can. J. Fish. Aquat. Sci. 55(Suppl. 1):281–287. [2]

Pärt, T. 1994. Male philopatry confers a mating advantage in the migratory collared flycatcher, *Ficedula albicollis*. Anim. Behav. 48:401–409. [2]

Partridge, L. 1988. The rare-male effect: what is its evolutionary significance? Phil. Trans. R. Soc. Lond. B 319:525–539. [5]

Pascual, M., P. Bentzen, C. R. Rossi, G. Mackay, M. T. Kinnison, and R. Walker. 2001. First documented case of anadromy in a population of introduced rainbow trout in Patagonia, Argentina. Trans. Am. Fish. Soc. 130:53–67. [3, 13, A3]

Pascual, M. A., T. P. Quinn, and H. Fuss. 1995. Factors affecting the homing of fall chinook salmon from Columbia River hatcheries. Trans. Am. Fish. Soc. 124:308–320. [I, 2, A1]

Paulik, G. J., A. S. Hourston, and P. A. Larkin. 1967. Exploitation of multiple stocks by a common fishery. J. Fish. Res. Board Can. 24:2527–2536. [13]

Payne, R. H., A. Forrest, and A. R. Child. 1972. Existence of natural hybrids between European trout and Atlantic salmon. J. Fish Biol. 4:233–236. [8]

Pearman, P. B. 2001. Conservation value of independently evolving units: Sacred cow or testable hypothesis. Cons. Biol. 15:780–783. [12]

Pearson, K. 1903. Mathematical contributions to the theory of evolution. III. Regression, heredity, and panmixia. Phil. Trans. R. Soc. Lond. A 200:1–66. [11]

Pella, J., and M. Masuda. 2001. Bayesian methods for analysis of stock mixtures from genetic characters. Fishery Bull. 99:151–167. [2]

Pen, I., and F. J. Weissing. 2000. Towards a unified theory of cooperative breeding: the role of ecology and life history re-examined. Proc. R. Soc. Lond. B 267:2411–2418. [2]

Penn, D. J., K. Danjanovich, and W. K. Potts. 2002. MHC heterozygosity confers a selective advantage against multiple-strain infections. Proc. Natl. Acad. Sci. USA 99:11260–11264. [9]

Pennock, D. S., and W. W. Dimmick. 1997. Critique of the evolutionarily significant unit as a definition for "distinct population segments" under the U.S. Endangered Species Act. Cons. Biol. 11:611–619. [12]

Perkins, D. L., C. C. Krueger, and B. May. 1993. Heritage brook trout in northeastern USA: genetic variability within and among populations. Trans. Am. Fish. Soc. 122:515–532. [A2]

Perrin, N., and J. Goudet. 2001. Inbreeding, kinship, and the evolution of natal dispersal. Pp. 123–142 *in* J. Clobert, E. Danchin, A. A. Dhondt, and J. D. Nichols, eds. Dispersal. Oxford Univ. Press, Oxford, U.K. [2]

Perrin, N., and V. Mazalov. 1999. Dispersal and inbreeding avoidance. Am. Nat. 154:282–292. [2]

Perrin, N., and V. Mazalov. 2000. Local competition, inbreeding, and the evolution of sex-biased dispersal. Am. Nat. 155:116–127. [2]

Peterman, R. M. 1984. Density-dependent growth in early ocean life of sockeye salmon (*Oncorhynchus nerka*). Can. J. Fish. Aquat. Sci. 41:1825–1829. [13]

Peterman, R. M., M. J. Bradford, and J. L. Anderson. 1986. Environmental and parental influences on age at maturity in sockeye salmon (*Oncorhynchus nerka*) from the Fraser River, British Columbia. Can. J. Fish. Aquat. Sci. 43:269–274. [13]

Peterson, N. P., and T. P. Quinn. 1996. Persistence of egg pocket architecture in redds of chum salmon, *Oncorhynchus keta*. Envir. Biol. Fishes 46:243–253. [2]

Petersson, E., and T. Järvi. 1997. Reproductive behaviour of sea trout (*Salmo trutta*)— the consequences of sea-ranching. Behaviour 134:1–22. [9]

Petersson, E., and T. Järvi. 2001. 'False orgasm' in female brown trout: trick or treat? Anim. Behav. 61:497–501. [9]

Petersson, E., T. Järvi, H. Olsén, I. Mayer, and M. Hedenskog. 1999. Male-male competition and female choice in brown trout. Anim. Behav. 57:777–783. [I, 9]

Pethon, P. 1974. Naturally occurring hybrids between whitefish (*Coregonus lavaretus* L.) and cisco (*Coregonus albula* L.) in Orrevann. Norw. J. Zool. 22:287–293. [8]

Pettersson, J. C. E., M. M. Hansen, and T. Bohlin. 2001. Does dispersal from landlocked trout explain the coexistence of resident and migratory trout females in a small stream? J. Fish Biol. 58:487–495. [3, A3]

Phelps, S. R., L. L. LeClair, S. Young, and H. L. Blankenship. 1994. Genetic diversity patterns of chum salmon in the Pacific northwest. Can. J. Fish. Aquat. Sci. 51(Suppl. 1):65–83. [A2]

Phillips, R. B., and T. H. Oakley. 1997. Phylogenetic relationships among the Salmoninae based on nuclear and mitochondrial DNA sequences. Pp. 145–162 *in* T. Kocker and C. Stepien, eds. Molecular systematics of fishes. Academic Press, San Deigo, CA. [7]

Phillips, R. B., K. A. Pleyte, and M. R. Brown. 1992. Salmonid phylogeny inferred from ribosomal DNA restriction maps. Can. J. Fish. Aquat. Sci. 49:2345–2353. [7, 8]

Pianka, E. R. 1970. On *r*- and *K*-selection. Am. Nat. 104:592–597. [1]

Pianka, E. R., and W. S. Parker. 1975. Age-specific reproductive tactics. Am. Nat. 109:453–464. [1]

Pielou, E. 1991. After the ice age. The return of life to glaciated North America. Univ. of Chicago Press, Chicago, IL. [8]

Pigeon, D., A. Chouinard, and L. Bernatchez. 1997. Multiple modes of speciation involved in the parallel evolution of sympatric morphotypes of lake whitefish (*Coregonus clupeaformis*, Salmonidae). Evolution 51:196–205. [6]

Pigliucci, M. 2001a. Phenotypic plasticity. Pp. 58–69 *in* C. W. Fox, D. A. Roff, and D. J. Fairbairn, eds. Evolutionary ecology: concepts and case studies. Oxford Univ. Press, Oxford, U.K. [5]

Pigliucci, M. 2001b. Phenotypic plasticity: beyond nature and nurture. John Hopkins, Baltimore. [5]

Pitcher, T. J., and P. J. B. Hart. 1982. Fisheries ecology. Croom Helm, London, U.K. [13]

Pitcher, T. J., P. J. B. Hart, and D. Pauley, eds. 1998. Reinventing fisheries management. Kluwer Academic Publishers, Dordrecht, The Netherlands. [13]

Plantegenest, M., and P. Kindlmann. 1999. Evolutionarily stable strategies of migration in heterogeneous environments. Evol. Ecol. 13:229–244. [2]

Podolsky, R. H., and T. P. Holtsford. 1995. Population structure of morphological traits in *Clarkia dudleyana* I. Comparison of Fst between allozymes and morphological traits. Genetics 140:733–744. [6]

Policansky, D. 1993. Fishing as a cause of evolution in fishes, Pp. 2–18 *in* T. K. Stokes, J. M. McGlade and R. Law, eds. The exploitation of evolving resources. Springer-Verlag, Berlin, Germany. [5, 13]

Porter, A. H. 2003. A test for deviation from island-model population structure. Mol. Ecol. 12:903–915. [2]

Poteaux, C., R. Guyomard, and P. Berrebi. 2000. Single and joint gene segregation in intraspecific hybrids of brown trout (*Salmo trutta* L.) lineages. Aquaculture 186:1–12. [8]

Potter, E. C. E., and I. C. Russell. 1994. Comparison of the distribution and homing of hatchery-reared and wild Atlantic salmon, *Salmo salar* L., from north-east England. Aquacul. Fish. Manage. 25(Suppl. 2):31–44. [A1]

Potts, W. K., C. J. Manning, and E. K. Wakeland. 1991. Mating patterns in seminatural populations of mice influenced by MHC genotype. Nature 352:619–621. [2]

Potvin, C., and L. Bernatchez. 2001. Lacustrine spatial distribution of landlocked Atlantic salmon populations assessed across generations by multilocus individual assignment and mixed-stock analyses. Mol. Ecol. 10:2375–2388. [I, 6]

Power, G. 1969. The salmon of Ungava Bay. Tech. Pap. Arctic Inst. N.A. 22:1–72. [1]

Power, G. 1981. Stock characteristics and catches of Atlantic salmon (*Salmo salar*) in Québec, and Newfoundland and Labrador in relation to environmental variables. Can. J. Fish. Aquat. Sci. 38:1601–1611. [5]

Press, W. H., Teukolsky, S. A., Vetterling, W. T. and B. P. Flannery. 1992. Numerical recipes in FORTRAN. Cambridge Univ. Press, Cambridge, U.K. [1]

Prévost, E., E. M. P. Chadwick, and R. R. Claytor. 1992. Influence of size, winter duration and density on sexual maturation of Atlantic salmon (*Salmo salar*) juveniles in little Codroy River (southwest Newfoundland). J. Fish Biol. 41:1013–1019. [3]

Primmer, C. R., T. Aho, J. Piironen, A. Estoup, J.-M. Cornuet, and E. Ranta. 1999. Microsatellite analysis of hatchery stocks and natural populations of Arctic charr, *Salvelinus alpinus*, from the Nordic region: implications for conservation. Hereditas 130:277–289. [A2]

Primmer, C. R., M. T. Koskinen, and J. Piironen. 2000. The one that did not get away: individual assignment using microsatellite data detects a case of fishing competition fraud. Proc. R. Soc. Lond. B. 267:1699–1704. [7]

Pritchard, A. L. 1937. Variation in the time of run, sex proportions, size and egg content of adult pink salmon (*Oncorhynchus gorbuscha*) at McClinton Creek, Masset Inlet, B.C. J. Biol. Board Can. 3:403–416. [9]

Prodöhl, P. A., A. F. Walker, R. Hynes, J. B. Taggart, and A. Ferguson. 1997. Genetically monomorphic brown trout (*Salmo trutta* L.) populations, as revealed by mitochondrial DNA, multilocus and single-locus minisatellite (VNTR) analyses. Heredity 79:208–213. [2]

Pulliam, H. R. 1988. Sources, sinks, and population regulation. Am. Nat. 132:652–661. [2]

Quinn, T. P. 1984. Homing and straying in Pacific salmon. Pp. 357–362 *in* J. D. McCleave, G. P. Arnold, J. J. Dodson, and W. H. Neill, eds. Mechanisms of migration in fishes. Plenum Press, New York. [2]

Quinn, T. P. 1993. A review of homing and straying of wild and hatchery-produced salmon. Fish. Res. 18:29–44. [I, 2, 4, 12, A1]

Quinn, T. P., and D. J. Adams. 1996. Environmental changes affecting the migratory timing of American shad and sockeye salmon. Ecology 77:1151–1162. [I]

Quinn, T. P., and C. A. Busack. 1985. Chemosensory recognition of siblings in juvenile coho salmon (*Oncorhynchus kisutch*). Anim. Behav. 33:51–56. [2]

Quinn, T. P., and A. H. Dittman. 1990. Pacific salmon migrations and homing: mechanisms and adaptive significance. Trends Ecol. Evol. 5:174–177. [I]

Quinn, T. P., and C. J. Foote. 1994. The effects of body size and sexual dimorphism on the reproductive behaviour of sockeye salmon, *Oncorhynchus nerka*. Anim. Behav. 48:751–761. [I, 3, 9]

Quinn, T. P., and K. Fresh. 1984. Homing and straying in chinook salmon (*Oncorhynchus tshawytscha*) from Cowlitz River Hatchery, Washington. Can. J. Fish. Aquat. Sci. 41:1078–1082. [2, A1]

Quinn, T. P., and M. T. Kinnison. 1999. Size-selective and sex-selective predation by brown bears on sockeye salmon. Oecologia 121:273–282. [2, 3, 7]

Quinn, T. P., and N. P. Peterson. 1996. The influence of habitat complexity and fish size on over-winter survival and growth of individually marked juvenile coho salmon (*Oncorhynchus kisutch*) in Big Beef Creek, Washington. Can. J. Fish. Aquat. Sci. 53:1555–1564. [13]

Quinn, T. P., and R. F. Tallman. 1987. Seasonal environmental predictability and homing in riverine fishes. Environ. Biol. Fishes 18:155–159. [2]

Quinn, T. P., M. D. Adkison, and M. B. Ward. 1996. Behavioral tactics of male sockeye salmon (*Oncorhynchus nerka*) under varying operational sex ratios. Ethology 102:304–322. [I, 9]

Quinn, T. P., A. H. Dittman, N. P. Peterson, and E. Volk. 1994. Spatial distribution, survival, and growth of sibling groups of juvenile coho salmon (*Oncorhynchus kisutch*) in an experimental stream channel. Can. J. Zool. 72:2119–2123. [2]

Quinn, T. P., A. P. Hendry, and G. B. Buck. 2001c. Balancing natural and sexual selection in sockeye salmon: interactions between body size, reproductive opportunity and vulnerability to predation by bears. Evol. Ecol. Res. 3:917–937. [2, 3, 7, 9]

Quinn, T. P., A. P. Hendry, and L. A. Wetzel. 1995. The influence of life history trade-offs and the size of incubation gravels on egg size variation in sockeye salmon (*Oncorhynchus nerka*). Oikos 74:425–438. [I, 2, 4]

Quinn, T. P., M. T. Kinnison, and M. J. Unwin, 2001b. Evolution of chinook salmon (*Oncorhynchus tshawytscha*) populations in New Zealand: pattern, rate, and process. Genetica 112/113:493–513. [I, 2, 4, 7, 12]

Quinn, T. P., R. S. Nemeth, and D. O. McIsaac. 1991. Homing and straying patterns of fall chinook salmon in the lower Columbia River. Trans. Am. Fish. Soc. 120:150–156. [2, A1]

Quinn, T. P., M. J. Unwin, and M. T. Kinnison. 2000. Evolution of temporal isolation in the wild: genetic divergence in timing of migration and breeding by introduced chinook salmon populations. Evolution 54:1372–1385. [I, 2, 3, 7, 9, 12, 13]

Quinn, T. P., E. C. Volk, and A. P. Hendry. 1999. Natural otolith microstructure patterns reveal precise homing to natal incubation sites by sockeye salmon (*Oncorhynchus nerka*). Can. J. Zool. 77:766–775. [I, 2, A1]

Quinn, T. P., L. Wetzel, S. Bishop, K. Overberg, and D. E. Rogers. 2001a. Influence of breeding habitat on bear predation and age at maturity and sexual dimorphism of sockeye salmon populations. Can. J. Zool. 79:1782–1793. [2, 7, 9]

Quinn, T. P., C. C. Wood, L. Margolis, B. E. Riddell, and K. D. Hyatt. 1987. Homing in wild sockeye salmon (*Oncorhynchus nerka*) populations as inferred from differences in parasite prevalence and allozyme allele frequencies. Can. J. Fish. Aquat. Sci. 44:1963–1971. [I, 2, A1]

Raleigh, R. F. 1971. Innate control of migrations of salmon and trout fry from natal gravels to rearing areas. Ecology 52:291–297. [2, 3, 12]

Ralph, S. C., G. C. Poole, L. L. Conquest, and R. J. Naiman. 1994. Stream channel morphology and woody debris in logged and unlogged basins of western Washington. Can. J. Fish. Aquat. Sci. 51:37–51. [13]

Rand, D. M., and R. G. Harrison. 1989. Ecological genetics of a mosaic hybrid zone: mitochondrial, nuclear, and reproductive differentiation of crickets by soil type. Evolution 43:432–449. [8]

Rand, P. S., and S. G. Hinch. 1998. Swim speeds and energy use of upriver-migrating sockeye salmon (*Oncorhynchus nerka*): simulating metabolic power and assessing risk of energy depletion. Can. J. Fish. Aquat. Sci. 55:1832–1841. [3]

Ratner, S., and R. Lande. 2001. Demographic and evolutionary responses to selective harvesting in populations with discrete generations. Ecology 82:3093–3104. [11]

Redenbach, Z., and E. B. Taylor. 2002. Evidence for historical introgression along a contact zone between two species of char (Pisces: Salmonidae) in northwestern North America. Evolution 56:1021–1035. [8]

Redenbach, Z., and E. B. Taylor. 2003. Evidence for bimodal hybrid zones between two species of char (Pisces: *Salvelinus*) in northwestern North America. J. Evol. Biol. In review. [8]

Reeves, G. H., F. H. Everest, and J. R. Sedell. 1993. Diversity of juvenile anadromous salmonid assemblages in coastal Oregon basins with different levels of timber harvest. Trans. Am. Fish. Soc. 122:309–317. [13]

Refstie, T., and T. A. Steine. 1978. Selection experiments with salmon. III. Genetic and environmental sources of variation in length and weight of Atlantic salmon in the freshwater phase. Aquaculture 14:221–234. [13]

Regan, C. T. 1914. The systematic arrangement of the fishes of the family Salmonidae. Ann. Mag. Nat. Hist. 13(series 8):405–408. [7]

Reich, T., C. A. Morris, and J. W. James. 1972. Use of multiple thresholds in determining mode of transmission of semi-continuous traits. Ann. Hum. Genet. 36:163–184. [11]

Reisenbichler, R. R. 1997. Genetic factors contributing to declines of anadromous salmonids in the Pacific Northwest. Pp. 223–244 *in* D. J. Stouder, P. A. Bisson, and R. J. Naiman, eds. Pacific salmon & their ecosystems: status and future options, Chapman and Hall, New York. [13]

Reisenbichler, R. R., and J. D. McIntyre. 1977. Genetic differences in growth and survival of juvenile hatchery and wild steelhead trout, *Salmo gairdneri*. J. Fish. Res. Board Can. 34:123–128. [13]

Reisenbichler, R. R., and S. R. Phelps. 1987. Genetic variation in chinook, *Oncorhynchus tshawytscha*, and coho, *O. kistuch*, salmon from the north coast of Washington. Fishery Bull. 85:681–701. [A2]

Reisenbichler, R. R., and S. R. Phelps. 1989. Genetic variation in steelhead (*Salmo gairdneri*) from the north coast of Washington. Can. J. Fish. Aquat. Sci. 46:66–73. [A2]

Reisenbichler, R. R., and S. P. Rubin. 1999. Genetic changes from artificial propagation of Pacific salmon affect the productivity and viability of supplemented populations. ICES J. Mar. Sci. 56:459–466. [I, 2, 7]

Reisenbichler, R. R., J. D. McIntyre, M. F. Solazzi, and S. W. Landino. 1992. Genetic variation in steelhead of Oregon and northern California. Trans. Am. Fish. Soc. 121:158–169. [A2]

Reist, J. D. 1989. Genetic structuring of allopatric populations and sympatric life history types of charr, *Salvelinus alpinus/malma*, in the western Arctic, Canada. Physiol. Ecol. Japan, Spec. Vol. 1:405–420. [3, A3]

Reist, J. D., L. D. Maiers, R. A. Bodaly, and T. J. Carmichael. 1998. The phylogeny of new- and old-world coregonine fishes as revealed by sequence variation in a portion of the d-loop mitochondrial DNA. Adv. Limnol. 50:323–339. [6]

Reist, J. D., J. Vuorinen, and R. A. Bodaly. 1992. Genetic and morphological identification of coregonid hybrid fishes from Arctic Canada. Pol. Arch. Hydrobiol. 39:551–561. [8]

Remington, C. L. 1968. Suture-zones of hybrid interaction between recently joined biotas. Evol. Biol. 2:321–428. [8]

Repka, J., and M. R. Gross. 1995. The evolutionarily stable strategy under individual condition and tactic frequency. J. theor. Biol. 176:27–31. [I, 3]

Reshetnikov, Y. S. 1988. Coregonid fishes in recent conditions. Finnish Fish. Res. 9:11–16. [6]

Reynolds, J. D. 1996. Animal breeding systems. Trends Ecol. Evol. 11:68–72. [9]

Reynolds, J. D., and M. J. G. Gage. 2002. Animal mating systems. Pp. 696–699 in M. Pagel, ed. Encyclopedia of evolution. Volume 1. Oxford Univ. Press, Oxford, U.K. [9]

Reynolds, J. D., and M. R. Gross. 1990. Costs and benefits of female mate choice: is there a lek paradox? Am. Nat. 136:230–243. [9]

Reynolds, J. D., and M. R. Gross. 1992. Female mate preference enhances offspring growth and reproduction in a fish, *Poecilia reticulata*. Proc. R. Soc. Lond. B 250:57–62. [9]

Reynolds, J. D., and J. C. Jones 1999. Female preference for preferred males is reversed under low oxygen conditions in the common goby (*Pomatoschistus microps*). Behav. Ecol. 10:149–154. [9]

Reynolds, J. D., M. A. Colwell, and F. Cooke. 1986. Sexual selection and spring arrival times of red-necked and Wilson's phalaropes. Behav. Ecol. Sociobiol. 18:303–310. [9]

Reynolds, J. D., N. B. Goodwin, and R. P. Freckleton. 2002. Evolutionary transitions in parental care and live-bearing in vertebrates. Phil. Trans. R. Soc. Lond. B 357:269–281. [9]

Reznick, D. N. 1993. Norms of reaction in fishes. Pp. 72–90 in T. K. Stokes, J. M. McGlade, and R. Law, eds. The exploitation of evolving resources. Springer-Verlag, Berlin, Germany. [5, 13]

Reznick, D. N., and C. K. Ghalambor. 2001. The population ecology of contemporary adaptations: what do empirical studies reveal about the conditions that promote adaptive evolution? Genetica 112/113:183–198. [I, 2, 4, 7]

Reznick, D. A., H. Bryga, and J. A. Endler. 1990. Experimentally induced life-history evolution in a natural population. Nature 346:357–359. [7]

Reznick, D. N., F. H. Shaw, F. H. Rodd, and R. G. Shaw. 1997. Evaluation of the rate of evolution in natural populations of guppies (*Poecilia reticulata*). Science 275:1934–1937. [7]

Rhodes, J. S., and T. P. Quinn. 1998. Factors affecting the outcome of territorial contests between hatchery and naturally reared coho salmon parr in the laboratory. J. Fish Biol. 53:1220–1230. [2]

Rhymer, J. M., and D. Simberloff. 1996. Extinction by hybridization and introgression. Annu. Rev. Ecol. Syst. 27:83–109. [8]

Rice, W. R. 1987. Speciation via habitat specialization: the evolution of reproductive isolation as a correlated character. Evol. Ecol. 1:301–314. [6]

Rice, W. R., and E. E. Hostert. 1993. Laboratory experiments on speciation: what have we learned in 40 years? Evolution 47:1637–1653. [I, 6, 8]

Ricker, W. E. 1938. 'Residual' and kokanee salmon in Cultus Lake. J. Fish. Res. Board Can. 4:192–218. [3]

Ricker, W. E. 1940. On the origin of kokanee, a fresh-water type of sockeye salmon. Proc. Trans. R. Soc. Canada 34:121–135. [I, 3]

Ricker, W. E. 1954. Stock and recruitment. J. Fish. Res. Board Can. 11:559–623. [11, 13]

Ricker, W. E. 1958. Maximum sustained yields from fluctuating environments and mixed stocks. J. Fish. Res. Board Can. 15:991–1006. [11, 13]

Ricker, W. E. 1972. Hereditary and environmental factors affecting certain salmonid populations. Pp. 19–160 in R. C. Simon and P. A. Larkin, eds. The stock concept in Pacific salmon. H. R. MacMillan lectures in fisheries. Vancouver, BC, Canada. [1, 2, 7]

Ricker, W. E. 1976. Review of rate of growth and mortality of Pacific salmon in salt water, and non-catch mortality caused by fishing. J. Fish. Res. Board Can. 33:1483–1524. [11]

Ricker, W. E. 1981. Changes in the average size and average age of Pacific salmon. Can. J. Fish. Aquat. Sci. 38:1636–1656. [5, 7, 11, 13]

Ricker, W. E. 1995. Trends in the average size of Pacific salmon in Canadian catches. Can. Spec. Pub. Fish. Aquat. Sci. 121:593–602. [11]

Riddell, B. E. 1986. Assessment of selective fishing on age at maturity in Atlantic salmon (*Salmo salar*): a genetic perspective. Can. Spec. Pub. Fish. Aquat. Sci. 89:102–109. [11]

Riddell, B. E., and W. C. Leggett. 1981. Evidence of an adaptive basis for geographic variation in body morphology and time of downstream migration of juvenile Atlantic salmon (*Salmo salar*). Can. J. Fish. Aquat. Sci. 38:308–320. [3, 13]

Riddell, B. E., W. C. Leggett, and R. L. Saunders. 1981. Evidence of adaptive polygenic variation between two populations of Atlantic salmon (*Salmo salar*) native to tributaries of the S.W. Miramichi River, N.B. Can. J. Fish. Aquat. Sci. 38:321–333. [7, 13]

Rieman, B. E., R. C. Beamesderfer, S. Vigg, and T. P. Poe. 1991. Estimated loss of juvenile salmonids to predation by northern squawfish, walleyes, and smallmouth bass in John Day Reservoir, Columbia River. Trans. Am. Fish. Soc. 120:448–458. [3]

Rieman, B. E., D. L. Myers, and R. L. Nielsen. 1994. Use of otolith microchemistry to discriminate *Oncorhynchus nerka* of resident and anadromous origin. Can. J. Fish. Aquat. Sci. 51:68–77. [3]

Rieseberg, L. H. 1998. Molecular ecology of hybridization. Ad. Mol. Ecol. 306:243–265. [8]

Rieseberg, L. H., B. Sinervo, C. R. Linder, M. C. Ungerer, and D. M. Arias. 1996. Role of gene interactions in hybrid speciation: evidence from ancient and experimental hybrids. Science 272:741–745. [6]

Rieseberg, L. H., J. Whitton, and K. A. Gardner. 1999. Hybrid zones and the genetic architecture of a barrier to gene flow between two sunflower species. Genetics 152:713–727. [6]

Rijnsdorp, A. D. 1993. Fisheries as a large-scale experiment on life-history evolution: disentangling phenotypic and genetic effects in changes in maturation and reproduction of North Sea plaice, *Pleuronectes platessa* L. Oecologia 96:391–401. [5, 7, 11]

Rijnsdorp, A. D., and A. Jaworski. 1990. Size-selective mortality in plaice and cod eggs: a new method in the study of egg mortality. J. Cons. Int. Explor. Mer. 47:256–263. [4]

Rikardsen, A. H., and J. M. Elliott. 2000. Variations in juvenile growth, energy allocation and life-history strategies of two populations of Arctic charr in North Norway. J. Fish Biol. 56:328–346. [3]

Robertson, A. 1921. Further proof of the parent stream theory. Trans. Am. Fish. Soc. 51:87–90. [2]

Robinson, B. W., and D. S. Wilson. 1994. Character release and displacement in fishes: a neglected literature. Am. Nat. 144:596–627. [6]

Robinson, B., and D. Schluter. 2000. Natural selection and the evolution of adaptive genetic variation in northern freshwater fishes. Pp. 65–94 *in* T. A. Mousseau, B. Sinervo, and J. Endler, eds. Adaptive genetic variation in the wild. Oxford Univ. Press, Oxford, U.K. [6]

Robison, O. W., and L. G. Luempert. 1984. Genetic variation in weight and survival of brook trout (*Salvelinus fontinalis*). Aquaculture 38:155–170. [5]

Rodríguez, M. A. 1995. Habitat-specific estimates of competition in stream salmonids: a field test of the isodar model of habitat selection. Evol. Ecol. 9:169–184. [13]

Roff, D. A. 1975. Population stability and the evolution of dispersal in a heterogeneous environment. Oecologia 19:217–237. [2]

Roff, D. A. 1992. The evolution of life histories; theory and analysis. Chapman and Hall, New York. [1, 4, 9, 11]

Roff, D. A. 1994. Habitat persistence and the evolution of wing dimorphism in insects. Am. Nat. 144:772–798. [2]

Roff, D. A. 1995. The estimation of genetic correlations from phenotypic correlations: A test of Cheverud's conjecture. Heredity 74:481–490. [6]

Roff, D. A. 1996a. The evolution of genetic correlations: an analysis of patterns. Evolution 50:1392–1403. [6]

Roff, D. A. 1996b. The evolution of threshold traits in animals. Quart. Rev. Biol. 71:3–35. [13]

Roff, D. A. 1997. Evolutionary quantitative genetics. Chapman and Hall, New York. [3, 11]

Roff, D. A. 1998. The maintenance of phenotypic and genetic variation in threshold traits by frequency dependent selection. J. Evol. Biol. 11:513–529. [5]

Roff, D. A. 2002. Life history evolution. Sinauer Assoc., Sunderland, MA. [I, 1, 5, 11]

Roff, D. A., and D. J. Fairbairn. 2001. The genetic basis of dispersal and migration, and its consequences for the evolution of correlated traits. Pp. 191–202 *in* J. Clobert, E. Danchin, A. A. Dhondt, and J. D. Nichols, eds. Dispersal. Oxford Univ. Press, Oxford, U.K. [2]

Roff, D. A., and T. A. Mousseau. 1987. Quantitative genetics and fitness: lessons from *Drosophila*. Heredity 58:103–118. [6]

Rogers, S. M., D. Campbell, S. J. E. Baird, R. G. Danzmann, and L. Bernatchez. 2001. Combining the analyses of introgressive hybridisation and linkage mapping to investigate the genetic architecture of population divergence in lake whitefish (*Coregonus clupeaformis*, Mitchill). Genetica 111:25–41. [6, 8]

Rogers, S. M., V. Gagnon, and L. Bernatchez. 2002. Genetically based phenotype-environment associations for swimming behavior in lake whitefish ecotypes (*Coregonus clupeaformis* Mitchill). Evolution 56:2322–2329. [6]

Rohwer, F. C. 1988. Interspecific and intraspecific relationships between egg size and clutch size in waterfowl. Auk 105:161–176. [4]

Rojas, M. 1992. The species problem and conservation: What are we protecting? Cons. Biol. 6:170–178. [12]

Ronce, O., S. Gandon, and F. Rousset. 2000b. Kin selection and natal dispersal in an age-structured population. Theor. Pop. Biol. 58:143–159. [2]

Ronce, O., F. Perret, and I. Olivieri. 2000a. Evolutionary stable dispersal rates do not always increase with local extinction rates. Am. Nat. 155:485–496. [2]

Rose, M. R., and G. V. Lauder, eds. 1996. Adaptation. Academic Press, New York. [I, 4]

Rosenau, M. L., and J. D. McPhail. 1987. Inherited differences in agonistic behavior between two populations of coho salmon. Trans. Am. Fish. Soc. 116:646–654. [13]

Rosenfield, J. A., T. Todd, and R. Greil. 2000. Asymmetric hybridization and introgression between pink salmon and chinook salmon in the Laurentian Great Lakes. Trans. Am. Fish. Soc. 129:670–679. [8]

Rosenzweig, M. L. 1978. Competitive speciation. Biol. J. Linn. Soc. 10:275–289. [6]

Rounsefell, G. A. 1958. Anadromy in North American Salmonidae. U.S. Fish Wild Serv. Fish. Bull. 58:171–185. [3]

Rounsefell, G. A. 1962. Relationships among North American Salmonidae. U.S. Fish Wild. Serv. Fish. Bull. 62:235–270. [7]

Routledge, R. D., and J. R. Irvine. 1999. Chance fluctuations and the survival of small salmon stocks. Can. J. Fish. Aquat. Sci. 56:1512–1519. [13]

Rowan, D. J., and J. B. Rasmussen. 1997. Measuring the bioenergetic cost of fish activity in situ using a globally dispersed radiotracer (^{137}Cs)—reply. Can. J. Fish. Aquat. Sci. 54:1955–1956. [6]

Rowe, D. K., J. E. Thorpe, and A. M. Shanks. 1991. Role of fat stores in the maturation of male Atlantic salmon (*Salmo salar*) parr. Can. J. Fish. Aquat. Sci. 48:405–413. [3]

Rubidge, E., P. Corbett, and E. B. Taylor. 2001. A molecular analysis of hybridization between native westslope cutthroat trout and introduced rainbow trout in southeastern British Columbia, Canada. J. Fish Biol. 59(Suppl. A):42–54. [8]

Ruckelshaus, M. H., P. Levin, J. B. Johnson, and P. M. Kareiva. 2002. The Pacific salmon wars: What science brings to the challenge of recovering species. Annu. Rev. Ecol. Syst. 33:665–706. [7]

Ruggerone G. T., R. Hanson, and D. E. Rogers. 2000. Selective predation by brown bears (*Ursus arctos*) foraging on spawning sockeye salmon (*Oncorhynchus nerka*). Can. J. Zool. 78:974–981. [2]

Rundle, H. D., and M. C. Whitlock. 2001. A genetic interpretation of ecologically dependent isolation. Evolution 55:198–201. [I, 6, 8]

Rundle, H. D., L. Nagel, J. W. Boughman, and D. Schluter. 2000. Natural selection and parallel speciation in sympatric sticklebacks. Science 287:306–308. [I, 6]

Ruzzante, D. E., M. M. Hansen, and D. Meldrup. 2001. Distribution of individual inbreeding coefficients, relatedness and influence of stocking on native anadromous brown trout (*Salmo trutta*) population structure. Mol. Ecol. 10:2107–2128. [2, A2]

Ryder, O. A. 1986. Species conservation and systematics: the dilemma of subspecies. Trends Ecol. Evol. 1:9–10. [12]

Ryman, N. 1970. A genetic analysis of recapture frequencies of released young of salmon (*Salmo salar* L.). Hereditas 65:159–160. [2]

Ryman, N., and L. Laikre. 1991. Effects of supportive breeding on the genetically effective population size. Cons. Biol. 5:325–329. [I, 10]

Ryman, N., P. E. Jorde, and L. Laikre. 1995. Supportive breeding and variance effective population size. Cons. Biol. 9:1619–1628. [10]

Sabo, J. L., and G. B. Pauley. 1997. Competition between stream-dwelling cutthroat trout (*Oncorhynchus clarki*) and coho salmon (*Oncorhynchus kisutch*): effects of relative size and population origin. Can. J. Fish. Aquat. Sci. 54:2609–2617. [13]

Sajdak, S. L., and R. B. Phillips. 1997. Phylogenetic relationships among *Coregonus* species inferred from the DNA sequence of the first internal transcribed spacer (ITS1) of ribosomal DNA. Can. J. Fish. Aquat. Sci. 54:1494–1503. [6]

Sakai, S., and Y. Harada. 2001. Why do large mothers produce large offspring? Theory and a test. Am. Nat. 157:348–359. [4]

Sánchez, J. A., C. Clabby, D. Ramos, G. Blanco, F. Flavin, E. Vázquez, and R. Powell. 1996. Protein and microsatellite single locus variability in *Salmo salar* L. (Atlantic salmon). Heredity 77:423–432. [A2]

Sandercock, F. K. 1991. Life history of coho salmon (*Oncorhynchus kisutch*). Pp. 397–445 *in* C. Groot and L. Margolis, eds. Pacific salmon life histories. UBC Press, Vancouver, BC, Canada. [13]

Sandlund, O. T., T. F. Næsje, and R. Saksgård. 1995. Ecological diversity in whitefish *Coregonus lavaretus*: ontogenetic niche shifts and polymorphism. Arch. Hydr. Spec. Issues Adv. Limnol. 46:49–59. [6]

Sanz, N., J.-L. García-Marín, and C. Pla. 2000. Divergence of brown trout (*Salmo trutta*) within glacial refugia. Can. J. Fish. Aquat. Sci. 57:2201–2210. [A2]

Sargent, R. C., M. R. Gross, and E. P. van den Berghe. 1986. Male mate choice in fishes. Anim. Behav. 34:545–550. [9]

Sargent, R. C., P. D. Taylor, and M. R. Gross. 1987. Parental care and the evolution of egg size in fishes. Am. Nat. 129:32–46. [4]

Savvaitova, K. A. 1980. Taxonomy and biogeography of charrs in the Palearctic. Pp. 281–294 *in* E. K. Balon, ed. Charrs: salmonid fishes of the genus *Salvelinus*. Dr. W. Junk Publishers, Dordrecht, The Netherlands. [8]

Sawamura, K., A. Davis, and C.-L. Wu. 2000. Genetic analysis of speciation by means of introgression into *Drosophila melanogaster*. Proc. Natl. Acad. Sci. USA 97:2652–2655. [6]

Scarnecchia, D. L. 1983. Age at maturity in Icelandic stocks of Atlantic salmon (*Salmo salar*). Can J. Fish. Aquat. Sci. 40:1456–1468. [1]

Schaffer, W. M. 1974a. Selection for optimal life histories: the effects of age structure. Ecology 55:291303. [I, 1]

Schaffer, W. M. 1974b. Optimal reproductive effort in fluctuating environments. Am. Nat. 108:783–790. [I, 1]

Schaffer, W. M. 1977. Some observations on evolution of reproductive rate and competitive ability in flowering plants. Theor. Pop. Biol. 11:90–104. [1]

Schaffer, W. M. 1979a. The theory of life-history evolution and it application to Atlantic salmon. Proc. Zool. Soc. Lond. 44:307–326. [1]

Schaffer, W. M. 1979b. Equivalence of maximizing reproductive value and fitness in the case of reproductive strategies. Proc. Natl. Acad. Sci. USA. 76:3567–3569. [1]

Schaffer, W. M. 1981. On reproductive value and fitness. Ecology 62:1683–1685. [1]

Schaffer, W. M. 1983. The application of optimal control theory to the general life history problem. Am. Nat. 121:418–431. [1]

Schaffer, W. M. 1985. Can nonlinear dynamics help us infer mechanisms in ecology and epidemiology? IMA J. Math. Appl. Med. Biol. 2:221–252. [1]

Schaffer, W. M. 2003. Forward and retrospective: life histories, evolution and salmonids. Pp. 20–51 *in* A. P. Hendry and S. C. Stearns, eds. Evolution illuminated: salmon and their relatives. Oxford Univ. Press, New York, NY—*this volume* [I, 5, 7]

Schaffer, W. M., and P. F. Elson. 1975. The adaptive significance of variations in life history among local populations of Atlantic salmon in North America. Ecology 56:577–590. [I, 1, 3]

Schaffer, W. M., and M. D. Gadgil. 1975. Selection for optimal life histories in plants. Pp. 142–157 *in* M. Cody and J. M. Diamond, eds. Ecology and Evolution of Communities. Belknap Press, Cambridge, MA. [1]

Schaffer, W. M., and M. L. Rosenzweig. 1977. Selection for optimal life histories. II: Multiple equilibria and the evolution of alternative reproductive strategies. Ecology 58:60–72. [1]

Schaffer, W. M., and M. V. Schaffer. 1977. The adaptive significance of variation in reproductive habit in Agavaceae. Pp. 261–276 *in* B. Stonehouse and C. M. Perrins, eds. Evolutionary Ecology. MacMillan, London, U.K. [1]

Schaffer, W. M., and M. V. Schaffer. 1979. Adaptive significance of variations in reproductive habit in the Agavaceae. II: Pollinator foraging behavior and selection for increased reproductive expenditure. Ecology 60:1051–1069. [1]

Schaffer, W. M., and R. H. Tamarin. 1973. Changing reproductive rates and population cycles in lemmings and voles. Evolution 27:114–124. [1]

Schaffer, W. M., Inouye, R. S., and T. S. Whittam. 1982. Energy allocation when the effects of seasonality on growth and reproduction are decoupled. Am. Nat. 120:787–815. [1]

Scheer, B. T. 1939. Homing instinct in salmon. Quart. Rev. Biol. 14:408–430. [2]

Schemske, D. W., and H. D. Bradshaw. 1999. Pollinator preference and the evolution of floral traits in monkeyflowers (*Mimulus*). Proc. Natl. Acad. Sci. USA 96:11910–11915. [6]

Schiefer, K. 1971. Ecology of Atlantic salmon, with special reference to the occurrence and abundance of grilse, in North Shore Gulf of St. Lawrence Rivers. Ph.D. dissertation, Univ. Waterloo, Waterloo, ON, Canada. [1]

Schlichting, C. D., and M. Pigliucci. 1998. Phenotypic evolution: a reaction norm perspective. Sinauer Assoc., Sunderland, MA. [3, 5]

Schliewen, U., K. Rassmann, M. Markmann, J. Markert, T. Kocher, and D. Tautz. 2001. Genetic and ecological divergence of a monophyletic cichlid species pair under fully sympatric conditions in Lake Ejagham, Cameroon. Mol. Ecol. 10:1471–1488. [I]

Schliewen, U. K., D. Tautz, and S. Pääbo. 1994. Sympatric speciation suggested by monophyly of crater lake cichlids. Nature 368:629–632. [I]

Schluter, D. 1988. Estimating the form of natural selection on a quantitative trait. Evolution 42:849–861. [11]

Schluter, D. 1994. Experimental evidence that competition promotes divergence in adaptive radiation. Science 266:798–801. [6]

Schluter, D. 1995. Adaptive radiation in sticklebacks: trade-offs in feeding performance and growth. Ecology 76:82–90. [6]

Schluter, D. 1996. Adaptive radiation along genetic lines of least resistance. Evolution 50:1766–1774. [6]

Schluter, D. 2000. The ecology of adaptive radiation. Oxford Univ. Press, Oxford, U.K. [I, 2, 6, 7, 8]

Schluter, D. 2001a. Ecological character displacement in adaptive radiation. Am. Nat. 156:S4-S16. [6]

Schluter, D. 2003. Frequency dependent natural selection during character displacement in sticklebacks. Evolution 57:1142–1150. [6]

Schluter, D. 2001b. Ecology and the origin of species. Trends Ecol. Evol. 16:372–380. [6]

Schluter, D., and P. R. Grant. 1984. Determinants of morphological patterns in communities of Darwin's finches. Am. Nat. 123:175–196. [6]

Schluter, D., T. D. Price, and L. Rowe. 1991. Conflicting selection pressures and life history trade-offs. Proc. R. Soc. Lond. B 246:11–17. [4]

Schmalhausen, I. I. 1949. Factors of evolution. Blakiston, Philadelphia, PA. [5]

Schmidt, D. C., S. R. Carlson, G. B. Kyle, and B. P. Finney. 1998. Influence of carcass-derived nutrients on sockeye salmon productivity of Karluk Lake, Alaska: importance in the assessment of an escapement goal. N.A. J. Fish. Manage. 18:743–763. [2]

Schmidt-Nielsen, K. 1984. Scaling: why is animal size so important? Cambridge Univ. Press. Cambridge, U.K. [1]

Schoener, T. W. 1970. Non-synchronous spatial overlap of lizards in patchy habitats. Ecology 58:408–418. [6]

Scholz, A. T., R. M. Horrall, J. C. Cooper, and A. D. Hasler. 1976. Imprinting to chemical cues: basis for homestream selection in salmon. Science 192:1247–1249. [2]

Scholz, N. L., N. K. Truelove, B. L. French, B. A. Berejikian, T. P. Quinn, E. Casillas, and T. K. Collier. 2000. Diazinon disrupts antipredator and homing behaviors in chinook salmon (*Oncorhynchus tshawytscha*). Can. J. Fish. Aquat. Sci. 57:1911–1918. [2]

Schroder, S. L. 1981. The role of sexual selection in determining overall mating patterns and mate choice in chum salmon. Ph.D. dissertation, Univ. of Washington, Seattle, WA. [3, 9]

Schroder, S. L. 1982. The influence of intrasexual competition the distribution of chum salmon in a experimental stream. Pp. 275–285 *in* E. L. Brannon and E. O. Salo, eds. Salmon and trout migratory behavior symposium. School of Fisheries, Univ. Washington, Seattle, WA. [9]

Schroder, S. L., C. M. Knudsen, and E. C. Volk. 1995. Marking salmon fry with strontium chloride solutions. Can. J. Fish. Aquat. Sci. 52:1141–1149. [2]

Schroeder, R. K., R. B. Lindsay, and K. R. Kenaston. 2001. Origin and straying of hatchery winter steelhead in Oregon coastal rivers. Trans. Am. Fish. Soc. 130:431–441. [2, A1]

Schulte, P. M. 2001. Environmental adaptations as windows on molecular evolution. Comp. Biochem. Physiol. B 128:597–611. [8]

Schwartz, F. J. 1982. World literature to fish hybrids with an analysis by family, species, and hybrid. Suppl. 1. U.S. Nat. Mar. Ser. Spec. Sci. Rept. Fish. 750. [8]

Schwartz, J. H. 1999. Sudden origins: fossils, genes, and the emergence of species. Wiley, New York. [7]

Schwartz, M. K., D. A. Tallmon, and G. H. Luikart. 1998. Review of DNA-based census and effective population size estimators. Anim. Cons. 1:293–299. [10]

Scott, W. B., and E. J. Crossman. 1973. Freshwater fishes of Canada. Fish. Res. Board Can. Bull. 180. [I, 6]

Scribner, K. T., J. W. Arntzen, and T. Burke. 1997. Effective number of breeding adults in *Bufo bufo* estimated from age-specific variation at minisatellite loci. Mol. Ecol. 6:701–712. [10]

Scribner, K. T., P. A. Crane, W. J. Spearman, and L. W. Seeb. 1998. DNA and allozyme markers provide concordant estimates of population differentiation:

analyses of U.S. and Canadian populations of Yukon River fall-run chum salmon (*Oncorhynchus keta*). Can. J. Fish. Aquat. Sci. 55:1748–1758. **[A2]**

Scribner, K. T., K. S. Page, and M. L. Barton. 2000. Hybridization in freshwater fishes: a review of case studies and cytonuclear methods of biological inference. Rev. Fish Biol. Fish. 10:293–323. **[8]**

Seeb, L. W., and P. A. Crane. 1999. High genetic heterogeneity in chum salmon in western Alaska, the contact zone between northern and southern lineages. Trans. Am. Fish. Soc. 128:58–87. **[A2]**

Seeb, J. E., C. Habicht, W. D. Templin, L. W. Seeb, J. B. Shaklee, and F. M. Utter. 1999. Allozyme and mitochondrial DNA variation describe ecologically important genetic structure of even-year pink salmon inhabiting Prince William Sound, Alaska. Ecol. Freshwater Fish 8:122–140. **[2, A2]**

Seeb, L. W., C. Habicht, W. D. Templin, K. E. Tarbox, R. Z. Davis, L. K. Brannian, and J. E. Seeb. 2000. Genetic diversity of sockeye salmon of Cook Inlet, Alaska, and its application to management of populations affected by the *Exxon Valdez* oil spill. Trans. Am. Fish. Soc. 129:1223–1249. **[A2]**

Seehausen, O., J. J. M. van Alphen, and F. Witte. 1997. Cichlid fish diversity threatened by eutrophication that curbs sexual selection. Science 277:1808–1811. **[8]**

Shaklee, J. B., and N. V. Varnavskaya. 1994. Electrophoretic characterization of odd-year pink salmon (*Oncorhynchus gorbuscha*) populations from the Pacific coast of Russia, and comparison with selected North American populations. Can. J. Fish. Aquat. Sci. 51(Suppl. 1):158–171. **[A2]**

Shapovalov, L., and A. C. Taft. 1954. The life histories of the steelhead rainbow trout (*Salmo gairdneri gairdneri*) and silver salmon (*Oncorhynchus kisutch*) with special reference to Waddell Creek, California, and recommendations regarding their management. Cal. Dep. Fish Game, Fish Bull. 98. **[1, 2, A1]**

Sharma, R., and R. Hilborn. 2001. Empirical relationships between watershed characteristics and coho salmon (*Oncorhynchus kisutch*) smolt abundance in 14 western Washington streams. Can. J. Fish. Aquat. Sci. 58:1453–1463. **[13]**

Sharov, A. F., and A. V. Zubchenko. 1993. Influence of human activity on properties of Atlantic salmon populations. Pp. 62–69 *in* T. K. Stokes, J. M. McGlade and R. Law eds. The exploitation of evolving resources, Springer-Verlag, Berlin, Germany. **[13]**

Shaw, R. G. 1987. Maximum-likelihood approaches to quantitative genetics of natural populations. Evolution 41:812–829. **[11]**

Shearer, W. M. The Atlantic salmon. Natural history, exploitation and future management. Fishing News Books, Oxford, U.K. **[9]**

Shed'ko, S. V., L. K. Ginatulina, I. Z. Parpura, and A. V. Ermolenko. 1996. Evolutionary and taxonomic relationships among Far-Eastern salmonid fishes inferred from mitochondrial DNA divergence. J. Fish Biol. 49:815–829. **[7]**

Shelton, P. A., and B. P. Healey. 1999. Should depensation be dismissed as a possible explanation for the lack of recovery of the northern cod (*Gadus morhua*) stock? Can. J. Fish. Aquat. Sci. 56:1521–1524. **[13]**

Sheridan, A. K. 1988. Agreement between estimated and realized genetic parameters. Anim. Breed. Abstr. 56:877–889. **[11]**

Shine, R. 1978. Propagule size and parental care: the "safe harbor" hypothesis. J. theor. Biol. 75:417–424. **[4]**

Sholes, W. H., and R. J. Hallock. 1979. An evaluation of rearing fall-run chinook salmon, *Oncorhynchus tshawytscha*, to yearlings at Feather River Hatchery, with a comparison of returns from hatchery and downstream releases. Calif. Fish Game 65:239–255. [A1]

Shykoff, J. A., and A. Widmer. 1998. Eggs first. Trends Ecol. Evol. 13:158. [4]

Sibly, R., and K. Monk. 1987. A theory of grasshopper life cycles. Oikos 48:186–194. [4]

Sih, A. 1994. Predation risk and the evolutionary ecology of reproductive behaviour. J. Fish Biol. 45(Suppl. A):111–130. [9]

Siitonen, L., and G. A. E. Gall. 1989. Response to selection for early spawn date in rainbow trout, *Salmo gairdneri*. Aquaculture 78:153–161. [9]

Silverstein, J. T., and W. K. Hershberger. 1992. Precocious maturation in coho salmon (*Oncorhynchus kisutch*): estimation of the additive genetic variance. J. Hered. 83:282–286. [13]

Simon, R. C., J. D. McIntyre, and A. R. Hemmingsen. 1986. Family size and effective population size in a hatchery stock of coho salmon (*Oncorhynchus kisutch*). Can. J. Fish. Aquat. Sci. 43:2434–2442. [10]

Simpson, G. G. 1953. The major features of evolution. Columbia Univ. Press, New York. [6]

Simpson, G. G. 1944. Tempo and mode in evolution. Columbia Univ. Press, New York. [7]

Sinervo, B., and P. Licht. 1991. Proximate constraints on the evolution of egg size, number, and total clutch mass in lizards. Science 252:1300–1302. [4]

Sinervo, B., and E. Svensson. 1998. Mechanistic and selective causes of life history trade-offs and plasticity. Oikos 83:432–442. [4]

Sinervo, B., P. Doughty, R. B. Huey, and K. Zamudio. 1992. Allometric engineering: a causal analysis of natural selection on offspring size. Science 258:1927–1930. [4]

Sinervo, B., E. Svensson, and T. Comendant. 2000. Density cycles and an offspring quantity and quality game driven by natural selection. Nature 406:985–988. [4]

Skaala, Ø., and G. Nævdal. 1989. Genetic differentiation between freshwater resident and anadromous brown trout, *Salmo trutta*, within watercourses. J. Fish Biol. 34:597–605. [3]

Skarstein, F., and I. Folstad. 1996. Sexual dichromatism and the immunocompetence handicap: an observational approach using Arctic charr. Oikos 76:359–367. [9]

Skrochowska, S. 1969. Migrations of the sea-trout (*Salmo trutta* L.), brown trout (*Salmo trutta* m. *fario* L.) and their crosses. Part 1. Problem, methods and results of tagging. Pol. Arch. Hydrobiol. 16:125–140. [3]

Skúlason, S., and T. B. Smith. 1995. Resource polymorphisms in vertebrates. Trends Ecol. Evol. 10:366–370. [I]

Skúlason, S., S. S. Snorrason, and B. Jonsson. 1999. Sympatric morphs, populations and speciation in freshwater fish with an emphasis on arctic charr. Pp. 160–183 *in* A. E. Magurran and R. M. May, eds. Evolution of biological diversity. Oxford Univ. Press, Oxford, U.K. [6]

Slaney, P. A., and T. G. Northcote. 1974. Effects of prey abundance on density and territorial behavior of young rainbow trout (*Salmo gairdneri*) in laboratory stream channels. J. Fish. Res. Board Can. 31:1201–1209. [13]

Slaney, T. L., K. D. Hyatt, T. G. Northcote, and R. J. Fielden. 1996. Status of anadromous salmon and trout in British Columbia and Yukon. Fisheries (Bethesda) 21(10):20–35. [13]

Slatkin, M. 1985. Rare alleles as indicators of gene flow. Evolution 39:53–65. [2]

Slatkin, M. 1987. Gene flow and the geographic structure of natural populations. Science 236:787–792. [2]

Slatkin, M. 1993. Isolation by distance in equilibrium and non-equilibrium populations. Evolution 47:264–279. [2]

Small, M. P., T. D. Beacham, R. E. Withler, and R. J. Nelson. 1998. Discriminating coho salmon (*Oncorhynchus kisutch*) populations within the Fraser River, British Columbia, using microsatellite DNA markers. Mol. Ecol. 7:141–155. [A2]

Smith, C. C., and S. D. Fretwell. 1974. The optimal balance between size and number of offspring. Am. Nat. 108:499–506. [1, 4]

Smith, G. R. 1992. Introgression in fishes: significance for paleontology, cladisitics, and evolutionary rates. Syst. Biol. 41:41–57. [8]

Smith, G. R., and R. F. Stearley. 1989. The classification and scientific names of rainbow and cutthroat trouts. Fisheries (Bethesda) 14(1):4–10. [7]

Smith, G. R., and T. N. Todd. 1984. Evolution of species flocks of fishes in north temperate lakes. Pp. 45–68 *in* A. A. Echelle and I. Kornfield, eds. Evolution of fish species flocks. Univ. of Maine at Orono Press, Orono, ME. [6]

Smith, P. J., R. I. C. C. Francis, and M. McVeagh. 1991. Loss of genetic diversity due to fishing pressure. Fish. Res. 10:309–316. [11]

Smoker, W. W., and F. P. Thrower. 1995. Homing propensity in transplanted and native chum salmon. Am. Fish. Soc. Symp. 15:575–576. [A1]

Smoker, W. W., A. J. Gharrett, and M. S. Stekoll. 1998. Genetic variation of return date in a population of pink salmon: a consequence of fluctuating environment and dispersive selection? Alaska Fishery Res. Bull. 5:46–54. [12, 13]

Smoker, W. W., A. J. Gharrett, M. S. Stekoll, and J. E. Joyce. 1994. Genetic analysis of size in an anadromous population of pink salmon. Can. J. Fish. Aquat. Sci. 51 (Suppl.1):9–15. [13]

Snucins, E. J., R. A. Curry, and J. M. Gunn. 1992. Brook trout (*Salvelinus fontinalis*) embryo habitat and timing of alevin emergence in a lake and a stream. Can. J. Zool. 70:423–427. [5]

Sober, E., ed. 1984. Conceptual issues in evolutionary biology. The MIT Press, Cambridge, MA. [13]

Sokal, R. R., and F. J. Rohlf. 1995. Biometry, 3rd ed. W. H. Freeman, New York. [6]

Solomon, D. J. 1973. Evidence for pheromone-influenced homing by migrating Atlantic salmon, *Salmo salar* (L.). Nature 244:231–232. [A1]

Soto, D., F. Jara, and C. Moreno. 2001. Escaped salmon in the inner seas, southern Chile: facing ecological and social conflicts. Ecol. Appl. 11:1750–1762. [I]

Soulé, M. E., ed. 1986. Conservation biology: the science of scarcity and diversity. Sinauer Assoc., Sunderland, MA. [10, 13]

Spidle, A. P., W. B. Schill, B. A. Lubinski, and T. L. King. 2001. Fine-scale population structure in Atlantic salmon from Maine's Penobscot River drainage. Cons. Genet. 2:11–24. [A2]

Spitze, K. 1993. Population structure in *Daphnia obtusa*: quantitative genetic and allozymic variation. Genetics 135:367–374. [6]

Spruell, P., B. E. Rieman, K. L. Knudsen, F. M. Utter, and F. W. Allendorf. 1999. Genetic population structure within streams: microsatellite analysis of bull trout populations. Ecol. Freshwater Fish 8:114–121. [A2]

Stabell, O. B. 1984. Homing and olfaction in salmonids: a critical review with special reference to the Atlantic salmon. Biol. Rev. 59:333–388. [2, A1]

Stacey, P. B., and J. D. Ligon. 1987. Territory quality and dispersal options in the acorn woodpecker, and a challenge to the habitat-saturation model of cooperative breeding. Am. Nat. 130:654–676. [2]

Ståhl, G. 1987. Genetic population structure of Atlantic salmon. Pp. 121–140 *in* N. Ryman and F. Utter, eds. Population genetics & fishery management, Univ. of Washington Press, Seattle, WA. **[A2]**

Ståhl, G., and K. Hindar. 1988. Genetisk struktur hos norsk laks: status og perspektives. Rapport fra Fiskeforskningen No. 1. Direktoratet for Naturforvaltning, Trondheim. **[A2]**

Stanley, S. M. 1979. Macroevolution: pattern and process. W. H. Freeman, San Fancisco, CA. **[7]**

Staurnes, M., G. Lysfjord, and O. K. Berg. 1992. Parr–smolt transformation of a nonanadromous population of Atlantic salmon (*Salmo salar*) in Norway. Can. J. Zool. 70:197–199. **[3]**

Staurnes, M., G. Lysfjord, L. P. Hansen, and T. G. Heggberget. 1993. Recapture rates of hatchery-reared Atlantic salmon (*Salmo salar*) related to smolt development and time of release. Aquaculture 118:327–337. **[3]**

Stearley, R. F. 1992. Historical ecology of Salmoninae, with special reference to *Oncorhynchus*. Pp. 622–658 *in* R. Mayden, ed. Systematics, historical ecology, and North American freshwater fishes. Stanford Univ. Press, Stanford, CA. **[7]**

Stearley, R. F., and G. R. Smith. 1993. Phylogeny of the Pacific trouts and salmons (*Oncorhynchus*) and genera of the family Salmonidae. Trans. Am. Fish. Soc. 122:1–33. **[1, 7, 8]**

Stearns, S. C. 1976. Life history tactics: a review of ideas. Quart. Rev. Biol. 51:3–47. **[I, 1, 9]**

Stearns, S. C. 1977. The evolution of life history traits: a critique of the theory and a review of the data. Annu. Rev. Ecol. Syst. 8:145–171. **[1]**

Stearns, S. C. 1987. The selection arena hypothesis. Pp. 299–311 *in* S. C. Stearns, ed. The evolution of sex and its consequences. Birkhaeuser, Verlag, Berlin, Germany. **[I]**

Stearns, S. C. 1992. The evolution of life histories. Oxford Univ. Press, Oxford, U.K. **[1, 5, 9, 11, 13]**

Stearns, S. C. 2000. Daniel Bernoulli (1738): evolution and economics under risk. J. Bioscience 25:221–228. **[I]**

Stearns, S. C., and A. P. Hendry. 2003. The salmonid contribution to key issues in evolution. Pp. 3–19 *in* A. P. Hendry and S. C. Stearns, eds. Evolution illuminated: salmon and their relatives. Oxford Univ. Press, New York, NY—*this volume* **[1, 4, 7, 12]**

Stearns, S. C., and J. C. Koella. 1986. The evolution of phenotypic plasticity in life-history traits: predictions of reaction norms for age and size at maturity. Evolution 40:893–913. **[5, 13]**

Steen, R. P., and T. P. Quinn. 1999. Egg burial depth by sockeye salmon (*Oncorhynchus nerka*): implications for survival of embryos and natural selection on female body size. Can. J. Zool. 77:836–841. **[3, 9]**

Stein, R. A., P. E. Reimers, and J. D. Hall. 1972. Social interaction between juvenile coho (*Oncorhynchus kisutch*) and fall chinook (*O. tshawytscha*) in Sixes River, Oregon. J. Fish. Res. Board Can. 29:1737–1748. **[13]**

Steinmann, P. 1950. Monographie der schweizerischen Koregonen. Beitrag zum Problem der Entstehung neur Arten. Spezieller Teil. Schweizerische Zeitschrift für Hydrobiologie 12:340–491. **[6]**

Stirling, G., D. Roff, and D. Fairbairn. 1999. Four characters in a trade-off: dissecting their phenotypic and genetic relations. Oecologia 120:492–498. **[2]**

Stockley, P. 1999. Sperm selection and genetic incompatibility: does relatedness of mates affect male success in sperm competition. Proc. R. Soc. Lond. B 266:1663–1669. [9]

Stockley, P., M. J. G. Gage, G. A. Parker, and A. P. Møller. 1997. Sperm competition in fishes: the evolution of testis size and ejaculate characteristics. Am. Nat. 149:933–954. [9]

Stockwell, C. A., and S. C. Weeks. 1999. Tranlocations and rapid evolutionary responses in recently established populations of western mosquitofish (*Gambusia affinis*). Anim. Cons. 2:103–110. [1]

Stockwell, C. A., A. P. Hendry, and M. T. Kinnison. 2003. Contemporary evolution meets conservation biology. Trends Ecol. Evol. 18:94–101. [I, 4, 7, 13]

Stokes, K., and R. Law. 2000. Fishing as an evolutionary force. Mar. Ecol. Prog. Ser. 208:307–309. [5]

Stokes, T. K., J. M. McGlade, and R. Law, eds. 1993. The exploitation of evolving resources. Springer-Verlag, Berlin, Germany. [11, 13]

Storfer, A. 1996. Quantitative genetics: a promising approach for the assessment of genetic variation in endangered species. Trends Ecol. Evol. 11:343–348. [11]

Streelman, J. T., and T. D. Kocher. 2000. From phenotype to genotype. Evol. Dev. 2:166–173. [6]

Su, G.-S., L. E. Liljedahl, and G. A. E. Gall. 1996. Effects of inbreeding on growth and reproductive traits in rainbow trout (*Oncorhynchus mykiss*). Aquaculture 142:139–148. [2]

Su, G.-S., L. E. Liljedahl, and G. A. E. Gall. 1997. Genetic and environmental variation of female reproductive traits in rainbow trout (*Oncorhynchus mykiss*). Aquaculture 154:115–124. [I, 4]

Sutherland, W. J. 1996. From individual behaviour to population ecology. Oxford Univ. Press, Oxford, U.K. [9]

Sutterlin, A. M., and D. MacLean. 1984. Age at first maturity and the early expression of oocyte rectruitment processes in two forms of Atlantic salmon (*Salmo salar*) and their hybrids. Can. J. Fish. Aquat. Sci. 41:1139–1149. [A3]

Svärdson, G. 1949. Natural selection and egg number in fish. Rep. Inst. Freshwater Res., Drottningholm 29:115–122. [4]

Svärdson, G. 1979. Speciation of Scandinavian *Coregonus*. Rep. Inst. Freshwater Res., Drottningholm 57. [6, 8]

Svärdson, G. 1998. Postglacial dispersal and reticulate evolution of Nordic coregonids. Nordic J. Freshwater Res. 74:3–32. [6, 8]

Svedäng, H. 1992. Observations on interbreeding between dwarf and normal arctic charr, *Salvelinus alpinus*, from Stora Rösjön, central Sweden. Environ. Biol. Fishes 33:293–298. [8]

Swain, D. P., and L. B. Holtby. 1989. Differences in morphology and behavior between juvenile coho salmon (*Oncorhynchus kisutch*) rearing in a lake and in its tributary stream. Can. J. Fish. Aquat. Sci. 46:1406–1414. [13]

Swain, D. P., and B. E. Riddell. 1990. Variation in agonistic behavior between newly emerged juveniles from hatchery and wild populations of coho salmon, *Oncorhynchus kisutch*. Can. J. Fish. Aquat. Sci. 47:566–571. [13]

Swain, D. P., B. E. Riddell, and C. B. Murray. 1991. Morphological differences between hatchery and wild populations of coho salmon (*Oncorhynchus kisutch*): environmental versus genetic origin. Can. J. Fish. Aquat. Sci. 48:1783–1791. [13]

Székely, T., J. N. Webb, A. I. Houston, and J. M. McNamara. 1996. An evolutionary approach to offspring desertion in birds. Current Ornithol. 13:271–330. [9]

Szymura, J. M., and N. H. Barton. 1986. Genetic analysis of a hybrid zone between the fire-bellied toads, *Bombina bombina* and *B. variegata*, near Cracow in southern Poland. Evolution 40:1141–1159. [8]

Szymura, J. M., and N. H. Barton. 1991. The genetic structure of the hybrid zone between the fire-bellied toads *Bombina bombina* and *B. variegata*: comparisons between transects and between loci. Evolution 45:237–261. [8]

Taborsky, M. 1998. Sperm competition in fish: 'bourgeois' males and parasitic spawning. Trends Ecol. Evol. 13:222–227. [8, 9]

Taggart, J. B., I. S. McLaren, D. W. Hay, J. H. Webb, and A. F. Youngson. 2001. Spawning success in Atlantic salmon (*Salmo salar* L.): a long-term DNA profiling-based study conducted in a natural stream. Mol. Ecol. 10:1047–1060. [I, 3, 8, 9, 10]

Takahata, N. 1983. Gene identity and genetic differentiation of populations in the finite island model. Genetics 104:497–512. [2]

Tallman, R. F. 1986. Genetic differentiation among seasonally distinct spawning populations of chum salmon, *Oncorhynchus keta*. Aquaculture 57:211–217. [12]

Tallman, R. F., and M. C. Healey. 1994. Homing, straying, and gene flow among seasonally separated populations of chum salmon (*Oncorhynchus keta*). Can. J. Fish. Aquat. Sci. 51:577–588. [2, A1, A2]

Tamate, T., and K. Maekawa. 2000. Interpopulation variation in reproductive traits of female masu salmon, *Oncorhynchus masou*. Oikos 90:209–218. [4]

Tanksley, S. D. 1993. Mapping polygenes. Annu. Rev. Genet. 27:205–233. [6]

Tautz, A. F., and C. Groot. 1975. Spawning behavior of chum salmon (*Oncorhynchus keta*) and rainbow trout (*Salmo gairdneri*). J. Fish. Res. Board Can. 32:633–642. [9]

Tave, D. 1993. Genetics for fish hatchery managers, 2nd ed. AVI [12]

Taylor, E. B. 1990a. Environmental correlates of life-history variation in juvenile chinook salmon, *Oncorhynchus tshawytscha* (Walbaum). J. Fish Biol. 37:1–17. [7]

Taylor, E. B. 1990b. Phenotypic correlates of life-history variation in juvenile chinook salmon, *Oncorhynchus tshawytscha*. J. Anim. Ecol. 59:455–468. [7]

Taylor, E. B. 1991a. Behavioral interaction and habitat use in juvenile chinook, *Oncorhynchus tshawytscha*, and coho, *O. kisutch*, salmon. Anim. Behav. 42:729–744. [13]

Taylor, E. B. 1991b. A review of local adaptation in Salmonidae, with particular reference to Pacific and Atlantic salmon. Aquaculture 98:185–207. [I, 2, 4, 7, 8, 12, 13]

Taylor, E. B. 1999. Species pairs of north temperate freshwater fishes: evolution, taxonomy, and conservation. Rev. Fish Biol. Fish. 9:299–324. [6, 7, 8]

Taylor, E. B. 2003. Evolution in mixed company: evolutionary inferences from studies of natural hybridization in Salmonidae. Pp. 232–263 *in* A. P. Hendry and S. C. Stearns, eds. Evolution illuminated: salmon and their relatives. Oxford Univ. Press, New York, NY—*this volume* [I, 7]

Taylor, E. B., and C. J. Foote. 1991. Critical swimming velocities of juvenile sockeye salmon and kokanee, the anadromous and non-anadromous forms of *Oncorhynchus nerka* (Walbaum). J. Fish Biol. 38:407–419. [3, A3]

Taylor, E. B., and J. D. McPhail. 1985a. Variation in body morphology among British Columbia populations of coho salmon, *Oncorhynchus kisutch*. Can. J. Fish. Aquat. Sci. 42:2020–2028. [3]

Taylor, E. B., and J. D. McPhail. 1985b. Variation in burst and prolonged swimming performance among British Columbia populations of coho salmon, *Oncorhynchus kisutch*. Can. J. Fish. Aquat. Sci. 42:2029–2033. [3]

Taylor, E. B., and J. D. McPhail. 2000. Historical contingency and ecological determinism interact to prime speciation in sticklebacks, *Gasterosteus*. Proc. R. Soc. Lond. B. 267:2375–2384. [7]

Taylor, E. B., C. J. Foote, and C. C. Wood. 1996. Molecular genetic evidence for parallel life-history evolution within a Pacific salmon (sockeye salmon and kokanee, *Oncorhynchus nerka*). Evolution 50:401–416. [I, 3, 7]

Taylor, E. B., S. Harvey, S. Pollard, and J. Volpe. 1997. Postglacial genetic differentiation of reproductive ecotypes of kokanee *Oncorhynchus nerka* in Okanagan Lake, British Columbia. Mol. Ecol. 6:503–517. [A2]

Taylor, E. B., A. Kuiper, P. M. Troffe, D. J. Hoysak, and S. Pollard. 2000. Variation in developmental biology and microsatellite DNA in reproductive ecotypes of kokanee, *Oncorhynchus nerka*: implications for declining populations in a large British Columbia lake. Cons. Genet. 1:231–249. [8, A2]

Taylor, E. B., Z. Redenbach, A. B. Costello, S. M. Pollard, and C. J. Pacas. 2001. Nested analysis of genetic diversity in northwestern North American char, Dolly Varden (*Salvelinus malma*) and bull trout (*S. confluentus*). Can. J. Fish. Aquat. Sci. 58:406–420. [I, 8, A2]

Taylor, H. M., R. S. Gourley, C. E. Lawrence, and R. S. Kaplan. 1974. Natural selection of life history attributes: analytical approach. Theor. Pop. Biol. 5:104–122. [1]

Taylor, P. D. 1988. An inclusive fitness model for dispersal of offspring. J. theor. Biol. 130:363–378. [2]

Tchernavin, V. 1938. Changes in the salmon skull. Trans. Zool. Soc. Lond. 24:103–184. [9]

Tchernavin, V. 1944. The breeding characters of salmon in relation to their size. Proc. Zool. Soc., Lond. B 113:206–232. [9]

Teel, D. J., G. B. Milner, G. A. Winans, and W. S. Grant. 2000. Genetic population structure and origin of life history types in chinook salmon in British Columbia, Canada. Trans. Am. Fish. Soc. 129:194–209. [7, 12, A2]

Tessier, N., and L. Bernatchez. 1999. Stability of population structure and genetic diversity across generations assessed by microsatellites among sympatric populations of landlocked Atlantic salmon (*Salmo salar* L.). Mol. Ecol. 8:169–179. [2]

Tessier, N., and L. Bernatchez. 2000. A genetic assessment of single versus multiple origin of landlocked Atlantic salmon (*Salmo salar*) from Lake Saint-Jean, Québec, Canada. Can. J. Fish. Aquat. Sci. 57:797–804. [7, A2]

Tessier, N., L. Bernatchez, and J. M. Wright. 1997. Population structure and impact of supportive breeding inferred from mitochondrial and microsatellite DNA analyses in land-locked Atlantic salmon *Salmo salar* L. Mol. Ecol. 6:735–750. [2, A2]

Thedinga, J. F., A. C. Wertheimer, R. A. Heintz, J. M. Maselko, and S. D. Rice. 2000. Effects of stock, coded-wire tagging, and transplant on straying of pink salmon (*Oncorhynchus gorbuscha*) in southeastern Alaska. Can. J. Fish. Aquat. Sci. 57:2076–2085. [2, A1]

Thomas, W. K., and A. T. Beckenbach. 1989. Variation in salmonid mitochondrial DNA: evolutionary contraints and mechanisms of substitution. J. Mol. Evol. 29:233–245. [8]

Thomaz, D., E. Beall, and T. Burke. 1997. Alternative reproductive tactics in Atlantic salmon: factors affecting mature parr success. Proc. R. Soc. Lond. B 264:219–226. [3, 5, 9]

Thorne, R. E., and J. J. Ames. 1987. A note on variability of marine survival of sockeye salmon (*Oncorhynchus nerka*) and effects of flooding on spawning success. Can. J. Fish. Aquat. Sci. 44:1791–1795. [2]

Thorpe, J. E. 1986. Age at first maturity in Atlantic salmon, *Salmo salar*: freshwater period influences and conflicts with smolting. Can. Spec. Pub. Fish. Aquat. Sci. 89:7–14. [5, 9]

Thorpe, J. E. 1993. Impacts of fishing on genetic structure of salmonid populations. Pages 68–81 *in* J. G. Cloud and G. H. Thorgaard, eds. Genetic conservation of salmonid fishes. Plenum Press, New York. [3]

Thorpe, J. E. 1994. An alternative view of smolting in salmonids. Aquaculture 121:105–113. [3]

Thorpe, J. E., and K. A. Mitchell. 1981. Stocks of Atlantic salmon (*Salmo salar*) in Britain and Ireland: discreteness, and current management. Can. J. Fish. Aquat. Sci. 38:1576–1590. [1]

Thorpe, J. E., M. Mangel, N. B. Metcalfe, and F. A. Huntingford. 1998. Modelling the proximate basis of salmonid life-history variation, with application to Atlantic salmon, *Salmo salar* L. Evol. Ecol. 12:581–599. [3, 5]

Thorpe, J. E., M. S. Miles, and D. S. Keay. 1984. Developmental rate, fecundity and egg size in Atlantic salmon, *Salmo salar* L. Aquaculture 43:289–305. [5]

Thorpe, J. E., R. I. G. Morgan, C. Talbot, and M. S. Miles. 1983. Inheritance of developmental rates in Atlantic salmon, *Salmo salar* L. Aquaculture 33:119–128. [3, 5]

Tilzey, R. D. J. 1977. Repeat homing of brown trout (*Salmo trutta*) in Lake Eucumbene, New South Wales, Australia. J. Fish. Res. Board Can. 34:1085–1094. [2]

Ting, C. T., S. C. Tsaur, M. L. Wu, and C. I. Wu. 1998. A rapidly evolving homeobox at the site of a hybrid sterility gene. Science 282:1501–1504. [6, 8]

Travis, J. M. J., and C. Dytham. 1998. The evolution of dispersal in a metapopulation: a spatially explicit, individual-based model. Proc. R. Soc. Lond. B 265:17–23. [2]

Travis, J. M. J., and C. Dytham. 1999. Habitat persistence, habitat availability and the evolution of dispersal. Proc. R. Soc. Lond. B 266:723–728. [2]

Travis, J. M. J., D. J. Murrell, and C. Dytham. 1999. The evolution of density-dependent dispersal. Proc. R. Soc. Lond. B 266:1837–1842. [2]

Trippel, E. A. 1995. Age at maturity as a stress indicator in fisheries. BioScience 45:759–771. [5, 7, 11, 13]

Trivers, R. L. 1974. Parent-offspring conflict. Am. Zool. 14:249–264. [4]

Trudel, M., A. Tremblay, A. Schetagne, and J. B. Rasmussen. 2000. Estimating food consumption rates of fish using a mercury mass blance model. Can. J. Fish. Aquat. Sci. 57:414–428. [6]

Trudel M., A. Tremblay, A. Schetagne, and J. B. Rasmussen. 2001. Why are dwarf fish so small? An energetic analysis of polymorphism in lake whitefish (*Coregonus clupeaformis*). Can. J. Fish. Aquat. Sci. 58:394–405. [6]

Tsiger, V. V., V. I. Skirin, N. I. Krupyanko, K. A. Kashkin, and A. Yu. Semenchenko. 1994. Life history forms of male masu salmon (*Oncorhynchus masou*) in South Primor'e, Russia. Can. J. Fish. Aquat. Sci. 51:197–208. [I, 1, 3, 9]

Tsuyuki, H., and E. Roberts. 1966. Inter-species relationships within the genus *Oncorhynchus* based on biochemical systematics. J. Fish. Res. Board Can. 23:101–107. [7]

Turelli, M. 1984. Heritable genetic variation via mutation-selection balance: Lerch's zeta meets the abdominal bristle. Theor. Pop. Biol. 25:138–193. [11]

Turelli, M. 1988. Phenotypic evolution, constant covariances, and the maintenance of additive variance. Evolution 42:1342–1347. [5]

Turgeon, J., and L. Bernatchez. 2001a. Mitochondrial DNA phylogeography of lake cisco (*Coregonus artedi*): evidence supporting extensive secondary contacts between two glacial races. Mol. Ecol. 10:987–1001. [6]

Turgeon, J., and L. Bernatchez. 2001b. Clinal variation at microsatellite loci reveals historical secondary intergradation between glacial races of *Coregonus artedi* (Teleostei: Coregoninae). Evolution 11:2274–2286. [6, 8, A2]

Turgeon, J., A. Estoup, and L. Bernatchez. 1999. Species flock in the North American Great Lakes: molecular ecology of Lake Nipigon Ciscoes (Teleostei: Coregonidae: *Coregonus*). Evolution 53:1857–1871. [6]

Turner, T. F., and J. C. Trexler. 1998. Ecological and historical associations of gene flow in darters (Teleostei: Percidae). Evolution 52:1781–1801. [4]

Turner, T. F., L. R. Richardson, and J. R. Gold. 1999. Temporal genetic variation of mitochondrial DNA and female effective population size of red drum (*Sciaenops ocellatus*) in the northern Gulf of Mexico. Mol. Ecol. 8:1223–1230. [10]

Underwood, T. J., M. J. Millard, and L. A. Thorpe. 1996. Relative abundance, length frequency, age, and maturity of Dolly Varden in nearshore waters of the Arctic National Wildlife Refuge, Alaska. Trans. Am. Fish. Soc. 125:719–728. [8]

Unwin, M. J., and T. P. Quinn. 1993. Homing and straying patterns of chinook salmon (*Oncorhynchus tshawytscha*) from a New Zealand hatchery: spatial distribution of strays and effects of release date. Can. J. Fish. Aquat. Sci. 50:1168–1175. [2, A1]

Unwin, M. J., M. T. Kinnison, N. C. Boustead, and T. P. Quinn. 2003. Genetic control over survival in Pacific salmon (*Oncorhynchus* spp.): experimental evidence between and within New Zealand chinook salmon (*O. tshawytscha*). Can. J. Fish. Aquat. Sci. 60:1–11. [I, 7, 10]

Unwin, M. J., M. T. Kinnison, and T. P. Quinn. 1999. Exceptions to semelparity: postmaturation survival, morphology, and energetics of male chinook salmon (*Oncorhynchus tshawytscha*). Can. J. Fish. Aquat. Sci. 56:1172–1181. [I, 1, 3, 7, 9]

Unwin, M. J., T. P. Quinn, M. T. Kinnison, and N. C. Boustead. 2000. Divergence in juvenile growth and life history in two recently colonized and partially isolated chinook salmon populations. J. Fish Biol. 57:943–960. [7]

Utter, F. M. 2000. Patterns of subspecific anthropogenic introgression in two salmonid genera. Rev. Fish Biol. Fish. 10:265–279. [2, 7, 8, 12]

Utter, F., P. Aebersold, J. Helle, and G. Winans. 1984. Genetic characterization of populations in the southeastern range of sockeye salmon. Pp. 17–32 *in* J. M. Walton and D. B Houston, eds. Proc. Olympic Wild Fish Conference. Peninsula College and Olympic National Park, Port Angeles, WA. [A2]

Utter, F. M., F. W. Allendorf, and H. O. Hodgins. 1973. Genetic variability and relationships in Pacific salmon and related trout based on protein variations. Syst. Zool. 22:257–270. [8]

Utter, F. M., D. W. Chapman, and A. R. Marshall. 1995. Genetic population structure and history of chinook salmon of the Upper Columbia River. Am. Fish. Soc. Symp. 17:149–168. [12]

Utter, F., G. Milner, G. Ståhl, and D. Teel. 1989. Genetic population structure of chinook salmon, *Oncorhynchus tshawytscha*, in the Pacific northwest. Fishery Bull. 87:239–264. **[12, 13, A2]**

Vamosi, S. M., and D. Schluter. 1999. Sexual selection against hybrids between sympatric stickleback species: evidence from a field experiment. Evolution 53:874–879. **[8]**

van den Berghe, E. P., and M. R. Gross. 1984. Female size and nest depth in coho salmon (*Oncorhynchus kisutch*). Can. J. Fish. Aquat. Sci. 41:204–206. **[9]**

van den Berghe, E. P., and M. R. Gross. 1989. Natural selection resulting from female breeding competition in a Pacific salmon (coho: *Oncorhynchus kisutch*). Evolution 43:125–140. **[3, 4, 9, 13]**

van den Berghe, E. P., F. Wernerus, and R. R. Warner. 1989. Female choice and mating cost of peripheral males. Anim. Behav. 38:875–884. **[9]**

Van Valen, L. 1965. Selection in natural populations. III. Measurement and estimation. Evolution 19:514–528. **[11]**

Van Valen, L. 1971. Group selection and the evolution of dispersal. Evolution 25:591–598. **[2]**

Varnavskaya, N. V., and T. D. Beacham. 1992. Biochemical genetic variation in odd-year pink salmon (*Oncorhynchus gorbuscha*) from Kamchatka. Can. J. Zool. 70:2115–2120. **[A2]**

Varnavskaya, N. V., C. C. Wood, R. J. Everett, R. L. Wilmot, V. S. Varnavsky, V. V. Midanaya, and T. P. Quinn. 1994. Genetic differentiation of subpopulations of sockeye salmon (*Oncorhynchus nerka*) within lakes of Alaska, British Columbia, and Kamchatka, Russia. Can. J. Fish. Aquat. Sci. 51(Suppl. 1):147–157. **[2, A2]**

Varvio, S. L., R. Chakraborty, and M. Nei. 1986. Genetic variation in subdivided populations and conservation genetics. Heredity 57:189–198. **[2]**

Veen, T., T. Borge, S. C. Griffith, G.-P. Sætre, S. Bures, L. Gustafsson, and B. C. Sheldon. 2001. Hybridization and adaptive mate choice in flycatchers. Nature 411:45–50. **[I]**

Verhulst, S., and H. M. van Eck. 1996. Gene flow and immigration rate in an island population of great tits. J. Evol. Biol. 9:771–782. **[2]**

Verspoor, E., and L. J. Cole. 1989. Genetically distinct sympatric populations of resident and anadromous Atlantic salmon, *Salmo salar*. Can. J. Zool. 67:1453–1461. **[3, A3]**

Verspoor, E., and J. Hammar. 1991. Introgressive hybridization in fishes: the biochemical evidence. J. Fish Biol. 39(Suppl. A):309–334. **[I, 8]**

Via, S. 2001. Sympatric speciation in animals: the ugly duckling grows up. Trends Ecol. Evol. 16:381–390. **[6]**

Via, S., and R. Lande. 1985. Genotype-environment interaction and the evolution of phenotypic plasticity. Evolution 39:505–522. **[5, 6, 11]**

Via, S., A. C. Bouck, and S. Skillman. 2000. Reproductive isolation between divergent races of pea aphids on two hosts. II. Selection against migrants and hybrids in the parental environments. Evolution 54:1626–1637. **[2]**

Visman, V., S. Pesant, J. Dion, B. Shipley, and R. H. Peters. 1996. Joint effects of maternal and offspring sizes on clutch mass and fecundity in plants and animals. EcoScience 3:173–182. **[4]**

Vladič, T. V., and T. Järvi. 2001. Sperm quality in the alternative reproductive tactics of Atlantic salmon: the importance of the loaded raffle mechanism. Proc. R. Soc. Lond. B 268:2375–2381. **[I, 9]**

Vladykov, V. D. 1954. Taxonomic characters of the eastern North American chars (*Salvelinus* and *Cristivomer*). J. Fish. Res. Board Can. 11:904–932. [9]

Vladykov, V. D. 1962. Osteological studies on Pacific salmon of the genus *Oncorhynchus*. Fish. Res. Board Can. Bull. 136:1–172. [9]

Vogler, A. P., and R. Desalle. 1994. Diagnosing units of conservation management. Cons. Biol. 8:354–363. [12]

Volk, E. C., S. L. Schroder, and J. J. Grimm. 1999. Otolith thermal marking. Fish. Res. 43:205–219. [2]

Vøllestad, L. A., and J. H. L'Abée-Lund. 1994. Evolution of the life-history of Arctic charr, *Salvelinus alpinus*. Evol. Ecol. 8:315–327. [9]

Vøllestad, L. A., E. M. Olsen, and T. Forseth. 2002. Growth-rate variation in brown trout in small neighbouring streams: evidence for density-dependence? J. Fish Biol. 61:1513–1527. [2]

Volpe, J. P., E. B. Taylor, D. W. Rimmer, and B. W. Glickman. 2000. Evidence of natural reproduction of aquaculture-escaped Atlantic salmon in a coastal British Columbia river. Cons. Biol. 14:899–903. [1]

Vucetich, J. A., T. A. Waite, and L. Nunney. 1997. Fluctuating population size and the ratio of effective to census population size. Evolution 51:2017–2021. [10]

Vuorinen, J. 1988. Enzyme genes as interspecific hybridization probes in Coregonine fishes. Finnish Fish. Res. 9:31–37. [8]

Vuorinen, J., and O. K. Berg. 1989. Genetic divergence of anadromous and nonanadromous Atlantic salmon (*Salmo salar*) in the River Namsen, Norway. Can. J. Fish. Aquat. Sci. 46:406–409. [3, A3]

Vuorinen, J. A., R. A. Bodaly, J. D. Reist, and M. Luczynski. 1998. Phylogeny of five *Prosopium* species with comparisons with other Coregonine fishes based on isozyme electrophoresis. J. Fish Biol. 53:917–927. [6]

Wade, M. J. 1976. Group selection among laboratory populations of *Tribolium*. Proc. Natl. Acad. Sci. USA 73:4604–4607. [13]

Wade, M. J. 1978. A critical review of the models of group selection. Quart. Rev. Biol. 53:101–114. [13]

Wade, M. J. 1985. Soft selection, hard selection, kin selection, and group selection. Am. Nat. 125:61–73. [13]

Wade, M. J., and S. Kalisz. 1990. The causes of natural selection. Evolution 44:1947–1955. [4, 13]

Waknitz, F. W., G. M. Matthews, T. Wainwright, and G. Winans. 1995. Status review for Mid-Columbia River summer chinook salmon. NOAA Tech. Memo. NMFS-NWFSC-22. [12]

Wallace, B. 1968. Polymorphism, population size, and genetic load. Pp. 87–108 *in* R. C. Lewontin, ed. Population biology and evolution, Syracuse Univ. Press, Syracuse, NY. [13]

Wallace, B. 1975. Hard and soft selection revisited. Evolution 29:465–473. [13]

Walters, C. 1999. Variation in productivity of southern British Columbia coho salmon (*Oncorhynchus kisutch*) stocks and implications for mixed-stock management. unpublished report. [13]

Walters, C. J. 1983. Mixed-stock fisheries and the sustainability of enhancement production for chinook and coho salmon. Pp. 109–115 *in* W. J. McNeil, ed. Salmon production, management, and allocation. Biological, economic, and policy issues. Oregon State Univ. Press, Corvallis, OR. [11]

Walters, C. J. 1985. Bias in the estimation of functional relationships from time series data. Can. J. Fish. Aquat. Sci. 42:147–149. [13]

Walters, C. J. 1986. Adaptive management of renewable resources. MacMillan, New York. [13]

Walters, C. J., and P. Cahoon. 1985. Evidence of decreasing spatial diversity in British Columbia salmon stocks. Can. J. Fish. Aquat. Sci. 42:1033–1037. [13]

Walters, C., and J. F. Kitchell. 2001. Cultivation/depensation effects on juvenile survival and recruitment: implications for the theory of fishing. Can. J. Fish. Aquat. Sci. 58:39–50. [13]

Walters, C. J., and D. Ludwig. 1981. Effects of measurement errors on the assessment of stock-recruitment relationships. Can. J. Fish. Aquat. Sci. 38:704–710. [13]

Wang, J. L., and N. Ryman. 2001. Genetic effects of multiple generations of supportive breeding. Cons. Biol. 15:1619–1631. [I, 10]

Wang, S., J. J. Hard, and F. Utter. 2002. Genetic variation and fitness in salmonids. Cons. Genet. 3:321–333. [11]

Wang, S., J. J. Hard, and F. Utter. 2001. Salmonid inbreeding: a review. Rev. Fish Biol. Fish. 11:301–319. [2]

Waples, R. S. 1989. A generalized approach for estimating effective population size from temporal changes in allele frequency. Genetics 121:379–391. [10]

Waples, R. S. 1990a. Conservation genetics of Pacific salmon. II. Effective population size and the rate of loss of genetic variability. J. Hered. 81:267–276. [I, 10]

Waples, R. S. 1990b. Conservation genetics of Pacific salmon. III. Estimating effective population size. J. Hered. 81:277–289. [I, 10]

Waples, R. S. 1991a. Pacific salmon, Oncorhynchus spp., and the definition of "species" under the Endangered Species Act. Mar. Fish. Rev. 53:11–22. [I, 7, 12, 13]

Waples, R. S. 1991b. Genetic methods for estimating the effective size of cetacean populations. Rep. Int. Whaling Commission (Special Issue 13):279–300. [10]

Waples, R. S. 1995. Evolutionarily significant units and the conservation of biological diversity under the Endangered Species Act. Am. Fish. Soc. Symp. 17:8–27. [I, 12, 13]

Waples, R. S. 1998a. Evolutionarily significant units, distinct population segments, and the Endangered Species Act: reply to Pennock and Dimmick. Cons. Biol. 12:718–721. [12]

Waples, R. S. 1998b. Separating the wheat from the chaff: patterns of genetic differentiation in high gene flow species. J. Hered. 89:438–450. [2]

Waples, R. S. 2002a. Effective size of fluctuating salmon populations. Genetics 161:783–791. [10]

Waples, R. S. 2002b. Evaluating the effect of stage-specific survivorship on the N_e/N ratio. Mol. Ecol. 11:1029–1037. [10]

Waples, R. S. 2002c. Definition and estimation of effective population size in the conservation of endangered species. Pp. 147–168 in S. R. Beissinger and D. R. McCullough, eds. Population viability analysis. Univ. Chicago Press, Chicago, IL. [10]

Waples, R. S. 2003. Salmonid perspectives into effective population size. Pp. [update pages] in A. P. Hendry and S. C. Stearns, eds. Evolution illuminated: salmon and their relatives. Oxford Univ. Press, New York, NY—this volume [I, 2, 13]

Waples, R. S., and C. Do. 1994. Genetic risk associated with supplementation of Pacific salmonids: captive broodstock programs. Can. J. Fish. Aquat. Sci. 51(Suppl. 1):310–329. [10]

Waples, R. S., and P. E. Smouse. 1990. Gametic disequilibrium analysis as a means of identifying mixtures of salmon populations. Am. Fish. Soc. Symp. 7:439–458. [10]

Waples, R. S., O. W. Johnson, P. B. Aebersold, C. K. Shiflett, D. M. VanDoornik, D. J. Teel, and A. E. Cook. 1993. A genetic monitoring and evaluation program for supplemented populations of salmon and steelhead in the Snake River Basin. Annual Report of Research, Bonneville Power Administration, Portland, OR. [10]

Waples, R. S., R. P. Jones, B. R. Beckman, and G. A. Swan. 1991. Status review for Snake River fall chinook salmon. NOAA Tech. Memo. NMFS F/NWC-201. [12]

Waples, R. S., and 15 other authors. 2001. Characterizing diversity in salmon from the Pacific Northwest. J. Fish Biol. 59(Suppl. A):1–41. [I, 12]

Ward, B. R., and P. A. Slaney. 1988. Life history and smolt-to-adult survival of Keogh River steelhead trout (*Salmo gairdneri*) and the relationship to smolt size. Can. J. Fish. Aquat. Sci. 45:1110–1122. [13]

Ward, R. D., M. Woodwark, and D. O. F. Skibinski. 1994. A comparison of genetic diversity levels in marine, freshwater and anadromous fishes. J. Fish Biol. 44:213–232. [12]

Ware, D. M. 1975. Relation between egg size, growth, and natural mortality of larval fish. J. Fish. Res. Board Can. 32:2503–2512. [4]

Ware, D. M. 1978. Bioenergetics of pelagic fish: theoretical change in swimming speed and ration with body size. J. Fish. Res. Board Can. 35:220–228. [9]

Warner, R. R. 1991. The use of phenotypic plasticity in coral reef fishes as tests of theory in evolutionary ecology. Pp. 387–398 *in* P. F. Sale, ed. The ecology of fishes on coral reefs. Academic Press, New York. [5]

Waser, P. M., and W. T. Jones. 1983. Natal philopatry among solitary mammals. Quart. Rev. Biol. 58:355–390. [2]

Waser, P. M., S. N. Austad, and B. Keane. 1986. When should animals tolerate inbreeding? Am. Nat. 128:529–537. [2]

Wehrhahn, C. F., and R. Powell. 1987. Electrophoretic variation, regional differences, and gene flow in the coho salmon (*Oncorhynchus kisutch*) of southern British Columbia. Can. J. Fish. Aquat. Sci. 44:822–831. [A2, 13]

Weir, B. S., and C. C. Cockerham. 1984. Estimating *F*-statistics for the analysis of population structure. Evolution 38:1358–1370. [6]

Weitkamp, L. A., T. C. Wainwright, G. J. Bryant, G. B. Milner, D. J. Teel, R. G. Kope, and R. S. Waples. 1995. Status review of coho salmon from Washington, Oregon and California. NOAA Tech. Memo. NMFS-NWFSC-24. [12]

Wenburg, J. K., and P. Bentzen. 2001. Genetic and behavioral evidence for restricted gene flow among coastal cutthroat trout populations. Trans. Am. Fish. Soc. 130:1049–1069. [A1, A2]

Wenburg, J. K., P. Bentzen, and C. J. Foote. 1998. Microsatellite analysis of genetic population structure in an endangered salmonid: the coastal cutthroat trout (*Oncorhynchus clarki clarki*). Mol. Ecol. 7:733–749. [A2]

West, S. A., C. M. Lively, and A. F. Read. 1999. A pluralist approach to sex and recombination. J. Evol. Biol. 12:1003–1012. [1]

White, K. P. 2001. Functional genomics and the study of development, variation and evolution. Nature Rev. Genet. 2:528–537. [6]

Whitlock, M. C., and D. E. McCauley. 1999. Indirect measures of gene flow and migration: $F_{ST} \neq 1/(4Nm + 1)$. Heredity 82:117–125. [2]

Wiklund, C., and T. Fagerström. 1977. Why do males emerge before females? A hypothesis to explain the incidence of protandry in butterflies. Oecologia 31:153–158. [9]

Wilder, D. G. 1952. A comparative study of anadromous and freshwater populations of brook trout (*Salvelinus fontinalis* (Mitchill)). J. Fish Res. Board Can. 9:169–203. [3, A3]

Williams, G. C. 1966a. Adaptation and natural selection. A critique of some current evolutionary thought. Princeton Univ. Press, Princeton, NJ. [1, 9, 11, 13]

Williams, G. C. 1966b. Natural selection, the cost of reproduction, and a refinement of Lack's principle. Am. Nat. 100:687–690. [4]

Williams, T. H., K. P. Currens, N. E. Ward III, and G. H. Reeves. 1997. Genetic population structure of coastal cutthroat trout. Pp. 16–17 in J. D. Hall, P. A. Bisson, and R. E. Gresswell, eds. Sea-run cutthroat trout: biology, management, and future conservation. Oregon Chapter American Fisheries Society, Corvallis, OR. [A2]

Willson, M. F. 1997. Variation in salmonid life histories: patterns and perspectives. Res. Paper PNW-RP-498. USDA Forest Service. Pacific NW Res. Station. Portland, OR. [1]

Willson, M. F., and K. C. Halupka. 1995. Anadromous fish as keystone species in vertebrate communities. Cons. Biol. 9: 489–497. [13]

Willson, M. F., S. M. Gende, and B. H. Marston. 1998. Fishes and the forest. Bioscience 48:455–462. [13]

Wilmot, R. L., and C. V. Burger. 1985. Genetic differences among populations of Alaskan sockeye salmon. Trans. Am. Fish. Soc. 114:236–243. [A2]

Wilmot, R. L., R. J. Everett, W. J. Spearman, R. Baccus, N. V. Varnavskaya, and S. V. Putivkin. 1994. Genetic stock structure of Western Alaska chum salmon and a comparison with Russian Far East stocks. Can. J. Fish. Aquat. Sci. 51(Suppl. 1):84–94. [A2]

Wilson, C. C., and L. Bernatchez. 1998. The ghost of hybrids past: fixation of arctic charr (*Salvelinus alpinus*) mitochondrial DNA in an introgressed population of lake trout (*S. namaycush*). Mol. Ecol. 7:127–132. [I, 8]

Wilson, C. C., and P. D. N. Hebert. 1993. Natural hybridization between Arctic char (*Salvelinus alpinus*) and lake trout (*S. namaycush*) in the Canadian Arctic. Can. J. Fish. Aquat. Sci. 50:2652–2658. [8]

Wilson, C. C., and P. D. N. Hebert. 1998. Phylogeography and postglacial dispersal of lake trout (*Salvelinus namaycush*) in North America. Can. J. Fish. Aquat. Sci. 55:1010–1024. [8]

Wilson, D. S. 1975. Theory of group selection. Proc. Natl. Acad. Sci. USA 72:143–146. [13]

Wilson, H. B. 2001. The evolution of dispersal from source to sink populations. Evol. Ecol. Res. 3:27–35. [2]

Winans, G. A., P. B. Aebersold, S. Urawa, and N. V. Varnavskaya. 1994. Determining continent of origin of chum salmon (*Oncorhynchus keta*) using genetic stock identification techniques: status of allozyme baseline in Asia. Can. J. Fish. Aquat. Sci. 51(Suppl. 1):95–113. [A2]

Winans, G. A., P. B. Aebersold, and R. S. Waples. 1996. Allozyme variability of *Oncorhynchus nerka* in the Pacific Northwest, with special consideration to populations of Redfish Lake, Idaho. Trans. Am. Fish. Soc. 125:645–663. [A2]

Wipfli, M. S., J. P. Hudson, J. P. Caouette, and D. T. Chaloner. 2003. Marine subsidies in freshwater ecosystems: salmon carcasses increase the growth rate of stream-resident salmonids. trans. Am. Fish. Soc. 132:371–381. [2, 13]

Wirtz, P. 1999. Mother species-father species: unidirectional hybridization in animals with female choice. Anim. Behav. 58:1–12. [8]

Withler, F. C. 1982. Transplanting Pacific salmon. Can. Tech. Rep. Fish. Aquat. Sci. 1079. [2, 7, 12]

Withler, R. E. 1985. *Ldh-4* allozyme variability in North American sockeye salmon (*Oncorhynchus nerka*) populations. Can. J. Zool. 63:2924–2932. [A2]

Withler, R. E., K. D. Le, R. J. Nelson, K. M. Miller, and T. D. Beacham. 2000. Intact genetic structure and high levels of genetic diversity in bottlenecked sockeye salmon (*Oncorhynchus nerka*) populations of the Fraser River, British Columbia, Canada. Can. J. Fish. Aquat. Sci. 57:1985–1998. [A2]

Wolf, J. B., and M. J. Wade. 2001. On the assignment of fitness to parents and offspring: whose fitness is it and when does it matter? J. Evol. Biol. 14:347–356. [4]

Wollenberg, K., J. Arnold, and J. C. Avise. 1996. Recognizing the forest for the trees: testing temporal patterns of cladogenesis using a null model of stochastic diversification. Mol. Biol. Evol. 13:833–849. [6]

Woltereck, R. 1909. Weiterer experimentelle Untersuchuingen über Artveranderung, Speziell über das Wessen Quantitativer Artunterschiede bei Daphniden. Versuch. Deutsch Zoologiche Geselleschaft 19:110–172. [5]

Wood, C. C. 1995. Life history variation and population structure in sockeye salmon. Am. Fish. Soc. Symp. 17:195–216. [I, 2, 3, 7, 12]

Wood, C. C., and C. J. Foote. 1990. Genetic differences in the early development and growth of sympatric sockeye salmon and kokanee (*Oncorhynchus nerka*), and their hybrids. Can. J. Fish. Aquat. Sci. 47:2250–2260. [I, 3, A3]

Wood, C. C., and C. J. Foote. 1996. Evidence for sympatric genetic divergence of anadromous and nonanadromous morphs of sockeye salmon (*Oncorhynchus nerka*). Evolution 50:1265–1279. [I, 3, 4, 7, 8, A3]

Wood, C. C., C. J. Foote, and D. T. Rutherford. 1999. Ecological interactions between juveniles of reproductively isolated anadromous and non-anadromous morphs of sockeye salmon, *Oncorhynchus nerka*, sharing the same nursery lake. Environ. Biol. Fishes 54:161–173. [3]

Wood, C. C., B. E. Riddell, D. T. Rutherford, and R. W. Withler. 1994. Biochemical genetic survey of sockeye salmon (*Oncorhynchus nerka*) in Canada. Can. J. Fish. Aquat. Sci. 51(Suppl. 1):114–131. [A2]

Woody, C. A., J. Olsen, J. Reynolds, and P. Bentzen. 2000. Temporal variation in phenotypic and genotypic traits in two sockeye salmon populations, Tustumena Lake, Alaska. Trans. Am. Fish. Soc. 129:1031–1043. [2, A2]

Wootton, R. J. 1984. Introduction: strategies and tactics in fish reproduction. Pp. 1–12 *in* G. W. Potts and R. J. Wootton, eds. Fish reproduction: strategies and tactics. Academic Press, New York. [4]

Wright, S. 1931. Evolution in Mendelian populations. Genetics 16:97–159. [I, 2, 10]

Wright, S. 1938. Size of population and breeding structure in relation to evolution. Science 87:430–431. [10]

Wright, S. 1951. The genetical structure of populations. Ann. Eugenics 15:323–354. [6]

Wright, S. 1968. Evolution and the genetics of populations. Univ. of Chicago Press, Chicago, IL. [1, 11]

Yamada, Y. 1977. Evaluation of the culling variate used by breeders in actual selection. Genetics 86:885–899. [11]

Yamahira, K., and D. O. Conover. 2002. Intra- vs. interspecific latitudinal variation in growth: adaptation to temperature or seasonality? Ecology 83:1252–1262. [3]

Yodzis, P. 1981. Concerning the sense in which maximizing fitness is equivalent to maximizing reproductive value. Ecology 62:1681–1682. [1]

Young, K. A. 1999. Environmental correlates of male life history variation among coho salmon populations from two Oregon coastal basins. Trans. Am. Fish. Soc. 128:1–16. [I, 9, 13]

Young, K. A. 2001. Defining units of conservation for intraspecific biodiversity: reply to Dimmick et al. Cons. Biol. 15:784–787. [12, 13]

Young, K. A. 2003. Toward evolutionary management: lessons from salmonids. Pp. 358–376 in A. P. Hendry and S. C. Stearns, eds. Evolution illuminated: salmon and their relatives. Oxford Univ. Press, New York, NY—this volume [I, 1, 7, 9]

Young, K. A. In press. Asymmetric competition, habitat selection and niche overlap in juvenile salmonids. Ecology. In Press. [13]

Young, W. P., C. O. Ostberg, P. Keim, and G. H. Thorgaard. 2001. Genetic characterization of hybridization and introgression between anadromous rainbow trout (Oncorhynchus mykiss irideus) and coastal cutthroat trout (O. clarki clarki). Mol. Ecol. 10:921–930. [8]

Yund, P. O., and M. A. McCartney. 1994. Male reproductive success in sessile invertebrates: competition for fertilizations. Ecology 75:2151–2167. [9]

Zahavi, A. 1975. Mate selection—selection for a handicap. J. theor. Biol. 53:205–214. [9]

Zeng, Z.-B. 1988. Long-term correlated response, interpopulation covariation, and interspecific allometry. Evolution 42:363–374. [11]

Zimmerman, C. E., and G. H. Reeves. 2000. Population structure of sympatric anadromous and nonanadromous Oncorhynchus mykiss: evidence from spawning surveys and otolith microchemistry. Can. J. Fish. Aquat. Sci. 57:2152–2162. [3]

Zimmerman, C. E., and G. H. Reeves. 2002. Identification of steelhead and resident rainbow trout progeny in the Deschutes River, Oregon, revealed with otolith microchemistry. Trans. Am. Fish. Soc. 131:986–993. [3]

Zinn, J. L., K. A. Johnson, J. E. Sanders, and J. L. Fryer. 1977. Susceptibility of salmonid species and hatchery strains of chinook salmon (Oncorhynchus tshawytscha) to infections by Ceratomyxa shasta. J. Fish. Res. Board Can. 34:933–936. [12]

Zouros, E., and D. W. Foltz. 1987. The use of allelic isozyme variation for the study of heterosis. Pp. 1–59 in M. C. Rattazzi, J. G. Scandalios, and G. S. Whitt, eds. Isozymes. Alan R. Lish, New York. [11]

INDEX